The Gluten Proteins

The Gluten Proteins

Edited by

D. Lafiandra, S. Masci and R. D'Ovidio
Universita' degli Studi della Tuscia, Viterbo, Italy

advancing the chemical sciences

The proceedings of the 8th Gluten Workshop held on 8-10 September 2003 in Viterbo, Italy.

Special Publication No. 295

ISBN 0-85404-633-X

A catalogue record for this book is available from the British Library

© The Royal Society of Chemistry 2004

All rights reserved

Apart from any fair dealing for the purpose of research or private study for non-commercial purposes, or criticism or review as permitted under the terms of the UK Copyright, Designs and Patents Act, 1988 and the Copyright and Related Rights Regulations 2003, this publication may not be reproduced, stored or transmitted, in any form or by any means, without the prior permission in writing of The Royal Society of Chemistry, or in the case of reprographic reproduction only in accordance with the terms of the licences issued by the Copyright Licensing Agency in the UK, or in accordance with the terms of the licences issued by the appropriate Reproduction Rights Organization outside the UK. Enquiries concerning reproduction outside the terms stated here should be sent to The Royal Society of Chemistry at the address printed on this page.

Published by The Royal Society of Chemistry,
Thomas Graham House, Science Park, Milton Road,
Cambridge CB4 0WF, UK

Registered Charity Number 207890

For further information see our web site at www.rsc.org

Printed by Athenaeum Press Ltd, Gateshead, Tyne and Wear, UK

Preface

The end use characteristics of flours and semolina obtained from hexaploid bread wheat and tetraploid durum wheat are most strongly influenced by the gluten proteins. Because of their importance, the heterogeneous group of proteins that make up gluten have been studied extensively in order to facilitate development of products that are more satisfying to consumers. The most recent contributions to the study of gluten proteins were presented at the VIII International Gluten Workshop held in Viterbo, Italy, on September 8-10, 2003. The workshop was attended by 161 people from thirty different Countries and was organized in sessions dealing with quality, genetics, biotechnology, environmental aspects, and non food-uses of gluten.

The wide ranging presentations included papers describing the use of recently developed sophisticated tools, such as proteomics and genomics, which are contributing importantly to the unravelling of the complexity and heterogeneity of gluten proteins and their corresponding genes. These new techniques are accelerating progress toward development of a full understanding of the molecular basis for quality trait variation. The introduction at the VIIIth Gluten Workshop of a session dedicated to gluten allergies and intolerances is noteworthy. Greater attention is being paid to the involvement of these proteins in triggering intolerances and allergies.

Finally, we would like to thank the colleagues D. Bhandari, J.M. Bietz, B.J. Butow, F. Clarke, J.M. Clarke, B. Dobraszczyk, F. Dupont, J. Goodwin, J. Jenkins, H. Jones, D.D. Kasarda, C. Mills, A. Perdon, K. Preston, J. Robertson, P.R. Shewry, P. Wiley, who helped us in reviewing the papers here presented.

Sponsors of the 8th Gluten Workshop

Comune di Viterbo

Presidenza del Consiglio regionale del Lazio

Provincia di Viterbo

Consorzio " Gian Pietro Ballatore"

Contents

Biotechnology, Transcriptomics and Proteomics

The use of biotechnology to study wheat endosperm development and improve grain quality
P.R. Shewry, H.D. Jones, M.J. Holdworth, J.R. Lenton, K.J. Edwards ... 3

Agronomic, biochemical and quality characteristics of wheats containing HMW-glutenin transgenes
A.E. Blechl, P. Bregitzer, K. O'Brien, J. Lin, S.B. Nguyen, O.D. Anderson ... 6

Characterization of glutenin polymers in a transgenic bread wheat line over-expressing a LMW-GS
S. Masci, R. D'Ovidio, F. Scossa, C. Patacchini, D. Lafiandra, O.D. Anderson, A.E. Blechl ... 10

The potential of biotechnology to produce novel gluten phenotypes
H.D. Jones, H. Wu, C. Sparks, A. Doherty, M. Cannell, J. Goodwin, G. Pastori, P.R. Shewry ... 14

The use of ESTs to analyze the spectrum of wheat seed proteins
O.D. Anderson, V. Carollo, S. Chao, D. Laudencia-Chinguanco, G.R. Lazo ... 18

Transgenic expression of epitope-tagged LMW glutenin subunits in durum wheat: effect on polymer size distribution and dough mixing properties
P. Tosi, R. D'Ovidio, F. Békés, J.A. Napier, P.R. Shewry ... 22

Heterotypic aggregation of an *in vitro* synthesised low-molecular-weight glutenin subunit
A. Orsi, , F. Sparvoli, P.A. Della Cristina, A. Ceriotti ... 26

Proteomic analysis of wheat storage proteins: a promising approach to understand the genetic and molecular bases of gluten components
G. Branlard, J. Dumur, E. Bancel, M. Merlino, M. Dardevet ... 30

Mass spectrometric approaches for the characterization of low-molecular weight glutenin subunits
V. Cunsolo, S. Foti, V. Muccilli, R. Saletti, D. Lafiandra, S. Masci ... 34

Comparison of different peptidic hydrolyses to identify wheat storage proteins using MALDI-TOF
E. Bancel, J. Dumur, C. Chambon, G. Branlard ... 38

Characterization of omega gliadins encoded on chromosome 1A and evidence for post translational cleavage of omega-gliadins by an asparginyl endoprotease
F.M. DuPont, W.H. Vensel, D.D. Kasarda ... 42

DNA marker linked to low-molecular-weight glutenins that are resolved by two-dimensional polyacrylamide gel electrophoresis and are associated with bread-making quality 46
W.M. Funatsuki, K. Takata, A. Kato, K. Saito, T. Tabiki, Z. Nishio, H. Saruyama, E. Yahata, H. Funatsuki, H. Yamauchi

A monoclonal antibody specific for a unique epitope of HMW glutenin subunit 1 allows immunological discrimination of *Glu-A1* alleles 50
H. Gruber, M. Sedlmeier, B. Killermann

Use of recombinant peptides to explore the molecular mechanism of gluten protein viscoelasticity 54
N.G. Halford, A. Savage, N. Wellner, E.N.C. Mills, P.S. Belton, P.R. Shewry

Bacterial expression purification and functional studies of a y-type high molecular weight glutenin present *T. tauschii* 58
M.E. Hassani, M.C. Gianibelli, R.G. Solomon, L. Tamas, M.R. Shariflou, P.J. Sharp

Information hidden in the low molecular weight glutenin gene sequences 62
A. Juhász, M.C. Gianibelli

Identification and characterisation of new chimeric storage protein genes from an old hungarian wheat variety 66
I.J. Nagy, I. Takács, A. Juhász, L. Tamás, Z. Bedő

Wheat protein inhbitors of insect digestive proteinases: molecular characterization and potential biotechnological use 70
E. Poerio, I. Bellavita, S. Di Gennaro, F. Farisei, D. Panichi, A.G. Ficca

Molecular modeling of peptide sequences of gliadins and LMW-glutenin subunits 74
F. Yaşar, S. Çelik, H. Köksel

Genetics and Quality

Effect of the introduction of novel high Mr glutenin subunits on quality of durum wheat semolina 81
M.C. Gianibelli, O.R. Larroque, E. De Ambrogio, D. Lafiandra

Use of segregating double haploid populations to investigate the effects of *Glu-B1* and *Glu-D1* alleles on dough strength 85
B.J. Butow, K.R. Gale, W. Ma, O.R. Larroque, M.K. Morell, F. Békés

Biochemical analysis of gluten-forming proteins in a double haploid population of bread wheat and their relation to dough extensibility 89
A. Juhàsz, O.R. Larroque, H. Allen, J. Oliver, M.K. Morell, M.C. Gianibelli

Contents

The transmission route through which the common wheat (*Triticum aestivum* L.) has reached the Far-East, Japan — 93
H. Nakamura

Allelic variation at the *Glu-1* and *Glu-3* loci, presence of 1B/1R translocation and their effect on dough properties of Chinese bread wheat — 97
Z.H. He, L. Liu, R.J. Peña

Characterisation of LMW-GS genes and the corresponding proteins in common wheats — 101
T.M. Ikeda, T. Nagamine, H. Yano

The use of SE-HPLC for quality prediction in two African countries — 105
M.T. Labuschagne, E. Koen, T. Dessalegn

Biochemical and functional studies of wheat isolines containing single type high molecular weight glutenin subunits (HMW-GS) — 109
O.R. Larroque, B. Margiotta, M.C. Gianibelli, F. Bekes, P.J. Sharp, D. Lafiandra

Influence of glutenins and puroindolines composition on the quality of bread wheat varieties grown in Portugal — 113
A.S. Bagulho, M.C. Muacho, J.M. Carrillo, C. Brites

Variability and genetic diversity for endosperm storage proteins in Spanish spelt wheat — 117
L. Caballero, L.M. Martín, J.B. Alvarez

Biochemical and technological indicators of pasta quality — 121
M. Carcea, N. Guerrieri, E. Marconi, S. Salvatorelli, E. Franchi, M.C. Messia

Genetic variation of the storage proteins in the structural mutant forms of *T. aestivum* L. — 125
T. Dekova, S. Georgiev, K. Gecheff

Genetic variation of the storage proteins in the *sphaerococcum* mutant forms of hexaploid wheat — 129
T. Dekova, Y. Yordanov, S. Georgiev

Durum wheat dough strength relationship to polymeric protein quantity and composition — 132
N.M. Edwards, M.C. Gianibelli, N.P. Ames, J.M. Clarke, J.E. Dexter, O.R. Larroque, T.N. McCaig

LMW-i type subunits are expressed in wheat endosperm and belong to the glutenin fraction — 136
P. Ferrante, C. Patacchini, S. Masci, R. D'Ovidio, D. Lafiandra

Qualitative and quantitative analysis of gliadin composition in an old Hungarian wheat population 140
A. Juhász, R. Haraszi, O.R. Larroque, M.C. Gianibelli, F. Békés, Z. Bedő

Additive and epistatic effects of *Glu-1*, *Glu-3* and *Gli-1* alleles on characteristics of baking-quality and agronomic performance in four doubled haploid wheat populations 144
B. Killermann, G. Zimmermann, W. Friedt

QTL association with measures of gluten strength across environments in durum wheat 148
R.E. Knox, S. Houshmand, F.R Clarke, J.M.Clarke, N.A. Ames

Development of NILs of bread wheat differing in High Molecular Weight Glutenin Subunits and their use in quality related studies 152
B. Margiotta, G. Colaprico, M. Aramini, S. Masci, C. Patacchini, D. Lafiandra

Relationship between some *Glu-D1/Glu-B3* allelic combinations and bread-making quality related parameters commonly used in wheat breeding 156
R. J. Peña, H. Gonzalez-Santoyo, F. Cervantes

Importance of HMW and LMW glutenin subunits and their interactions on bread-making quality 158
L.A. Pflüger, D. Lafiandra, S. Benedettelli

Understanding the functionality of wheat high molecular weight glutenin subunits (HMW-GS) in chapati making quality 162
G. Sreeramulu, R. Banerjee, A. Bharadwaj, P.P.Vaishnav

Environmental Effects

Proteomic analysis of wheat endosperm proteins: changes in response to development and high temperature 169
W.J. Hurkman, C.K. Tanaka, W.H. Vensel, J.H. Wong, Y. Balmer, N. Cai, B.B. Buchanan

Analysis of the effect of heat shock on wheat storage proteins 173
T. Majoul, E. Bancel, E. Triboi, J. Ben Hamida, G. Branlard

Glutenin particles are affected by growing conditions 177
C. Don, G. Lookhart, H. Naeem, F. MacRitchie, R.J. Hamer

Grain protein polymer formation: influences of cultivar, environment and dough treatment 180
E. Johansson, M.L. Prieto-Linde, R. Kuktaite, A. Andersson, H. Larsson

Relationship between some prolamin variants and quality parameters at different levels of nitrogenous fertilizer in durum wheat 184
J.M. Carrillo, M. Ruiz, M.C. Martinez, M. Rodrìguez-Quijano, J.F. Vàzquez

Influence of sulphur fertilisation on the quantitative composition of gluten protein types in wheat flour *P. Koehler, H. Wieser, R. Gutser, S. von Tucher*	188
Enviromental effects on measurement of gluten index and SDS-sedimentation volume in durum wheat *F.R. Clarke, J.M.Clarke, N.A. Ames, R.E. Knox*	192
A dynamic model of the effects of nitrogen fertilization, water deficit, and temperature on grain protein level and composition for bread wheat (*Triticum aestivum* L.) *P. Martre, J. R. Porter, P. D. Jamieson, S.M. Henton, E. Triboï*	196

Gluten Rheology and Functionality

What makes a good theory of gluten viscoelasticity? *P.S. Belton*	203
Viscoelastic and flow behaviour of doughs from transgenic wheat lines differing in HMW glutenin subunits *J. Lefebvre, C. Rousseau, Y. Popineau*	207
Large deformation properties of wheat flour and gluten dough in uni- and biaxial extension *E.L. Sliwinski, P. Kolster, T. van Vliet*	211
Extensional rheology measurements as predictors of wheat quality *G. Mann, F. Békés, M.K. Morell*	215
Dough mixing studies on the micro Z-arm mixer *R. Haraszi, F. Békés, M.L. Bason, J.M.C. Dang, J.L. Blakeney*	219
The effects of dough mixing on GMP re-aggregation and dough elasticity during dough rest *C. Don, W.J. Lichtendonk, J.J. Plijter, R.J. Hamer*	223
Polymer concepts applied to gluten behaviour in dough *F. MacRitchie, H. Singh*	227
Rheological mechanisms of stability of bubble expansion in breadmaking doughs *B.J. Dobraszczyk, J.D. Schofield, J. Smewing, M. Albertini, G. Maesmans*	231
Effect of high-pressure and temperature on the functional and chemical properties of gluten *R. Kieffer, H. Wieser*	235
Carbon atom and the thermal stability of wheat gluten proteins: effect on dough properties and noodle texture *M.I.P. Kovacs, B.X. Fu, S.M. Woods, K. Khan*	239

Why dough inflation should have to modify the rate of gluten protein thermosetting
M. Pommet, A. Redl, S. Domenek, S. Guilbert, M.H. Morel — 243

On the swelling properties of gluten in the presence of salts and reduced pH
H. Larsson, U. Hedlund — 247

Studies on the differing properties of gluten and dough and on the formation of the gluten network
R. Kieffer — 251

The dynamic development and distribution of gas cells in bread-making dough during proving and baking
W. Li, B.J. Dobraszczyk — 255

Vital wheat gluten: chemical and functional aspects
E. Marconi, M.C. Messia, M.F. Caboni, M.C. Trivisonno, G. Iafelice, R. Cubadda — 259

Confocal visualisation of MDD dough development
M.P. Newberry, L.D. Simmons, M.P. Morgenstern — 263

Determination of breadmaking quality of wheat flour dough with different macro and micro mixers
S. Tömösközi, Á. Kindler, J. Varga, M. Rakszegi, L. Láng, Z. Bedő, O. Baticz, R. Haraszi, F. Békés — 267

Rheological properties and microstructure of mixed and unmixed flour-water systems
L. Unbehend, M.G. Lindhauer, F. Meuser — 271

Evaluation of durum wheat quality using micro-scale and basic rheological tests
S. Uthayakumaran, Lafiandra D., M.C. Gianibelli — 275

Gluten extensibility: a key factor in Uruguayan wheat quality
D. Vázquez, B. Watts — 279

Gluten Polymers and Tools to Investigate their Structure

Linking glutenin particle size to HMW glutenin subunit composition
C. Don, G. Mann, F. Békés, R.J. Hamer — 285

Gluten macropolymer in wheat flour doughs: structure and function for wheat quality
R. Kuktaite, H. Larsson, S. Marttila, K. Brismar, M. Prieto-Linde, E. Johansson — 288

Critical factors governing gluten protein agglomeration on a micro-scale
B. Schurgers, W.S. Veraverbeke, E. Dornez, J.A. Delcour — 292

Application of flow field-flow fractionation and multiangle laser light scattering for size determination of polymeric wheat proteins K.R. Preston, S.G. Stevenson, S. You, M.S. Izydorczyc	296
Confocal scanning laser microscopy of gluten proteins and lipids in bread dough W. Li, B.J. Dobraszczyk, P.J. Wilde	300
Dye ligand chromatography in the purification of wheat flour proteins G. Alberghina, M.E. Amato, S. Fisichella, D. Lafiandra, D. Mantarro, A. Palermo, A. Savarino, G. Scarlata	304
Structural studies of wheat flour glutenin polymers and high molecular weight glutenin subunits by circular dichroism spectroscopy M.E. Amato, S. Fisichella, D. Lafiandra, D. Mantarro, A. Palermo, A. Savarino, G. Scarlata	308
Prediction of grain protein content through portable FT-NIR measurement of developing wheat D.G. Bhandari, S.J. Millar, J.C. Richmond	312
Structural analyses of two heterologously expressed native and mutated low molecular weight glutenin subunits V. Consalvi, R. Chiaraluce, C. Patacchini, D. Lafiandra, R. D'Ovidio, S. Masci	316
Gluten surface hydrophobicity of bread and durum wheat N. Guerrieri	320
Relationship between functional properties of wheat dough and the relative proportion of the polymeric fraction I. Kiràly, O. Baticz, O.R. Larroque, A. Juhàsz, S. Tömösközi, F. Békés, A. Guòth, T. Abonyi, Z. Bedő	323
Analysis of sulphur in gluten proteins by X-ray Absorption Near Edge structure (XANES) spectroscopy A. Prange, B. Birzele, H. Modrow, J. Hormes, P. Koehler	327
Molecular characterisation of wheat gluten and barley protein A.A. Tsiami, D.L Pyle, J.D. Schofield	331

Gluten Interactions with Exogenous and Endogenous Components

Association of non-protein components in wheat gluten with its quality L. Day, M. Augustin, I.L. Batey, C.W. Wrigley	337
Ascorbate improver effects, dough mixing properties and bread quality interdependence on flour protein composition D. Every, W.B. Griffin, L.D. Simmons, K.H. Sutton	341

Utilization of Rapid Visco Analyzer for assessing the effect of different levels of transglutaminase on gluten quality *A. Basman, H. Köksel, P.K.W. Ng*	345
Gliadins and polysaccharides interaction *N. Guerrieri, P. Cerletti, F. Secundo*	349
Structure-function relationships of phospholipids in breadmaking *G. Helmerich, P. Koehler*	353
Effect of ascorbic acid in dough: reaction of oxidised glutathione with reactive thiol groups of wheat glutelin *P. Koehler*	357
Effect of pentosans on gluten formation and properties *M. Wang, T. van Vliet, R.J. Hamer*	361

Nutritional Aspects, Intolerances and Allergies

The structural and biological relationships of cereal proteins involved in type I allergy *E.N.C. Mills, J.A. Jenkins, S. Griffiths-Jones, P.R. Shewry*	367
Coeliac disease-specific toxicological and immunological studies of peptides from α-gliadins *H. Wieser, W. Engel, J. Fraser, E. Pollok, H.J. Ellis, P.J. Ciclitira*	371
Mass spectrometry as a tool for probing gluten peptide modifications relevant to celiac disease *G. Mamone, P. Ferranti, D. Melck, F. Tafuro, F. Addeo*	375
Formulation of gluten-free bread using response surface methodology *D.F. McCarthy, E. Gallagher, T.R. Gormley, T.J. Schober, E.K. Arendt*	379
Involvement of lipid transfer proteins in food allergy to wheat *F. Battais, J.P. Douliez, D. Marion, Y. Popineau, G. Kanny, D.A. Moneret-Vautrin, S. Denery-Papini*	383
Properties of proteins in immature wheat grains relevant to their trasformation *F. Bonomi, S. Iametti, M. Zardi, M.A. Pagani, M.G. D'Egidio*	387
Binding of gluten peptides to the coeliac disease-associated HLA-DQ2 molecule by computational methods *S. Costantini, G. Colonna, M. Rossi, A.M. Facchiano*	391
Evidence of *Panicum miliaceum* as a safe food for coeliac patients *F. Fusari, A. Petrini, M. De Vincenzi, N.E. Pogna*	395
Effects of rye and barley in coeliac disease *L. Sabbatella, S. Vetrano, M. Di Tola, C. Casale, M.C. Anania, A. Picarelli*	398

Contents

Nutritional components of mill stream fractions L.D. Simmons, K.H. Sutton, M.S. Noorman	402
Fermented wheat germ extract in the treatment of colorectal cancer R. Tömösközi-Farkas, M. Hidvégi	406

Modifications Due to Parasite Attacks

Specificity of action of the wheat bug (*Nysius huttoni*) proteinase D. Every, K.H. Sutton, P.R. Shewry, A.S. Tatham, T. Coolbear	413
Inhibition effects of plant extracts on the bug (*Eurygaster* spp.) damaged wheat B. Olanca, D. Sivri	417
Effects of intercultivar variation of the gluten proteins and rheological properties of bug (*Eurygaster* spp.) damaged wheats D. Sivri, B. Olanca, A. Atlı, H. Köksel	421
Relationships between timing of *Eurygaster maura* attacks and gluten degradation in two bread wheat cultivars P. Vaccino, M. Corbellini, A. Curioni, G. Zoccatelli, M. Migliardi, L. Tavella	425
Effect of *Fusarium* proteases on breadmaking properties D. Vázquez, S. Gonnet, M. Nin, O. Bentancur	429
Breakdown of glutenin polymers during dough mixing by *Eurigaster maura* protease G. Zoccatelli, S. Vincenzi, M. Corbellini, P. Vaccino, L. Tavella, A. Curioni	433

Non-Food Uses

A way to improve the water resistance of gluten-based biomaterials: plasticization with fatty acids M. Pommet, A. Redl, M.H. Morel, S. Guilbert	439
Wheat gluten based biomaterials: environmental performance, degradability and physical modifications S. Guilbert, M.-H. Morel, S. Domenek	443
Gluten films: effect of composition and processing on properties and structure Y. Popineau, C. Mangavel, J. Guéguen	447

Non-Gluten Components

A lipid transfer protein from farro (*Triticum dicoccum* Schrank) and common wheat (*Triticum aestivum* L. cv. Centauro) A. Capocchi, D. Fontanini, L. Galleschi, L. Lombardi, R. Lorenzi, F. Saviozzi, M. Zandomeneghi	453

Evidence of a gene coding for grain softness protein (*GSP-1a*) on the long arm of chromosome 5D in *Triticum aestivum* 457
L. Gazza, A. Niglio, F. Nocente, N.E. Pogna

Relationship between sequence polymorphism of *GSP-1* and puroindolines in *Triticum aestivum* and *Aegilops tauschii* 461
A. Massa, C.F. Morris

Author Index 465

Subject Index 469

ns and Proteomics
Biotechnology, Transcriptomics and Proteomics

THE USE OF BIOTECHNOLOGY TO STUDY WHEAT ENDOSPERM DEVELOPMENT AND IMPROVE GRAIN QUALITY

P.R. Shewry[1], H.D. Jones[1], M.J. Holdworth[2], J.R. Lenton[3] and K.J. Edwards[4]

1. Rothamsted Research, Harpenden, Herts, AL5 2JQ, UK
2. University of Nottingham, Sutton Bonington, Leicestershire, LE12, 5RD, UK
3. Long Ashton Research Station, Long Ashton, Bristol, BS41 9AF, UK.
4. University of Bristol, Biological Science Building, Woodland Road, Bristol, BS8 1UG, UK

1. INTRODUCTION

Pollination of bread wheat results in a double fertilization event within the embryo sac. One pollen nucleus fuses with the egg cell which subsequently gives rise to the zygote which has the normal hexaploid constitution of 42 chromosomes. At the same time the second pollen nucleus fuses with two polar nuclei in the embryo sac to give rise to the endosperm which consequently has three copies of each chromosome. Therefore, although the endosperm of bread wheat is often referred to as triploid, it actually has 63 chromosomes, nine of each homoeologous group.

The primary endosperm nucleus divides mitotically with the products of the first two divisions establishing the right and left halves and the distal and proximal poles of the endosperm, respectively. Further nuclear divisions then occur, which are initially synchronous, to give a syncytium which, in wheat may have 1000 to 2000 nuclei. This stage is usually reached by about 72 hours, after which cell wall formation occurs. The newly formed cells then divide and differentiate with division becoming restricted to the outer layer of cells which form the aleurone.

Storage products, starch and protein, are first observed in the starchy endosperm cells at about 14 days after anthesis with maximum accumulation occurring over the following 14 days. Subsequently, the grain dries down, the starchy endosperm cells become disorganised and die and the embryo and aleurone enter a state of dormancy. Consequently, the development of wheat under UK conditions can be divided into three phases of approximately equal duration. Phase 1 (0 – 14 days) is when the patterns of cell division and differentiation essentially establish the basic structure and organisation of the tissue. This will include genetically determined differences in size, shape and architecture (e.g. crease structure). Phase 2 (15 – 28 days) is grain filling which determines the final yield and quality of the grain. Phase 3 (29 - 42 days) is desiccation and dormancy development[1]. However, it must be noted that the duration of these phases will be greatly affected by environmental conditions, being shortened under high temperatures

---

[1] The website http://www.wheatbp.net gives further information on grain development including downloadable images of tissue and cells.

It is clear that we need to understand how events taking place during grain development are regulated if we wish to manipulate the yield and quality of the grain

2. TRANSCRIPTOMICS

The identification and quantitization of the whole range of transcripts expressed in specific cells, tissues and stages of development has become a standard tool for molecular biologists as it allows transcripts which are associated with specific characteristics (events, mutations, environmental impacts etc.) to be identified. The standard system is to use arrays of DNA sequences corresponding to specific genes for hybridization against cDNA fractions from the tissue of interest.

In order to generate a resource for transcriptional analysis of wheat development we constructed 35 cDNA libraries, using mRNA fractions from various stages of grain development as well as from vegetative tissues grown under normal and stress conditions. Over 26,000 of these cloned "expressed sequence tags" (ESTs) have been subjected to single pass sequencing (ie, one strand only) and their sequences made publicly available in the IGF (Investigating Gene Function) database (http://www.cerealsdb.uk.net).

10,000 of these EST have also been arrayed on glass slides to give a unigene set which is publicly available for high throughput gene expression studies. We are currently using this array for several projects, including determining the effects of crop nutrition and environmental factors (temperature, water availability) on grain development and quality and comparison of the "substantial equivalence" of GM and non-GM wheat.

3. TRANSFORMATION AND GENE IDENTIFICATION

High throughput transformation is an essential prequisite for determining the functions of transcripts identified by transcriptome analysis and confirming the identities of genes identified by tagging or mutagenesis.

We have focused on developing a routine biolistics (particle bombardment) system which can be applied to a wide range of wheat genotypes[1,2], as discussed in a separate chapter in this volume. In addition, we are focusing on two lines of wheat as tools for functional genomics studies.

Firstly, the cultivar Cadenza, which was grown recently in the UK as a winter wheat, although vernalisation is not strictly required. It is hard with moderate breadmaking quality and is classed as NABIN Group 2 with a NIAB score of 6. We have found that Cadenza gives higher rates of regeneration and transformation than other commercial cultivars which have been grown in the UK in recent years, averaging about 10 % but ranging up to 20 %. We therefore routinely use Cadenza as a "model" commercial bread wheat for transformation.

Secondly, the diploid cultivated species *Triticum monococcum* (einkorn)which is related to the ancestral donor of the A genome of polyploid wheats but is free threshing and has plump seeds and good agronomic performance. As a diploid it is more appropriate for mutagenesis, gene tagging and transformation to determine gene function than is hexaploid bread wheat. We have therefore screened a number of accessions of *T monococcum* from the collections held by the John Innes Centre (Norwich, UK) and the Vavilov Institute (St. Petersburg, Russia) and selected one line which exhibits good regeneration capacity for transformation. The first transformed plants in this line have recently been generated by particle bombardment.

We are currently using *T. monococcum* for two projects aimed at discovering new genes; mutagenesis using chemical and physical mutagens and gene tagging using a system based on the Ac/Ds transposable element system of maize.

4. APPLICATION OF TRANSFORMATION TO IMPROVING GRAIN PROCESSING QUALITY

Much of our work over the past 10 years has focused on understanding the molecular basis for wheat gluten visco-elasticity. This has included the transformation of bread and durum wheats to express additional genes for the quality-associated HMW subunits 1Ax 1 and 1Dx5, including analysis of their effects on dough mixing characteristics and breadmaking quality[3-7]. These studies have shown that the two subunits have dramatically different effects on dough properties Whereas subunit 1Ax1 gave the expected increase in dough strength and gluten elasticity, the expression of subunit 1Dx5 often resulted in low water absorption and the failure of the flour to form a normal dough, giving a dense loaf of low volume. Fractionation of the gluten proteins demonstrated that this was associated with a high proportion of highly cross-linked insoluble glutenin polymers[6]. This could have resulted from an additional cysteine residue present within the repetitive domain of subunit 1Dx5 (when compared with subunit 1Ax1 and all other characterised x-type subunits).

Current work with HMW subunit transgenes is focusing on determining their impact on quality parameters of "general purpose" cultivars grown in the UK and western Europe. This is being achieved by direct transformation of cultivars (notably cvs. Cadenza, Canon and Imp) and by introgression of transgenes unto cultivars by crossing with the transgenic lines reported by Barro *et al*[3].

References

1. S. Rasco-Gaunt, A. Riley, M. Cannell, P. Barcelo, P.A. Lazzeri 2001 *J. Exp. Bot.* **52**, 865
2. G. Pastori, M. Wilkinson, S. Steele, C. Sparks, H.D. Jones, M.A.J. Parry 2001 *J. Exp. Bot.* 52, 857
3. F. Barro, L. Rooke, F. Bekes, P. Gras, A.S. Tatham, R.J. Fido, P. Lazzeri, P.R. Shewry and P. Barcelo, 1997 *Nature Biotech.* **15**, 1295
4. G. He, L. Rooke, S. Steele, F. Bekes, P. Gras, A.S. Tatham, R. Fido, P. Barcelo, P.R. Shewry and Lazzeri, P. 1999 *Molecular Breeding* **5**, 377
5. L. Rooke, F. Bekes, R.J. Fido, F. Barro, P. Gras, A.S. Tatham, P. Barcelo, P. Lazzeri and P.R. Shewry, 1999 *J. Cer. Sci.* **30**, 115
6. Y. Popineau, G. Deshayes, J. Lefebvre, R.J Fido, A.S. Tatham and Shewry P.R 2001 *J. Agric. Food Chem.* **49**, 395
7. H. Darlington, R.J Fido, A.S. Tatham, H. Jones, S.E. Salmon, P.R. Shewry, 2003 *J. Cer. Sci.* In Press

Acknowledgements

Rothamsted Research and Long Ashton Research Station receive grant-aided support from the Biotechnology and Biological Sciences Research Council (BBSRC) of the UK. Work described here is also supported by the BBSRC under the Investigating Gene Function (IGF) and Exploiting Genomics (Ex-Gen) programmes.

AGRONOMIC, BIOCHEMICAL AND QUALITY CHARACTERISTICS OF WHEATS CONTAINING HMW-GLUTENIN TRANSGENES

A.E. Blechl[1], P. Bregitzer[2], K. O'Brien[3], J. Lin[1], S.B. Nguyen[1] and O.D. Anderson[4]

[1]ARS-USDA, Crop Improvement and Utilization Research Unit, Western Regional Research Center, 800 Buchanan St., Albany, CA, 94710, ablechl@pw.usda.gov
[2]ARS-USDA, National Small Grains Germplasm Research Facility, P.O. Box 307, Aberdeen, ID, 83210
[3]University of Idaho, Aberdeen, ID 83210
[4]ARS-USDA, Genomics and Gene Discovery Unit, 800 Buchanan St., Albany, CA, 94710

1 INTRODUCTION

Bread dough strength is primarily dependent on its composition of high-molecular-weight glutenin subunits (HMW-GS), a class of storage proteins that typically comprises 5-10% of flour proteins[1]. We have made a set of transgenic wheats that differ both quantitatively and qualitatively in their HMW-GS compositions. All the transgenics were derived from the cultivar Bobwhite, a hard white spring wheat that contains HMW-GS Ax2*, Bx7, Dx5, By9 and Dy10, and also the 1BL/1RS rye translocation. The introduced transgenes included native wheat genes for Dx5 and/or Dy10[2], or modified genes that encode a Dx5 variant with an extra-long repeat region[3] or a Dy10:Dx5 hybrid subunit[4]. The majority of 28 lines exhibit increases in HMW-GS levels due to additive expression of the transgenes and endogenous genes. However, five lines show transgene-mediated suppression of endogenous HMW-glutenin genes[5]. In this paper, we present results of a single field trial for 32 different transgenic wheats and mixing characteristics for transgenic wheats that over-express native subunits Dx5 and/or Dy10.

2 METHODS AND RESULTS

2.1 Field Evaluations

Thirty-two different transgenic wheats and their non-transformed parent Bobwhite were planted in Aberdeen, Idaho, in the spring of 2001. The planting was a randomized complete block design with four replicates of each entry. All the transgenic wheats contained the transformation marker gene, bialaphos resistance encoded by the *BAR* gene under control of the maize *Ubiquitin1* promoter[6]. Twenty-eight also contained transgenes that encoded HMW-glutenin subunits. Figure 1 shows the average yields of these 33 entries plotted as a frequency diagram. None of the four lines that contained only the marker gene had significantly different yields compared to the non-transformed parent. Most of the lines were phenotypically similar to Bobwhite, although several showed variability for height, vigor, and/or time of maturity (Figure 2). Eleven transgenics with added HMW-glutenin genes had reduced yields (significant at the 5% level in a Dunnett's t test). Four of the latter lines were shorter than the control (significant at the 1% level in a Dunnett's t test) and one of those was also delayed in maturing (Figure 2). Another low-yielding line had a lower test weight than non-transformed Bobwhite.

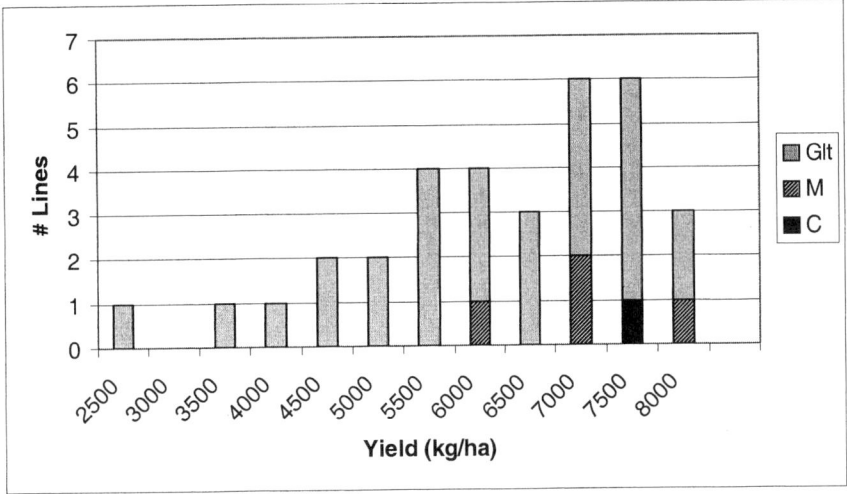

Figure 1 *Frequency diagram showing the number of wheat lines with the plotted yields. C = non-transformed control; M = transgenic wheats containing only the bialaphos-resistance marker gene; Glt = transgenic wheats containing the marker gene and added HMW-glutenin transgenes. Yields below 6000 kg/ha were significantly different from the control at the 5% level.*

Figure 2 *Transgenic wheats growing in field plots just before harvest. The plot on the left contains the transgenic wheat line with the lowest yield, shortest stature and late maturity. The plot on the right contains a transgenic wheat line that had the same yield and appearance as the non-transformed parent in this trial.*

2.2 Mixing and Baking Tests

Flours from these field-grown wheats were milled in a Quadramat Senior mill and subjected to quality analyses. Figure 3 shows traces from 10-gram mixographs[7] for the non-transformed control and five of the transgenic lines with increased levels of the natural Dx5 and/or Dy10 subunits. All such lines had improved mixing tolerance compared to the control, as indicated by decreased slope and/or increased bandwidth after peak resistance was achieved. Lines that contained more than a doubling of Dy10 (A) or Dx5 (B) or both (D) had mixing curves characterized by more rapid development and lower peak resistances, compared to those of the parental cultivar (C), and long stabilities (beyond the seven minutes over which the experiments were conducted). Lines with lesser amounts of transgene-encoded Dx5 and Dy10 increases (E and F) had curves characterized by longer dough development times than the control. Mixograms of flours with unbalanced amounts of Dy10 and Dx5 (A and B) had narrow bandwidths while those with subunit ratios near 1 (D, E and F) had broad bandwidths. This may indicate that more energy can be applied to doughs with more balanced x and y compositions before they break or disengage from the mixing pins. This property may reflect differences in the structure and/or size of the gluten macropolymer when x- and y-type subunits levels are balanced.

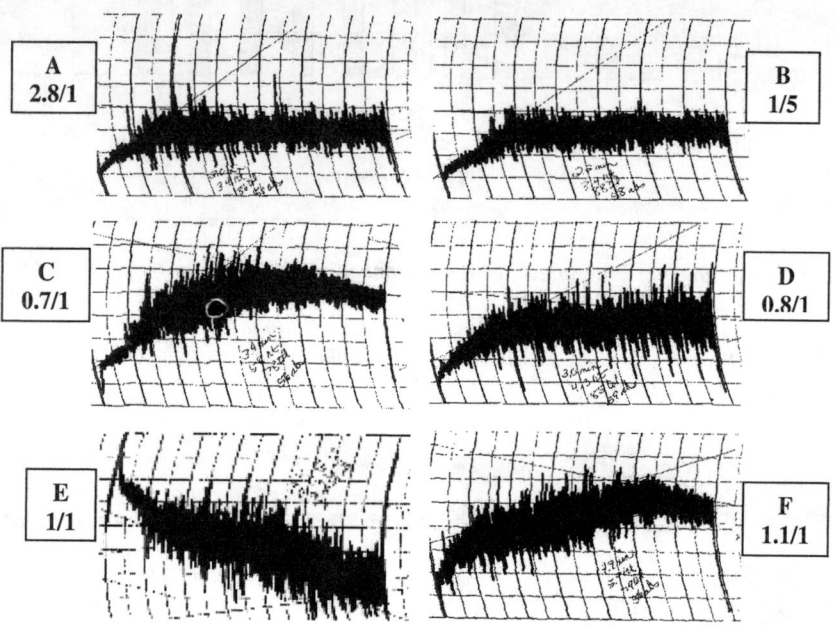

Figure 3 *Mixograms of flour from control (C) and transgenic wheats transformed with Dy10 (A) or Dx5 (B) or both (D-F). The numbers in the boxes are the ratios of subunits Dy10 to Dx5 determined by RP-HPLC.*

None of the transgenic wheats exhibited significant improvements in loaf volume compared to their parent. Of the lines depicted in Figure 3, baking results[8] were obtained for flours C (control), D, E and F (extra Dx5 and Dy10). The control had a loaf volume of 1025 cc, as did line F, while the volume of a loaf made from the flour in E was slightly reduced (1000 cc). The loaf made from flour D was less than 500 cc. The decreases in loaf volumes for D and E could not be accounted for by their flour protein contents, which were greater than that of the control (12.2 and 11.4, respectively, vs 11%).

3 CONCLUSION

We have used genetic transformation to change mixing and baking characteristics of wheat flour. Our field data show that such changes can be achieved without necessarily changing the agronomic characteristics of the wheat plant. Increases in HMW-GS content increase dough mixing tolerances. This is desirable because a wide range of mixing times will produce doughs with optimum development. However, overly strong doughs may require more work input to achieve even hydration and mixing. They also typically produce smaller loaves. As more transgenic lines are characterized by biochemical and quality analyses, we will learn more about the relationship between dough functionality and HMW-glutenin subunit composition.

References

1 P.R. Shewry, N.G. Halford and A.S. Tatham, *J. Cereal Sci.*, 1992, **15**, 105.
2 O.D. Anderson, F.C. Greene, R.E. Yip, N.G. Halford, P.R. Shewry and J.-M. Malpica-Romero, *Nucl. Acids Res.*, 1989, **17**, 3156.
3 R. D'Ovidio, O.D. Anderson, S. Masci, J. Skerritt and E. Porceddu, *J. Cereal Sci.*, 1997, **25**, 1.
4 A. Blechl and O.D. Anderson, *Nature Biotechnology*, 1996, **14**, 875.
5 A.E. Blechl, H.Q. Le and O.D. Anderson, *J. Plant Physiol.*, 1998, **152**, 703.
6 A.H. Christensen and P.F. Quail, *Transgenic Res.*, 1996, **5**, 213.
7 Mixograph Method 54-40, *Approved Methods of the American Association of Cereal Chemists*, 10th edition, AACC, St. Paul, MN, 2000.
8 Optimized Straight-Dough Bread-Baking Method 10-10B, *Approved Methods of the American Association of Cereal Chemists*, 10th edition, AACC, St. Paul, MN, 2000.

CHARACTERIZATION OF GLUTENIN POLYMERS IN A TRANSGENIC BREAD WHEAT LINE OVER-EXPRESSING A LMW-GS

S. Masci[1], R. D'Ovidio[1,2], F. Scossa[1], C. Patacchini[1], D. Lafiandra[1], O.D. Anderson[2], A.E. Blechl[2]

[1]Dipartimento di Agrobiologia e Agrochimica, Università della Tuscia, Via S.C. de Lellis, 01100 Viterbo, Italy (masci@unitus.it)
[2]US Department of Agriculture, Agricultural Research Service, Western Regional Research Center, 800 Buchanan Street, Albany, CA 94710

1 INTRODUCTION

Technological properties of wheat flour depend mainly on the proteins that make up the polymeric network called gluten. High and low molecular weight glutenin subunits (HMW-GS and LMW-GS, respectively) are the most abundant components of gluten and both contribute to the formation of the glutenin polymers, whose size is directly correlated with flour rheological properties. HMW-GS play the most prominent role in bread wheat, whereas the LMW-GS are relatively more important for durum wheat properties. However, LMW-GS also have a role in bread-making quality.

Because LMW-GS are the most common polypeptides in the glutenin polymer and because their relative amount and/or allelic forms are known to influence dough visco-elasticity properties, we are investigating the effects of increasing the LMW-GS fraction. Here we present the results of analyses of flours from a transgenic bread wheat line that over-expresses a LMW-GS gene controlled by its own promoter. We compare these results to those obtained from transgenic wheats that over-express HMW-GS.

2 MATERIALS AND METHODS

Immature embryos of the bread wheat cultivar Bobwhite were used for transformation experiments[1]. UBI:BAR and pLMWF23A plasmid DNAs, the latter containing a LMW-GS gene cloned from the *Glu-D3* locus of the bread wheat cultivar Cheyenne[2], were used for wheat transformation. The alcohol-soluble and insoluble seed protein fractions were analysed by one and two dimensional SDS-PAGE under reducing and non-reducing conditions, by RP-HPLC, and by SE-HPLC. Southern blotting of genomic DNA were probed with a cloned LMW-GS gene. Protein content was determined by the Kjeldahl method (N X 5.7) and the SDS sedimentation test was performed according to Dick and Quick[3].

3 RESULTS

One out of the eleven bialaphos-resistant transformed plants showed detectable over-expression of the transgenic LMW-GS (Figure 1). Plants were grown from T_1 seeds that showed over-expression. The integration of the transgene in these wheat plants was demonstrated by Southern blotting which showed the presence of the expected 3.7 kbp fragment derived from the plasmid used in the transformation experiment (data not shown). Plants over-expressing the transgenic LMW-GS were propagated by selfing for three further generations.

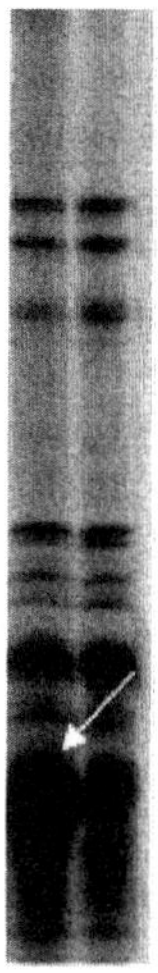

Figure 1 *SDS-PAGE of protein extracts from a T_1 seed (BW_T) belonging to the line showing over-expression of the transgenic LMW-GS and from a seed protein extract from untransformed cv. Bobwhite (BW_{WT}). The arrow shows the position of the transgenic LMW-GS.*

Seed proteins were fractionated by ethanol solubility. Comparing the soluble and insoluble fractions showed that the transgenic LMW-GS is incorporated into the glutenin polymers, and it is not present as a monomer (data not shown). Moreover, two-dimensional SDS-PAGE analysis showed that the transgenic LMW-GS is the main component of the oligomers present in the alcohol-soluble fraction (data not shown).

Densitometric analysis performed on the SDS-PAGE patterns of total endosperm protein extracted from T_3 seeds indicated a twelve-fold increase of the transgenic product compared to native LMW-GS with similar molecular weights, whereas RP-HPLC of the reduced insoluble fraction indicated a sixteen-fold increase (Figure 2).

To test whether over-production of the transgene-encoded LMW-GS changes the molecular size distribution of glutenin polymers, SE-HPLC was performed on both the soluble and insoluble (residue) fractions of expressor and non-expressor seeds of the same generation (Figure 3). The chromatograms were divided into three areas, roughly corresponding to glutenin polymers greater than 500,000 kDa in molecular weight and mainly containing HMW-GS and LMW-GS (P1), smaller glutenin polymers plus higher molecular weight monomers (P2), and oligomers plus monomers less than 30,000 kDa

(P3), respectively. The total soluble fraction obtained from the expressor genotype (dashed line in Figure 3A) is 65% lower than that from the non-expressor. Conversely, the total area occupied by the insoluble fraction from the expressor genotype (Figure 3B) is 1.2 times larger than that of the non-expressor genotype, indicating that the expressors have a larger amount of glutenin polymers, and a lower content of smaller oligomers and gliadins, compared to the non-expressor.

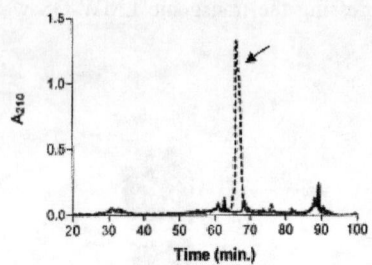

Figure 2: *Comparison between the RP-HPLC patterns of cv. Bobwhite (solid line) and the transformed genotype (dotted line). Arrow shows the peak corresponding to the transgene-encoded LMW-GS*

The same flour samples used for SE-HPLC were submitted to the SDS-sedimentation test. The transgenic genotype (expressor) shows a considerably lower SDS sedimentation value (55 mm) compared to the control (74.3 mm), even though the former's seed protein content is higher (17.3% in the expressor and 15.7% in the non-expressor). Since increased protein content generally increases SDS sedimentation values, these results suggest that the quality of the protein in the LMW-GS transgene expressors, as measured in this test, has been changed.

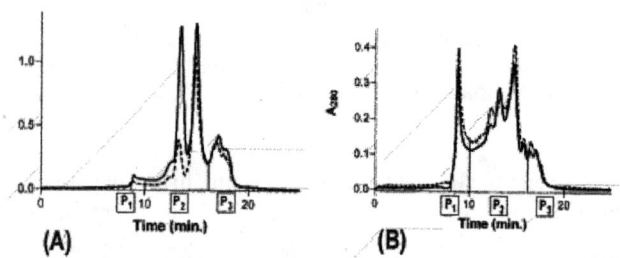

Figure 3 *SE-HPLC patterns of the alcohol-soluble (A) and insoluble (B) fractions of expressor (dashed lines) and non expressor (solid lines) genotypes. P_1, P_2, and P_3 define areas corresponding to different molecular weight ranges.*

4 CONCLUSION

Our experiments comprise the first report of transformation of bread wheat with a LMW-GS gene and allow us to test, for the first time, the effect of an increase in the amount of this protein class on gluten visco-elastic properties. Previous work has indicated that transformation of wheat with HMW-GS genes modifies dough properties, as measured by mixographic and rheological analyses and baking tests. Effects varied, depending on the type of subunit and the recipient genotype$^{4\text{-}6}$ and 1Dx5^{4,5,7}. Worthy of note is that the presence of transgenic 1Dx5 consistently resulted in either an over-strong dough^{4,7}, or in an inability to form a cohesive dough^5. The behaviour of doughs containing elevated levels of Dx5 could be due to the presence of the extra cysteine residue, compared to the other x-type subunits, that very likely increases cross-linking. In comparison to HMW-GS, the smaller size of the repetitive domain and the different cysteine distribution in LMW-GS could be expected to contribute in a different manner to the glutenin architecture and, consequently, to gluten visco-elastic properties. At the same time, glutenin polymers containing the transgenic LMW-GS could be smaller than those formed by HMW-GS, thereby allowing a finer tuning of glutenin polymer size. However, the lower sedimentation volumes of the transgenic flour measured by the SDS sedimentation test (indicative of lower gluten strength), show that over-expression of the transgenic LMW-GS, even when found in a higher protein seed, negatively affects gluten visco-elastic properties. This observation could suggest that the existing correlations between the size and/or amount of glutenin polymers and quality parameters are not valid under all conditions. If this hypothesis is valid, then reducing the level of expression of the transgenic LMW-GS, either by decreasing the number of active genes (through segregation, if the multiple insertions are not located at closed linked loci) or by finding transformants with lower expression of the transgene, might result in improved strength.

Our results demonstrate that gluten polymer composition can be altered by over-expression of a LMW-GS and that such changes affect wheat end-use properties.

References

1. A.E. Blechl and O.D. Anderson, *Nat. Biotechnol.*, 1996, **14**, 875
2. B.G.Cassidy, J. Dvorak and O.D. Anderson, *Theor. Appl. Genet.*, 1998, **96**, 743
3. J.W. Dick and J.S. Quick, *Cereal Chem.*, 1983, **60**, 315
4. G.Y. He, L. Rooke, S. Steele, F. Békés, P. Gras, A.S. Tatham, R. Fido, P. Barcelo, P.R. Shewry and P.A. Lazzeri, *Mol. Breed.*, 1999, **5**, 377
5. Y. Popineau, G. Deshayes, J. Lefebvre, R. Fido, A.S. Tatham and P.R. Shewry, *J. Agric. Food Chem.*, 2001, **49**, 395
6. I.K. Vasil, S. Bean, J. Zhao, P. McCluskey, G. Lookhart, H.-P. Zhao, F. Altpeter and V. Vasil, *J. Plant Physiol.*, 2001, **158**, 521
7. L. Rooke, F. Bekes, R. Fido, F. Barro, P. Gras, A.S. Tatham, P. Barcelo, P. Lazzeri and P.R. Shewry, *J. Cereal Sci.*, 1999, **30**, 115

Acknowledgements

Research supported by the Italian Ministry for University and Research (MIUR), projects "Aspetti biochimici, genetici e molecolari delle proteine della cariosside dei frumenti in relazione alle caratteristiche nutrizionali e tecnologiche dei prodotti derivati" (PRIN 2002), and "Espressione genica ed accumulo di proteine d'interesse agronomico nella cellula vegetale: meccanismi trascrizionali e post-trascrizionali" (FIRB RBNE01TYZF).

THE POTENTIAL OF BIOTECHNOLOGY TO PRODUCE NOVEL GLUTEN PHENOTYPES

H.D. Jones, H. Wu, C. Sparks, A. Doherty, M. Cannell[1], J. Goodwin, G. Pastori and P.R. Shewry

huw.jones@bbsrc.ac.uk CPI Division, Rothamsted Research, Harpenden, Herts, AL5 2JQ, UK. [1][Present address: Syngenta, Jealotts Hill International Research Centre, UK]

1 INTRODUCTION

The introduction of foreign genes into wheat is a powerful tool for research and to improve commercial wheat germplasm. The successful application of genetic modification depends upon an effective transformation system, the availability of genes for target traits and the use of regulatory sequences capable of driving appropriate levels of expression in the tissues and developmental stages required. Robust and relatively efficient transformation methods using biolistics to deliver DNA into regenerable immature embryos are now available for wheat. These protocols have proved valuable to analyse and validate the function of many genes, and recently used to demonstrate the production of 'clean' wheat lines with no ampicillin-resistance genes. However, biolistic DNA delivery can result in high copy number insertions with rearrangements that can make analysis difficult and lead to unpredictable silencing. The advantages of transformation via *Agrobacterium* have been recognised for several years but wheat has proved recalcitrant and lagged behind other plants in its ability to be transformed in this way. The first report of *Agrobacterum*-mediated transformation of wheat was by Cheng *et al.* using the model wheat variety Bobwhite.[1] We have developed protocols for the transformation of bread wheat varieties via *Agrobacterium* and are currently analysing transgene expression and inheritance in lines generated by this method and developing tools to target and regulate transgene expression.[2]

Production of 'clean' transgenic lines.

The presence of antibiotic resistance genes, the origin of replication and other sequences involved in maintenance of plasmids in bacteria are undesirable in transgenic plants, particularly when environmental release is anticipated. Our aim was to produce transgenic plants containing the 1Ax1 or 1Dx5 fragment with no other recombinant DNA. The plasmid backbone and other unnecessary sequences were separated from the DNA to be transferred by digestion and purification. The linear fragments containing the trait gene (eg. genomic 1Dx5 fragment) and the plant selectable marker (*bar* expression cassette) were transferred to plants using conventional biolistic wheat transformation[3,4]. More than thirty independent lines were obtained, the majority of which expressed the *bar* and 1Dx5 genes. We are currently analysing the transgene insertions in this population for lines in which the *bar* selectable marker and 1Ax1 or 1Dx5 fragment have integrated at unlinked loci. Previous experience indicates that this occurs in approximately 5% of lines. These plants will represent an important demonstration that a genomic fragment of functional wheat DNA can be moved from one variety to another with no unnecessary associated sequences.

Agrobacterium-mediated wheat transformation

We have used freshly isolated immature embryos of cv. Florida and *Agrobacterium* strain AGL1 harbouring pAL154/pAL156 (shown previously to effectively transfer T-DNA to wheat embryos[5]) containing a T-DNA incorporating the *bar* gene and a modified *uid*A (GUS) gene, to investigate and optimise major T-DNA delivery and tissue culture variables for the efficient generation of stable transformants. Transient expression of GUS was used to assess *Agrobacterium*-mediated DNA delivery and frequency of callus formation or regeneration. A summary of the results follow. The addition of acetosyringone to the inoculation and co-cultivation media increased the efficiency of T-DNA delivery both in terms of the number of embryos that displayed at least one focus of GUS staining and the average numbers of GUS foci per explant. Both scores were higher when using larger embryos (L and XL). However, the response of these embryos in tissue culture showed the opposite trend, with the smaller-sized class showing significantly higher regeneration frequency.

Embryos of size 0.8mm to 2.0 mm were chosen for a compromise between good T-DNA delivery and recovery of regenerants. DNA-delivery was improved at higher Silwet L-77 concentrations, presumably due to lowering of surface tension allowing better of penetration of bacteria into the plant cells. However, at the Silwet L-77 concentrations where this effect was most marked (0.02 – 0.04%), phytotoxicity of the surfactant killed embryos and/or prevented callus induction. A total of 44 transgenic lines were generated using these optimised parameters. Of the seven checked so far by Southerns, four had a single transgene copy and three had multi-copy insertions. Expression studies on the two genes in the T-DNA revealed that half the 44 lines showed a 3:1 segregation pattern for at least one of the transgenes.

Tools to provide desired levels of transgene transcription
Promoter catalogue

One of the current major limitations in wheat transformation is the lack of suitable promoters to target expression of identified genes of interest. There is a need for a range of well-analysed promoters with known strength and different expression characteristics. Current work involves analysing promoters from a variety of genes and species in transient and stable transformation systems. We used *Uid*A (GUS), GFP and anthocyanin genes scorable marker genes. The aim is to accumulate a "catalogue" of constitutive, tissue-specific, developmentally regulated and inducible promoters of different strength, from which to select those appropriately targeted for specific transformation projects. The rice Actin and maize Ubiquitin promoters were generally constitutive, showing expression in all major organs and in the majority of tissue types assayed. The rice tungro bacilliform virus promoter gave expression in most tissue types, but was notably absent in roots and pollen. Other promoters clearly showed spatial and/or temporal specificity. For example, 1Dx5 and Bx17 expression was evident only in seed endosperm and α-Amylase 1 only in the scutellum region of the embryo. Expression from the tissue-specific promoters was tightly regulated demonstrating their potential to express genes of interest in a highly controlled manner. Although *uid*A is a reliable reporter gene, the destructive nature of the assay means it has limited use in monitoring real-time developmental expression. Both GFP and anthocyanin offer promising alternative non-destructive assay systems as shown by transient expression studies. The efficacy of these reporter genes is currently being assessed in stably transformed plants (table 1 and figure 1).

Lines available	Promoter
6	Rice Actin
>50	Maize Ubiquitin
7	Rice tungro bacilliform virus
>50	Wheat HMW glutenin (1Dx5)
3	Wheat HMW glutenin (Bx17)
12	Wheat α-Amylase 1 and 2

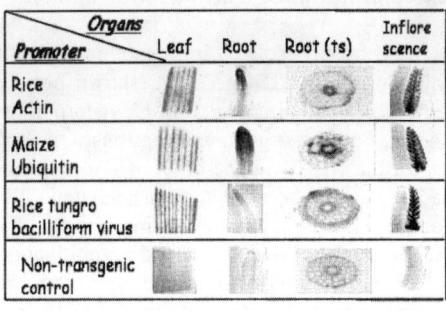

Table 1. *Transgenic plants containing the uidA gene under the control of various promoters*

Figure 1. *GUS expression (dark staining) driven by various promoters in leaf, root and immature inflorescence of wheat*

Matrix attachment regions

Several studies have shown that Matrix attachment regions (MARs) enhance transgene expression and reduce the variability of expression caused by position effects.[6,7,8] The mechanism by which MARs exert this influence is not fully understood, although there is evidence to suggest that they are able to reduce the effectiveness of transcriptional gene silencing and that they may promote the attachment of chromatin loops to the nuclear matrix, which protects transgenes on this loop from negative effects on gene expression. The objective of this project was to investigate the effect of cereal MAR sequences on the transgenic expression of the wheat high molecular weight glutenin (HMWG) subunit gene (1Dx5). In addition to basic plus and minus MARs-HMWG constructs, transposable elements were cloned, flanking the MARs-HMWG expression cassette and be used to generate new insertion sites in subsequent generations from a selection of primary transformation events. Total protein from ten T1 seeds of each primary transgenic line were separated by SDS-PAGE. The gels were digitally imaged and protein bands quantified using Phoretix 1D software (Nonlinear Dynamics, UK). Clear differences between the MAR and non-MAR lines were evident, with levels of expression from the 1Dx5 transgene flanked by MARs significantly higher (Figure 2).

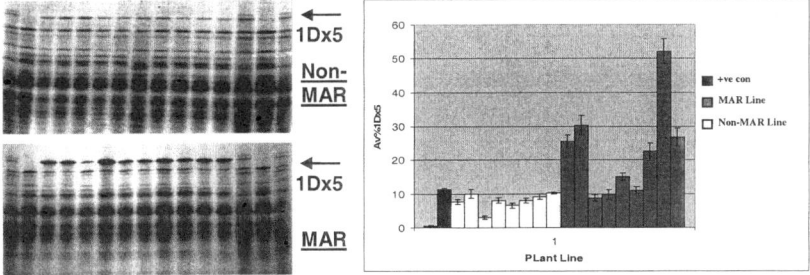

Figure 2 *SDS-PAGE separation of total proteins from ten T1 seeds from a non-MAR and a MAR transgenic line in the background L88-31 (left). Quantitation (in arbitary units) of the 1Dx5 transgene either flanked by MARs or not.*

References

1. M. Cheng, J.E. Fry, S.Z. Pang, H.P. Zhou, C.M. Hironaka, D.R. Duncan, T.W. Conner, Y.C. Wan, *Plant Physiol.*, 1997, **115**, 971
2. H. Wu, C. Sparks, B. Amoah, H.D. Jones, *Plant Cell Rep*, 2003, **21**, 659.
3. G.M. Pastori, M.D. Wilkinson, C. Sparks, H.D. Jones, M.A.J. Parry, *J Exp Bot*, 2001, **52**, 857.
4. S. Rasco-Gaunt, A. Riley, M. Cannell, P. Barcelo, P.A. Lazzeri. *J. Exp. Bot.*, 2001, **52**, 865.
5. B.K. Amoah, H. Wu, C. Sparks, H.D. Jones, *J Exp Bot*, 2001, **52**: 1135
6. G.C. Allen, G. Hall, S. Michelowski, W. Newman, S. Spiker. A.K. Weisinger and W.F. Thompson, *Plant Cell*, 1996, **8**, 899
7. G.C. Allen, G.E. Hall, L.C. Childs, A.K. Weisinger, S. Spiker and W.F. Thompson, *Plant Cell*, 1993, **5**, 603
8. P. Breyne, M. van Montagu, A. Depicker and G. Gheysen, *Plant Cell*, 1992, **4**, 463.

Acknowledgements

Rothamsted receives grant-aided support from the Biotechnological and Biological Sciences Research Council UK. Thanks go to Carmen Lamacchia, Sue Steele and Marlon Stone for provision of 1Dx5, Bx17 and α-Amylase lines respectively. The pBx17::GUS construct was kindly provided by Dr. Laszlo Tamas, ELTE University, Budapest, Hungary.

THE USE OF ESTS TO ANALYZE THE SPECTRUM OF WHEAT SEED PROTEINS

O.D. Anderson, V. Carollo, S. Chao, D. Laudencia-Chinguanco, G.R. Lazo

Genomics and Gene Discovery Research Unit, Western Regional Research Center, Agricultural Research Service, U.S. Department of Agriculture, 800 Buchanan Street, Albany, CA, USA, oandersn@pw.usda.gov

1. INTRODUCTION

In the past few years the number of Triticeae ESTs (Expressed-Sequence-Tags) has gone from less than 100 to over 800,000 wheat, barley, rye and related grass ESTs in the public NCBI (National Center for Biotechnology Information: http://www.ncbi.nlm.nih.gov) database. In fact, wheat now has the greatest EST resource of all plants - more than model systems such as rice and *Arabidopsis*. ESTs are valuable resources for candidate gene identification, gene structure characterization, evolutionary studies, informatics and physical resources for DNA microarray development, gene expression profiling, generation of DNA markers (SNPs, SSRs, RFLP), and genome physical mapping.

Among the cDNA libraries used to generate wheat ESTs are many from developing seed, endosperm, and whole grain - all of which would be expected to contribute to a large number of ESTs of the developing endosperm proteins. By analysis of the public ESTs, it is possible to develop detailed information on many aspects of seed genes, their encoded proteins, and relative expression levels by *in silico* (bioinformatics) methods. Two examples are given for "mining" the wheat EST resource for novel information on gene structure.

2 METHODS AND RESULTS

2.1 EST Analysis

ESTs (Expressed-Sequence-Tags) are short sequence reads, usually from either the 5' or 3' end of a cDNA clone randomly picked from cDNA recombinant libraries constructed from mRNA preparations. For a cDNA project, typically thousands of clones from each of many cDNA libraries are sequenced at one or both ends. The short (100-700 base) DNA sequences become "tags" for the genes from which the mRNA originated. ESTs can be analyzed to determine if there are ESTs that contain overlapping sequences from which a consensus contiguous sequence, or "contig" can be derived. Such overlapping partial sequences can allow reconstruction of complete original sequences, although a complication is the potential for erroneous assembly of sequences from different genes, particularly from paralogous multigene families and orthologous genes.

Assembling large numbers of ESTs gives a collection of contigs (sets of overlapping ESTs) and singletons (non-overlapping with other ESTs). The sum of the contigs and singletons approximate the total individual genes represented in the EST collection and are termed a unigene set. The results of an assembly are dependent on the specific algorithm used for assembly and the settings of relevant parameters. For example, in the recently completed U.S. NSF EST/Mapping project, the last assembly of project ESTs on January 8, 2003 used 115,510 ESTs and resulted in 18,876 contigs and 23,034 singletons, or 41,910 unigenes.

Two examples of using wheat ESTs to study seed proteins will be summarized. Once a complete gene coding region sequence is known, ESTs can allow a reliable assembly of allelic genes from other germplasm using the known sequence to "seed" the EST assembly. A reconstruction can also be accomplished without the known sequence, but then uncertainty as to the accuracy of the final result increases. Figure 1 shows the full-length amino acid sequence of the cultivar Chinese Spring high-molecular-weight (HMW) glutenin Bx7 subunit as derived from an EST assembly and using the cv Cheyenne Bx7 sequence[1] as the seed. Throughout the reconstructed DNA sequence, the only difference found was one complete repeat motif found in Chinese Spring but not Cheyenne. An apparent neutral single base change (leucine residue indicated by *) was an error in the Cheyenne sequence[1]. The region containing the repeat motif difference is supported by 28 Chinese Spring EST sequences, 27 of which are identical throughout this region and one of which contains poor quality reads and was discarded as not being representative of the major Bx sequence. As a check on the Cheyenne Bx7 sequence, it was also reconstructed

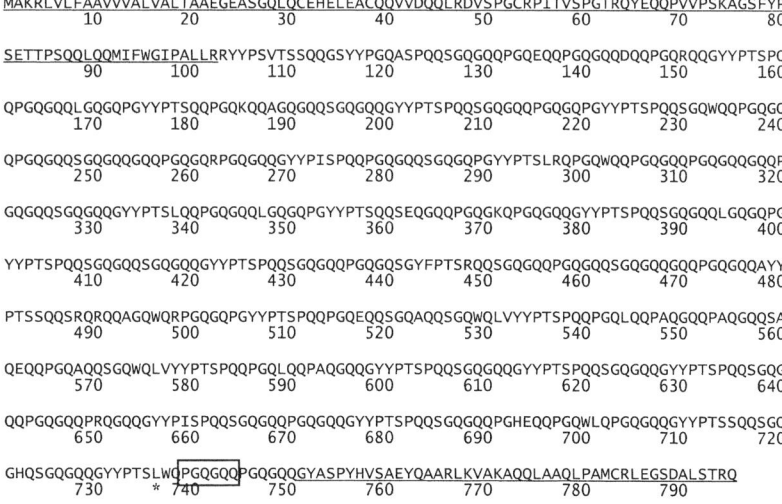

Figure 1 *Sequence of the EST-derived Chinese Spring Bx7 subunit; * indicates error in original Cheyenne sequence; box indicates repeat motif not present in cv Cheyenne Bx7 subunit; nonrepetitive terminal domains underlined*

from ESTs and found identical to the published sequence except the mentioned single error. But, the Chinese Spring and Cheyenne Bx7 HMW-glutenin coding regions are not identical in spite of their common nomenclature.

One of the most elusive wheat seed protein families has been the ω-gliadins. The first DNA sequences were reported for two ω-gliadin genes[2] likely from the D-genome. However, although the B-genome ω-gliadins are correlated with quality, no gene sequences have been reported, and only limited peptide fragment sequences are known[3]. By searching the wheat EST collection, candidate B-genome ω-gliadin clones can be identified. This was accomplished using the N- and C-terminal peptide sequences[3] and the tBLASTn algorithm at NCBI (http://www.ncbi.nlm.nih.gov) or the GrainGenes wEST site (http://wheat.pw.usda.gov/wEST). The identified ESTs were assembled into contigs and some of the results shown in Figure 2 and include N- and C-terminal portions of sequences, giving the first extensive pictures of the structure of B-genome ω-gliadin genes

A) C-terminus

```
EST Contig                  PSGSDIISISGL
ω-gliadin gene Fg20b²       PYGSSLTSIGGQ
ω-gliadin gene G3²          PSGSSLTSIGGQ
Consensus                   P GS    SI G
```

B) N-terminus

```
EST Contig (DNA)            SRLLSPRGKELHTPQEQFPQQQ
1B2 ω-gliadin³ (Peptide)    SRLLSPRGKELHTPQEQFPQQQ
```

C) N-terminus Repeats

```
FP  QQQQ
FP  QPQQ
FP  QQQ
IP  QQHQ
IP  QQPQQ
FP  QQQQ
FL  QQQQ
IP  QQQ
IP  QQHQ
IP  QQPQQ
FP  QQQ
FP  QQQQ
FP  QQHQ
```

D) C-terminus Repeats

```
FP   QQQ          IP   QQPQQ
LGG  QQQ          FP   QQQ
IP   QQQQ         FP   QQQQ
IP   QQPQQ        FP   QQQE
IP   QQQQ         FP   QQQ
IP   QQPKQ        FP   QQQ
FP   QQQ          FH   QQQ
FP   QQQ          LP   QQQ
FP   QQQ          FP   QQQ
FP   QQQ          FP   QQQ
FP   QQ           FP   QQQQ
FP   QQQQ         FP   QQQQ
FP   QQQ          LT   QQQ
IA   QQPQQ        FPR  PQQ
LP   QQQQ         SP   EQQQ
IP   QQPQL        FP   QQQ
FP   QQQQ         FP   QQ
FP   QQQ          PP   QQ
SP   QQQQ         FP   QQQ
FP   QQQ          FP   IP
FP   QQQQ         PP   QQ
LP   QQQ          SQ   IP
FP   QPQQ         SP   YQQ
IP   QQQQ         YP   QQQ
```

Figure 2 *B-genome ω-gliadin amino acid sequence derived from wheat EST contigs* and proteins. Figure 2A shows the C-terminal region of a B-genome ω-gliadin sequence encoded by a contig compared to the N-termini of the two known full-length ω-gliadins[2]. The N-terminal sequence of another contig encodes the identical peptide as found for a B-genome ω-gliadin[3] (Figure 2B). Boxed sequences in Figure 2D indicate identical peptide sequences to those determined for B-genome ω-gliadins[3]. As suggested[3], the repetitive domain (Figure 2C, 2D) of the B-genome ω-gliadins (consensus repeat motif of FPQ_{2-5}) is structured different from the ω-gliadins of other genomes (consensus motif of $PFPQ_{1-2}PQQ$). It is interesting to note that the two previously reported ω-gliadin genes[2] were originally isolated during a probe for γ-gliadin genes – presumably because their repeat motif pattern (consensus motif of $PFPQ_{1-2}PQQ$) was similar to that of γ-gliadins[4] ($PFPQ_{1-2}(PQQ)_{1-2}$). Researchers attempting to isolate ω-gliadin genes from different Triticeae germplasms should be aware that this difference in repeat pattern should be taken into consideration when choosing probes, or they may not succeed depending on the specific ω-gliadin probe used. If confirmed, this would be a unique situation for the wheat prolamin gene families.

A general word of caution is warranted when analyzing derived contig sequences – since they originate from different clones, the accidental merging of ESTs from orthologous or paralogous genes must be considered. Such a possibility can often be reasonably discounted by checking the contig sequence through full-length sequencing of strategically placed EST members of the contig, or if sufficient distinctive EST sequences rule out merging artifacts.

3 CONCLUSION

The good news is that Triticeae ESTs are a tremendous resource, and perhaps half of the total wheat genes have been identified via ESTs. The bad news is that the other half of the wheat gene complement is still to be identified (and are likely to be among the lesser expressed, and more interesting genes), and the degree of complexity in EST analysis caused by multi-gene families and orthologs is still to be determined. Nevertheless, efforts to analyze the increasing numbers of Triticeae ESTs is a worthy task certain to lead to new knowledge and insights.

References

1 O.D. Anderson and F.C. Greene, *Theor. Appl. Genet.*, 1989, **77**, 689.
2 C.C. Hsia and O.D. Anderson, *Theor. Appl. Genet.*, 2001, **103**, 37.
3 F.M. Dupont, W.H. Vensel, R. Chan, and D.D. Kasarda, *Cereal Chem.*, 2000, **77**, 607.
4 O.D. Anderson, C.C. Hsia, and V. Torres, *Theor. Appl. Genet.*, 2001, **103**, 323.

TRANGENIC EXPRESSION OF EPITOPE-TAGGED LMW GLUTENIN SUBUNITS IN DURUM WHEAT: EFFECT ON POLYMER SIZE DISTRIBUTION AND DOUGH MIXING PROPERTIES.

P. Tosi[1], R. D'Ovidio[2], F. Bekes[3], J.A. Napier[1], P.R. Shewry[1]

1. Rothamsted Research, Harpenden, AL5 2JQ, UK.
2. Dipartimento di Agrobiologia e Agrochimica, Università della Tuscia, Via San Camillo de Lellis, 01100, Viterbo, Italy.
3. CSIRO Plant Industry, PO Box 7, North Ryde, NSW 1670, Australia.

1. INTRODUCTION

Low molecular weight (LMW) subunits are important determinants of dough quality in both durum and bread wheats, being major components of the highly cross-linked polymers that confer viscoelastic properties to gluten. They comprise a complex mixture of proteins with similar structures and properties and have both quantitative and qualitative effects on dough viscoelasticity[1].

In order to determine the relationship between dough viscoelasticity and the amounts and properties of the LMW subunits, we have transformed the pasta wheat cultivars Ofanto and Svevo with three genes encoding proteins which differ in their numbers and positions of cysteine residues. The transgenic proteins were tagged at their C-termini with a 14 amino acid residue c-myc epitope to facilitate their detection. This has allowed the determination of transgene expression without compromising the correct folding of the transgenic subunits and their incorporation into the glutenin polymers was possible.

Detailed analyses were carried out of two lines expressing a LMW subunit with two cysteine residues available for inter-molecular disulphide bonds. In one line this resulted in increases in the proportion of high molecular mass glutenin polymers and dough strength. In contrast, the second line exhibited down-regulation of endogenous LMW subunit genes and both polymer size and dough strength were negatively affected.

2. MATERIALS AND METHODS

2.1 Materials

2.1.1 Plant material Immature inflorescence and scutellum explants of cultivars Svevo and Ofanto were used for transformation.
2.1.2 Plasmid DNAs Plasmids pRDPT$_5$1B*, pRDPT$_5$1B*_ and pRDPT$_5$1A3*[2] contained the epitope tagged LMW subunit transgenes under the control of the high molecular weight (HMW) subunit 1Dx5 gene promoter and terminator regions.

2.2 Methods

2.2.1 Durum wheat transformation Immature inflorescence and scutellum explants were co-transformed with a plasmid containing the gene of interest and plasmid pAHC25. The transformation and regeneration/selection procedure was as for Lamacchia *et al.*(2001)[3].

2.2 2 DNA analysis Total genomic DNA was isolated from leaves of plants surviving selection and PCR and Southern blotting analysis were performed as described[4].

2.2.3 Protein analysis Total proteins were extracted from single half grains and separated by SDS PAGE using a Tris-borate buffer system and by western blotting using the anti-*c-myc* polyclonal antibody A-14 with an anti-rabbit alkaline phosphatase conjugated secondary antibody.

Reversed-phase HPLC of glutenin proteins was performed as in Marchylo *et al*[5]. Samples were extracted in duplicate with a single analysis of each extract being performed.

Size-exclusion HPLC was performed on total unreduced proteins extracted from flour by sonication for 15 sec in 0.5% (w/v) SDS in 0.05M sodium phosphate buffer, pH6.9, on a Phenomenex BiosepTM SEC-400 column (Phenomenex) with an eluant of acetonotrile/water (50/50) containing 0.05% (v/v) TFA.

Mixograph analysis Grain samples were milled on a FQC-2000 Micro scale laboratory mill. Mixing was carried out in triplicate in a 2-g MixographTM (TMCO, Lincoln, NE, USA). Parameters recorded were: mean time to peak dough development (PDD); mixing time (MT); Mixograph peak resistance (MPR); resistance breakdown (RBD); bandwidth breakdown (BWBD); maximum bandwidth (MBW) and time to maximum bandwidth (TMBW).

3. RESULTS AND DISCUSSION

3.1 Production of transgenic durum wheat lines expressing epitope-tagged LMW glutenin subunits

PCR analysis of putative transformants led to the identification of eleven transgenic plants (Table 1) containing both the *bar* selectable marker gene and the gene of interest. Transformation efficiencies varied between 0.6 and 3.1% while the overall co-transformation efficiency was of 64%. Southern blotting using the restriction endonuclease *Xba*I, which linearises the plasmid containing the genes of interest, showed that all the primary transformants contained multiple inserts. Hybridising fragments larger than the size of the plasmid used for transformation (6.4 kb), which were presumed to correspond to full-length versions of the genes, were detected in DNA from all of the transgenic lines. However, hybridising bands of the same size or smaller than the transformation constructs, indicating the presence of truncated and/or rearranged forms of the genes, were also detected in the same lines.

3.2 Detection of transgenic LMW glutenin subunit in seeds of transformed plants

The availability of a polyclonal antibody against the *c-myc* epitope-tag allowed the identification of transgene expression by immunoblotting. Analysis of 25-50 seeds of each primary transformant showed expression of the transgenes in all eleven PCR positive lines. The identification of the transgenic protein by SDS-PAGE and staining with Coomassie Blue, however, was only possible in eight of the transgenic lines, either because they were expressed at very low levels or because they co-migrated with endogenous subunits.

3.3 Analysis of the glutenin subunit composition by RP-HPLC

More detailed analyses were carried out on the two independent lines of cultivar Ofanto expressing lmw1B*, which is the longest of the two LMW glutenin genes used in the transformation experiments and contains two cysteine residues that are thought to be available for intermolecular disulphide bond formation.

Glutenin extracts from flour of homozygous transgenic and control plants were analysed by RP-HPLC in order to identify differences in their glutenin composition. Eight main peaks were separated and each peak area expressed as a percentage of the total peak area to compensate for differences in the amounts of storage proteins.

Comparison of the 1061 null and 1061 transgenic lines showed no significant differences between peaks 3, 4, 5, 6 and 7, all corresponding to LMW subunits, while peaks 1 and 2, corresponding to HMW subunits, were significantly lower in the transgenic samples and peak 8, corresponding to LMW subunits, was significantly higher.

Comparison of the 1093 null and the 1093 transgenic lines showed a more complex pattern of differences. Significant differences were observed for all eight peaks, with the transgenic line having higher values for peaks 1, 2 and 8 and lower values for all the other peaks.

The fact that peak 8 gave consistently higher values in both transgenic lines when compared to the Ofanto control and the null lines, strongly suggests that this peak contains the transgenic subunit.

The transgenic LMW glutenin subunit contributed about 7-8 % of the total glutenins in line 1061 while in the transgenic line 1093 it represented just over 15 % of the total glutenin fraction and was the most abundant of the glutenin subunits. However, the higher proportion of the transgenic protein in line 1093 probably relates to the reduced expression of the endogenous LMW glutenin subunits rather than to a higher level of expression of the transgene, the absolute areas of peak 8 indicating similar levels of expression in the two lines.

3.4 Comparison of the molecular size distribution of protein polymers in transgenic and control lines by SE-HPLC.

For comparison, the SE-HPLC elution profiles were divided into four main parts. The first two parts corresponded to protein polymers of large and medium size, respectively, which are enriched in HMW subunits and polymeric LMW subunits (B-type). Part 3 corresponded to ω-gliadins and to oligomers enriched in D-type and C-type LMW subunits, while part 4 contained monomeric α-type and γ-type gliadins and low molecular weight albumins and globulins. The areas of the four groups were expressed as percentages of the total area and analysis of variance (ANOVA) was carried out on the mean values. Variation between extractions and between replicate runs was combined to give an overall estimate of residual variation against which to test for differences between lines.

Transgenic line 1061 had the highest proportion of high M_r polymers (area 1), the highest proportion of total polymeric proteins (area 1+area 2+area 3) and the highest ratio of large size polymers: total polymeric protein (area 1/area 1+area 2 +area 3). The transgenic line 1093 had a lower value for area 1 when compared to the transgenic line 1061, but also correspondently lower values for areas 2 and 3, so that the proportion of the total polymeric protein present in large size polymers was very similar in the two lines. In contrast, the ratio of polymeric : monomeric protein (area 1+area 2+area 3/area 4) was very different in the two transgenic lines (1.7:1 for line 1061, and 1.2:1 in 1093) but only slightly different between the two transgenic lines and their respective null lines (1.5:1 for 1061 null, 1.2:1 for 1093 null).

3.5 Mixograph analysis

Flour samples with a higher percentage of large size-polymers are expected to form dough with greater resistance than those with a smaller proportion of such polymers[6]. Comparison of the Mixograph properties of the transgenic 1061 line with those of the null and control lines confirmed this association between large size polymers and dough properties. Dough from the transgenic 1061 line showed higher resistance to mixing (PR) and lower loss of resistance on overmixing (RBD) which together indicated an increase in elasticity (strength). No significant increases in the mixing time (MT) requirement were recorded. However, recent studies have shown that MT is not always appropriate for assessing dough strength[7, 8] since it seems to be greatly influenced by the hydratation properties of the flour, which in turn depend on the degree of starch damage. The same studies indicated that the bandwidth of the Mixograph curve is a key characteristic that relates to dough strength and the transgenic 1061 line did indeed show significantly higher values for both the maximum band width (MBW) and the bandwidth at peak resistance (BWPR).

4. CONCLUSIONS

RP-HPLC showed no clear changes in the pattern of endogenous storage proteins in line 1061. The changes in the size distribution of polymeric protein and in the Mixograph properties can therefore be explained solely in terms of the increased amount of LMW glutenin subunits resulting from expression of the lmw1B* transgene.

In contrast, the expression of the lmw1B* transgene in line1093 is accompanied by the suppression of some endogenous LMW subunits. Differences in the amount and size distribution of the polymeric protein between the null and transgenic lines and their mixing properties may therefore result from a combination of these two effects.

References

1. S. Masci, E. J-L. Lew, D. Lafiandra, E. Porceddu and D. D. Kasarda, *Cereal Chem*, 1995, **72**, 100.
2. P. Tosi, J. A. Napier, R. D'Ovidio, H. D. Jones and P. R. Shewry, In *Wheat Gluten*, eds P. R. Shewry and A. S. Tatham, Royal Society of Chemistry, p93.
3. C. Lamacchia, P .R. Shewry, N. Di Fonzo, J. L. Forsyth, N. Harris, P. Lazzeri, J. A. Napier, N. G. Halford and P. Barcelo, *J. Exp. Bot.*, 2001, **52**, 243.
4. P. Tosi, R. D'Ovidio, J. A. Napier, F. Bekes and P. R. Shewry, *Theor. Appl. Genet.*, 2003, In press.
5. B. A. Marchylo, J. E. Kruger and D. W. Hatcher, *J. Cereal Sci.*, 1988, **9**, 113.
6. R. B. Gupta, K. Khan and F. MacRichie, *J. Cereal Sci.*, 1993, **18**, 23.
7. J-P. Martinant, Y. Nicolas, A. Bouguennec, Y. Popineau, L. Saulnier and G. Branlard, *J. Cereal Sci*, 1998, **27**, 179.
8. P.W. Gras, H.C. Carpenter, and R.S. Anderssen, *J. Cereal Sci*, 2000, **31**, 1.

HETEROTYPIC AGGREGATION OF AN *IN VITRO* SYNTHESIZED LOW-MOLECULAR-WEIGHT GLUTENIN SUBUNIT

A. Orsi, F. Sparvoli, P. A. Della Cristina, and A. Ceriotti

Istituto di Biologia e Biotecnologia Agraria, Consiglio Nazionale delle Ricerche, Via Bassini 15, 20133 Milano, Italy.

1 INTRODUCTION

We have previously described an *in vitro* system based on plant components that allows the synthesis, translocation and folding of a low-molecular-weight glutenin subunit (LMW-GS)[1,2]. The system is composed of a wheat germ extract supplemented with microsomal membranes isolated from developing bean cotyledons.

LMW-GS are characterized by the presence of three conserved intrachain disulfide bonds[3]. We previously found that when *in vitro* synthesized mRNA coding for the B11-33 LMW-GS[4] is translated in the presence of sufficient levels of oxidized glutathione (GSSG), the protein accumulates as a monomer containing intrachain disulfide bonds. Conversely, when translation is performed under reducing conditions the *in vitro* synthesized protein misfolds and enters high molecular weight aggregates[2].

Here we present a further characterization of the aggregates formed in this *in vitro* system.

2 METHODS AND RESULTS

The Binding Protein BiP is a molecular chaperone belonging to the Hsp70 family that assists the structural maturation of proteins that are inserted in the endoplasmic reticulum (ER)[5]. BiP binds transiently to unfolded proteins, possibly preventing their aggregation and thus allowing them to reach their native conformation. For this reason BiP is often found in prolonged association with terminally misfolded proteins, from which it can be released *in vitro* by treatment with ATP.

To investigate whether misfolded B11-33 polypeptides synthesized *in vitro* interact with BiP, synthetic mRNA coding for this LMW-GS was translated in a wheat germ extract in the presence of bean cotyledon microsomes. Synthesis was performed either in the absence or in the presence of GSSG. Aliquots of the translation reactions were then used for immunoselection using anti-BiP antibodies (Figure 1, lanes 1, 2, 5 and 6) or a pre-immune serum (lanes 3 and 4). Half of each sample was treated with ATP before analysis.

When synthesis was performed under oxidizing conditions no B11-33 polypeptides were evident among the material immunoselected by anti-BiP antibodies (lane 5). Conversely, a band corresponding to signal processed, translocated B11-33 polypeptides was immunoselected by the anti-BiP antiserum when translation was performed under conditions that do not support disulfide bond formation (lane 1). Treatment with ATP failed to release a large fraction of these co-selected B11-33 polypeptides (lane 2).

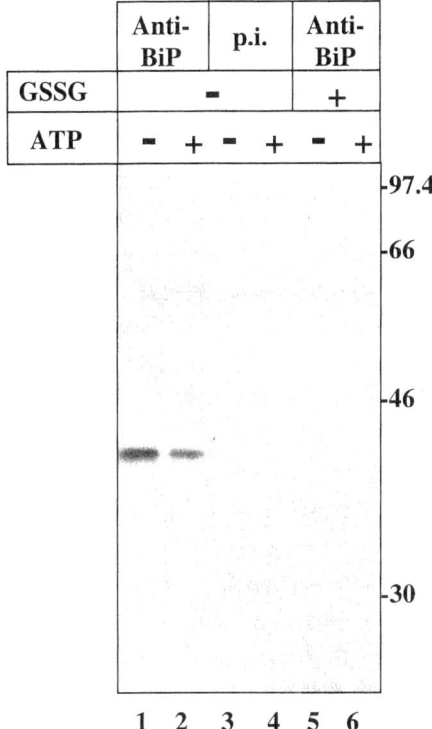

Figure 1 *Synthetic mRNA coding for the B11-33 protein was translated in vitro for 30 minutes in the presence of bean microsomal membranes as previously described[2] (1 µL of microsomal membrane preparation/12.5 µL reaction). Synthesis was performed either in the absence or in the presence of GSSG (7 mM). At the end of the translation period cycloheximide (2 mM final concentration) was added and the samples were treated with apyrase. Samples were immunoprecipitated with either anti-BiP[10] or pre-immune serum (p.i.) and then split into two aliquots, one of which was incubated in the presence of ATP. Analysis was by SDS-PAGE under reducing conditions and fluorography. The position of molecular mass markers (kDa) is indicated on the right.*

B11-33 polypeptides were not selected by the pre-immune serum (lane 3) confirming the specificity of the immunoprecipitation procedure.

The inability of ATP to fully release B11-33 polypeptides suggests that BiP might be somehow trapped into the large B11-33 aggregates that are formed when synthesis occurs under conditions that do not favour disulfide bond formation[2]. We therefore decided to examine whether another protein, which does not have any known chaperone activity, could be trapped into B11-33-containing aggregates. Phaseolin is a trimeric protein that after segregation in the ER is transported to the storage vacuoles in bean cotyledons. Phaseolin does not contain cysteine residues, and therefore its folding should not be directly affected by the redox conditions of the system. It should be noted that, since the microsomes utilized in this study are isolated from bean cotyledons, unlabelled phaseolin polypeptides would be expected to be present within the microsomes.

Messenger RNAs coding for phaseolin and for the B11-33 protein were translated together *in vitro*. In addition to untranslocated and translocated B11-33 polypeptides, four phaseolin polypeptides were evident among the total translation products (Fig. 2, lanes 5 and 6). A characterisation of the *in vitro* synthesized phaseolin polypeptides has been previously presented[6]. While immunoprecipitation with anti-phaseolin antibodies specifically selected the four phaseolin polypeptides from the translation products synthesized under oxidizing conditions (Figure 2, lane 1), B11-33 polypeptides were also selected from the translation products synthesized under conditions not allowing the formation of disulphide bonds (lane 2). As control, the immunoselection was also performed using an irrelevant antiserum (lanes 3 and 4). These results indicate that a cysteine-free protein can be bound into B11-33 containing aggregates formed when synthesis is performed under conditions that do not allow correct folding of the LMW-GS.

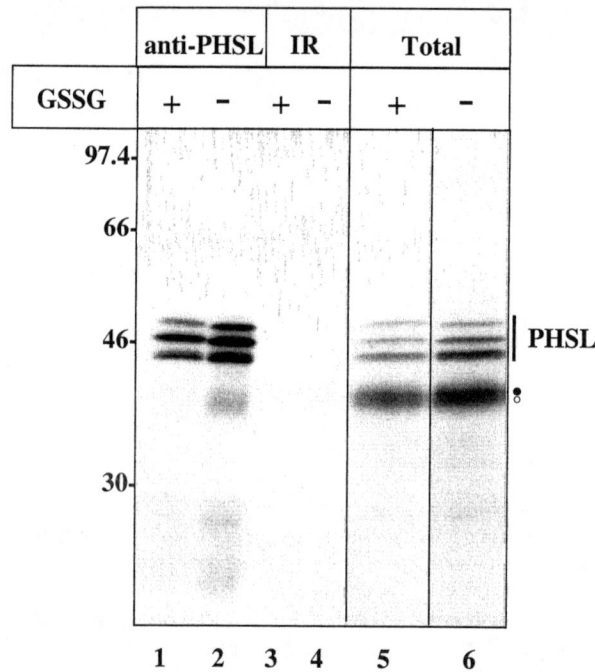

Figure 2 *Synthetic mRNAs coding for the B11-33 protein and for β-phaseolin [11] were translated together in vitro as described in Fig. 1. Synthesis was performed either in the absence or in the presence of GSSG, as indicated. Aliquots of the translation reactions were saved for direct analysis (lanes 5 and 6), while the rest was used for immunoprecipitation with either anti-phaseolin (PHSL) (lanes 1 and 2) or an irrelevant rabbit antiserum (IR, lanes 3 and 4). Polypeptides were analysed by SDS-PAGE under reducing conditions and fluorography. The position of molecular mass markers (kDa) is indicated on the left. Dot: untranslocated B11-33. Empty circle: translocated B11-33. PHSL: phaseolin polypeptides.*

3 CONCLUSIONS

Unrelated proteins can form mixed heteroaggregates when their folding in the ER is perturbed[7-9]. In the case of the B11-33 protein, rapid formation of intrachain disulfide bonds is essential to prevent the aggregation of newly synthesized polypeptides imported into bean microsomes[2]. Clearly, the formation of such aggregates indicates that ER chaperones are unable, at least in this *in vitro* situation, to maintain reduced B11-33 polypeptides in soluble form. Our data indicate that some interaction between the chaperone BiP and reduced B11-33 polypeptides does occur, but BiP appears to be at least in part trapped in these aggregates, rather than bound while performing its typical chaperone function. A similar phenomenon has been previously observed *in vivo*[9].

Our inability to co-select oxidized *in vitro* synthesized B11-33 polypeptides using anti-BiP antiserum could have different explanations and should be interpreted with caution. *In vivo*, BiP can be found in an ATP-sensitive association with an epitope-tagged form of the B11-33 protein (A. Orsi and A. Ceriotti, unpublished observation) and further work will be required to determine whether BiP plays any role in the productive folding/assembly pathway(s) followed by LMW-GS.

References

1 A. Ceriotti, E. Pedrazzini, M. de Silvestris and A. Vitale, in *Methods in Cell Biology*, eds. D. W. Galbraith, D. P. Bourque and H. J. Bohnert, Academic Press, San Diego, 1995, 50B, ch. 21, 295.
2 A. Orsi, F. Sparvoli and A. Ceriotti. *J Biol. Chem.*, 2001, **276**, 32322.
3 P.R. Shewry and A.S. Tatham, *J. Cereal Sci.*, **25**, 207.
4 T.W. Okita, V. Cheesbrough and C.D. Reeves, *J. Biol. Chem.*, **260**, 8203.
5 I.G. Haas, *Experientia*, 1994, **50**, 1012.
6 F. Lupattelli, E. Pedrazzini, R. Bollini, A. Vitale and A. Ceriotti. *Plant Cell*, 1997, **9**, 597.
7 T. Marquardt and A. Helenius, *J. Cell Biol.*, 1992, **117**, 505.
8 J.T. Sawyer, T. Lukaczyk and M. Yilla, *J. Biol. Chem.*, 1994, **269**, 22440.
9 F. Sparvoli, F. Faoro, M. Daminati, A. Ceriotti and R. Bollini, *Plant J.*, 2000, **24**, 825.
10 E. Pedrazzini, G. Giovinazzo, A. Bielli, M. de Virgilio, L. Frigerio, M. Pesca, F. Faoro, R. Bollini, A. Ceriotti and A. Vitale. *Plant Cell*, 1997, **9**, 1869.
11 A. Ceriotti, E. Pedrazzini, M.S. Fabbrini, M. Zoppè, R. Bollini and A. Vitale, *Eur. J. Biochem.*, 1991, **202**, 959.

Acknowledgements

This work was supported by grant RBNE01TYZF (FIRB) from the Ministry of Education, University and Research.

PROTEOMIC ANALYSIS OF WHEAT STORAGE PROTEINS: A PROMISING APPROACH TO UNDERSTAND THE GENETIC AND MOLECULAR BASES OF GLUTEN COMPONENTS

G. Branlard [1], J. Dumur [1,2], E. Bancel [1], M. Merlino [1] and M. Dardevet [1]

1- INRA UMR Amélioration et Santé des Plantes, 234 Avenue du Brezet, 63039 Clermont Ferrand Cedex 02, France.
2- INRA UMR Amélioration des Plantes et Biotechnologies Végétales, BP 35323, 35653 Le Rheu Cedex, France.

1 INTRODUCTION

The proteome is composed of the whole set of proteins present in a given tissue, cell, or in sub-cellular components of a living organism at a given time[1]. The identification of all the proteins expressed, the study of their genetic determinism (gene location and sequence), the study of their possible polymorphism through post-translational and post-traductional modifications, and the identification of their function in physiological metabolism are the major tasks that comprise the proteomic approach[2]. Wheat storage proteins (WSP) have been extensively analysed in the two last decades (for a review, see Kasarda[3], Lafiandra[4], Shewry[5]). These early findings on genetic determinism, allelic diversity, protein sequences, and functional properties were in fact the main steps of proteomic analysis long before this approach became a powerful technique, thanks to progress in mass spectrometry. Several studies have been carried out on the endosperm proteins using proteomic tools. For example, total endosperm proteins extracted from mature wheat kernels have been analysed both in normal and heat-stressed growing conditions[6-11]. Two dimensional electrophoresis (2-DE), which is currently the most powerful method for protein separation[12], has not often been used to analyse either gliadins or glutenins. This technique was recently used to characterise wheat endosperm proteins by determining their N-terminal amino acid sequences[10,13]. Proteomic analysis offers new opportunities to investigate important questions concerning the gluten proteins (gliadins and glutenins) which are not usually separated together on the same gel. Some examples of proteomic approaches are described below.

2 MATERIAL AND METHODS

The experiment shown in Figure 1 was performed using the cultivar "Courtot", a French semi-dwarf wheat grown at INRA Clermont Ferrand, France in 2002.
The main steps of a proteomic approach i.e. protein extraction, 2DE, gel staining, gel scanning, image analysis, statistical comparisons, trypsic digestion, mass spectrometry and database interrogation, have been described elsewhere. Specific details about the above steps will be found in the following studies on albumins and globulins[9], on amphiphilic proteins[14] and on total endosperm protein[8].
Some brief remarks may help avoiding low reproducibility of 2-DE and wasted time:

- The conditions used for protein extraction are crucial as they considerably influence the diversity of proteins revealed on 2-DE. In all cases they need to be carefully controlled and kept constant throughout the series of gels.
- The quantity of the loaded proteins on the focussing strip must be controlled.
- The parameters of 2-DE must be accurately controlled and kept constant over the series of runs; particularly water conductivity, brand and purity of the chemicals, room temperature, electric parameters, temperature of migration, etc..
- The detection of the protein depends on the staining procedure. Some staining protocols are better suited for quantitative analysis and others for mass spectrometry.
- 2-DE has some limits: many proteins are not necessarily always revealed, particularly the less abundant, i.e. those of pI < 3 and pI > 11 or of MW > 250 kDa.
- Spots may contain a mixture of proteins, and the proteins that we are interested in may not be present in the data bases.

3 RESULTS

3.1 Preliminary observations on wheat kernel proteomic

Due to different flowering dates and locations of the kernel on the spike, the kernels may display differences in size, as well as in starch and protein content, consequently the sampling material must be homogenous to avoid undesired variability.

Depending on the milling process, some components of the aleurone layer and the embryo may or may not contaminate the starchy endosperm components. It is recommended that proteins be extracted from known kernel parts.

Not all the wheat protein fractions (albumins, globulins, amphiphilic, gliadins, glutenins) are stable in solution over time due to protein refolding, protease activities, ionic interactions, hydrophobicity, etc. It is recommended to run protein samples with the same history of extraction and conditioning.

3.2 What can the gluten proteomic approach be used for?

The answers to many questions concerning the genetics, composition and properties of gluten can be investigated using the proteomic approach. The list of examples below is not exhaustive:

Characterisation of the gene products. The use of high resolution 2-DE allows gliadins and reduced glutenins to be separated together on the same gel. Most of these proteins have been characterised through one dimensional SDS-PAGE and HPLC techniques. 2-DE allows the pI and apparent MW to be simultaneously determined for the more than 100 components that form the WSP of any classic bread and durum wheat cultivars. In this way, the many alleles[15] registered so far will be better characterised thus rending their links with quality parameters more reliable.

Study of the expression of WSP genes. Although loci encoding WSP are known, their DNA sequences are unknown in most cases, and their promoter sequences are not yet analysed. The proteomic approach will be very useful in characterising the transcription factors, and both the qualitative (presence / absence) and quantitative expression of the complex WSP loci.

Study of interactions between chromosomes. Genetic regulation in response to polyploidy, and chromosome dosage and their consequences for WSP polymerisation and aggregation are questions that require the whole set of WSP to be simultaneously separated and analysed. For example, in the Courtot cultivar, the percentage volume of 45 WSP significantly changed (25 up-regulated, and 20 down-regulated) in its monosomic 1D line compared to the normal euploid Courtot (Figure 1). Interestingly, although chromosome 1D was the only one that lost its homologous chromosome, 6 spots identified as HMW-GS encoded on *Glu-B1* and 6 α-gliadins encoded at *Gli-A2* were up-regulated[16].

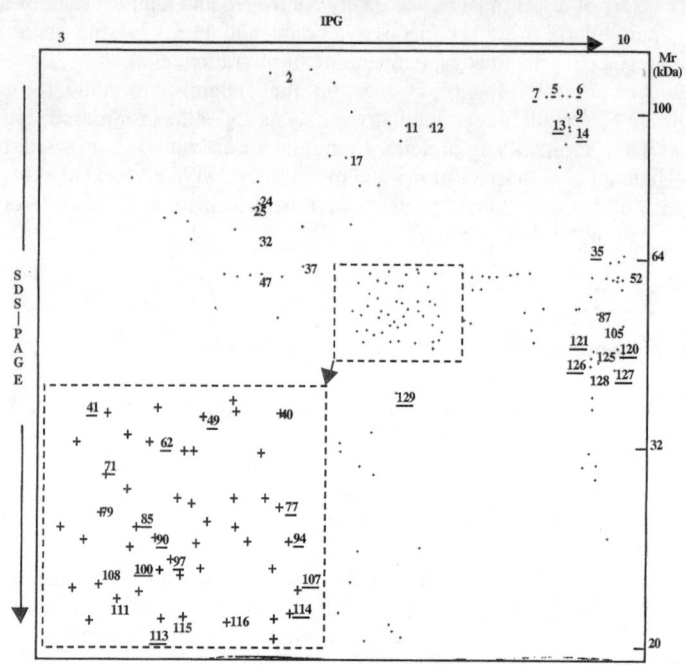

Figure 1 *Spots of WSP of Courtot significantly changed (underlined = up-regulated, not underlined = down-regulated spots) in a monosomic 1D line.*

This approach will be useful both for understanding the expression of homoeologous genes resulting from interactions between genomes[17] and also for breeding future synthetic wheats.

Understanding the accumulation of WSP in protein bodies. The mechanisms involved in the polymerisation of glutenin subunits and their aggregation together with gliadins in the protein bodies remain largely unknown. Specific extractions of endoplasmic reticulum, and of light and dense protein bodies offer key materials for the proteomic approach.

Analysis of post-translational modifications. The peptide fragmentation of WSP obtained using MALDI-TOF provides many peptidic masses, many of which do not match virtual masses deduced from database interrogations[18]. Preliminary results indicate that some peptides may be modified either as a result of genetic and environmental influence or of chemical treatment. Software will allow the identification of amino-acids that could carry the modifications[19].

Analysis of the environmental influences on WSP. Several studies have shown that increases in temperature induce the synthesis of gliadins at the expense of glutenins. Further studies are required to understand both transcriptomic and proteomic responses to environmental factors (heat stress, water stress, nitrogen deficiency, etc.).The regulation factors that control the balance between WSP synthesis and starch synthesis also remain to be identified.

4 CONCLUSION

The proteomic approach is complementary to genome analysis as it allows the study of functional rather than informational molecules. It offers additional information for the analysis of gluten components. This tool is expected to be very efficient in association with transcriptome analysis in providing answers to the above questions in the search for future genetic improvements in the stability of wheat quality.

References
1. M.R. Wilkins, J. C. Sanchez, A. A. Gooley, R. D. Appel, I. Humphery-Smith, D. F. Hochstrasser and K.L. Williams. *Biotechnol. Genet. Eng Rev.* 1995, **13**, 19.
2. H. Thiellement, N. Bahrman, C. Damerval, C. Plomion, M. Rossignol, V. Santoni, D.de Vienne and M. Zivy, *Electrophoresis* 1999, **20**, 2013.
3. D.D. Kasarda, *Cereal Food World.* 1999, **44**, 566.
4. D. Lafiandra, S. Masci, C. Blumenthal and C.W. Wrigley. *Cereal Food World* 1999, **44**, 572.
5. P. R. Shewry, N. G. Halford and D. Lafiandra. Adv Genet. 2003, **49**, 111.
6. N. Islam, S.H. Woo, H. Tsujimoto, H. Kawasaki and H. Hirano. *Proteomics* 2002, **2**, 1146.
7. N. Islam, H. Tsujimoto and H.Hirano. *Proteomics*, 2003, **3**, 307
8. T. Majoul, E. Bancel, E. Triboï, J. Ben Hamida and G. Branlard. *Proteomics* 2003a, **3**, 175.
9. T. Majoul, E. Bancel, E. Triboï, J. Ben Hamida and G. Branlard. *Proteomics* 2003b, in press.
10. D. J. Skylas, J. A. Mackintosh, S. J. Cordwell, D. J. Basseal, D. J., Walsh, J., Harry, C. Blumenthal, L. Copeland, C. W. Wrigley and W. Rathmell. *J. Cereal Sci.* 2000, **32**,169.
11. D.J. Skylas, S. J. Cordwell, P. G. Hains, M.R. Larsen, D. J. Basseal, B. J., Walsh, C. Blumenthal W. Rathmell, L. Copeland and C. W. Wrigley. *J. Cereal Sci.* 2002, **35**, 175.
12. T. Rabilloud, *Proteomics* 2002, **2**, 3.
13. H. Nakamura, *Cereal Chem.* 2001, **78**, 79.
14. G. Branlard, G., N. Amiour, G. Igrejas, T. Gaborit, S. Herbette, M. Dardevet and D. Marion. *Proteomics* 2003, **3**, 168.
15. R. A. McIntosh, Y. Yamazaki, K. M. Devos, J. Dubcovsky, W. J. Rogers and R. Appels.. *Proceedings of the 10th International Wheat Genetics Symposium*, (on CD), Paestum, Italy, 1-6 September, 2003, Volume 4.
16. J. Dumur, J. Jahier, E. Bancel, A. M. Tanguy, M. Bernard and G. Branlard *Proteomics*, 2003, submitted.
17. N. Islam, H. Tsujimoto and H. Hirano. *Proteomics,* 2003, **3**, 549.
18. E. Bancel, J. Dumur, C. Chambon and G. Branlard Proceeding 8th Inter. Gluten Workshop, Viterbo, Italy, 7-10th September, 2003
19. M.R. Wilkins, E. Gasteiger, A. A. Gooley, B. R. Herbert, M. P. Molloy, P-A. Binz, K. Ou, J-C. Sanchez, A. Bairoch, K. L. Williams and D.F. Hochstrasser. *J. Mol. Biol.* 1999, **289**, 645.

MASS SPECTROMETRIC APPROACHES FOR THE CHARACTERIZATION OF LOW-MOLECULAR WEIGHT GLUTENIN SUBUNITS

V. Cunsolo[1], S. Foti[1], V. Muccilli[1], R. Saletti[1], D. Lafiandra[2] and S. Masci[2]

[1]Dipartimento di Scienze Chimiche, Università di Catania, V.le A. Doria 6, 95125 Catania, Italy
[2]Dipartimento di Agrobiologia ed Agrochimica, Università degli Studi della Tuscia, Via San Camillo de Lellis, 01100 Viterbo, Italy

1 INTRODUCTION

Low molecular weight glutenin subunits (LMW-GS) are the most abundant components of the glutenin macropolymer, which plays a major role in determining the viscoelastic properties of wheat flour. They are subdivided into B-, C- and D-type, according to SDS-PAGE mobility. B-type subunits are considered typical LMW-GS, whereas C- and D-type correspond mostly to modified gliadins (monomeric) that become part of the glutenin polymer because of mutations in the number and/or position of Cys residues[1,2]. The number of Cys residues and their availability for the formation of inter-chain disulphide bonds is in fact one major point in the structural characterisation of these subunits.

In this investigation, an enriched B-type LMW-GS fraction from *Triticum aestivum* L. cultivar Chinese Spring has been investigated by use of separation techniques as 2D electrophoresis and RP-HPLC coupled with ESI and MALDI mass spectrometry, in order to determine the number and molecular weight of the components of the fraction examined and the number of Cys residues present in some subunits.

2 METHOD AND RESULTS

2.1 Isolation of B LMW-GS Fraction

The B LMW-GS fraction was obtained by fractionated precipitation from hydro-alcoholic extracts of *Triticum aestivum* L. cultivar Chinese Spring[2]. The reduced B LMW-GS fraction was alkylated, as described by Lew et al.,[3] using 4-vinyylpyridine, which increases the M_r by 105.1 for each free thiol group.

2.2 RP-HPLC and Mass Spectrometry Analysis

On-line RP-HPLC/ESI-MS analysis was obtained by injection of 50 µg of B-type LMW-GS solution (CH_3CN 0.05 % TFA / H_2O 0.05 % TFA (25:75), conc. 2 µg/µL) onto a C_4 narrow-bore column (2.1x250 mm) coupled with an electrospray ion source of a DECA ion trap mass spectrometer (ThermoFinnigan). Proteins were eluted with a linear gradient of A,

CH$_3$CN 0.05 % TFA in H$_2$O and B, 0.05 % TFA, from 25% to 60% A in 80 min at a flow rate of 200 µL/min at 50 °C. The B-type LMW-GS fraction was also separated by RP-HPLC, applying 25 µL of solution onto a C$_4$ narrow-bore column coupled with an UV detector (λ=224 nm), under the same conditions used for the on-line LC analysis. Fractions were collected manually and directly used for MALDI-TOF mass spectrometric analysis (Voyager DE-PRO, PerSeptive Biosystems). On-line RP-HPLC/ESI-MS analysis was also carried out for the alkylated B LMW-GS in order to determine the number of Cys residues present in some subunits. The on-line RP-HPLC/ESI mass spectrometric analysis of unalkylated B-type LMW-GS allowed separation of 24 different fractions (Figure 1). Analogously, the off-line RP-HPLC (UV trace not shown) resulted in the separation of 24 fractions, which were analysed by MALDI-TOF MS. While HPLC/ESI-MS was able to provide accurate mass measurements for most components, it presented limitation for the resolution of multi-components chromatographic peaks in comparison with the off-line MALDI-TOF mass analysis. In summary, 42 proteins were detected by HPLC/ESI-MS and 65 by HPLC followed by MALDI-MS. Their experimentally determined molecular masses and the HPLC fraction in which they were detected are reported in Table 1.

Except for some minor components eluting at the beginning of the chromatogram, the molecular masses of the detected proteins were in the range of 26,000-43,000, which is lower than that determined by SDS-PAGE, thus confirming that this technique tends to overestimate the molecular masses of wheat prolamins. Most of the components co-eluting in the same HPLC peak present minor differences in the molecular masses, supporting the existence of extensive micro-heterogeneity in prolamines. For the determination of the number of Cys residues present in the detected proteins, the fraction was alkylated and analysed by HPLC/ESI-MS.

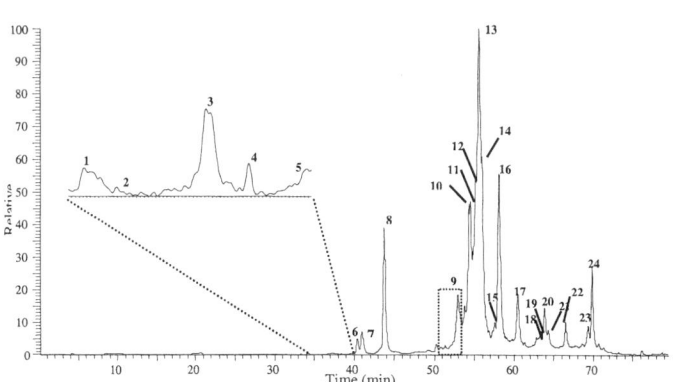

Figure 1 *Total Ion Current (TIC) of on line RP-HPLC/ESI-MS analysis of unalkylated B LMW-GS fraction*

The number of Cys residues was deduced from the difference in the molecular masses of the alkylated and unalkylated protein. By this analysis 25 components (Table 2) could be correlated with the corresponding unalkylated proteins. The correlation was very reliable as indicated by the precise correspondence of the molecular masses of alkylated and unalkylated proteins

Table 1

HPLC Peaks	Retention Time (min)	Ident. N.	M_r (Da) ESI	M_r (Da) MALDI
1	34.27-34.82	1	18351	
2	35.30-35.46	2	18464	
3	37.40-37.79	3	9606	9605
		4	10608	
4	38.04-38.78	5	9475	9745
		6	10921	10921
5	40.00-40.20	7	19487	
6	40.45-40.54	8	41866	41880
		9	42741	42764
7	40.99-41.37	10	41651	
8	43.72-44.01	11		22763
		12	40953	40963
		13	41835	41845
9	50.45-53.42	14		21696
		15		22030
		16		22500
		17		22847
		18	25492	25495
		19		27390
		20		29199
		21		30170
		22		30204
		23		30331
		24		30678
		25	31989	32000
		26	36348	36310
10	53.71-54.11	27	39483	39479
		28	39648	39647
11	54.21-54.47	29		30811
		30		31423
		31		31537
		32		31581
		33		32047
		34		32476
		35	32474	32780
		36		33973
12	54.71-55.10	37	31191	31182
		38	37755	37750
		39		37789
13	55.18-55.38	40	37494	37482
		41	41118	41113
14	55.62-55.98	42		31209
		43		31559
		44	33280	33287
		45	33407	33405
		46		33810
		47	41215	41204
15	57.10-57.56	48	30503	30506
		49	30636	30621
16	57.83-58.23	50		32745
		51	32824	32823
		52	32939	32976
17	60.37-60.48 60.62-60.81	53		24697
		54		32195
		55		32822
		56		35193
		57	35182	39211
		58	39223	39243
		59		39263
18	62.86-63.10	60	31644	
19	63.37-63.60	61	30982	30983
20	63.75-64.08	62		20972
		63	31685	31671
		64	32012	32021
		65	32425	32432
21	64.33-64.54	66	31484	31490
22	66.39-66.86	67		29197
		68		31721
		69		31990
		70		32787
		71	36926	
23	69.21-69.73	72	30651	30641
		73	32654	
24	69.80-70.10	74	32603	32594

(Table 2). Nine proteins, eluting in the middle part of the chromatogram showed an even number (six or eight) of Cys residues. Whereas typical LMW-GS (mostly present in the B group) present eight Cys residues, the components with six Cys might be α-gliadins (typically showing six Cys residues) that might be part of the glutenin polymer because the position of Cys residues is changed, thus affecting polymerisation properties. Alternatively, they might be typical α-gliadins that are present as contaminants of the glutenin preparation. Four components present at the beginning of the chromatogram have molecular masses of 9,000-11,000 and eight Cys. Six components eluting at the end of the chromatogram with an odd number (seven or nine) of Cys might likely be modified gliadins (α-gliadins that have acquired a Cys residue or γ-gliadins that have lost one, respectively, thus entering into the glutenin fraction) such as the majority of C subunits[1]. Four components with one Cys could correspond to D-type LMW-GS, namely modified ω-gliadins[1]. Finally, two components of the fraction with M_r of 41,866 and 42,741 do not contain any Cys. These components might correspond to ω-gliadins, which were fortuitously co-isolated in the fraction examined.

2. Component Identification

In order to identify the proteins, the unalkylated B-type LMW-GS fraction was also analysed by a proteomic approach involving two-dimensional (2D) gel electrophoresis, mass spectrometry, and bioinformatics screening[4]. Although very good MALDI spectra were obtained for all the tryptic digests of the major spots present in the 2D electrophoresis, attempts to identify proteins by database search using the MALDI determined molecular masses of these tryptic fragments were unsuccessful because of the lack, in the databases, of the sequences corresponding to the proteins present in the fraction examined. Future attempts will be based on the determination of peptide sequences using tandem mass spectrometry. This approach has been shown[5] to be able to identify proteins by homology with proteins present in other cereals as barley and especially rice, for which the genome is completely known.

Table 2

HPLC Peaks	Retention Time (min)	Ident. N.	Unalkylated Ident. N.	M_r (Da) ESI	Calculated N. of Cys Residues
1	29.44-29.94	1	4	11448	10608 + 8 Cys (7.99)
2	30.03-30.70	2	5	10318	9475 + 8 Cys (8.02)
		3	3	10447	9606 + 8 Cys (8.00)
3	31.05-31.33	4	6	11763	10921 + 8 Cys (8.01)
4	33.82-34.30	5	2	18457	18351 + 1 Cys (1.01)
5	40.13-40.34	6	8	41861	41866 + 0 Cys
		7	9	42744	42741 + 0 Cys
6	40.42-40.69	8	10	41765	41651+ 1 Cys (1.02)
7	43.08-43.52	9	12	41055	40953 + 1 Cys (0.97)
		10	13	41939	41835 + 1 Cys (0.99)
8	44.12-44.42	11	12	36966	36348 + 6 Cys (5.88)
9	46.75-48.16	12	25	32834	31989 + 8 Cys (8.03)
		13	34	33213	32474 + 7 Cys (7.03)
		14	41	41961	41118 + 8 Cys (8.02)
10	48.32-48.71	15	44	33920	33280 + 6 Cys (6.09)
		16	45	34042	33407 + 6 Cys (6.04)
11	48.89-50.35	17	40	38346	37494 + 8 Cys (8.10)
		18	38	38601	37755 + 8 Cys (8.04)
		19	27	40323	39483 + 8 Cys (7.99)
12	50.43-50.66	20	39	38757	37789 + 9 Cys (9.20)
13	52.53-53.31	21	51	33670	32824 + 8 Cys (8.04)
14	53.94-54.36	22	64	32964	32012 + 9 Cys (9.05)
		23	65	33380	32425 + 9 Cys (9.08)
15	54.61-54.81	24	66	32439	31484 + 9 Cys (9.08)
16	56.83-57.23	25	74	33561	32603 + 9Cys (9.10)

3 CONCLUSION

The use of RP-HPLC coupled with ESI and MALDI mass spectrometry made it possible to detect about 70 components in a fraction of B-type LMW-GS extracted from the bread wheat cultivar Chinese Spring and to determine their molecular masses. The presence of micro-heterogeneity was suggested by the detection of several co-eluting proteins with minor differences in the molecular masses. Analysis of the alkylated fraction allowed determination of the number of Cys present in 25 subunits. The results obtained suggest that the classification in B, C, and D-type LMW-GS does not strictly reflect structural peculiarities of these groups, and that LMW-GS are even more complex than expected.

References

1 S. Masci, T.A. Egorov, C. Ronchi, D.D. Kuzmicky, D.D. Kasarda and D. Lafiandra. *J. Cereal Sci.*, 1999, **29**, 17
2 S. Masci, L. Rovelli, D.D. Kasarda, W.H. Vensel and D. Lafiandra. *Theor. Appl. Genet.*, 2002, **104**, 422
3 E. Lew, D. Kuzmicky and D. Kasarda, *Cereal Chem.*, 1992, **68**, 122.
4 O. Jensen, M Larsen and P. Roepstorff, *Prot.: Struct., Funct. and Gen. Suppl.*, 1998, **2**, 74.
5 N. Andon, S. Hollingworth, A. Koller, A. Greenland, J. Yates III and P. Haynes, *Proteomics*, 2002, **2**, 1156.

Acknowledgments

This work was supported by MIUR, projects "Aspetti biochimici, genetici e molecolari delle proteine della cariosside dei frumenti in relazione alle caratteristiche nutrizionali e tecnologiche dei prodotti derivati" (PRIN 2002) and "Espressione genica ed accumulo di proteine d'interesse agronomico nella cellula vegetale: meccanismi trascrizionali e post-trascrizionali" (FIRB RBNE01TYZF)

COMPARISON OF DIFFERENT PEPTIDIC HYDROLYSES TO IDENTIFY WHEAT STORAGE PROTEINS USING MALDI-TOF

E. Bancel [1], J. Dumur [1,2], C. Chambon [3] and G. Branlard [1]

1- INRA UMR Amélioration et Santé des Plantes, 234 Avenue du Brezet, 63039 Clermont Ferrand Cedex 02, France.
2- INRA UMR Amélioration des Plantes et Biotechnologies Végétales, BP 35323, 35653 Le Rheu Cedex, France.
3- INRA Centre de Clermont-Theix, Plateforme Protéomique, 63122 Saint Genès Champanelle, France.

1 INTRODUCTION

Proteomic analysis is an extremely useful analytical tool for the separation, the quantification and identification of proteins from complex mixtures[1]. This approach was used for -(i) analysing the effects of heat stress on wheat kernel[2], -(ii) mapping genes of amphiphilic proteins[3,4], and (iii) analysing the endosperm proteins varying in some aneuploid lines[5] or in some wheat related species[6].

Two-dimensional electrophoresis (the first step of proteomic analysis) is currently the most powerful tool for protein separation. Spots of interest are usually extracted from pieces of gel and then subjected to hydrolysis giving different peptides whose masses, when accurately assessed using mass spectrometer, provide the basis for identify the protein via database interrogation.

Trypsin is the most widely used enzyme for protein hydrolysis. Trypsin cuts the protein after each arginine (R) and lysine (K) amino acid. Because these two amino acids are rather rare among wheat storage proteins (WSP) only a few peptides result from trypsic hydrolysis of WSP (gliadins and glutenins) rendering the identification of these proteins almost impossible. To overcome such problems, several methods of protein hydrolysis were compared: enzymatic hydrolysis using trypsin or endoproteinase Glu-C V8 and chemical hydrolysis using CNBr. The different peptides were subjected to MALDI-Tof and the measured masses used for database interrogation.

2 MATERIAL AND METHODS

All the experiments reported here were performed using the cultivar "Courtot", a French dwarf wheat of good bread making qualities grown in 2002 at INRA Clermont Ferrand, France.

Two-dimensional electrophoresis (immobilized pH gradient IPG (pH 3-10) x SDS-PAGE was performed on WSP. Coomassie Blue stained gels were analysed using Melanie-3 software (GeneBio, Switzerland). Spots of interest (gliadins, glutenins) were selected, excised from three to four replicates, digested in gels using either trypsin as previously reported[2], endoproteinase Glu-C (Sigma) or CNBr[7,8]. Conditions of use of these hydrolysis agents are reported on Table 1. The peptides were subjected to MALDI-Tof mass spectrometry (Voyager DE super STR, Applied Biosystems) according to previously described procedure[3]. To identify candidate proteins, the peptide masses were used to search for the SWISS-PROT and NCBInr databases using Profound or Mascot softwares as previously described[2,3]. Search parameters were as follows : viridiplantae as taxonomic category, protein mass range of 0-200 kDa, S-pyridylethyl modified cysteine, oxidized methionine, one maximum missing cleavage site and a monoisotopic mass tolerance of 25 ppm. The criteria used to accept the protein identification were the followings: extent of sequence coverage, agreement between theoretical and observed Mr and pI values and high probabilistic scores.

Table 1 *Characteristics of the peptidic hydrolyses used for WSP identification*

Name	Buffer	Time/Temp.	cleave	Ref
Trypsin (porcine Promega)	25 mM NH$_4$HCO$_3$ pH 7.8	16h/37°C	K, R	1
Glu-C (V8-E, bicarbonate) (Sigma)	100 mM NH$_4$HCO$_3$ pH 7.8	16h/37°C	E	sigma
Trypsin/CNBr (Promega/Fluka)	25 mM NH$_4$HCO$_3$ pH 7.8 HCOOH 70%	16h/37°C 16h/20°C	K,R M	7, 8

3 RESULTS

3.1 2 dimensional gel electrophoresis

Two-dimensional gel electrophoresis IPG x SDS-PAGE of WSP extracted from Courtot kernels followed by Coomassie Brillant Blue staining allowed more than 150 spots to be resolved (Figure 1).

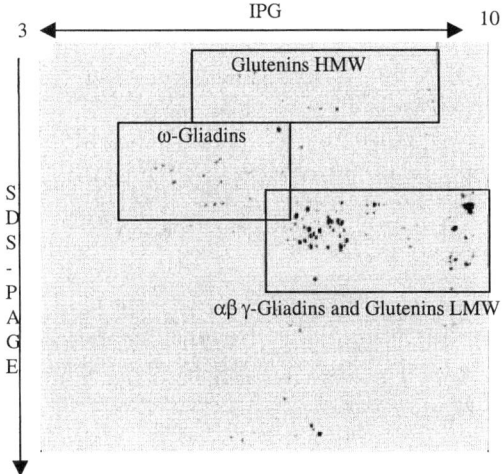

Figure 1 *IPG x SDS-PAGE of the "Courtot" endoperm proteins*

Several spots of known glutenin subunits (HMW and LMW) and gliadins (αβ–, γ- and ω-) were excised from three or four replicates. The quantity of protein collected from the 4 replicates was about 1.5 μg per spot. The 3 peptidic hydrolyses were tested for each spot. To increase the possibility of peptidic hydrolysis combination of trypsic digestion and subsequent CNBr cleavage were employed for some recalcitrant spots.

3.2 HMW glutenin subunits (HMW-GS)

MALDI-Tof spectrum obtained from spot of HMW-GS, hydrolysed using trypsin, gave numerous peptides in the m/z [600-3500] range of fragmentation (Figure 2). Many peaks were clearly resolved and the spectrum of measured peptide masses allowed to unambiguously identify the original protein through database interrogation. The use of the endoproteinase Glu-C V8 gave similar results. Methionine (M) being a rather rare amino acid in the HMW-GS, CNBr which cut the protein at the M location, was not used.

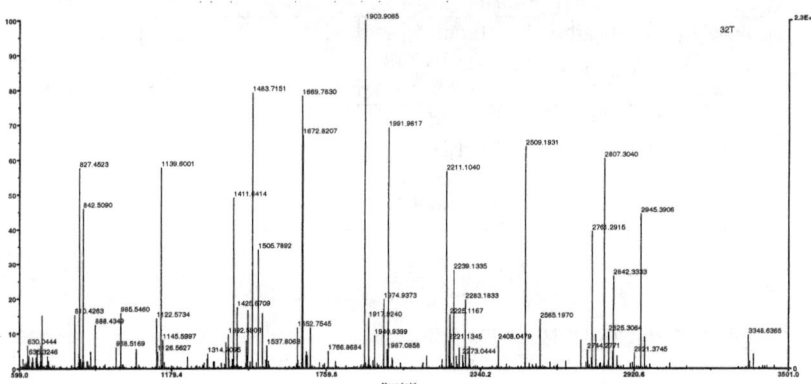

Figure 2 *MALDI-Tof spectrum of peptidic masses of the HMW glutenin subunit 2 hydrolysed using trypsin*

3.3 LMW glutenin subunits (LMW-GS)

Peptide mass fingerprinting, after trypsic digestion of the LMW-GS spots, gave partial results. Although numerous peaks were observed only few peptide masses (3 or 4) matched with those resulting from virtual fragmentation of LMW-GS occurring in the databases. The sequence coverage was generally lower than 10%. No identification was obtained after Glu-C V8 or CNBr hydrolysis.

3.4 Gliadins

The tryptic hydrolysis of the αβ-gliadins resulted in only few masses (<4) matching those present in the databases. The combination of trypsin digestion and subsequent CNBr cleavage on the same αβ-gliadin enhanced the sequence coverage as compared to the individual cleavage methods. Hydrolysis using endoproteinase Glu-C V8 didn't improve the number of masses matching the α- or β-gliadin sequences in the databases.

For the γ-gladin spot tested only one peptide matched after tryptic or Glu-C V8 hydrolysis. The sequence coverage was 7 or 6 % respectively. This was simply resulting of the low occurrence of the R, K or E amino acids in the γ-gliadins (Figure 3).

MKTLLILTILAMAITIGTANIQVDPSGQVQWLQQQLVPQLQQPLSQQPQQTFPQPQQTFPHQPQQQV
PQPQQPQQPFLQPQQPFPQQPQQPFPQTQQPQQPFPQQPQQPFPQTQQPQQPFPQQPQQPFPQTQQPQ
QPFPQLQQPQQPFPQPQQQLPQPQQPQQSFQQRPFIQPSLQQQLNPCKNILLQQSKPASLVSSLWSIIW
PQSDCQVMRQQCCQQLAQIPQQLQCAAIHSVVHSIIMQQQQQQQQQQGIDIFLPLSQHEQVGQGSLV
QGQGIIQPQQPAQL E**AIRSLVLQTLPSMCNVYVPPE**CSIMRAPFASIVAGIGGQ *
            EAIR**SLVLQTLPSMCNVYVPPECSIMR**APFASIVAGIGGQ**

Figure 3 : *Sequence of the wheat γ-gliadin precursor [P08453]. Bolded and underlined amino acids correspond to the matched peptides using : (*) Glu-C V8 ; (**) trypsin.*

The trials carried out with ω-gliadin spots hydrolysed using either trypsin or Glu-C V8 or CNBr, didn't result of peptide mass fingerprint matching with the database masses. Here again this was directly resulting of the too low occurrence of hydrolysis sites in the considered protein.

4 CONCLUSION

MALDI-Tof analysis of proteins, although considered fast, specific and with high resolution presents in our study limits resulting of the particularity of WSP sequences which do not possess enough cleavage sites for trypsin (R, K) or endoproteinase Glu-C V8 (E). The high frequency of prolin (P) in the protein sequences was also a factor limiting the peptidic cleavages. As a consequence, the size of the generated fragments (<800 Da or> 3000 Da) became incompatible with MALDI-Tof analysis.
In order to identify, after MALDI-Tof MS, LMW glutenins and γ-or ω-gliadins, other cleavage agents must be considered. Moreover, to improve the probability of identification, a complementary approach can be considered in studying the potential post-translational modifications of these WSP[9].

References

1   H. Thiellement, N. Bahram, C. Damerval, C. Plomion, M. Rossignol, V. Santoni, D. de Vienne and M. Zivy, *Electrophoresis*, 1999, **20**, 2013.
2   T. Majoul, E. Bancel, E. Triboï, J. Ben Hamida and G. Branlard, *Proteomics*, 2003, **3**, 175.
3   N. Amiour, M. Merlino, P. Leroy and G. Branlard, 2002. *Proteomics*, 2002, **2**, 632.
4   N. Amiour, M. Merlino, P. Leroy and G. Branlard, 2003. *Theor. Applied Genet*, 2003, DOI 10.1007/s00122-003-1411-0
5   J. Dumur, J. Jahier and G. Branlard, In : Proceeding of Eucarpia Cereal Section Meeting (M. Stanca ed), Salsomaggiore, Italy, 21-25 November 2002. In press.
6   N. Islam, H. Tsujimoto and H. Hirano, *Proteomics*, 2003, 4, 549.
7   B. A. Van Montfort, M.K. Doevena, B. Canas, L.M. Veenhoff, B. Poolman and G.T. Robillard, 2002. *Biochim Biophys Acta*, 2002, **1555**, 111.
8   U. Hellman, C. Wernstedt, J. Gonez and C.H. Heldin, *Anal. Biochem.*, 1997, **224** , 451.
9   M.R. Wilkins, E. Gasteiger, A. Gooley, B. Herbert, M.P. Molloy, P.A. Binz, K. Ou, J.-C. Sanchez, A. Bairoch, K.L. Williams and D.F. Hochstrasser, *J. Mol Biology* 1999, **289** : 645.

CHARACTERIZATION OF OMEGA GLIADINS ENCODED ON CHROMOSOME 1A AND EVIDENCE FOR POST-TRANSLATIONAL CLEAVAGE OF OMEGA GLIADINS BY AN ASPARAGINYL ENDOPROTEASE

F.M. Dupont, W. Vensel, and D.D. Kasarda

U.S.D.A. Agricultural Research Service, Western Regional Research Center, 800 Buchanan Street, Albany, CA, USA

1. INTRODUCTION

Three types of wheat ω-gliadins were identified previously, based on N-terminal sequences. Types beginning ARQ or RQ were separated from types beginning KEL by acid PAGE and ion exchange chromatography and referred to as ω-2 and ω-1 types, respectively[1,2]. N-terminal sequences of ARQ and KEL types were similar except for the first 8 amino acids of the ARQ type. ARQ and KEL types were associated with chromosomes 1A and 1D of bread wheat. A third type beginning SRL was associated with chromosome 1B, and referred to as ω-5 type. Since it was not clear how to distinguish between the ω-gliadins from chromosomes 1A and 1D based on protein or gene sequence data, ω-gliadins were purified from *T. urartu*, the probable donor of the A-genome[3], and compared with 1A ω-gliadins that were purified from the bread wheats Butte86 and Chinese Spring nullisomic 1D tetrasomic 1A (CS N1DT1A). N-terminal sequences and masses of the purified ω-gliadins suggested that the KEL types (ω-1) are derived from ARQ types (ω-2) by cleavage between an asparagine and a lysine. Data presented elsewhere indicate that 1A and 1D ω-gliadins can be separated by HPLC and that 1A ω-gliadins are generally somewhat smaller that 1D ω-gliadins but are similar in terms of N-terminal sequence.

2. METHODS AND RESULTS

Gliadins were extracted with 40% ethanol, precipitated with 1.5 M NaCl, dissolved in 0.1 N acetic acid and separated on a BioGel P100 column to obtain a fraction rich in ω-gliadins. The ω-gliadins were then separated by RP-HPLC[4]. Proteins in each RP-HPLC peak were analyzed by SDS-PAGE, electrospray ionization (ESI) mass spectrometry and N-terminal Edman sequence analysis.

HPLC elution profiles of ω-gliadins from *T. urartu* were complex. Six peaks were resolved (Fig 1A) with a retention time similar to that of the single 1A ω-gliadin peak from bread wheat (not shown). When proteins in the six RP-HPLC peaks were analyzed further by SDS-PAGE, at least nine different protein bands were distinguished (Fig. 1B).

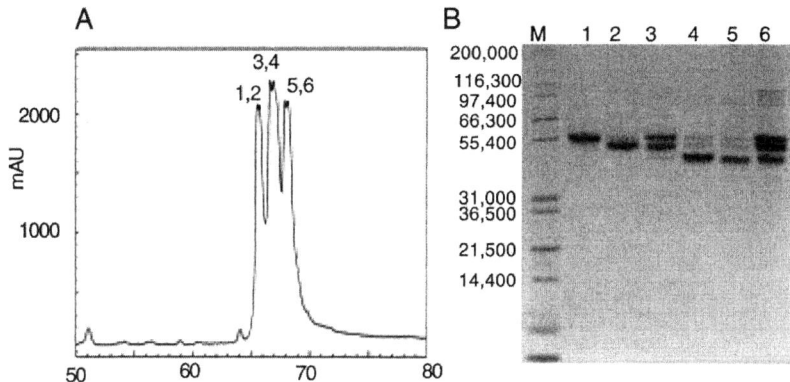

Figure 1 *Purification of ω-gliadins from T.urartu. A. RP-HPLC of the ω-gliadin-enriched fraction from a P100 column. X-axis is elution time in min. B. SDS-PAGE of the RP-HPLC peaks. M, size markers.*

Masses of all ω-gliadins determined by mass spectroscopy were lower than estimated by SDS-PAGE (Table 1). Although peaks 1, 2, 4 and 5 had only one major band in SDS-PAGE, mass spectroscopy revealed that each peak contained a pair of ω-gliadins differing in mass by 881 to 883 Daltons, corresponding to the mass of the first 8 amino acids of the ARQ type protein (Table I). The N-terminal sequence for one member of each pair began ARQ and the other began KEL. The simplest explanation for this result is that each pair resulted from partial cleavage of the longer ARQ-type protein, between the asparagine and the lysine at amino acids 8 and 9, as indicated in bold in Table 1. Results are more complex for peak 3, which had 2 bands in SDS-PAGE (data not shown). Peak 6, with 3 bands, was not analyzed.

Table 1. *Masses and N-terminal amino acid sequences of ω-gliadins from T. urartu*

RP-HPLC peak	Mr by SDS-PAGE	Mass^a (amu)	Δ Mass^b (amu)	Aligned N-terminal Amino Acid Sequences
1	58000	43363		ARQLNPS**NK**ELQSPQQS
		42479	883	**K**ELQSPQQSFSHQQQPF
2	55000	41241		ARQLNPS**NK**ELQSPQQ
		40358	883	**K**ELQ
4	45000	35440		ARQLNPS**NK**ELQSPQQS
		34559	881	**K**ELQSPQQSFSH?Q??F?
5	45000	35449		?RQ?NP?**NK**ELQ?PQ
		34566	883	**K**ELQSPQQSFSHQQQPF

^{a}Determined by mass spectrometry, +/- 2 amu.
^{b}Difference in mass between the two proteins in a single RP-HPLC peak, +/- 2 amu.

The 1A ω-gliadin compositions of CS and Butte86 were simpler than that of *T. urartu*, with a single broad peak and shoulder in RP-HPLC and a single band in SDS-PAGE (not shown). The 1A ω-gliadins from CS N1D/T1A lacked the asparagine/lysine motif described above, and the KEL sequence was not detected (Table 2). The 1A ω-gliadins from Butte86 had RQ/KEL pairs similar to the ARQ/KEL pairs of *T. urartu*, plus an additional protein beginning AR (Table 2). Two proteins differed in mass by 810, corresponding to the mass of the peptide that would be removed by cleaving the RQ protein at the NK site at amino acids 7 and 8 to form the KEL protein. A third protein was considerably longer than the other two. It was previously reported that the 1D ω-gliadins from Butte86 and CS had ARE/KEL pairs[1,4]. They were otherwise similar to the 1A ω-gliadins from *T. urartu*.

Table 2. *Masses and N-terminal amino acid sequences of 1A ω-gliadins*

Source	Mr by SDS-PAGE	Mass^a (amu)	Δ Mass^b (amu)	Aligned N-terminal Amino Acid Sequences
CS N1DT1A 1A1ω	46000	39347		ARQLNPSKQELQ PQQL RQLNPSKQ
CS N1DT1A 1A2ω	46000	39450		ARQL R?
Butte 86 1A1ω	47500	39447		AR??NP?
		35342	4105	RQLNPSNKELQSPQ
		34530	812	KELQSPQQSFS?H?P
Butte 86 1A2ω	47500	39345		ARQ?N
		35334	4011	RQLNPSNKELQSPQQ
		34523	811	KELQSPQQSFSHQQQ

ESI gave much higher resolution of ω-gliadin masses than were obtained in an earlier study that utilized MALDI-TOFF[4]. In order to solubilized the ω-gliadins for analysis by ESI, it was necessary to dissolve them in guanidine HCl. An interesting phenomenon noted during mass spectrometry using ESI was detection of possible guanidine residues in the spectrum of each single protein. The ESI technique resolved each ω-gliadin into a discrete set of multiple peaks separated by the approximate size of a guanidine ion (Figure 2). In contrast, MALDI-TOFF failed even to resolve individual members of ω-gliadin pair such as those in Tables 1 and 2. Instead, broad peaks were detected with estimated masses intermediate between those of the two members of each pair.

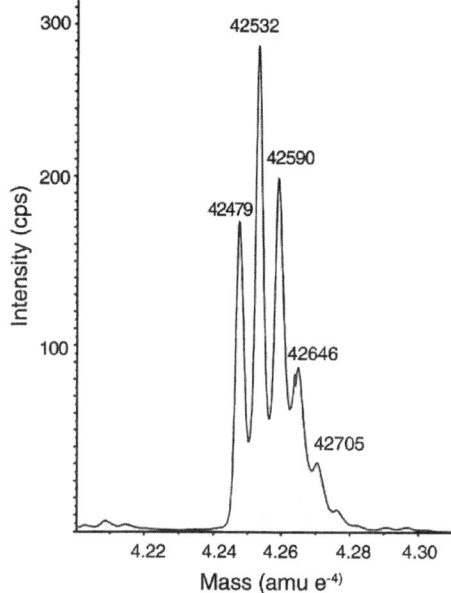

Figure 2 ESI *spectrum of an ω-gliadin from T.urartu. The principal mass is 42,479. The additional peaks probably resulted from addition of 59 Dalton guanidine ions.*

3. DISCUSSION

We suggest that those 1A and 1D ω-gliadins with N-terminal sequences beginning KEL are produced by the action of an asparaginyl endoprotease[5] that cleaves between the asparagine and lysine at amino acids 8 and 9 (NK motif) of the ARQ, ARE and RQ ω-gliadin types. Partial cleavage results in the presence of both the longer and shorter type. Not all ω-gliadins have the NK motif, and it is absent from known sequences of secalins and hordeins. Also, it is likely that the proline at amino acid 6 prevents cleavage after the asparagine at amino acid 5. Asparaginyl endoproteases, also known as legumains, are reported to be involved in post-translational processing of other seed storage proteins. It is not known whether cleavage at the NK motif occurs at the same time as post-translational cleavage of the ω-gliadin signal peptide or later.

References

1. D.D. Kasarda, J.-C. Autran, E.J.L. Lew, C.C. Nimmo, and P.R. Shewry, Biochim. Biophys. Acta 1983 **74**, 138.
2. D. Lafiandra, D.D. Kasarda, and R. Morris, Theor. Appl. Genet. 1984, **68**, 53.
3. J. Dvorak, P. di Terlizzi, H.B. Zhang, and P. Resta, Genome, 1993, **36**, 21.
4. F.M. DuPont, W.H. Vensel, R. Chan, and D.D. Kasarda, Cereal Chem., 2000, **77**, 607.
5. K. Müntz, and A.D. Shutov, (2002). Trends in Plant Science 2002, **7**, 340.

DNA MARKER LINKED TO LOW-MOLECULAR-WEIGHT GLUTENINS THAT ARE RESOLVED BY TWO-DIMENSIONAL POLYACRYLAMIDE GEL ELECTROPHORESIS AND ARE ASSOCIATED WITH BREAD-MAKING QUALITY

W.M. Funatsuki[1], K. Takata[1], A. Kato[1], K. Saito[1], T. Tabiki[1], Z. Nishio[1], H. Saruyama[2], E. Yahata[2], H. Funatsuki[1], H. Yamauchi[1]

[1] National Agricultural Research Center for Hokkaido Region, Sapporo, Hokkaido 062-8555, Japan
[2] Hokkaido Green-Bio Institute, Naganuma, Yubari-Gun, Hokkaido 069-1317, Japan

1 INTRODUCTION

Takata et al. (1) developed a near-isogenic line (NIL) in which some low-molecular-weight glutenin subunits (LMW-GSs) were substitutively introduced from a Canada Western Extra-Strong (CWES) wheat cultivar, 'Glenlea', into a Japanese spring wheat cultivar, 'Harunoakebono', and showed that the NIL has much better bread-making quality than does Harunoakebono. In the NIL, two LMW-GSs of Glenlea with sizes of approximately 42 kDa (LMW-GS 42K) and 40 kDa (LMW-GS 40K) replaced one LMW-GS of Harunoakebono with a size of about 48 kDa (LMW-GS 48K). The LMW-GSs 42K and 40K are therefore thought to be associated with good bread-making quality. In this study, we monitored those LMW-GSs by two-dimensional polyacrylamide gel electrophoresis (2-D PAGE) in a segregating population. We then developed a PCR-amplified molecular marker linked to the allele corresponding to LMW-s glutenin components associated with good bread-making quality.

2 METHODS AND RESULTS

2.1 Allelism test of LMW-GSs

The NIL was developed by backcrossing Harunoakebono to F_1 of Harunoakebono and Glenlea and the progenies carrying LMW-GS 42K/40K and by self-crossing a B_5F_2 progeny carrying LMW-GS 42K/40K. SDS-PAGE showed that the NIL had two major LMW-GSs (42K and 40K) introduced from Glenlea instead of one major band (LMW-GS 48K) specific to Harunoakebono (Figure 1).

To determine the allelism of genes encoding the LMW-GSs responsible for good bread-making quality of Glenlea and the NIL, we analyzed SDS-PAGE patterns of the B_5F_2 population (Figure 2. A). Ninety-five seeds of the B_5F_2 population were divided three groups: 26 seeds carrying LMW-GS 42K/40K, 25 seeds carrying LMW-GS 48K, and 44 seeds carrying both. A chi-square test was applied to verify that the segregation ratio of B_5F_2 seeds carrying LMW-GS 42K/40K, LMW-GS 48K, and both is 1 : 1 : 2. The result was consistent with that obtained by using a monogenic model (p=0.7649), indicating that LMW-GS 42K/40K and LMW-GS 48K are coded by allelic genes at one *Glu-3* locus.

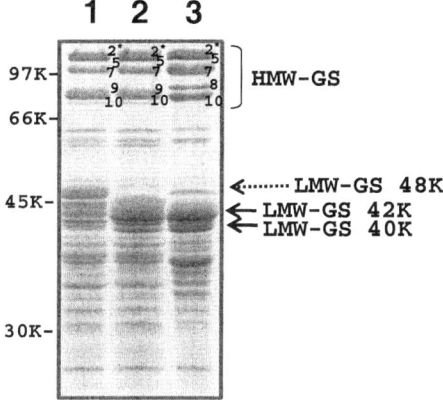

Figure 1. *SDS-PAGE separation of glutenin subunits: Harunoakebono (1), NIL (2), Glenlea (3).*

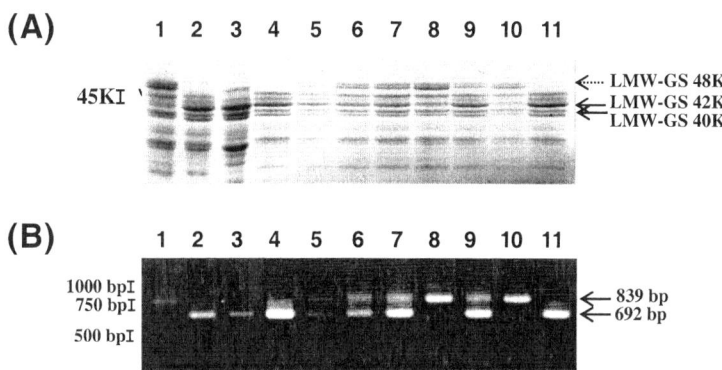

Figure 2. *SDS-PAGE separation and genomic PCR amplification of B_5F_2 progenies: (A) SDS-PAGE patterns of Harunoakebono (1), NIL (2), Glenlea (3) and B_5F_2 progenies with LMW-GS 42K/40K (4), 48K/42K/40K (5), 48K/42K/40K (6), 48K/42K/40K (7), 48K (8), 48K/42K/40K (9), 48K (10) and 42K/40K (11), (B) Results of PCR using s-F1/s-R2 primers performed for the same B_5F_2 progenies as those listed in (A).*

2.2 2-D PAGE resolution

To characterize LMW-GSs, the glutenin fractions of Glenlea and Harunoakebono were subjected to 2-D PAGE. The 2-D PAGE highly resolved the LMW-GSs, allowing one band visualized in SDS-PAGE to be resolved into two to five protein spots (Figure 3). LMW-GS 42K included spot nos. 1, 2 and 3. LMW-GS 40K consisted mainly of spot nos. 4 and 5. LMW-GS 48K in Harunoakebono consisted of protein spot nos. 6, 7, 8, 9 and 10. In the NIL, LMW glutenin components nos. 1 to 5 derived from Glenlea replaced nos. 6 to 10 in Harunoakebono, resulting in improved bread-making quality.

Glutenin extracted individually from 95 half-cut grains of B_5F_2 progenies were also resolved by 2-D PAGE. The results showed that LMW glutenin components nos. 1 to 5 or nos. 6 to 10 were cosegregated in all B_5F_2 seeds.

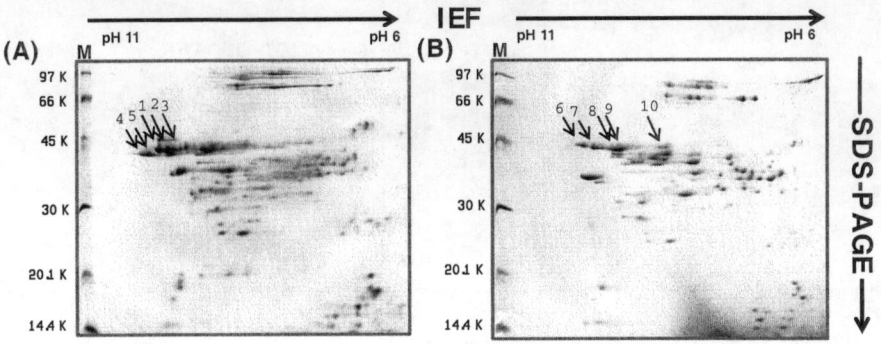

Figure 3. 2–D PAGE resolution of glutenin components of Glenlea (A) and Harunoakebono (B).

2.3 N-terminal amino acid sequences of LMW glutenin components

N-terminal amino acid sequences of LMW glutenin spot nos. 1, 2 and 3 corresponding to LMW-GS 42K and spot nos. 4 and 5 corresponding to LMW-GS 40K from Glenlea and spot no. 7 corresponding to LMW-GS 48K from Harunoakebono were determined. We could not determine the sequences of spot nos. 6, 8, 9 and 10 from Harunoakebono because amino acid peaks were not detected. The determined N-terminal amino acid sequences all corresponded fully or partially to SHIPGLERPSQQQPLPP (Table 1). They could all be classified as LMW-s (LMW glutenin component carrying a serine residue at the first position of the mature peptide). N-terminal sequence SHIPGLE is the most frequently analyzed protein-based sequence.

Table 1. N-terminal sequences of LMW glutenin components

Wheat cultivar	Spot no.	Sequences	Corresponding subunit
Glenlea	1	SHIPGLERPSQQQPL	LMW-GS
	2	SHIPGLERPSQQQ	LMW-GS
	3	SHIPGLERPSQQQ	LMW-GS
	4	SHIPGLERPSQQQ	LMW-GS
	5	SHIPGLERPSQQQ	LMW-GS
Harunoakebono	7	SHIPGLERPSQQ	LMW-GS

Each spot no. corresponds to the LMW glutenin spot in Fig. 3. The subunit in Fig. 1 corresponding to each spot is shown. Data are presented using the standard one-letter abbreviations for amino acids.

2.4 Molecular marker linked to the allele corresponding to LMW glutenin components associated with bread-making quality

To develop molecular markers linked to the alleles encoding LMW-GS 42K/40K and LMW-GS 48K, respectively, we performed genomic PCR using primers s-F1 5'-CCATCCAACAACAACCACACC 3' and s-R2 5'-CCCGAGTTGCTGTTGTGACTGC -3' for 65 B_5F_2 progenies. The results of PCR revealed that an amplicon of 692 bp (*GLEN42K*) was specific to the progeny carrying LMW-GS 42K/40K, that an amplicon of 839 bp (*HARU48K*) was specific to the progeny carrying LMW-GS 48K, and that both amplicons were amplified for the progeny carrying LMW-GS 48K/42K/40K (Figure 2). The amino acid sequence deduced from *GLEN42K* had a novel repetitive sequence (Figure 4). *HARU48K* showed 99.5% identity to the nucleotide sequence of an LMW-s glutenin gene, EMBL accs. Y17845 (2).

```
GLEN42K   IQQQPHQFPQQQFCSQQQQ---------------------------------QQQQ---  22
HARU48K   IQQQPHQFPQQQFCSQQQQQPPLSQQQQPPFSQQQQPPFSQQQQPVLPQQPSFSQQQLPP  60
          **** *************                                      ***
GLEN42K   -QQQQQQQ-PLSQQQQPP-------------FSQQQPPFSQQQQPVLPQQPSFSQQQLP   66
HARU48K   FSQQQ--QPPFSQQQQPVLPQQPSFSQQQLPPFSQQLPPFSQQQQPVLPQQPPFSQQQLP  118
          ***  * ******                  **** *************** *******
GLEN42K   PFSQQQPPFSQQQQPVLPQQPPFSQQQQQQPILPQQPSFSQQQQLVLPQQQIPFVHPSI  126
HARU48K   PFSQQLPPFSQQQQPVLPQQPPFSQQQQQ-PILPQQPPFSQQQQPVLLQQ-QIPFVHPSI  176
          ***** ******************* ******* ****** *  * * ********
GLEN42K   LQQLNHCKVFLQQQCSPVAMPQSLARSQMLQQSSCHVMQQQCCQQLPQIPQQSRYEAIRA  186
HARU48K   LQQLNHCKVFLQQCCSPVAMPQSLARSQMLQQSSCHVMQQCCQQLPQIPQQSRYEAIRA  236
          ************************************************************
GLEN42K   IIYSIVLQEQQQVRGSIQTQQQQPQQLGCCVSQPQQQSQQQLG                 229
HARU48K   IVYSIILQEQQQVQGSIQTQQQQPQQLGCCVSQPQQQSQQQLG                 279
          * *** ******* ****************************
```

Figure 4. *Alignment of the deduced amino acid sequences of DNA markers. Asterisks indicate identical residues, dashes are inserted to give maximum sequence alignment, sequences corresponding to primers are underlined, and cysteine residues are in squares.*

3 CONCLUSIONS

2-D PAGE showed that LMW-GS 42K/40K, which are associated with good bread-making quality, consist of five spotted LMW glutenin components that are cosegregated in the segregating population and that the allelic LMW-GS 48K derived from Harunoakebono is also composed of five spotted LMW glutenin components that are cosegregated.

The PCR-amplified marker *GLEN42K* is linked to the gene encoding LMW-GS 42K/40K, which are associated with good bread-making quality, and the marker *HARU48K* is linked to the allelic gene encoding LMW-GS 48K. These DNA markers would be useful for selection by wheat breeders of varieties carrying LMW glutenin components associated with good bread-making quality in a breeding program.

References
1 K. Takata, H. Yamauchi, Z. Nishio and T. Kuwabara, Breeding Sci., 2001, **51**, 143.
2 S. Masci, R D'Ovidio, D Lafiandra and D.D. Kasarda, Plant Physio., 1998, **118**, 1147.

A MONOCLONAL ANTIBODY SPECIFIC FOR A UNIQUE EPITOPE OF HIGH MOLECULAR WEIGHT GLUTENIN SUBUNIT 1 ALLOWS IMMUNOLOGICAL DISCRIMINATION OF *GLU-A1* ALLELES

H. Gruber, M. Sedlmeier and B. Killermann

Bavarian State Research Center for Agriculture, Institute for Crop Production and Plant Breeding, Lange Point 6/II, D-85354 Freising, Germany

1 INTRODUCTION

The high molecular weight glutenin subunits (HMW-GS), coded by the genes of the *Glu-1* loci are related to significant differences in breadmaking quality[1]. As a method with high sample throughput and high sensitivity an enzyme-linked immunosorbent assay (ELISA) can be a valuable tool to determine the allelic composition of HMW glutenin subunits in the selection process of breeding programs. Because the HMW glutenins show a high degree of homology in their amino acid sequences, it is difficult to develop specific antibodies (Abs) for single HMW glutenin subunits[2]. We therefore decided to generate specific monoclonal antibodies (mAbs) on the basis of peptides representing unique sequence sections of the glutenins of interest. One target protein was the HMW glutenin subunit 1 (1Ax1), which was repeatedly shown to be correlated with good dough quality[3]. We present the development and characterization of a mAb highly specific for this subunit and its application in a fast and simple *Glu-A1* Allele Assay.

2 MATERIALS AND METHODS

Wheat genetic stock and flour samples: For purification of defined HMW-GS wholemeal of the German wheat cultivars *Prinz*, *Basalt* and *Cadenza* was used. The investigated doubled haploid population derived from a cross between winter wheat cultivar *Flair* and line *CWW 92-6* (Bavarian State Research Center for Agriculture, Freising, Table 1).
Glutenin preparation and electrophoresis: Reduced and alkylated HMW-GS were extracted from wholemeal according to Marchylo et al.[4] and then purified by preparative A-PAGE[5]. For SDS-PAGE analysis reduced glutenin was extracted from crushed wheat kernels using 4,5 M urea, 10% (w/v) SDS, 3 % (v/v) ß-mercaptoethanol and separated on 12,5% SDS-polyacrylamide gels.
Production of monoclonal antibodies: MAbs were generated against synthetic peptide 592 conjugated to hemocyanin. Preparation of immunogen, immunization of mice, cell fusion, plating and preliminary screening was done by BioGenes GmbH, Berlin. Cell culture supernatants (CCS) were then screened for specific Abs by indirect ELISA and in Western Blots after SDS-PAGE of purified HMW-GS 1 (1Ax1) and 2 (1Dx2) and glutenin extracts of selected varieties.

Table 1 *HMW glutenin subunit composition of wheat cultivars for antibody screening (a) and parents of the investigated doupled haploid (DH) population (b). The HMW-GS corresponding to the Glu-1 alleles are shown in brackets.*

a Wheat cultivar	Glu-A1	Glu-B1	Glu-D1
Basalt	c [N]	a [7]	a [2+12]
Cadenza	c [N]	h [14+15]	d [5+10]
Prinz	a [1]	i [17+18]	d [5+10]
b DH parental cultivars			
Flair	c [N]	d [6+8]	d [5+10]
CWW 92-6	a [1]	i [17+18]	a [2+12]

Glu-A1 Allele Assay (indirect ELISA): Single wheat kernels were crushed and stirred with 800 µl of 2 M urea, 50 mM DTT, 50 mM carbonate buffer, pH 9,6 for 1 hour at 37 °C. Diluted supernatants were coated onto ELISA microwell plates. After standard blocking and washing steps mAb Anti-pep592 was added to the wells for 1 h incubation at room temperature. Bound mAb Anti-pep592 was detected by goat anti-mouse IgG (Fc specific) alkaline phosphatase conjugate and the enzyme substrate paranitrophenylphosphate was determined by measurement of absorbances at 405 nm.

3 RESULTS

The alignments of deduced published amino acid (AA)-sequences showed that HMW-GS 1 and HMW-GS 2* (1Ax2*) have an AA-sequence common in their repeat domains which does not occur in other wheat storage proteins sequenced up to date (Figure 1). Use of a

^{314}STQEQQLGQEQQDQQSGQGRQ^{334}

Figure 1 *The predicted unique amino acid sequence selected from the repeat domain of HMW-GS 1 (residues 314-334). This sequence served as pattern for the synthesis of the peptide immunogen peptide 592. Depiction in single-letter nomenclature for amino acids.*

peptide immunogen corresponding to this sequence led to hybridoma cell clone 4_2 producing mAbs, which are able to distinguish HMW-GS 1 from other HMW-GS in an indirect ELISA. MAbs from other clones show broad specificity for all investigated HMW glutenin subunits (Figure 2).

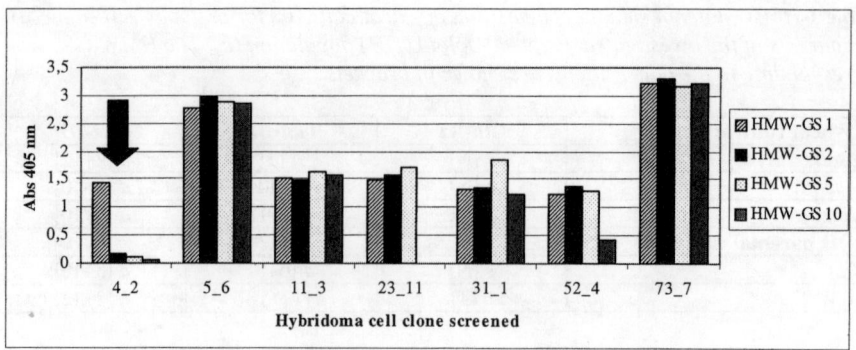

Figure 2 *ELISA-screening for hybridomas producing HMW-GS 1 specific mAbs: Reaction of mAbs using purified HMW glutenin subunits at 10 µg/ml as coating antigen and 1:2 diluted cell culture supernatants as first antibody. The arrow indicates the hybridoma cell line selected as antibody producer of mAb Anti pep592. MAbs from other clones show broad specificity for HMW glutenin subunits.*

From hybridoma cell clone 4_2 mAb was purified and further characterized as mAb Anti pep592. MAb Anti pep592 is able to recognize HMW-GS 1 in the complex mixture of wheat storage proteins. Assessment of different sample preparations showed that one extraction step from crushed kernels with 2 M urea, 50 mM DTT carbonate buffer is sufficient to successfully discriminate the alleles *Glu-A1a* and *Glu-A1c* in an indirect ELISA (Figure 3).

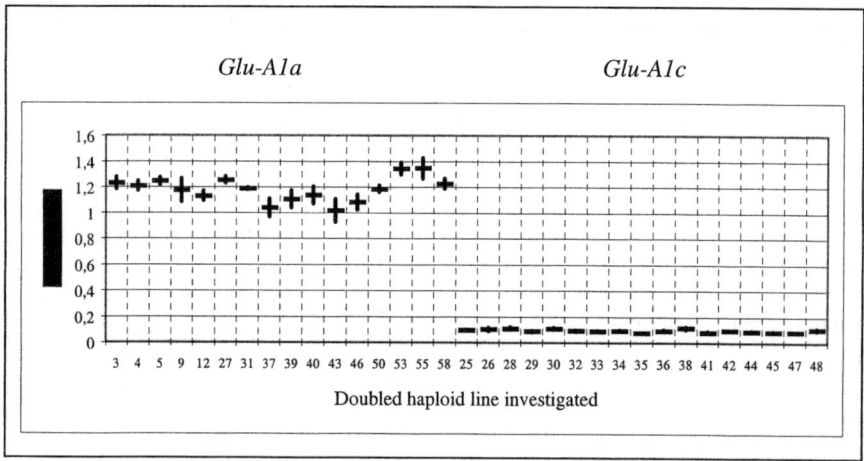

Figure 3 *Application of the Glu-A1 Allele Assay to doubled haploid lines from the cross Flair/CWW 92-6. Data are means ± standard deviation from five determinations for each line. Samples containing the Glu-A1a allele reach absorbances from 1,0 to 1,4 (16 DH-lines) whereas samples containing the Glu-A1c allele show absorbance values lower than 0,2 (17 DH-lines).*

4 DISCUSSION

The alignments of deduced published AA-sequences showed that the HMW-GS of interest very rarely form unique structures suitable as antigenic sites for recognition by antibodies. However, in HMW-GS 1 and HMW-GS 2* we found that a unique sequence which corresponds to the residues 314-334 of HMW-GS 1 has the potential to build out a specific epitope which does not occur in other HMW-GS. Our strategy to use the synthetic peptide 592 corresponding to this sequence as immunogen led to mAb Anti pep592 which is highly specific for HMW-GS 1. We propose that mAb Anti pep592 also recognizes HMW-GS 2* and we will present evidence for its specificity for HMW-GS 2* shortly. As the allele *Glu-A1b* responsible for expression of HMW-GS 2* very rarely occurs in German wheat cultivars, we concentrated our investigations onto the discrimination of the *Glu-A1a* from the *Glu-A1c* allele. Regarding the high AA-sequence homology of the known HMW-GS it is a great success for us to present a monoclonal antibody specific for two single HMW-GS only. To our knowledge all antibodies developed so far show crossreactivity with several other HMW-GS. We suppose that a part of the target sequence produces a continuous epitope in the glutenin subunit, as mAb Anti pep592 recognizes the epitope in the reduced and denatured protein under assay conditions. The simple *Glu-A1* Allele Assay enables us to perform a high throughput screening of wheat samples for the expression of HMW glutenin subunit 1, which makes an important contribution to improved selection for breadmaking quality.

References

1 P.I. Payne, K.G. Corfield, L.M. Holt, J.A. Blackman, *J. Sci. Food Agric.*, 1981, **32**, 51.
2 J.H. Skerrit, *AgBiotech News and Information*, 1998, **10**, 247.
3 N.G. Halford, J.M. Field, H. Blair, P. Urwin, K. Moore, L. Robert, R. Thompson, R.B. Flavell, A.S. Tatham and P.R. Shewry, *Theor. Appl. Genet.*, 1992, **83**, 373.
4 B.A. Marchylo, J.E. Kruger, D.W. Hatcher, *J. Cereal Sci.*, 1989, **9**, 113.
5 M.H. Morel, *Cereal Chem.*, 1994, **71**, 238.

Acknowledgements

We are grateful to Dr. G. Daniel (Bavarian State Research Center for Agriculture, Institute for Crop Production and Plant Breeding) for producing the doubled haploid populations. We also thank the GFP (Gemeinschaft zur Förderung der privaten deutschen Pflanzenzüchtung e.V.) and the BMVEL (Bundesministerium für Verbraucherschutz, Ernährung und Landwirtschaft) for financial support of the project G 87/00 HS.

USE OF RECOMBINANT PEPTIDES TO EXPLORE THE MOLECULAR MECHANISM OF GLUTEN PROTEIN VISCOELASTICITY

N.G. Halford[1], A. Savage[1], N. Wellner[2], E.N.C. Mills[2], P.S. Belton[3] and P.R. Shewry[1]

1. Rothamsted Research, Harpenden, Hertfordshire AL5 2JQ, UK
2. Institute of Food Research, Norwich Research Park, Colney, Norwich NR4 7US, UK
3. School of Chemical Sciences and Pharmacy, University of East Anglia, Norwich NR4 7TJ, UK

1. INTRODUCTION

Wheat gluten proteins are characterised by the presence of domains comprising repeated sequences based on one or more short peptide motifs. In the HMW subunits of glutenin these domains range from about 480 to 700 amino acids, accounting for 75-85% of the whole protein[1]. Two motifs, based on hexapeptides (PGQGQQ) and nonapeptides (GYYPTSP/LQQ), are present in all HMW subunits while the x-type subunits also contain tripeptide motifs (GQQ). Both the nonapeptide and tripeptide are always present in tandem with a hexapeptide.

It has been proposed that these repeat motifs contribute to the viscoelastic mechanism of gluten by forming an equilibrium between rigid β-sheet structures stabilised by inter-chain hydrogen bonds (trains) and β-turn rich structures (loops)[2]. If so it can be predicted that the length of the repetitive domain, the concensus sequence motifs and the presence of amino acid substitutions within the motifs will affect the viscoelastic properties via effects on the hydrogen bonding pattern and hence the loop to train ratio under different conditions of hydration.

We have therefore developed a strategy to explore this relationship in detail, based on the purification and characterisation of peptides expressed in *E.coli*.

2. RESULTS AND DISCUSSION

2.1 Strategy for Expression of Peptides in *E.coli*

The cloning strategy for the production of expression vectors encoding perfect repeat peptides has been described in detail by Feeney *et al*[3,4]. In brief, two double-stranded DNA molecules were synthesised and named linker A and insert B. Linker A encodes a protein of 23 amino acids, comprising methionine and alanine residues at the *N*-terminus followed by a hexapeptide in which the glutamine residue at position 6 of the motif is replaced with a cysteine residue. This is followed by a perfect nonapeptide except that the glutamine at position 9 of the motif is also replaced with a cysteine residue. There is a perfect hexapeptide repeat at the *C*-terminal end.

Insert B encodes a sequence of 30 residues comprising perfect hexapeptide and nonapeptide repeats (6+9+6+9). This was cloned into a *Pst*I site in the middle of linker A to create a synthetic gene, R1, that encoded a protein of 53 amino acids. Sequentially larger genes were made by cloning additional copies of insert B into a *Pst*I site in the middle of the central insert B sequence inserted in the previous step. This was possible because each cloning step produced a gene with a unique *Pst*I site.

The cysteines at positions 8 and –7 relative to the *N*- and *C*-termini respectively, of the expressed peptides allow their assembly into linear polymers by oxidation. The largest peptide of 'perfect' repeats that we have expressed and purified is R6 (Fig. 1a), which comprises 203 amino acids and has a M_r of 22005. Preliminary experiments have demonstrated that reoxidation of peptides based on R6 gives rise to a cohesive mass that can be used for biomechanical measurements.

2.2 Effects of Peptide Chain Length and Repeat Motif Conservation

Initial studies were carried out on R6 (203 amino acids), and a smaller peptide, R3 (110 amino acids), even smaller synthetic peptides (21 and 45 residues) and a larger (556 residue, M_r 58,000) recombinant peptide from the repetitive domain of subunit 1Dx5. All peptides were studied by FT-IR spectroscopy in the dry solid state, at hydration levels of 76% r.h. and 100% r.h., and either in solution (P21, P45, M_r 58,000 peptide), or fully hydrated (R3, R6).

The results (summarised in Table 1) showed that the dry peptides had similar spectra and that hydration resulted initially in increases in β-sheet but then decreases in β-sheet and increases in β-turn and hydrated extended structures. However, the magnitude of these effects and the precise hydration levels at which they occurred depended on the chain length and probably also on the degree of conservation of the motifs. Thus, the changes during the second phase of hydration were more marked in the P21, P45 (small, perfect repeat) peptides and the M_r 58,000 (large, poorly conserved repeat) peptide than in the R3 and R6 (larger, perfect repeat) peptides. The decreased extent of structural changes in these two peptides indicates that cumulative intermolecular hydrogen bonds that formed during the initial hydration provided a more stable inter-chain β-sheet structure which was less readily converted to β-turn and extended structure on further hydration.

Table 1 *Structural Changes During Hydration of Recombinant and Synthetic Peptides*

STATE	STRUCTURE
Dry solid	Mainly unordered structure and β-turns in all peptides.
Hydration: 76% r.h. ↓ 100% r.h. ↓ in water	Increases in β-sheet, β-turn and hydrated extended structures, but less marked with increasing M_r ↓ β-sheet content reaches maximum, further increases in β-turn and hydrated extended structures ↓ P21/P45/M_r 58,000 R3/R6 Forms solutions, high contents Insoluble, forms interchain β-sheet, of β-turns/hydrated extended decreased β-turn content structures

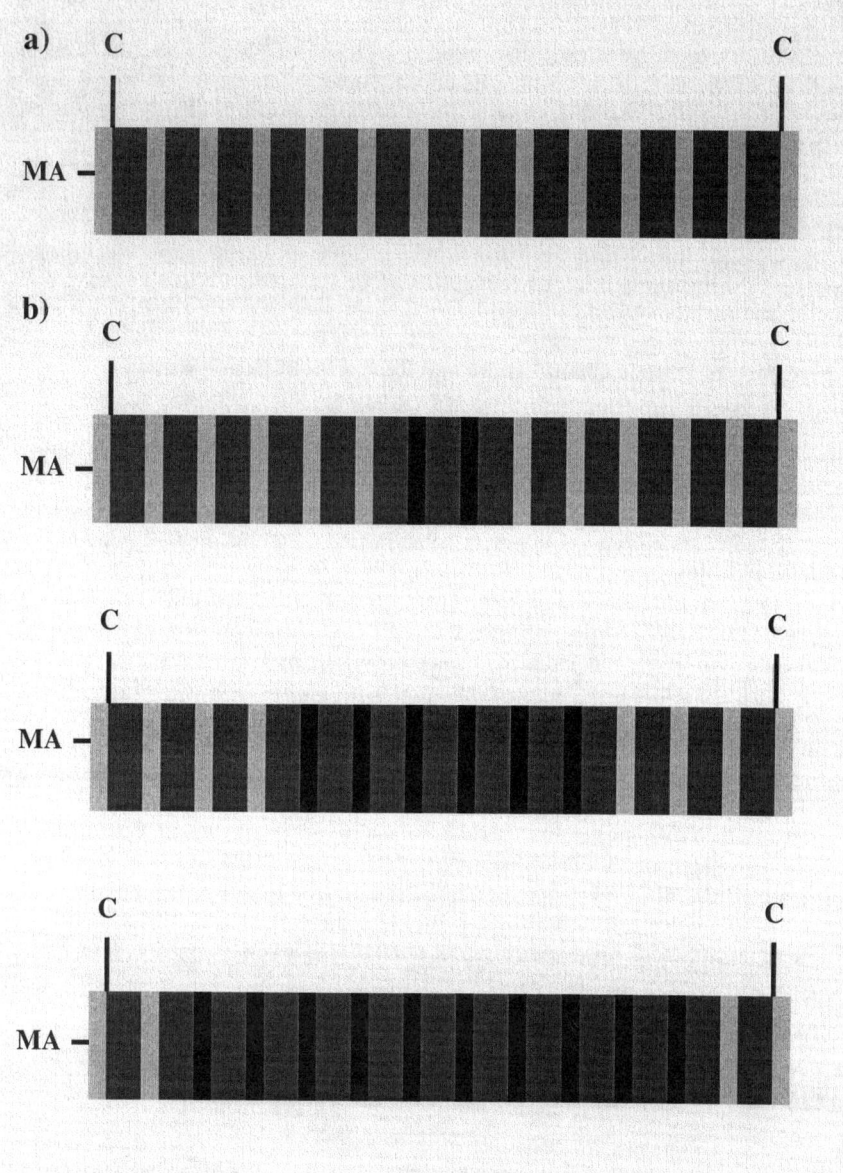

Figure 1 *a) Diagrammatic representation of R6, a 203 amino acid, M_r 22005 peptide comprised of `perfect` HMW subunit hexapeptides (PGQGQQ) (light grey) and nonapeptides (GYYPTSLQQ) (dark grey). b) Set of three variants of R6 containing 2, 6 or 10 copies of a modified hexapeptide motif (black)*

2.3 Effects of Amino Acid Substitution in R6 Peptides

Comparison of the patterns of amino acid substitutions within the repeat motifs of sequenced HMW subunits shows high frequencies of substitution of specific amino acids at some positions. For example, histidine is substituted for tyrosine at position 2 of over 40% of the nonapeptides present in y-type subunits[1]. Similarly, leucine is the most abundant amino acid at position 7 of the nonapeptide of y-type subunits but only occurs in about 10% of the nonapeptides of x-type subunits with proline being the most abundant amino acid in this position (approx. 70% of all peptides). Such natural variants have been used to design a series of peptides in which two, six or ten variant motifs are present within the R6 peptide structure, as shown diagrammatically for the hexapeptide motif in Fig. 1b.

The structures, interactions and biophysical properties of these peptides will be studied by subjecting single peptides and mixtures to FT-IR spectroscopy at varying levels of hydration and under compression using the system described by Wellner *et al* elsewhere in this volume. The information provided on the relationship between peptide motif sequence and biomechanical properties should allow a more rational approach to be adopted in identifying and designing novel subunits for exploitation to improve bread and pasta wheats.

References

1. P. R. Shewry, N. G. Halford, A. S. Tatham, Y. Popineau, D. Lafiandra and P. S. Belton, *Advances in Food and Nutrition Research,* 2003, **45**, 219.
2. P. S. Belton, On the elasticity of wheat gluten. *J. Cereal Sci.*, 1999, **29**, 103.
3. K. A. Feeney, A. S. Tatham, S. M. Gilbert, R. J. Fido, N. G. Halford and P. R. Shewry, *Biochim Biophys Acta*, 2001, **1546**, 346.
4. K. A. Feeney, N. Wellner, S. M. Gilbert, N. G. Halford, A. S. Tatham, P. R. Shewry and P. S. Belton, *Biopolymers (Biospectroscopy)*, 2003, **72**, 123.

Acknowlegments

Rothamsted Research and the Institute of Food Research receive grant-aided support from the Biotechnology and Biological Sciences Research Council (BBSRC) of the UK. The work described here is supported by BBSRC grant Ref. D14544.

BACTERIAL EXPRESSION, PURIFICATION AND FUNCTIONAL STUDIES OF A Y-TYPE HIGH MOLECULAR WEIGHT GLUTENIN SUBUNIT IN *T. tauschii*

M.E. Hassani[1,2], M.C. Gianibelli[3], R.G. Solomon[3], L. Tamas[4], M.R. Shariflou[1,2], P.J. Sharp[1,2]

[1]The University of Sydney, Plant Breeding Institute PMB11 Camden NSW 2570, Australia.
[2]Value Added Wheat CRC Limited, Locked Bag 1345, North Ryde NSW 1670, Australia.
[3]CSIRO Plant Industry, GPO Box 1600, Canberra ACT 2601, Australia.
[4] Eötvös Loránd University, Department of Plant Physiology, Budapest H-1117 Pázmány P. stny. 1/C

1 INTRODUCTION

Gluten is the essential determinant of bread-making and other wheat flour-based end use products quality. It consists of two components: glutenins and gliadins. Glutenins are divided in two groups according to their electrophoretic mobility: high molecular weight subunits (HMW-GS) and low molecular weight subunits (LMW-GS). They are linked together by covalent disulfide bonding and form very large polymeric structures. HMW-GS (x- and y-types) are encoded by two closely linked genes at the *Glu-1* loci (*GluA1*, *GluB1* and *GluD1*) that are located on the long arms of homologous chromosomes one. Bread wheat usually contains 3-5 HMW-GS arising from the expression of 1Ax, 1Bx or 1By, 1Dx and 1Dy. Studies on quality characteristics of wheat flour have shown that the HMW-GS composition strongly affects dough properties. Therefore, it is very important to determine the effect of individual polypeptides on dough functionality, particularly if testing unusual subunits that could be included in breeding programs for improving end product quality. Nevertheless, due to the number and amount of HMW-GS, purification of individual polypeptides is difficult because it is time consuming and very often the purified protein is contaminated with other accompanying polypeptides. This contamination prevents the accurate evaluation of functional properties of the individual polypeptides in dough. In order to avoid those drawbacks, recent testing developments have greatly contributed to properly study the functionality of individual polypeptides in dough. On one hand, the small scale testing system, using a 2 g Mixograph, extensograph and micro baking has facilitated the study of functional properties of small amounts of exogenous protein incorporated into different sample flours.[1] On the other hand, *in-vitro* expression techniques allowed the expression of large amounts of protein in heterologous systems.[2,3] The advantage of the latter is the production of large amounts of correctly folded polypeptides for functional studies. This paper mainly considers the expression, purification and functional studies of the smallest HMW-GS, subunit 12.4^t present in *Triticum tauschii*.

2 METHODS AND RESULTS

A typical SDS-PAGE of subunits 12.4^t is shown in Figure 1.

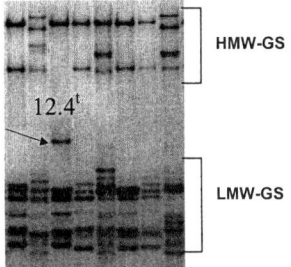

Figure 1: *SDS-PAGE of polymeric proteins of T. tauschii accessions (from Gianibelli et al., 2001)*

Primers were designed to amplify the entire encoding region of the gene for a mature y-type HMW-GS excluding signal peptides. *Escherichia coli* strain BL21-(DE3) pLysS and pET-3a expression vector were supplied by Novagen. PCR products were cloned and transformed in *E. coli*. Freshly transformed single colony was grown in 3 ml LB media at 37°C for 3 hours (Novagen handbook) with shaking. A small-scale expression, the 3 ml culture was used to inoculate 50 ml media in a 250 ml flask. LB media was supplemented with 30 mM of glycerol and glucose, 50 mg/ml ampicillin and 34 mg/ml chloramphenicol. In preliminary assessment and optimization of the expression of the HMW-GS, different parameters including media, host strains, level of IPTG and inoculation at different level of optical density (OD_{600}) were tested. The results have shown that different media had little impact on expression level while application of glycerol and glucose significantly improved the level of expression. Among host strains, BL21-CodonPlus™ (DE3)-RIL and BL21-(DE3) pLysS have improved expression level in comparison to BL21-CodonPlus™ (DE3)-RP and AD494-(DE3). There were no significant differences among IPTG concentration and level of OD_{600}. At the OD_{600} of 1.0, culture media was induced by IPTG to final concentration of 0.4 mM. A time course expression of 1, 2, 3, 4, 5, 6 and 16 hours was performed. The results indicated that bacterially expressed protein remarkably increased up to 5 hours incubation while further incubation leads to protein degradation (Figure 2a). The level of expression in small scale for y-type HMW-GS could be increased by up to 5%.[4] We have also observed a high level of expression in small-scale approach, where an appreciable amount of protein was obtained. The maximum expression after 5 hours of incubation showing that in *E. coli* system the expression of y-type HMW-GS will be prevented and protein degradation will occur if the incubation period is extended. This could be related to the toxicity of expressed protein.

2.1 SDS PAGE and Immunobloting Analysis of Expressed Protein

SDS-PAGE analysis of the expressed y-type Dy12.4^t was performed in 10% and 15% polyacrylamide gels.[5] Dy12.4^t subunit expressed in *E. coli* was readily identified as a unique band (Figure 2b). The level of expression of Dy12.4^t in *E. coli* was examined by analyzing the total cell proteins of uninduced and induced for 1, 2, 3, 4, 5, 6 and 16 hours. Monoclonal antibody was used for immunodetection of bacterially expressed protein. The result showed the strongest detection of Dy12.4^t in wholemeal flour and induced samples (Figure 2b).

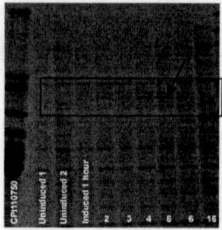 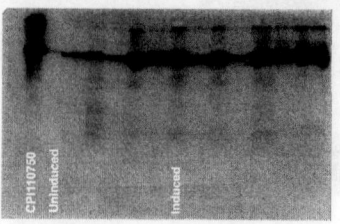

Figure 2: *a) Expression level of uninduced and induced bacteria showing the maximum expression after 5 hours of induction (arrow). b) Immunodetection of bacterial expressed protein using monoclonal antibody*

2.2 Purification of Expressed Dy12.4^t in *E. coli*

The method used for extraction and purification of bacterially expressed protein from insoluble inclusion bodies was based on the alcohol extraction method of Marchylo et al.[6] Bacterially expressed protein in *E. coli* that accumulated as inclusion bodies made difficult to purify the expressed protein. Extraction and purification method of HMW-GS described by Dowd and Bekes with some modification removed all the contaminants and allowed the purification of HMW-GS Dy12.4^t.[4] A substantial amount of purified protein was obtained when 70% ethanol and 100 mM DTT was used with two hours of incubation at 65°C, sonication on ice and vortexing. Purified proteins were pooled and dialyzed against 0.1 M acetic acid and freeze- dried. Although SDS-PAGE analysis of total cell proteins showed the strong expression of Dy12.4^t in *E. coli*, exposing cell pellet to 70% ethanol with 10 mM DTT at 65°C for one hour did not extract any protein. This indicates that the expressed protein was accumulated as inclusion bodies. The expression level and protein concentration in *E. coli* as well as the high content of amino acids such as cysteine and proline privileged the formation of inclusion bodies. Toxic proteins are accumulated as inclusion bodies by *E. coli* defence mechanism. However, it is not clear if the formation of inclusion bodies is correlated to the size of the protein, number of repetitive sequence or number of beta-turns structures.

RP-HPLC was used to assay bacterially expressed y-type HMW-GS.[7] Bacterially expressed Dy12.4^t showed earlier elution time (~ 1.00 minute) than the normal y-type HMW-GS obtained from flour of *T. tauschii*. No other differences were noted.

2.3 Functionality Studies

Purified bacterially expressed Dy12.4^t subunit was also incorporated into wheat flour samples using 2 g mixograph and mixing parameters were measured by software MDO6.[8] Bacterially expressed subunit Dy12.4^t has significantly improved mixing time and reduced resistance break down when was incorporated to a base flour (++-) where the *Glu-D1* encoding proteins were not expressed (Table 1)

Table 1: *Mixing properties of base flour, base flour plus 10 mg and 20 mg of bacterial expressed subunit Dy12.4^t. MT: mixing time (seconds); PR: peak resistance (arbitrary units); RBD: resistance breakdown (%).*

	MT	PR	RBD
Control	107	234	20
+10 mg	127	207	14
+20 mg	150	216	14

3 DISCUSSION

Two important features: 1-the location and number of cysteine residues and 2- the length and amino acid composition of repetitive domain affect the formation of the glutenin polymers. In terms of the number of cysteine, Dy12.4^t has the same number of cysteines as subunits Dy10 and Dy12 from bread wheat. In terms of the length of repetitive domain and amino acid composition they vary greatly[9]. Dy12.4^t is much smaller than Dy12 because of the deletions of 215 and 6 amino acid residues in its central repetitive domain. The deletion has reduced the number of hexapeptide to 29 from 49 and the number of nonapeptide to 10 from 21 compared with subunit Dy12.[9] Belton (1999)[10] has reported that the formation of hydrogen bonds between adjacent HMW-GS glutenin subunits could explain gluten elasticity. The same author has also described a model of train and loops to explain the constant formation of hydrogen bonds between central repetitive domains rich in glutamine residues. Therefore, the reduction in size of repetitive domain of subunits 12.4^t could reduce the number of hydrogen bonds and thus affect the dough properties.

The incorporation of 12.4^t subunit into wheat flour using 2 g mixograph showed a positive effect on functional properties of a base flour with an increase in mixing time and decrease resistance break down. Introgression of this subunit (*Glu-D1* allele from *T. tauschii*) in bread wheat is currently carried out to further evaluated its effect in dough quality.

References

1   F. Bekes, P.W. Gras and R.B. Gupta, *Cereal Chem.*, 1994, **1**, 44-50.
2   L. Tamas, J. Greenfield, N.G. Halford, A.S. Tatham and P.R. Shewry, *Prot. Expr. Purif.* 1994, **5**, 357-363.
3   R. D'Ovidio, O.D. Anderson, S. Masci, J. Skerritt and E. Porceddu, *J. Cereal Sci.*, 1997, **25**, 1-8.
4   C. Dowd and F. Bekes, *Prot. Expr. Purif.*, 2002, **25**, 97-104.
5   G.J. Lawrence and K.W. Shepherd, *Austr. J. Biol. Sci.*, 1980, **33**, 221-233.
6   B.A. Marchylo, J.E. Kruger, and D.W. Hatcher, *J. Cereal Sci.*, 1989, **9**, 113-130.
7. O.R. Larroque, F.Bekes, C.W. Wrigley and W.G. Rathmell, In *Gluten 2000*, Bristol, UK, April 2000. 136-139.
8   P. Gras and F. Bekes, *Proc. 6th Int. Gluten Workshop*, Sydney, Australia, 1996. 507-510
9   M.E. Hassani, M.C. Gianibelli, M.R. Shariflou and P.J. Sharp, *Proc. 10th Int. Wheat Genetic Symposium*, Paestum, Italy, 2003. 955-957.
10 P.S. Belton, *J. Cereal Sci.*, 1999, **29**, 103-107.

INFORMATION HIDDEN IN THE LOW MOLECULAR WEIGHT GLUTENIN GENE SEQUENCES

A. Juhász and M.C. Gianibelli

CSIRO Plant Industry, GPO Box 1600, Canberra, ACT 2601 Australia, angela.juhasz@csiro.au

1 INTRODUCTION

Low molecular weight glutenin subunits play an important role in determining dough rheological properties such as dough strength and extensibility. To define the mechanism through which LMW-GSs control these parameters a better understanding of *Glu-3* gene family is needed. In the last ten years the development of different biochemical analysis techniques, such as proteomics or the novel techniques used in biotechnology opened new possibilities for the study of these proteins. However, matching the results from the different techniques is not an easy task. Along with the protein biochemistry, biotechnology provides enormous amount of information from sequencing of cDNA libraries or genomic clones. The amino acid sequences derived from the nucleotide sequences may be easily analysed for different protein characteristics including Mr, pI or secondary and tertiary structure information. The information collected from the nucleic acid and protein databases may help the evaluation of biochemical results and the determining of structure –function relationship.

2 MATERIALS AND METHODS

Forty nine complete LMW glutenin gene sequences, mainly genomic clones (Table 1) and about 1000 ESTs were collected using NCBI and TIGR databases. The sequences were aligned and grouped based on their N terminal sequences (ClustalW). Physical and chemical properties (Mr and pI) of the different LMW glutenin types were calculated based on derived amino acid sequences using programs provided by ExPaSy Molecular Biology Server and matched with some experimental results. Frequency of some LMW-glutenin gene types in different genotypes was analysed.

3 RESULTS AND DISCUSSION

The LMW glutenin genes are grouped in three main groups based on the first amino acid of the mature peptide: i- (Ile), s- (Ser) and m- type (Met). However based on Ikeda et al. (2002) 12 subgroups of LMW-glutenin genes can be distinguished when both the N- and the C-terminal sequence are considered. Based on our analysis altogether 17 different subgroups (4 i-type, 7 s-type and 6 m-type subgroups)

Table 1. *LMW glutenin gene sequences analysed*

AN (NCBI)	Specie / cv	Clone type	Location	N terminal type	Mr	pI	N. of residues	Ref.
X07747	T. ae. Yamhill	G	Glu-A3	ISQQQQAPPFS	41020	8.63	356	1
Ab062876	T. ae. Norin 61	G	Glu-A3	ISQQQQPPFS	21936	8.43	192	2
Ab062878	T. ae. Norin 61	G	Glu-A3	ISQQQQPPFS	42847	7.31	369	2
Ab008497	T. ae. 1CW	G		ISQQQQPPPFS	42587	8.43	367	3
Aj293097	T. du. Langdon	G	Glu-A3	ISQQQQPPPFS	42549	8.63	367	4
Ab062877	T. ae. Norin 61	G	Glu-A3	ISQQQQPPFS	43035	7.60	371	2
U86030	T. ae. Cheyenne	C		ISQQQQPPFS	39289	8.05	339	5
Ab062863	T. ae. Norin 61	G	Glu-B3	(IEN)SHIPGLEK	36379	8.25	320	2
Ab062864	T. ae. Norin 61	G	Glu-B3	(IEN)SHIPGLEK	31414	8.06	277	2
Aj519838	T. ae. Neepawa	G		(MEN)SHIPGLER	31954	7.61	281	6
Ab062853	T. ae. Norin 61	G	Glu-B3	(MEN)SHIPGLER	42628	8.25	373	2
Ab062854	T. ae. Norin 61	G	Glu-B3	(MEN)SHIPGLER	40593	7.87	357	2
Ab062855	T. ae. Norin 61	G	Glu-B3	(MEN)SHIPGLER	40607	8.25	356	2
Ab062856	T. ae. Norin 61	G	Glu-B3	(MEN)SHIPGLER	40584	8.06	356	2
Ab062857	T. ae. Norin 61	G	Glu-B3	(MEN)SHIPGLER	37376	8.07	328	2
Ab062858	T. ae. Norin 61	G	Glu-B3	(MEN)SHIPGLER	35788	8.07	314	2
Ab062859	T. ae. Norin 61	G	Glu-B3	(MEN)SHIPGLER	33940	8.25	298	2
Ab062860	T. ae. Norin 61	G	Glu-B3	(MEN)SHIPGLER	32457	8.56	284	2
Ab062861	T. ae. Norin 61	G	Glu-B3	(MEN)SHIPGLER	25294	7.86	223	2
Ab062862	T. ae. Norin 61	G	Glu-B3	(MEN)SHIPGLER	27988	8.07	246	2
Aj007746	T. du. Lira	G	Glu-B3	(MEN)SHIPGLER	42242	8.26	369	7
Y17845	T. ae. Yecora Rojo	G	Glu-B3	(MEN)SHIPGLER	42485	8.06	372	8
Y18159	T. du. Lira	G	Glu-B3	(MEN)SHIPGLER	40847	8.44	357	7
Ab062851	T. ae. Norin 61	G	Glu-D3	METSHIPGLER	39662	8.25	345	2
Ab062852	T. ae. Norin 61	G	Glu-D3	METSHIPSLEK	36951	8.06	323	2
Y14104	T. du. Langdon	G	Glu-B3	METSHIPSLEK	37743	7.58	330	4
Ab062868	T. ae. Norin 61	G	Glu-A3	MDTSCIPGLER	32041	7.53	283	2
Ab062869	T. ae. Norin 61	G	Glu-A3	MDTSCIPGLER	31941	7.82	282	2
Ab062870	T. ae. Norin 61	G	Glu-A3	MDTSCIPGLER	30623	7.53	270	2
Ab062871	T. ae. Norin 61	G	Glu-A3	MDTSCIPGLER	29126	7.82	259	2
Aj293099	T. du. Langdon	G	Glu-A3	MDTSCIPGLER	31804	7.15	281	4
X62588	T. du. Langdon	G		MDTSCIPGLER	30706	7.56	272	12
Ab007763	T. ae. 1CW	G		METRCIPGLER	32928	7.82	288	3
Ab007764	T. ae. 1CW	G		METRCIPGLER	30521	8.34	267	3
Ab062875	T. ae. Norin 61	G	Glu-D3	METRCIPGLER	32567	8.04	284	2
Aj519835	T. ae. Neepawa	G		METRCIPGLER	32465	8.24	284	6
U86027	T. ae. Cheyenne	C		METRCIPGLER	32879	8.72	287	5
U86029	T. ae. Cheyenne	C		METRCIPGLER	32751	8.24	286	5
X13306	T. ae. Chinese Spring	G	Glu-D3	METRCIPGLER	32879	8.72	287	9
Ab062873	T. ae. Norin 61	G		METSCIPGLER	31773	8.43	278	2
Ab062874	T. ae. Norin 61	G		METSCIPGLER	31809	8.24	278	2
U86026	T. ae. Cheyenne	C		METSCIPGLER	31829	8.43	278	5
X51759	T. du. Mexicali	C	Glu-A3?	METSCIPGLER	31273	7.22	275	10
Ab062872	T. ae. Norin 61	G	Glu-D3	METSCISGLER	32047	7.53	283	2
M11077	T. ae. Cheyenne	C	Glu-D3	METSCISGLER	32175	7.53	284	11
U86028	T. ae. Cheyenne	C		METSCIGLER	32120	7.82	283	5
Ab062865	T. ae. Norin 61	G	Glu-D3	METSRVPGLEK	37752	7.84	330	2
Ab062866	T. ae. Norin 61	G	Glu-D3	METSRVPGLEK	37466	7.84	328	2
Ab062867	T. ae. Norin 61	G	Glu-D3	METSRVPGLEK	34294	8.06	301	2

were identified, differing not only in the N-terminal sequence but showing some specific features in all three domains (results not shown).

It is well known from the literature that the molecular weight of glutenin subunits determined on SDS-PAGE is overestimated [13]. A similar discrepancy can be observed between the iso-electric points calculated based on the protein sequences and the observed values in the gels (Figure 1).

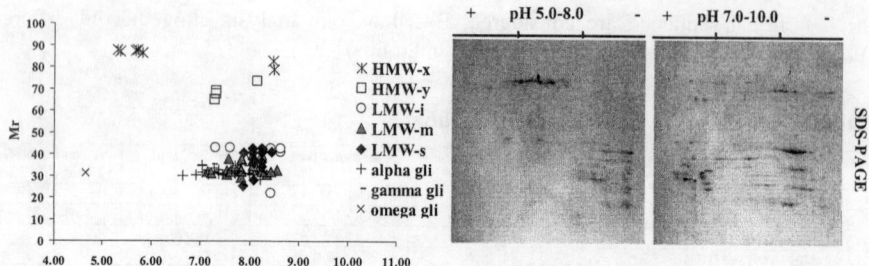

Figure 1. *IEF x SDS-PAGE gels present of reduced glutenin fraction of Glu-A3a, -,- null line*

The chromosomal location of genes was identified using their sequence annotations. The LMW-glutenin subgroups present on chromosomes A, B and D were evaluated using their calculated Mr and pI values (Figure 2). Based on our present knowledge i-type genes are located only on chromosome A. Analysing i-type ESTs several different i-type genes were identified in cv-s Chinese Spring, Cheyenne and Norin 61[14, 2]. S-type LMW glutenins are located mainly on locus *Glu-B3*, however there is a subgroup on *Glu-D3* locus. Most of the m-type sequences, especially those with cysteine residue at position five belong to LMW-C glutenins. Using these information for example some sub-groups of s-type genes, known as quality factors (Mr above 42kDa) can be identified and used to marker assisted selection (Figure 3)[15].

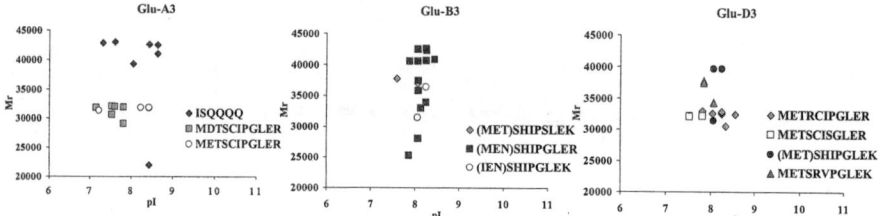

Figure 2. *LMW glutenin types of Glu-3 loci and their two dimensional (Mr x pI) map (deduced from their amino acid sequences)*

4 CONCLUSION

The increasing amount of sequence information provided by sequence and protein databases can be used not only for genomic purposes (e.g. primer design, gene sequencing, marker development). Using bioinformatics tools a better understanding of relationship between results arising from molecular genetic and proteomic studies is possible. It is especially true for proteins, like LMW glutenins where better understanding of structure-function relationships is important.

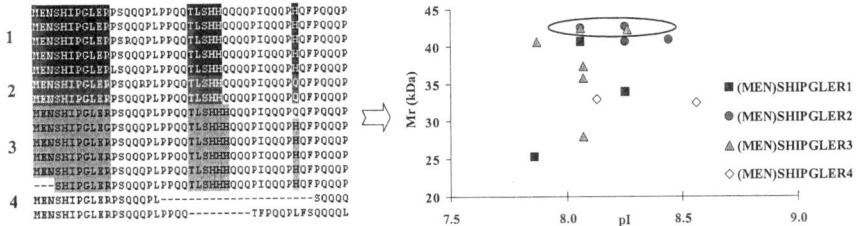

Figure 3. *Characterisation and identification s-type LMW glutenin genes which may have positive influence on rheological properties.*

References

1. E.G. Pitts, J.A. Rafalski and C. Hedgcoth. *Nucleic Acids Res,* 1988, **16**, 11376
2. T.M. Ikeda, T. Nagamine, H. Fukuoka and H. Yano. *Theor Appl Genet,* 2002, **104**, 680
3. N. Maruyama, K. Ichise, T. Katsube, T. Kishimoto, S. Kawase, Y. Matsumura, Y. Takeuchi, T. Sawada and S. Utsumi. *Eur. J. Biochem.* 1998, **255**, 739
4. R. D'Ovidio, S. Masci, C. Marchitelli, P. Tosi, M. Simeone and E. Porceddu 2000, direct submission.
5. O.D. Anderson, B. Cassidy, and J. Dvorak 1997, direct submission
6. T, Chardot, T. Do, L. Perret and M. Lauriere. In *Plant Biopolymer Science,* ed. D. Renard, G. Delle Valle and Y. Popineau, RCS, 2002, p24
7. R. D'Ovidio, C. Marchitelli, L. Ercoli Cardelli and E. Porceddu. *Theor. Appl. Genet* 1999, **98**, 455
8. S. Masci, R. D'Ovidio, D. Lafiandra and D.D. Kasarda. *Plant Physiol.* 1998, **118**, 1147
9. V. Colot, D. Bartels, R. Thompson and R. Flavell. *Mol. Gen. Genet.* 1989, **216**, 81
10. B.G. Cassidy, and J. Dvorak. unpublished 1995
11. T.W. Okita, V. Cheesbrough and C.D. Reeves. *J. Biol. Chem.* 1985, **260**, 8203
12. R. D'Ovidio and O. Tanzarella, E. *Plant Mol. Biol.* 1992, **18**, 781
13. N.A.C. Bunce, R.P. White and P.R. Shewry. *J. Cer. Sci* 1985. 131
14. A. Juhász, W. Zhang, K. R. Gale, M.K. Morell and C.M. Gianibelli, In *Proceedings of the 10th IWGS,* ed. N.E. Pogna, M. Romano, E.A. Pogna, G. Galterio, Instituto Sperimentale per la Cerealicoltura, Rome, 2003, 3, p. 965
15. S. Masci, R. D'Ovidio., D. Lafiandra and D.D. Kasarda *Theor Appl Genet,* 2000, **100**, 396

Acnowledgments

The authors wish to thank NSW Centre for Agricultural Genomics to support this work.

IDENTIFICATION AND CHARACTERISATION OF NEW CHIMERIC STORAGE PROTEIN GENES FROM AN OLD HUNGARIAN WHEAT VARIETY

I. J. Nagy[1], I. Takács[1], A. Juhász[1], L. Tamás[2], Z. Bedő[1]

[1] Agricultural Research Institute of the Hungarian Academy of Sciences, Martonvásár, Brunszvik u. 2, H-2462, Hungary
[2] Department of Plant Physiology, Eötvös Loránd University, Budapest, Pázmány P. stny. 1/C, H-1117, Hungary

1 INTRODUCTION

Old Hungarian wheat varieties resemble landraces with regard to their heterogeneous properties. It is interesting to note that although the Bánkúti 1201 population contains mainly subunits 2+12 or 3+12 encoded by chromosome 1D it has good breadmaking quality.

The aim of this project was to find new genes in one of the isogenic lines of Bánkúti 1201 which could be used for marker assisted selection. A cDNA library was prepared from the immature kernels. The library was screened with the amplified fragment of the C-terminal region of the LMW-1D1 glutenin subunit gene.

Amongst the positive clones a short gene with a very specific sequence was found. Its 5' end is identical to a gliadin sequence, while the 3' end is identical to the sequence of an LMW-GS. Screening the cDNA library by PCR using two oligonucleotides, specific for gliadin as forward, and for LMW-GS as reverse, more clones with similar sequences were found. The existence of this type of storage protein gene was also proved by PCR using genomic DNA as a template.

2 METHOD AND RESULTS

2.1. Method

2.1.1 PCR Primers and Reaction Conditions. The gliadin-specific primer was designed for the beginning of the N-terminal region of the M16064 gene[1] while the LMW-GS-specific primer for the very end (including the stop codons) of the LMW-1D1 gene[2].
GliNF1: 5'-TCATGAATATACAGGTCGACCCTAG-3'
LMWCR1: 5'-CCATGGGGATCCTTATCAGTAGGCACCAAC-3'
Reactions were carried both on cDNA clones and genomic DNA using Taq DNA polymerase (Promega) and Herculase DNA polymerase (Stratagene).

2.1.2 Southern Blot. The PCR products were blotted on a Hybond N membrane according to Maniatis et al.[3]. The membrane was hybridised with a radioactively labelled probe specific to the C-terminal region of LMW-GS 1D1[4].

2.1.3 Cloning of PCR Products. Prior to cloning, adenine addition (adenine "tailing") was carried out. A TOPO TA Cloning Kit (Invitrogen) was used for cloning the PCR products. DNA was transformed into EC TOP10 cells by electroporation.

2.1.4 Sequencing. Sequencing was carried out with an ABI Prism BigDye Terminator Cycle Sequencing Ready Reaction Kit and the DNA was run on an ABI PRISM™ 310 Genetic Analyser.

2.1.5 Sequence Analysis. Sequences were analysed using Wisconsin Package Version 10.0, Genetics Computer Group (GCG), Madison, Wisc. Software.

2.2. Results

In addition to the LMW-GS clones isolated by Nagy et al.[4] from the cDNA library, the same method was used to isolate the B21 clone (gene1). Its 5' end showed high similarity to the γ-gliadin sequences while its 3' end resembled the sequence of LMW-GS molecules. (Fig. 1). Analysis also revealed that a segment of 24 base pairs was deleted from the original gliadin allele. It seems that a frame shift occurred during the recombination process through which this gene evolved. Due to this frame shift the C-terminal region of the protein (31 amino acid residues) is not similar to any of the storage protein sequences. This new sequence however contains a cysteine residue, which is the only one in the molecule, suggesting that the protein may play a chain-terminating role in the gluten polymers.

Additional chimeric genes were amplified by PCR from 10 aliquots of the cDNA library containing 50 000 clones. PCR products were cloned and sequenced. Results of the sequence analysis are summarised in Table 1.

In order to prove that these chimeric genes were not artefacts arising during the preparation of the cDNA library, genomic DNA was also isolated from the same Bánkúti line from which the cDNA library was constructed. The genomic PCR products were blotted after agarose gel electrophoresis and hybridised with the same LMW-GS C-terminal probe used to isolate the B21 clone. The autoradiogram showed that genomic DNA contained several sequences produced by the chimera-specific primers. Since PCR of genomic DNA gave fragments with different sizes, all of them were cloned and transformed into *E. coli*.

2.2.1 Analysis of Clones Isolated from the cDNA Library. The longest clone so far, named gene 2, comprises 896 nucleotides. This sequence can be divided into 255 and 641 long fragments according to the result of sequence homology search. The first one is similar to γ-gliadins, while the second is similar to LMW-glutenins (see Table 1.). There is

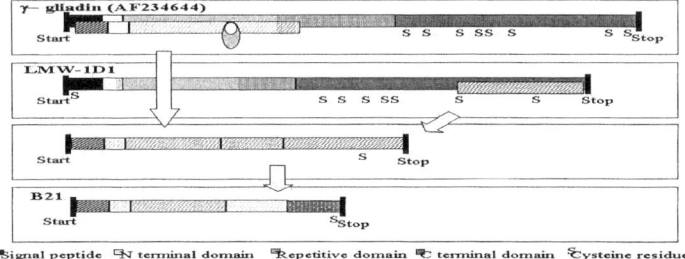

Figure 1 *A schematic drawing of the structure of the chimeric gene first isolated and of the protein it codes for.*

Designation	Length of isolated fragment (nt)	Gliadin–Glutenin border *	Full length of the protein **	No. of Cys res. in the protein	Homology Gliadin part	Homology Glutenin part
gene1^{cg}	580	+	360	152	1	af234644[5]	X13306[6]
gene2^c	896	−	253	317	7	af234644[5]	X13306[6]
gene3^c	266	−	44	108	1	af234644[5]	X13306[6]
gene4^c	863	−	42-45	307	7	af234644[5]	X13306[6]
gene5^c	281	−	102-113	113	1	M16064[1]	X13306[6]
gene6^c	315	+	81	82	1	M16064[1]	X13306[6]
gene7^c	588	+	360-363	177	1	af234644[5]	ab062863[8]
gene8^g	320	−	103-115	126	1	af234644[5]	X13306[6]
gene9^{cg}	432	+	123	122	1	af144104[7]	ab062851[8]

Table 1 *The results of sequence analysis of the chimeric genes. ^{c}isolated from cDNA library, ^{g}isolated from genomic DNA * Frame shift Present (+) Absent (−), **start–stop*

similar to γ-gliadins, while the second is similar to LMW-glutenins (see Table 1.). There is no frame shift in the gene. The length of the translated protein is 317 amino acids. It contains seven cysteine residues at the C terminal region.

Gene 3 comprises 266 base pairs, representing the shortest clone among the chimeric genes isolated from cDNA library. The first 43 nucleotides were clearly a γ-gliadin sequence. The rest of the gene was identical to the X13306[6] gene, with the deletion of a 588 base pair fragment. The protein consists of 108 amino acids and contains only one cysteine originating from the LMW glutenin gene.

Gene 4 consists of 863 base pairs. Only the first 44 base pairs derive from gliadin, while the rest is glutenin. No frame shift was observed. All seven cysteine residues in the 307 amino acid molecule were found in the glutenin part at their original positions.

The first 113 nucleotides out of the 281 bp long clone, named gene 5 can be regarded as gliadin. Only a single base pair change could be found in the first half of the primer binding site compared to the gene called M16064[1] in the database. The origin of a segment containing about 10 base pairs cannot be determined, as it is completely identical to the both the gliadin and the glutenin sequences. The last 179 nucleotides of gene 5 are identical to the sequence of the X13306[6] gene. There is no frame shift at the junction point, and the translated protein contains 113 amino acids with a single cysteine in the C-terminal region.

Clone, designated as gene 6, is 315 base pairs long. The first 43 nucleotides are similar to one of the γ-gliadin gene sequences. The rest of the cloned fragment seems to be derived from the X13306[6] gene. There is a frame shift at the junction point, which presumably results in a protein consisting of 63 amino acids. This molecule is 42 amino acids shorter than the protein without the frame shift, and contains a cysteine residue in the C-terminal region.

Gene 7 is 588 base pairs in length and resembles gliadin genes up to the 360th nucleotide. It seems that a short sequence, containing 69 base pairs, was deleted from the original gliadin gene. In the course of a recombination event this gene was fused with part of an LMW-GS gene. Due to the frame shift within the chimeric gene the translated protein is 39 amino acids shorter, consisting of 177 residues. It contains one cysteine residue in the C-terminal region of the molecule.

2.2.2 Analysis of the Sequence Amplified from Genomic DNA. The clone designated as gene 8 has 320 nucleotides. The sequence of the first 103 base pairs is similar to a γ-gliadin sequence (AF234644[5]), while the second part of the clone is identical to the 3' end of the X13306[6] LMW-GS gene. There is a short region between these two parts, containing about 12 nucleotides, which could originate either from a gliadin or from a glutenin gene according to sequence analysis. No frame shift was observed. The protein consists of 126 amino acids and contains one cysteine residue encoded by the LMW subunit part of the chimeric gene.

2.2.3 *Characterisation of Clones Isolated both from the cDNA Library and from Genomic DNA.* DNA was also amplified from genomic DNA by PCR using Herculase DNA polymerase. Sequence analysis of the cloned fragment revealed that it is almost completely identical (one bp substitution) to B21 (gene 1).

Another clone, called gene 9, was also identified by PCR and cloning of genomic DNA procedure. The first part (126 bp) of this 432 bp long clone is very similar to a γ-gliadin gene sequence (af144104[7]), however a 93 bp long fragment was missing, while the last part of the gene exhibits great similarity to the ab062851[8] LMW glutenin gene. Due to a frame shift at the gliadin-glutenin border the protein is 39 amino acids shorter. This protein has one cysteine residue in the C-terminal region.

3 CONCLUSIONS

In addition to the first γ-gliadin–LMW-glutenin chimeric clone identified in the cDNA library by means of radioactive hybridisation, further 8 DNA fragments with structures similar to the B21 (gene 1) were isolated from the cDNA library as well as from genomic DNA by PCR. The existence of this chimeric gene seems to be not an isolated event, and recombination between storage protein genes may have occurred several times leading to the formation of several functional chimeric genes. The chimeric genes isolated in the present work were obtained from line 11310 of Bánkúti 1201. The dough of Bánkúti wheat is well-known for its high extensibility, which can be attributed in part to the fact that the majority of the chimeric genes have an odd number of cysteines.

If γ-gliadin–LMW-glutenin chimeric genes exist, the possibility of LMW-glutenin–γ-gliadin chimeras or similar recombination processes between other storage protein gene families also arises. It would also be worth studying other wheat varieties in this respect.

The sequences of the isolated chimeric genes also allow conclusions to be drawn about the LMW-GS and γ-gliadin compositions of line 11310 of Bánkúti 1201. The part of gene which became linked in the chimeras obviously exist in original copies in the genome. It is highly probable that the γ-gliadins registered as af234644[5], M16064[1] and af144104[7] and the LMW-glutenins ab062863[8] and ab062851[8] are involved in the development of the gluten complex.

References

1 T. Sugiyama, A. Rafalski, D. Soell, *Plant Sci.*, 1986, **44**, 205
2 B.G. Cassidy, J. Dvorak, O.D. Anderson, *Theor. Appl. Genet.*, 1998, **96**, 743
3 T. Maniatis, E.F. Fritsch, J. Sambrook, *Molecular Cloning: A Laboratory Manual*, Cold spring Harbor Laboratory, New York, 1982
4 I.J. Nagy, I. Takács, L. Tamás, Z. Bedő, *Cereal Res. Comm.*, 2003, **31**, 25
5 O.D. Anderson, C.C. Hsia, V. Torres, *Theor. Appl. Genet.*, 2001, **103**, 323
6 V. Colot, D. Bartels, R. Thompson, R. Flavell, *Mol. Gen. Genet.*, 1989, **216**, 81
7 M. von Büren, J. Lüthy, P. Hübner, *Theor. Appl. Genet.*, 2000, **100**, 271
8 T.M. Ikeda, T. Nagamine, H. Fukuoka, H. Yano, 2001 Unpublished EMBL Database

Acknowledgement This research was funded by the Biotechnology 2000 project of the National Comitee for Technological Development, No. 2578.

WHEAT PROTEIN INIHBITORS OF INSECT DIGESTIVE PROTEINASES: MOLECULAR CHARACTERIZATION AND POTENTIAL BIOTECHNOLOGICAL USE

E. Poerio[1], I. Bellavita[1], S. Di Gennaro[1], F. Farisei[1], D. Panichi[1], and A.G. Ficca[2]

[1]Department of Agrobiologia e Agrochimica, Università degli Studi della Tuscia,via S. Camillo De Lellis snc. I-01100 Viterbo, Italy
[2]Department of Scienze Ambientali, Università degli Studi della Tuscia,via S. Camillo De Lellis snc. I-01100 Viterbo, Italy

1 INTRODUCTION

Several plant protein inhibitors of proteolytic enzymes have been widely investigated in order to evaluate their potential biotechnological use in agriculture. These proteins when expressed in transformed plants, could confer resistance to a number of insect pests by inhibiting their digestive proteolytic enzymes.[1-4]

Two monomeric proteins, capable to inhibit pancreatic chymotrypsin as well as chymotrypsin-like activities present in the digestive systems of some insect larvae (*Tenebrio molitor, Plodia interpunctella, Helicoverpa armigera*), have been isolated from endosperm of *Triticum aestivum*. Both inhibitors have been fully sequenced and extensively characterized.[5,6] One of these proteins, named WSCI (Wheat Subtilisin/Chymotrypsin Inhibitor) on the basis of its inhibitory specificity, is present in the aqueous protein extract of wheat endosperm at a concentration of 0.8-0.9% (w/w); it represents the first case of a wheat protein active in inhibiting both animal chymotrypsins and bacterial subtilisins. This inhibitor, which has been classified as member of the "Potato inhibitor I family", shows a high degree of homology with the barley chymotrypsin inhibitors CI-2A and CI-2B[7,8]. WSCI has a high content of essential amino acid residues; in particular, it contains six Lys and five Thr residues over a total of 72. The second inhibitor, named WCI (Wheat Chymotrypsin Inhibitor), is a typical CM-protein and is a member of the "cereal inhibitor superfamily". WCI consists of 119 amino acid residues and contains 10 cysteine residues all of which form disulphide bonds.

In order to explore the potential application of these two inhibitors in transforming plants of agronomic or ornamental interest, we decided to clone their cDNAs and to express them in a procaryotic system.

2 MATERIAL AND METHODS

Wheat seeds (*Triticum aestivum* cv S. Pastore) were originally supplied by the " Istituto Nazionale per la Cerealicoltura" (S. Angelo Lodigiano, Milano, Italy) and multiplied in the field at the University of Tuscia (Viterbo, Italy).

Total RNA was extracted from developing caryopses (collected 20 days post-anthesis) and purified by several steps of phenol/chloroform extraction. Two pairs of primers (one for each inhibitor) were designed on the basis of the N- and C- terminal regions of WSCI and WCI, respectively. 50 pmoles of each primer and 1 µg of total RNA were used for RT-PCR reactions.

Bacterial cells (*E. coli* BL21-DE3 strain), recombinant for the pGEX-2T/*wsci* and pGEX-2T/*wci* vectors, were growth at 30 °C in LB culture medium containing 0.1 mg/ml of ampicillin. Induction was performed with 1 mM IPTG.

3 RESULTS AND DISCUSSION

In order to clone WSCI and WCI cDNAs, total wheat RNA and the appropiate degenerate primers were used in RT-PCR experiments. The amplification products were then cloned in the pGEM-T easy vector. Among the transformants obtained, we isolated two plasmids, pGEM-*wsci* and pGEM-*wci*: the first plasmid contained the entire coding region (216 bp) of the native inhibitor WSCI; the second contained the coding region (357 bp) of the native inhibitor WCI. These cDNAs were then expressed in *E. coli*, after cloning them in pGEX-2T vector. Two bacterial cell cultures, one transformed with the recombinant plasmid pGEX-2T/*wsci* and the other with pGEX-2T/*wci*, expressed the products of interest fused with the enzyme glutathione-S-transferase (GST). The levels of expression were monitored by SDS-PAGE (Figures 1 and 2). The recombinant protein GST-WSCI was obtained in homogeneous form by affinity chromatography on glutathione-S-Sepharose column; the overall yield of the expression and purification of this protein was about 10 mg per liter of cell culture.

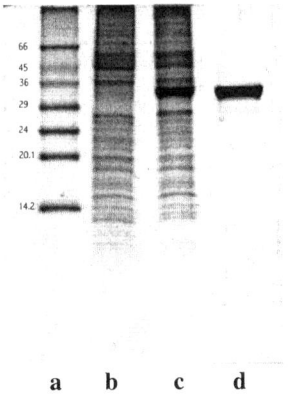

    a   b   c   d

Figure 1 *Expression and purification of recombinant GST-WSCI monitored by SDS-PAGE. a, markers (Mr expressed in kDa); b, non-induced bacterial cells extract; c, IPTG-induced bacterial cells extract; d, purified recombinant protein.*

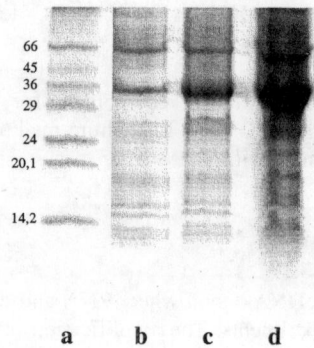

     a   b   c   d

Figure 2 *Expression of recombinant GST-WCI monitored by SDS-PAGE. a, markers (Mr expressed in kDa); b, non-induced bacterial cells extract; c, bacterial cells extract after 1h of IPTG-induction; d, bacterial cells extract after 4h of IPTG-induction.*

The inhibition properties of the pure recombinant GST-WSCI have been compared with those of the native protein purified from wheat endosperm. As shown in Table 1, both GST-WSCI and the native inhibitor are capable of interacting, *in vitro*, with pancreatic chymotrypsins and bacterial subtilisins. Although assayed as a protein fused with GST, the recombinant inhibitor is also active in inhibiting chymotrypsin-like activities detected in crude extracts of midguts dissected from coleopteran larvae *T. molitor* and from larvae of two Lepidoptera (*H. armigera* and *P. interpunctella*).

Table 1 *Inhibitory specificity of the native WSCI and of the recombinant product*

PROTEINASE ACTIVITY	NATIVE WSCI	GST-WSCI
$\alpha,\beta,\gamma,\delta$-chymotrypsin from bovine and porcine pancreas	+	+
Trypsin from bovine and porcine pancreas	-	-
Elastase from porcine pancreas	-	-
Subtilisin from *Bacillus subtilis* and from *Bacillus licheniformis*	+	+
Chymotrypsin-like activity from *Helicoverpa armigera*	+	+
Chymotrypsin-like activity from *Tenebrio molitor*	+	+
Chymotrypsin-like activity from *Plodia interpunctella*	+	+

The expression procedure of the recombinant product GST-WCI is still being optimised in order to improve the expression yield and to verify its inhibitory activity.

4 CONCLUSIONS

The results with GST-WSCI, although preliminary, strongly support the hypothesis that transformation of crop plants with the *wsci* cDNA could result in resistance to a number of insect pests. Over-expression in wheat of *wsci* or of a mutated form of *wsci*, could also result in improvement of the nutritional quality, whose low content of lysine and threonine is particularly significant.

Several studies have investigated the roles played by cereal proteins in human allergies[9-12]. The expression procedure reported here for the two inhibitors could therefore represent a rapid system for producing substantial amounts of protein to be used for toxicity and allergenicity tests. Such studies will be essential to establish the harmlessness or the potential risks associated with the ingestion of: i) foods prepared with wheat over-expressing the inhibitor WSCI; ii) foodstuffs derived from plants transformed with the cDNAs of the two wheat chymotrypsin inhibitors.

References.

1. J. Alfonso-Rubi, F. Ortego, P. Castanera, P. Carbonero and I. Diaz, *Transgenic Res.*, 2003, **12**, 23
2. M. Delledonne, G. Allegro, B. Belenghi, A. Balestrazzi, F. Picco, A. Levine, S. Zelasco, P. Calligari and M. Confalonieri, *Mol. Breed.*, 2001, **6**, 35.
3. X. Duan, X. Li, Q. Xue, M. Abo-El-Saad, D. Xu and R. Wu, *Nature Biotech.*, 1996, **14**, 494.
4. A.M.R. Gathehouse, G.M. Davidson and C.A. Newell, *Mol. Breed.*, 1997, **3**, 1.
5. E. Poerio, S. Di Gennaro, A. Di Maro, F. Farisei, P. Ferranti and A. Parente, *Biol. Chem.*, 2003, **384**, 295.
6. E. Poerio et al. *manuscript in preparation*
7. I.B. Svendsen, B. Martin and I.B. Jonassen, *Carlsberg Res. Commun.*, 1980, **45**, 79.
8. M.S. Williamson, J. Forde, B. Buxton and M. Kreis, *Eur. J. Biochem.*, 1987, **165**, 99.
9. G. Garcia-Casado, A. Armentia, R. Sanchez-Monge, J. M. Malpiga and G. Salcedo, *Clin. Exp. Allergy,* 1996, **26**, 42.
10. A. Curioni, B. Santucci, A. Cristaudo, C. Canistraci, M. Pietravalle, B. Simonato and M. Giannattasio, *Clin. Exp. Allergy,* 1999, **29**, 407.
11. M. Kusaba-Nakayama, M. Ki, M. Iwamoto, R. Shibata, M. Sato and K. Imaizumi, *Food Chem. Toxicol.*, 2000, **38**, 179.
12. B. Simonato, F. De Lazzari, G. Pasini, F. Polato, M. Giannattasio, C. Gemignani, A. D. Peruffo, B. Santucci, M. Plebani and A. Curioni, *Clin. Exp. Allergy,* 2001, **31**, 1771.

Acknowledgements

This research was supported by the grant Cofin 2002 (from MIUR and Università della Tuscia) to Elia Poerio.

MOLECULAR MODELING OF PEPTIDE SEQUENCES OF GLIADINS AND LMW-GLUTENIN SUBUNITS

F. Yaşar[1], S. Çelik[2], H. Köksel[2]

[1] Hacettepe University, Physics Engineering Department, 06532 Beytepe, Ankara, Turkey
[2] Hacettepe University, Food Engineering Department, 06532 Beytepe, Ankara, Turkey

1 INTRODUCTION

Gluten proteins are classified into two groups as gliadins and glutenins on the basis of their solubilities. Gliadins are divided into four subgroups, the α-, β-, γ-, and ω-gliadins based on their electrophoretic mobilites. Glutenins are subdivided into high molecular weight (HMW) and low molecular weight (LMW) subunits. Recent studies of amino acid compositions and sequence analysis suggest that the gliadins can be classified into two groups. The S-rich group includes the α/β- and γ-gliadins and S-poor group includes ω - gliadins. LMW glutenin subunits are related to S-rich prolamins and have been classified in this group.[1] In this study, we examined the possible contribution of one heptapeptide (PQPQPFP) and one pentapeptide (PQQPY) repeat sequence to the conformation of α/β-gliadins, one heptapeptide repeat sequence (PQQPFPQ) to the conformation of γ-gliadins and two heptapeptide repeat motifs (PQQPPFS and QQQQPVL; in one letter code) to the conformation of LMW glutenin subunits by using the recently developed multicanonical simulation procedure.[2] Effects of a tetrapeptide polyQ sequence (QQQQ) on the conformations of gliadins and LMW glutenin subunits were also investigated. The multicanonical simulation method allowed us to estimate the levels of different secondary structures in α/β-, γ-gliadins and LMW glutenin subunits at different temperatures.

2 METHODS

Conventional Monte Carlo (MC) methods in canonical ensemble are not well suited for obtaining a true sampling of the complete conformational space of proteins or peptides. In canonical ensemble, MC simulations such as Metropolis sample configurations with the Boltzmann probability of energy E, which is defined as

$$P^B(E) = n(E)\exp(-E/k_B T)/Z \qquad (1)$$

at a fixed temperature T, where n(E) is the density of states, k_B is the Boltzmann constant and Z is the partition function. In contrast to conventional MC, the Multicanonical (Muca) ensemble is based on probability function in which different energies are equally probable:

$$P^{MU}(E) \sim n(E)w^{MU}(E) = \text{constant} \qquad (2)$$

where w^{MU} (E)'s are Muca weight factors. Hence, a simulation with this weight factor, which has no temperature dependence, generates a one-dimensional random walk in the energy space, allowing itself to escape from getting trapped in any energy local minimum. The advantage of the Muca simulation procedure lies in the fact that it not only alleviates the multiple-minima problem but also allows the calculation of various thermodynamic quantities as functions of temperature from one simulation run.

In order to study the conformation of a sequence (e.g. PQQPFPQ; heptapeptide) in the tandem repeat region, the last residue of the previous repeat sequence was added to the N-terminus and the first residue of the next repeat sequence was added to the C-terminus of the sequence considered (i.e. Q-PQQPFPQ-P; nanopeptide). After the evaluation of the conformation, the central amino acid sequence (heptapeptide) was taken into account without counting the first and last residues (the added ones). A similar approach was used for the other sequences. The Muca method is investigated as applied to two α/β-gliadin peptide sequences (P-PQPQPFP-P, Y-PQQPY-P), a γ-gliadin sequence (Q-PQQPFPQ-P), LMW glutenin sequences (S-PQQPPFS-P, L-QQQQPVL-Q) and the polyQ sequence (Q-QQQQ-Q).[2] All molecules are modeled by potential energy function ECEPP/2 (Empirical Conformational Energy Program for Peptides), which assumes rigid geometry, and is based on the electrostatic term, 12-6 Lenard-Jones and hydrogen-bond term for all pairs of atoms in the peptide together with torsion term for all torsion angles.[3,4] For the present sequences, the peptide bond angles ω are kept fixed at 180°, which leaves dihedral angles φ, ψ in the backbone and χ in the side chain as independent degrees of freedom. We use the standard dielectric constant ε = 2 of ECEPP. This force is implemented in the package FANTOM,[5] which is used for the present simulations. For given sequences, Muca simulation started from completely random initial conformation. No *a-priori* information about groundstate is used in simulations

3 RESULTS AND DISCUSSION

The distribution of backbone φ and ψ angles were analysed for four heptapeptide, one pentapeptide and one tetrapeptide sequences investigated and the Ramachandran plots were prepared in the course of the multicanonical simulation. These Ramachandran plots were analysed to estimate the occurrence probabilities of various secondary structures in the simulated conformations.

The probabilities of different types of β-turns in the one heptapeptide (PQPQPFP) and one pentapeptide (PQQPY) sequences of α/β-gliadins, one heptapeptide (PQQPFPQ) of γ-gliadins and two heptapeptides (PQQPPFS and QQQQPVL) of LMW-glutenin subunits are calculated and presented in Table 1 except for the PQPQPFP sequence of α/β-gliadins. The heptapeptide repeat motif of α/β-gliadins did not show any significant predicted β-turn structures (data not presented). For the pentapeptide sequence, a βIII-turn is observed (3.5 %) when PQ sequence is placed in the 2nd and 3rd positions of the β-turn.

Table 1 Predicted percent probabilities of different β-turns in various repeat sequences and poly Q region of S-rich prolamins

Amino acid residues in the i+1, I+2 positions of β-turns		280-300 K	
		βI	βIII
PQQPY	PQ	0.65	3.46
PQQPFPQ	PQ	0.53	4.84
PQQPPFS	PQ	0.54	9.19
	PF	0.15	17.46
QQQQPVL	QQ	0.07	4.04
	QQ	0.01	1.72
	PV	-	8.93
	VL	-	0.18
QQQQ	QQ	0.09	15.37
	QQ	-	7.83
	QQ	0.09	2.62

Table 2 Predicted percent probabilities of different γ-turns in various repeat sequences of S-rich prolamins

Amino acid residue in the I+1 position of γ-turns		280 – 300 K
		Inverse γ-turns
PQPQPFP	P	12.36
	Q	-
	P	11.99
	Q	-
	P	6.09
	F	-
	P	-
PQQPY	P	6.51
	Q	1.62
	Q	-
	P	11.60
	Y	-
PQQPFPQ	P	0.23
	Q	0.33
	Q	-
	P	29.36
	F	-
	P	13.60
	Q	-

The repeats in the repetitive domain of γ-gliadins are based on the consensus PQQPFPQ.[6] Structural prediction at about room temperature indicates the presence of a βIII-turn with a probability of 4.8% if P and Q are placed in the 2nd and 3rd positions of the turn. Structural predictions of PQQPPFS sequence at about room temperature indicate the presence of βIII-turns with probabilities of 9.2% and 17.5% if PQ or PF are placed in the 2nd and 3rd positions of the turn, respectively. Structural predictions of QQQQPVL sequence at about room temperature indicate the presence of a βIII-turn with the highest probability of 8.9% if PV is placed in the 2nd and 3rd positions of the turn. There were also predicted βIII-turns if QQ was placed in the 2nd and 3rd positions of the turn. For the tetrapeptide sequence of the polyQ region present in gliadins and LMW-glutenin subunits,[7] the highest probability of βIII-turn was 15.4% (Table 1).

The probabilities of inverse γ-turns in the peptide sequences investigated are presented in Table 2. Structural predictions indicated that there are no classical γ-turns. The probability of inverse γ-turn was generally higher as P was placed in the i+1 and Q, Y or F in the i+2 positions of the inverse γ-turns in all sequences. In the PQQPFPQ sequence the highest probability (29.4%) was observed if P was placed in the i+1 and F in the i+2 position of the inverse γ-turn. Structural predictions indicated that the possibility of inverse γ-turns was around 12% if P was placed in the i+1 and Q in the i+2 positions of the γ-turn in the PQPQPFP sequence and the highest probability of inverse γ-turns was 11.6% if P was placed in the i+1 and Y in the i+2 positions of the γ-turn in the PQQPY sequence.

Structural predictions of the QQQQPVL sequence of LMW-glutenin subunits at about room temperature indicated the presence of helical structure with a probability of 20.6% at the first three amino acid residues (QQQ). When the temperature was increased to about 100°C, the probability of helical structure substantially decreased. Although, α/β- and γ-gliadins are reported to have helical structures, they were not detected in the sequences considered in the present study. This is probably due to the high P content of the repetitive sequences of α/β- and γ-gliadins investigated in this study. High contents of proline destabilizes α-helix structures.[8] Because of the lack of helical structures in the repetitive domain of gliadins, its probable occurence in the regions other than the repetitive domain (the polyQ region) was also investigated.[2] Structural predictions of the QQQQ sequence at about room temperature indicated the presence of helical structure with the highest probability of 22.2% at the first three amino acid residues (QQQ). When the temperature wasincreased to around 100°C, the probability of helical structure significantly decreased (data not presented).

4 CONCLUSIONS

The heptapeptide repeat motif of α/β-gliadins did not show β-turn structures while their pentapeptide sequence showed low level of βIII-turn. The probability of a βIII-turn in a γ-gliadin repeat motif was around 5%. The total β-turn probabilities are estimated to be about 27% and 15% of all possible conformations for each of the PQQPPFS and QQQQPVL sequences of LMW glutenin subunits, respectively. The probability of inverse γ-turn was generally higher than that of β-turns in all sequences investigated. The high level of inverse γ-turns found in these sequences suggests that γ-turns may contribute to the secondary structure of α/β-, γ-gliadins and LMW-glutenin subunits. Helical structures were not detected due to the high P content of the repetitive sequences. Structural predictions of the QQQQPVL sequence of LMW-glutenin subunits and the QQQQ sequence in the polyQ region of gliadins and LMW glutenin subunits at about room temperature indicate the presence of helical structure with the probability of >20%. When the temperature was increased to around 100°C, the probability of helical structure significantly decreased because of the increased thermal fluctuations.

References

1 J.D. Schofield, in *Wheat proteins: structure and functionality in milling and bread-making*, ed. W. Bushuk and V.F. Rasper, Chapman & Hall, Great Britain, 1994, p.66.
2 F. Yaşar, S. Çelik and H. Köksel, *Nahrung/Food*, 2003, **47**, 238.
3 F.A. Momany, R.F. McGuire, A.W. Burgess and H.A. Scheraga, *J. Phys. Chem.*, 1975, **79**, 2361.
4 M.J. Sippl, G. Nemethy and H.A. Scheraga, *J. Phys. Chem.*, 1984, **88**, 6231.
5 B. von Freyberg and W. Braun, *J. Comput. Chem.*, 1993, **14**, 510.
6 P.R. Shewry, M.J. Miles and A.S. Tatham, *Prog. Biophys. Molec. Biol.*, 1994, **61**, 37.
7 A.S. Tatham, J.M. Field, S.J. Smith, and P.R. Shewry, *J. Cereal Sci.*, 1987, **5**, 203.
8 A.S. Tatham, A.F. Drake and P.R. Shewry, *Biochemical Journal*, 1985, **226**, 557.

Acknowledgments: The support of Hacettepe University Research Fund (project 01.01.602.006) is acknowledged.

Genetics and Quality

EFFECT OF THE INTRODUCTION OF NOVEL HIGH M_r GLUTENIN SUBUNITS ON QUALITY OF DURUM WHEAT SEMOLINA.

M.C. Gianibelli[1*], O.R. Larroque[1], E. DeAmbrogio[2] and D. Lafiandra[3].

[1] CSIRO Plant Industry, GPO Box 1600, Canberra ACT 2601, Australia
[2] Società Produttori Sementi, Via Macero,1, 40050 Argelato (BO) Italy
[3] Università degli Studi della Tuscia, Dipartimento di Agrobiologia e Agrochimica, Via S. Camillo de Lellis, 01100 Viterbo, Italy.
* e-mail: Cristina.Gianibelli@csiro.au

1 INTRODUCTION

Durum wheat is mainly used for the manufacture of pasta which is made by extruding semolina, a coarse flour. Protein content and gluten composition are generally considered to be the main factors affecting dough properties and cooking quality of durum wheat pasta. Gluten proteins are divided into two groups, gliadins and glutenins, on the basis of their functional properties. Gliadins are monomeric proteins encoded at the complex loci *Gli-1* and *Gli-2*, located on the short arms of the group 1 and 6 chromosomes, respectively. Glutenins are formed by high molecular weight subunits (HMW-GS) and low molecular weight subunits (LMW-GS) encoded at the *Glu-1* and *Glu-3* loci, respectively, located on the long and short arms of the group 1 chromosomes. They are linked together by covalent disulfide bonding and form very large polymeric structures.

HMW-GS are the most studied proteins of the wheat endosperm. This is based on the fact that they are key components in terms of dough functionality and hence quality of the end-products. Being a tetraploid, durum wheat has two genomes: A and B. Each genome contains a *Glu-1* locus with two tightly linked HMW-GS genes (x- and y-type). However, the y-type gene at the *Glu-A1* is usually not expressed. From the three known allelic variants for the x-type genes, a null form without expressed protein, is widely spread in durum wheat cultivars. Therefore, HMW-GS expression is limited to those provided by the *Glu-B1* genes, with the allelic variants 7+8, 6+8 and 20 the most abundant. On the other hand, some alleles usually present in bread wheat and providers of good quality features (i.e. 17+18, high Payne's quality score) are not present in durum wheat.

The aim of this study was to increase the number and amount of HMW-GS expressed in durum wheat by means of the introduction of a *Glu-A1* allele (expressing x- and y-type subunits) from a wild relative, *Triticum dicoccoides*, into Svevo, an Italian durum cultivar. In addition, isogenic lines of Svevo carrying allele 17+18 instead of 7+8 for the *Glu-B1* were also produced. Biochemical determinations based on HPLC measurements and physical dough properties of the dough as determined by mixograph analysis were performed in order to establish if improvements in quality related parameters were achieved.

2 MATERIALS AND METHODS

2.1 Materials

The durum wheat cultivar "Svevo" was used in this study. Svevo has good quality for pasta making and good agronomic performance. A series of back-crosses were carried out to introduce: a) *Glu-B1* subunits 17+18 from the hexaploid cultivar Manital and b) *Glu-A1* subunits Ax and Ay from *Triticum dicoccoides*.

2.2 Methods

2.2.1 HMW-GS identification: Protein composition was analysed by electrophoresis using SDS-PAGE.[1] The HMW-GS alleles were classified according to Payne and Lawrence.[2]
2.2.2 Size Exclusion HPLC: samples were extracted using 0.5% sodium dodecyl sulfate - phosphate buffer (pH 6.9) according to standard procedures.[3,4] For reversed-phase HPLC, reduced and alkylated glutenin extracts were prepared according to Larroque *et al*.[5]
2.2.3. HPLC system and columns: Protein extracts were subjected to HPLC using a Phenomenex BIOSEP-SEC 4000 (5 µm, 500 Å, 7.8 mm x 300 mm) column (Phenomenex, Torrance, CA, USA) was utilized for size exclusion (SE-HPLC) and a Vydac C18 (10 µm, 4.6 mm x 250 mm) column (Vydac, Hesperia, CA, USA) was used for reversed-phase (RP-HPLC).
2.2.4 Mixograph: A 2 g CSIRO prototype Mixograph was used for the assessment of dough quality properties.[6]

3 RESULTS AND DISCUSSION

Electrophoretic analysis of the Svevo isogenic lines used in this study are shown in Figure 1.

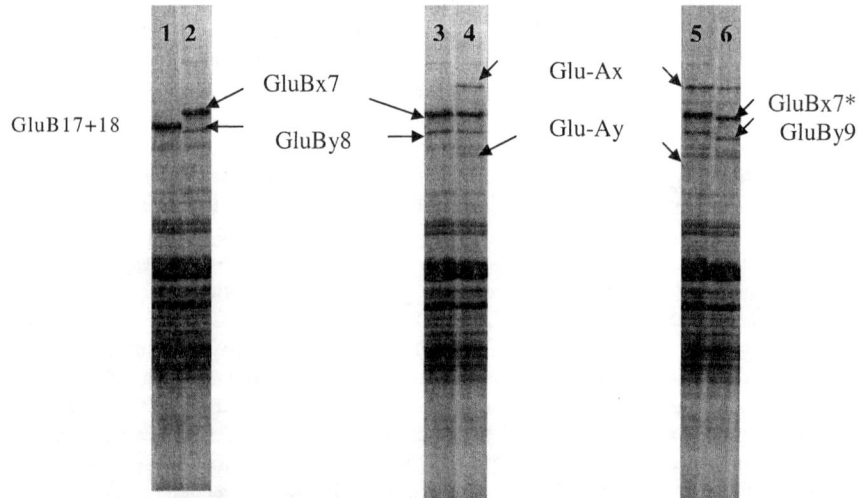

Figure 1: *Svevo isogenic lines. 1: subunits 17+18; 2-3: Svevo subunits 7+8. 4-5: subunits* Glu1Ax + Glu1Ay *and 7+8;* 6: Glu1Ax + Glu1Ay *and 7*+9.*

Isogenic lines with *Glu-B1* subunits 7*+9 (from *Triticum dicoccoides*) were also included in this study.

HPLC-based measurements showed %UPP values of 36.39 and 33.98 when Svevo and 17+18 were compared, respectively. They showed much even closer values when the polymeric peak percentage (from total protein extracts) was compared (50.26 versus 50.35 respectively). On the other hand, the total amount of polymeric protein in Svevo (as measured by the total chromatogram area in arbitrary units) was much higher than in the lines with subunit 17+18. When lines bearing four subunits were compared with Svevo (2 subunits, 7+8) the area for HMW-GS was higher in the former lines, as expected. Nevertheless, no differences where found between the total areas under the curve of the RP-HPLC chromatograms. In terms of %UPP, the presence of four subunits instead of two produced similarly low values (38.6 versus 36.38). A significant correlation between total area of HMW-GS (as measured by RP-HPLC) and %UPP was found (0.7418)

Mixing properties were evaluated with a 2 g Mixograph and the values are shown in Table 1. No changes in the mixing parameters were observed between isogenic lines with subunits 7+8 and 17+18 indicating that both alleles have a similar contribution to the mixing properties of semolina. When isogenic lines with 2 subunits (7+8) were compared with isolines with 4 subunits (*Glu1Ax* + *Glu1Ay* and 7+8) an increase in the mixing time was observed (Table 1). On the other hand, the presence of subunits 7*+9 reduced the mixing time when compared to lines with 7+8 subunits.

Table 1: *Mixing properties of Svevo isogenic lines.*

Svevo Isogenic lines	MT	PR	BWPR	RBD	BWBD	TMBW	MBW
7+8	126	834	857	16.5	49.5	91.0	960
17+18	125	847	878	16.0	48.0	90.5	970
Ax+Ay, 7*+9	136	770	648	14.5	34.7	89.5	799
Ax+Ay, 7+8	152	885	817	20.7	45.6	101.0	894

MT = Mixograph mixing time (seconds); PR = peak dough resistance (AU), RBD = resistance breakdown (%); BWPR = bandwidth at peak dough resistance (AU); BWBD = breakdown at peak bandwidth (%); TMBW = time to maximum bandwidth (s); MBW = maximum bandwidth (AU).

4 CONCLUSIONS

- Slight improvement in mixing properties was observed when the HMW-GS number increased to 4 by the addition of subunits from the *Glu-A1* locus.
- No significant differences were found between subunit 7+8 and 17+18 in terms of %UPP and mixing properties.
- Subunit 7*+9 showed mixing properties inferior to 7+8.

References

1. R.B. Gupta and F. MacRitchie, *J. Cereal Sci.*, 1991, **14**, 105.
2. P.I. Payne and G.J. Lawrence, *Cereal Res. Commun.*, 1984, **11**, 29.
3. I.L Batey, R.B. Gupta and F. MacRitchie, *Cereal Chem*, 1991, **68**, 207.
4. R.B. Gupta, K. Khan and F. MacRitchie, *J Cereal Sci*, 1993, **18**, 3.
5. O.R. Larroque, F. Bekes, C.W. Wrigley and W.G. Rathmell, in: *Wheat Gluten,* P.R. Shewry and A.S. Tatham eds., Royal Society of Chemistry Special Publications, Cambridge, 2000, p. 136.
6. F.Bekes, P.W. Gras, R.B.Gupta, D.R. Hickman and A.S. Tatham, *J. Cereal Sci*, **19**,

USE OF SEGREGATING DOUBLED HAPLOID POPULATIONS TO INVESTIGATE THE EFFECTS OF *GLU-B1* AND *GLU-D1* ALLELES ON DOUGH STRENGTH

B.J. Butow[1,2], K.R. Gale[1,2], W. Ma.[1,2], O. Larroque[1,2], M.K. Morell,[1,2] and F. Békés,[1,2]

[1]CSIRO Plant industry, GPO Box 1600, Canberra, ACT 2601, Australia.
[2]Graingene, 65 Canberra Avenue, Griffith ACT 2603 Australia.

1 INTRODUCTION

The balance between the viscous and the elastic properties of dough determine its strength during mixing and the baking process. In many breeding programs, high dough strength is synonymous with good quality bread wheat[1]. Dough strength has been attributed largely to the type of allele present at the *Glu-D1* locus; the HMW-GS allelic pair Dx2+Dy12 (i.e.Glu-1D a) has been found to give weaker, inferior doughs relative to wheat containing the allelic pair Dx5+Dy10 (i.e.Glu-1D d).[2] It has also been observed that cultivars and landraces with an over-expression of Glu1Bx7 ($Bx7^{OE}$) have improved dough strength.[3-7]

In order to estimate the contribution of different alleles to the genetic variance for dough strength, the use of a large segregating population is necessary. To this end, two different Doubled Haploid (DH) populations segregating at *Glu-B1* and *Glu-D1* have been developed from crosses between parents showing strong and weaker dough characteristics.

SDS PAGE has been found to be not wholly suitable to differentiate expression levels of Bx7 in the Australian cultivars, CD87 and Kukri. Alternatively, RP HPLC can be used, but is time consuming, expensive and requires dedicated equipment; moreover, both methods require an endosperm protein sample. The use of a PCR marker has the advantage for breeders of being more amenable to routine analysis of tissue and/or endosperm DNA and subsequent scoring allows the definition of sub-populations. Thus, the value of utilizing a new Bx PCR marker will also be assessed in this work, in addition to analysis of Bx7 expression by RP-HPLC.

2 MATERIALS & METHODS

2.1 Material

Two Doubled Haploid (DH) populations (parents: Kukri x Janz and CD87 x Katepwa) produced by the National Wheat Molecular Marker Program in Australia (NWMMP) were investigated.[8] Data was collected from two sites, Wongan Hills (1999) and Horsham (2000), for 156 progeny lines from the CD87 x Katepwa cross, and from one site, Roseworthy (2000), for 144 Kukri x Janz DH progeny lines.

2.2 DNA Extraction and PCR Analysis

Genomic DNA was extracted from 3-6 day old hypocotyls (10 mg) of germinating seeds using a rapid isolation technique.[9] PCR was performed and modified using the following Bx primers for the coding region (for the CD87 x Katepwa DH population): forward 5'-CAAGGGCAACCAGGGTAC-3', reverse 5'-AGAGTTCTATCACTGCCTGGT-3'.[10] A second set of primers were used for the Kukri x Janz population; the specific reverse primer was the same as before, but the new forward primer (5' - CGCAACAGCCAGGACAATT – 3') annealed with other repetitive sequences within the coding sequence and gave a more definitive difference between Bx7 and $Bx7^{OE}$ lines.[11]

2.3 Protein analysis of HMW-GS

Crushed grain extracts were subjected to one-dimensional SDS-PAGE.[12] Flour was also extracted sequentially for gliadins and glutenins and the HMW-GS composition was analysed by RP-HPLC.[13] SE-HPLC was also carried out to assess the unextractable polymeric protein (%UPP) content of all the flour samples.[14]

2.4 Dough Quality Testing

A two-gram Mixograph was used to evaluate functional dough properties of the DH lines.[15] Mixing tests were performed in duplicate and key parameters were measured, including mean time to peak dough development (Mixing Time, [s]).

3 RESULTS AND DISCUSSION

The use of a new DNA marker which differentiates $Bx7^{OE}$ from Bx7 coding region sequences, due to an 18 bp polymorphism present in $Bx7^{OE}$ – containing lines,[10] enabled quick and accurate differentiation of progeny from the Kukri x Janz DH population. Analysis of this population using RP-HPLC further enabled the quantification of Bx expression levels. The bimodal distribution of Bx expression level in the Kukri x Janz population (Figure 1A), was found to be similar to that for the CD87 x Katepwa DH population (Figure 1B). Both populations showed transgressive segregation whereby Bx values of the parent lines fell well within those of the extremes of the population. This is indicative of other factors influencing Bx expression levels.

We have recently shown, that the CD87 x Katepwa DH population segregates into four clear sub-populations according to dough strength, whereby those lines exhibiting an over-expression of Bx7 in addition to Glu-D1d (Dx5+Dy10), showed the highest dough strength.[10] Conversely, lines without the "over-expressing" Bx7 subunit in conjunction with Glu-D1a (Dx2+Dy12), gave the weakest dough strength. The effect of these Glu-B1 and Glu-D1 alleles on dough strength was validated in this current work using a second DH population, Kukri x Janz, and confirms the recent findings by Canadian,[16] and Hungarian research teams.[6] Statistical analysis has shown that the effect of *Glu-B1* and *Glu-D1* on dough strength is apparently an additive, and not an interactive effect. It is surmised that the unique, high expression levels of Bx7, possibly due to gene duplication,[3] exerts its influence on dough strength through a different mechanism to Dx5. Whilst not over-expressed, Dx5 influences its impact on dough strength due to differences in structure: an additional cysteine being thought to be responsible.[17]

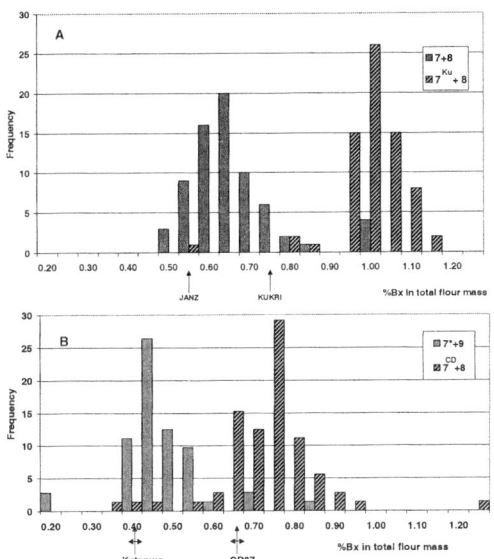

Figure 1 *Distribution of %Bx subunit in A) Kukri x Janz DH progeny (n=140) and B) CD87 x Katepwa DH progeny (n=156).*

The linear correlation found previously between dough strength and %UPP,[7] was upheld for both the DH populations tested. Analysis of the relationship between %UPP and MT (a dough strength parameter), in the Kukri x Janz progeny, for specific allelic combinations at Glu-B1 and Glu-D1, revealed that the correlation was most significant for progeny containing an over-expression of Bx7 ("7^{OE}+8") (Table 1). Furthermore, when the progeny were differentiated according to Glu-B1/Glu-D1 alleles, the sub-population "7^{OE}+8/5+10" showed the highest significant correlation, whereas the "7+8/2+12" group showed no significant correlation between MT and %UPP.

Table 1 *Linear correlation between mixing time and %UPP for the Kukri x Janz Doubled Haploid population. Significance level at P<0.05, 0.01, and 0.001 represented by *, ** or *** respectively. n.s., not significant.*

allelic composition at Glu-B1/Glu-D1	n	r^2	F	Significance
7+8 / 2+12	29	0.08	2.4	n.s.
7+8 / 5+10	40	0.14	6.11	*
7^{OE}+8 / 2+12	30	0.36	15.35	***
7^{OE}+8 / 5+10	45	0.323	20.5	***
all progeny	144	0.425	104.4	***

4 CONCLUSION

Dough strength was strongly correlated with both the over-expression of Bx7 and the presence of Glu-D1d alleles. However, the level of over-expression varied for different populations and appears to be influenced by other factors. A strong correlation between the amount of unextractable polymeric protein and mixing requirement of dough was verified, although the roles of the different Glu-B1 and Glu-D1 sub-units, within the polymeric matrix, appear to differ.

References

1. H.A Eagles, G.J. Hollamby, N.N. Gororo, and R.F. Eastwood, 2002, *Aust. J. Agric. Res.* **53**, 367.
2. P.I. Payne, M.A. Nightingale, A.F. Krattinger and L.M. Holt, 1987, *J. Sci. Food Agric*, **40**, 51.
3. B.A. Marchylo, O.M. Lukow and J.E. Kruger, 1992, *J. Cereal Science*, **15**, 29.
4. O.M. Lukow, S.A. Forsyth and P.I. Payne, *J. Genet. & Breed.*, 1992, **46**, 187.
5. R. D'Ovidio, S. Masci, E. Porceddu and D. Kasarda, *Plant Breeding*, 1997, **116**, 525.
6. A. Juhász, O.R. Larroque, L.Tamás, S.L.K. Hsam, F.J. Zeller, F. Békés, Z. Bedő, *Theor. Appl. Genet.*, 2003, (in press).
7. B.J Butow, P.W.Gras, R. Haraszi and F. Bekes, *Cereal Chem.*, 2002, **79**, 826.
8. S.J. Kammholz, A.W. Campbell, M.W. Sutherland, G.J. Hollamby, P.J. Martin, R.F. Eastwood, I. Barclay, R.E. Wilson, P.S. Brennan and J.A. Sheppard, *Aust. J. Agric. Res.*, 2001, **52**, 1079.
9. C.N. Stewart and L.E Via, *BioTechniques*, 1993, **15**, 748.
10. B.J.Butow, W Ma, K.R. Gale, G.B. Cornish, L. Rampling, O. Larroque, M.K. Morell, and F. Békés, *Theor. Appl. Genet*, 2003, (in press).
11. B.J.Butow, J. Ikea, W. Ma, M.K. Morell, F.Békés and K.R. Gale, 10th International Wheat Genetics Symposium, Paestum, Italy, 2003 (in press).
12. G.B. Cornish, F. Békés, H. Allen, and D.J. Martin, *Aust. J. Agric. Res.*, 2001, **52**, 1161.
13. O. R. Larroque, F. Békés, C.W. Wrigley and W.G. Rathmell, in *"Wheat Gluten"*, ed. P.R. Shewry and A.S. Tatham, Royal Society of Chemistry, Cambridge, 2001, p 136.
14. O.R. Larroque, and F. Bekes, *Cereal Chem.*, 2000, **77**, 451.
15. C. R. Rath., P.W. Gras, C.W. Wrigley and C.E. Walker, *Cereal Foods World*, 1990, **35**, 572.
16. N. Radovanovic, S. Cloutier, D. Brown, D.G. Humphreys, and O.M. Lukow, *Cereal Chem.*, 2002, **79** 843.
17. O.D. Anderson, F.C. Green, R.E. Yip, R.E. Halford, P.R. Shewry and J. M. Malpica-Romero, *Nucleic Acids Res.*, 1989, **17**, 461.

Acknowledgments

We thank the NWMMP for providing the Doubled Haploid samples and Mark Smith for excellent technical assistance. This work is supported by Graingene – a research joint venture between AWB Limited, CSIRO, GRDC and Syngenta Seeds.

BIOCHEMICAL ANANLYSIS OF GLUTEN-FORMING PROTEINS IN A DOUBLE HAPLOID POPULATION OF BREAD WHEAT AND THEIR RELATION TO DOUGH EXTENSIBILITY

A. Juhász[1], O. Larroque[1], H. Allen[2], J. Oliver[2], M.K. Morell[1], M.C. Gianibelli[1]

1 CSIRO Plant Industry, GPO Box 1600, Canberra, ACT 2601 Australia, angela.juhasz@csiro.au
2 NSW Agriculture, Wagga Wagga, Agricultural Institute, PMB, Wagga Wagga, NSW 2650 Australia

1 INTRODUCTION

The storage proteins of wheat play a key role in determining the unique viscoelastic properties of wheat flour dough. The effect on dough strength and stability of allelic composition and amounts of individual glutenin alleles is well known. Less information is available about the effect of storage proteins and especially those of LMW glutenin subunits on dough extensibility. Due to their complexity and heterogeneity, low molecular weight glutenin subunits have been studied to a lesser extent than HMW-GSs in hexaploid wheat. Genetic studies have indicated that LMW-GSs are encoded by genes located at the *Glu-3* loci on the short arm of the homoeologous group 1 chromosome (*Glu-A3*, *Glu-B3* and *Glu-D3*, respectively). Gliadin alleles located at the *Gli-1* loci show tight linkage to LMW-GSs. The improved protein separation techniques and biotechnological tools provide more information about the complexity of *Glu-3/Gli-1* complex loci. LMW-GS are present in the flour at about three times the level of HMW-GS. Earlier results indicate that the amount of polymeric glutenin fraction, especially those of the smaller polymeric proteins has strong effect on dough extensibility. Information provided by HPLC techniques may serve as important part in understanding the genetic background of dough extensibility.

2 MATERIALS AND METHODS

A doubled haploid population (DH) (n=190) developed from a cross between Chara and WW2449 was analysed for storage protein composition and rheological properties. Glutenin and gliadin composition of parental lines were analysed using SDS-PAGE, A-PAGE and A-PAGE x SDS-PAGE techniques[1]. HMW-, LMW-glutenin and gliadin allelic composition of DH lines were determined by SDS-PAGE. SE-HPLC analysis was used to determine the glutenin to gliadin ratio (Glu/Gli) and the unextractable polymeric protein (%UPP)[2]. Quantitative analysis of glutenin composition for parental lines was carried out by RP-HPLC[3]. The following parameters were defined after integration the glutenin chromatograms: HMW-GS composition (% of total HMW), HMW/LMW ratio, x/y ratio. Protein content was measured by NIR and dough extensibility was determined using Brabender Extensograph[TM4]. Relationship between allelic and quantitative storage protein composition with dough extensibility was identified. Statistical analyses were carried out using GENSTAT 4.1 (Lawes Agricultural Trust, Rothamsted).

3 RESULTS AND DISCUSSION

3.1 Characterisation of parental lines

Based on SDS-PAGE and A-PAGE results Chara and WW2449 differed on *Glu-B1*, *Glu-A3* and *Gli-A1* loci. Chara possess 2* 7^{OE}+8 2+12 HMW-GS composition and *b*, *b*, *b* for LMW-GS composition, while WW2449 possess 17+18 allele on *Glu-B1* and allele *Glu-A3c*. Based on 1D analysis of gliadins they have the same protein composition on *Gli-2* locus. Both parental lines express *Gli-B1b* and *Gli-D1f* alleles. A *Gli-A1* γ-gliadin band specific for Chara is missing in WW2449 and another band characteristic for WW2449 is missing from Chara. 2D-PAGE results however, revealed more differences for both LMW-glutenin and gliadin compositions (Figure 1).

Figure 1. *A-PAGE x SDS-PAGE gels of LMW glutenin and gliadin fractions of parental lines. Proteins different based on 1D PAGE are labelled by arrowheads, further differences are circled*

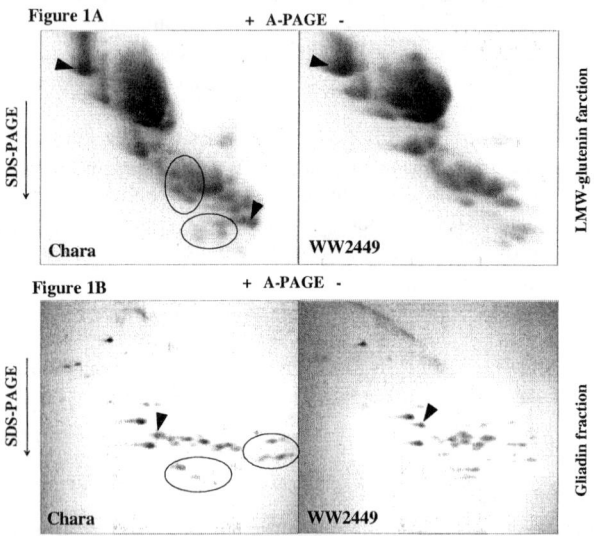

HPLC results show differences in most of the measured parameters (Table 1). Chara has higher glutenin and lower gliadin contents resulting in higher Glu/Gli values. The amount of unextractable polymeric protein was significantly higher. This result was in strong relation with the presence of over-expressing subunit $Bx7^{OE}$ in Chara ($Bx\%_{Chara}$=47.23±1.75, $Bx\%_{WW2449}$=34.86±1.45).

The ratio of amounts x-type subunits to y-type subunits was also significantly higher in Chara (x/y_{Chara}=4.86±0.57, x/y_{WW2449}=3.79±0.58). No significant difference was detected in HMW/LMW ratio between the two parental lines (HMW/LMW$_{Chara}$=0.49±0.02, HMW/LMW$_{WW2449}$=0.48±0.02). Both dough strength (Rmax) and extensibility (Ext) values were higher in Chara (Table 1).

Table 1. *Relationship between gliadin composition and amounts of gliadin fractions and sub-fractions (ANOVA)*

	Chara (n=20)	WW2449 (n=19)	Chara x WW2449 DH lines (n=190)		
	Mean±St.dev	Mean±St.dev	Mean±St.dev	Min	Max
Protein%	13.70±0.47	12.40±0.52	12.90±0.75	10.51	14.94
Glutenin%	51.16±0.75	50.04±0.49	50.51±0.96	47.92	52.98
Gliadin%	38.41±0.71	39.73±0.40	39.09±1.13	36.48	42.28
Glu/Gli	1.33±0.04	1.26±0.02	1.29±0.06	1.13	1.44
UPP%	48.37±2.39	37.83±2.46	43.32±4.15	33.37	54.20
Extensibility (cm)	23.49±1.28	18.40±1.13	20.69±1.74	16.00	25.15
Rmax (EU)	431.63±53.87	185.80±33.41	290.37±92.13	110	580

3.2 Characterisation of Chara x WW2449 DH population

23.9% of the DH lines showed HMW and LMW-glutenin pattern of Chara and 23.5% had the same composition as WW2449. 26.6% of the lines possess 2* 7^{OE}+8 2+12, *cbb* and 25.5% possess 2* 17+18 2+12, *bbb* glutenin composition. Two lines were identified expressing both the 7^{OE}+8 and 17+18 alleles. The DH lines were identical in *Gli-2* fraction and the allelic composition of *Gli-1* loci was 100% linked to the composition of *Glu-3* loci. The effect of allelic composition on dough extensibility was analysed using analysis of variance (ANOVA). The results show very strong positive effect of 7^{OE}+8 allele and less strong but significantly positive effect of *Glu-A3b* allele. No significant effect of *Glu-B1* x *Glu-A3* interaction was detected on dough extensibility values (Table 2).

Flour protein content shows strong and positive influence on dough extensibility (r=0.6272, p=0.01%). To compare the effect of the amount of different storage protein fractions, the measured values were expressed as a function of flour protein content. Both flour total monomeric content (FTMP%) and flour total polymeric protein content (FTPP%) have significant positive effect on extensibility ($r_{FTMP\%}$=0.4884, p=0.01% and $r_{FTPP\%}$=0.6874, p=0.01%), however the effect of amount of polymeric protein fraction in the flour is remarkably strong. The size distribution of the polymeric glutenin has an important effect on dough strength[5]. In this study the effect of polymer size distribution on extensibility was determined. Based on our results the amount of total extractable polymeric protein fraction in flour (FEPP%), it means the quantities of polymers under approx. 158kDa in size, did not show significant effect on dough extensibility, but strong significant effect of amount of large polymers above 158kDa (FUPP%) was detected (Figure 2). Analysing the impact of glutenin composition both on amounts of flour protein fractions and extensibility values significant differences were detected. In groups where *Glu-B1* allele 7^{OE}+8 was present stronger correlation between FUPP% and extensibility was detected, compared to those groups where subunits 17+18 were present (results not shown). Similar positive effect on polymer size was detected in presence of allele *Glu-A3b*. These results indicate that the strong effect of FUPP% on extensibility is mainly due to the presence of *Glu-B1* allele over-expressing subunit 7.

Table 2. *Unbalanced one way analysis of variance of effects of glutenin composition on extensibility*

N=188	Glu-B1		Glu-A3		Glu-B1 x Glu-A3	
ANOVA F values	91.22***		19.23***		1.61	
Means	7^{OE}+8	21.67 B	b	21.16 B	17+18 / c	19.37 A
	17+18	19.68 A	c	20.27 A	17+18 / b	20.00 A
					7+8 / c	21.11 B
					7+8 / b	22.25 C
LSD5%		0.40		0.40		0.69

Figure 2. *Effects of amounts of flour total monomeric proteins (FTMP%), total extractable (FEPP%) and total unextractable polymeric protein fractions(FUPP%) on dough extensibility (***: p=0.01%).*

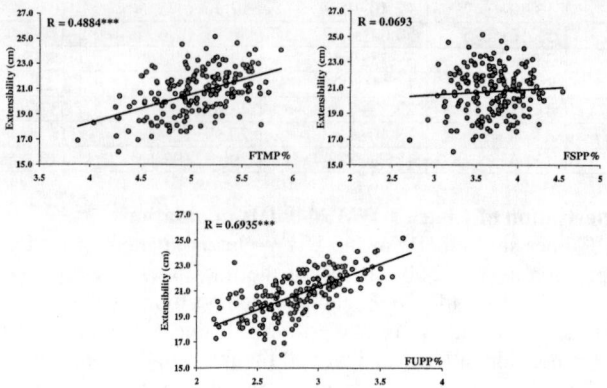

4 CONCLUSION

Genotypes Chara and WW2449 differing in dough strength and extensibility were used to develop DH population. Based on storage protein analysis the parental lines differ on *Glu-B1* and *Glu-A3/Gli-A1* loci. Chara has over-expressing subunit 7^{OE} combined with subunit 8, WW2449 has allele 17+18. Using SDS-PAGE *Glu-A3b* allele was identified in Chara and *Glu-B3c* allele in WW2449, however more differences were detected using 2D gel-electrophoresis technique. In DH lines over-expressing subunit 7^{OE} and expressing *Glu-A3b* allele higher amount of UPP was detected, resulting in higher extensibility values, however amount of extractable polymeric protein fraction did not show significant effect on dough extensibility. These results indicate that size distribution has effect not only on dough strength but on extensibility. The components of this unextractable polymeric protein fraction and the ratio of their amounts need further investigation.

References

1. E.A. Jackson, M.H. Morel, T. Sontag-Strohm, G. Branlard, E.V. Metakovsky and R. Redaelli *J. Genet & Breed.*, 1996, **50**, 321.
2. O.R. Larroque and F. Bekes *Cereal Chemistry*, 2000, **77**, 451.
3. O.R. Larroque, F. Bekes, C.W. Wrigley and W.G. Rathmell, in *Wheat gluten* eds.P.R. Shewry and A.S. Tatham, Royal Society of Chemistry Special Publications, Cambridge, 2000, p. 136.
4. American Association of Cereal Chemists 1983, Approved methods of the AACC. Method 54-10, approved 1961, revised 1982, The Association: St. Paul, MN
5. R.B.Gupta, K.Khan and F. MacRitchie *J. Cereal Sci*, 1993, **18**, 23.

THE TRANSMISSION ROUTE THROUGH WHICH THE COMMON WHEAT (*Triticum aestivum* L.) HAS REACHED THE FAR-EAST, JAPAN.

H. Nakamura

Japan International Research Center for Agricultural Sciences (JIRCAS),
Tsukuba 305-8686, JAPAN
hiro@jircas.affrc.go.jp

1 INTRODUCTION

It is well known that common wheat (*Triticum aestivum* L., $2n = 42$, AABBDD) was present about 10000 years ago in the Middle and Near East. It was then transmitted from its origin to Europe, Africa, southern Asia and China. Little is known, however, about the actual route of transmission of common wheat into Japan. I concentrated my study predominantly on the variation of the HMW-glutenin *Glu-D1f* allele, and the factors which affected its distribution in different parts of the world. The objective of this study was to analyze the distribution of alleles at the *Glu-D1* locus throughout Asia, and then to determine the route by which common wheat reached Japan, the most geographically remote region of common wheat production in the world.

2 METHODS AND RESULTS

In this study, I investigated the allelic composition of the high-molecular-weight (HMW) glutenin subunit loci from 305 Japanese, 353 Chinese, 150 Turkish, 3 Syrian, 6 Israeli, 4 Iranian, 1 Iraqi, 21 Afghanistan, 23 Indian, 15 Pakistani, 7 Bhutanese, 66 Nepalese, 1 Myanmar, 1 Filipino, 2 Thailand, 3 Indonesian, and 46 Taiwanese cultivars of common wheat. These were investigated by sodium dodecyl sulfate-polyacrylamide gel electrophoresis (SDS-PAGE). Gels were made up to 10% (w/v) acrylamide and 0.2% (w/v) bis-acrylamide containing 1.5 M Tris-HCl at pH 8.8 and 0.27% SDS. Published data for the worldwide distribution of the *Glu-1* alleles were available for 1380 cultivars and the frequency of the HMW glutenin alleles was available for the Japanese wheat varieties. These data sets were compared to the results for 1107 common Asian wheat varieties which were determined in this study. In total, 1380 published cultivars from 21 countries and 1107 Asian varieties were included in these comparisons.

The frequency of the HMW glutenin *Glu-D1* alleles in Japanese, Chinese and other Asian common wheat varieties was analyzed in order to investigate a possible transmission route for common wheat to Japan. Although the frequency varied among the areas, the allele *Glu-D1f* was present in wheat from northern and southern Japan, Xinjiang, Nanjing, Zhejiang, Beijing in China, and in Afghanistan (Table 1). However, a high frequency of the *Glu-D1f* allele was found predominantly in southern Japan. It was not detected in wheat from any other Asian region. It is said that there were four routes by which people moved across Asia in ancient times. The first of these routes, the so called Silk Road, ran through Afghanistan, Xinjiang (in north-west China), Gansu, Shanxi (in north-east China), Jiangsu, Zhejiang (in south-east China) and eventually reaching Japan (Figure 1). The second route ran through Pakistan, India, and Myanmar and then to Yunnan in China. The third ran through Nepal or Pamir, Tibet and into Sichuan in south-west China or Shanxi in north-east China. The final route was directly into southern China by boat from India. With regards to these routes across Asia, the distribution of the *Glu-D1* alleles is very interesting. The *Glu-D1f* allele has been regarded as a characteristic glutenin allele for Japanese wheat cultivars. In fact, while many hexaploid wheat cultivars in southern Japan possess this *Glu-D1f* allele, most of the northern Japanese cultivars do not. By comparison, it is known that consideration of β-amylase isozyme types shows that both types A and J are hexaploid wheat in Japan. Type A was found throughout Japan, whereas type J was present predominantly in southern regions. This distribution of β-amylase isozyme types is similar to that of *Glu-D1f* alleles in Japanese hexaploid wheat seed storage proteins. This distribution of an adaptively neutral character suggests a specific route of transmission for common wheat to eastern China and Japan. It was introduced from Afghanistan, carried to Xinjiang (in north-west China), Nanjing Zhejiang (in south-east China) and then to southern Japan along the "Silk Road". It is believed that cultivated common wheat originated in the Middle and Near East and was carried along "the Silk Road" through China to the Far East and Japan. Japan is the most geographically remote region from an origin of common wheat, the Middle and Near East. During the course of its long journey and its adaptation to diverse local environments, Japanese common wheat has developed a unique composition of glutenin alleles. In this study, the specific distribution of an adaptively neutral characteristic (*Glu-D1f* allele) suggested a transmission route for common wheat into eastern China, the Far-East, Japan.

Genetics and Quality

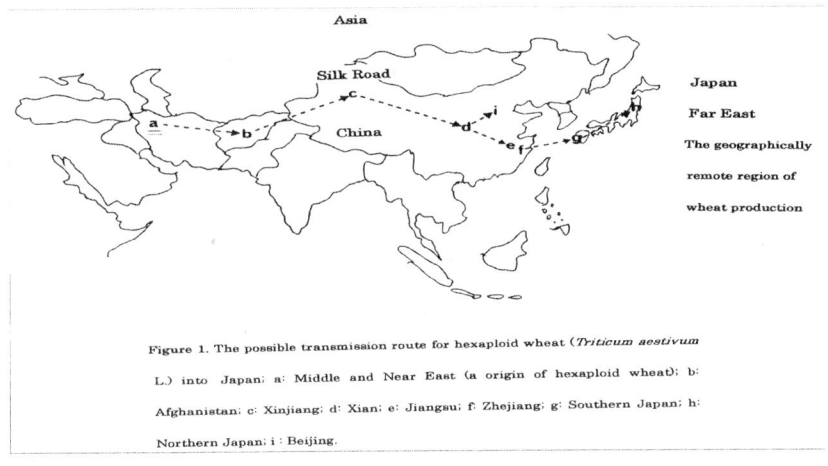

Figure 1. The possible transmission route for hexaploid wheat (*Triticum aestivum* L.) into Japan; a: Middle and Near East (a origin of hexaploid wheat); b: Afghanistan; c: Xinjiang; d: Xian; e: Jiangsu; f: Zhejiang; g: Southern Japan; h: Northern Japan; i : Beijing.

Glu-D1f allele

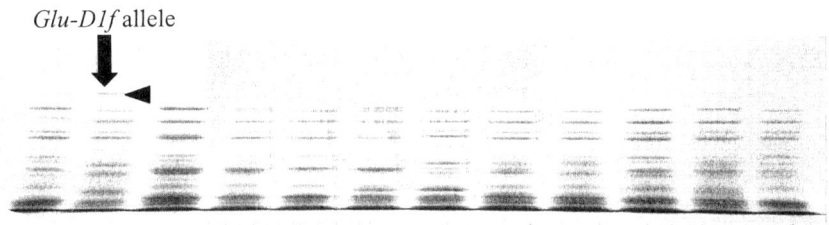

Figure 2 *SDS-PAGE gel electrophoresis separation of seed storage protein in common wheat*

Table 1 *Comparison of Glu-D1f Allele Frequency for Afghanistan, China, Japanese, and Other Asian Hexaploid Wheats (Triticum aestivum L.)*

COUNTRY	TOTAL NUMBER OF VARIETIES EXAMINED	NUMBER OF VARIETIES CARRYING *GLU-D1F* ALLELE	FREQUENCY (%)	X^2-VALUE
Other	428	0	0.0	1.4**
Afghanistan	21	2	9.5	46.86**
China	353	5	1.4	------
Japanese	305	90	29.5	564.00**

Other: Turkey, Syria, Israel, Iran, Iraq, India, Pakistan, Bhutan, Nepal, Myanmar, Philippine, Thailand, Indonesia, Taiwan.
** Significant at the 0.01 probability levels.
The *Glu-D1f* allele frequency of Chinese hexaploid wheats: the "expected" class.

3 CONCLUSION

1) In this study, the specific distribution of an adaptively neutral characteristic (the *Glu-D1f* allele) suggested a transmission route for hexaploid wheat into East China and Japan in Asia.
2) The common wheat was introduced from Afghanistan, moved through Xinjiang (in northwest China), into Jiangsu and Zhejiang (in southeast China), and then into southern Japan (Kyushu district) along the so-called "Silk Road".
3) The results presented here indicate that *Glu-D1* allele analysis is a powerful tool in the investigation of the real transmission route of common wheat across Asia and into the Far-East, JAPAN.

References

1 A.I. Morgunov, R.J. Pena, J. Crossa, and S. Rajarm, *J Genet. Breed.*, 1993, **47**,53.
2 H. Nakamura, A. Inazu, H. Hirano, *Euphytica*, 1999,**106**,131.
3 H. Nakamura, *J. Agric. Food Chem.*, 1999, **47**, 5273.
4 H. Nakamura, *Euphytica,* 2000, **112**,187.
5 H. Nakamura, *Aust. J. Agric. Res.*, 2000, **51**, 673.
6 H. Nakamura, *J. Agric. Food. Chem.,* 2000, **48**, 2648.
7 H. Nakamura, *Cereal Chem.*, 2001, **78**,79.
8 H. Nakamura, H. Fujimaki, *10th Australian Wheat Breeders Assembly,* 2001, 186.
9 H. Nakamura, and H. Fujimaki, *Cereal Chem.*, 2002, **79**, 486-490.
10 P.I. Payne, K.G. Corfield and J.A. Blackman, *Theor. Appl. Genet.*, 1979, **55**,153.
11 A.R. Templeton, *Genetics.,* 1980, **94**, 1011.
12 H. Tsujimoto, T. Yamada, and Sasakuma, *Breed. Sci.*, 1998, **48**, 287.
13 K. Tsunewaki, *Japan. J. Bot.*, 1966, **19**,175.

ALLELIC VARIATION AT THE *GLU-1* AND *GLU-3* LOCI, PRESENCE OF 1BL/1RS TRANSLOCATION, AND THEIR EFFECT ON DOUGH PROPERTIES OF CHINESE BREAD WHEATS

Z.H. He[1,2] L. Liu[1] and R.J. Peña[3]

[1]Institute of Crop Breeding and Cultivation, Chinese Academy of Agricultural Sciences (CAAS), No 12 Zhongguancun South Street, Beijing 100081, China
[2]CIMMYT China Office, C/O CAAS, No 12 Zhongguancun South Street, Beijing 100081, China
[3]CIMMYT, Apdo. Postal 6-641, 06600 Mexico, D.F., Mexico

1 INTRODUCTION

Chinese wheat has acceptable protein content, but weak gluten properties, which causes inferior pan bread and noodle quality. Therefore, improvement of gluten quality has been the major breeding objective of most wheat breeding programs in China[6]. Glutenin, a major storage protein in wheat, is polymeric and consists of high molecular weight (HMW) and low molecular weight (LMW) subunits. HMW and LMW glutenin subunits are encoded by the *Glu-1* and the *Glu-3* complex loci, respectively. The allelic variation at these loci is important in determining flour quality. The composition of HMW subunits in Chinese wheat and their association with bread making quality have been reported previously[2], but the composition of LMW gluten subunits and their potential effect on processing quality remain unclear.

Cultivars carrying the 1BL/1RS translocation, such as Kavkaz, Neuzucht, Lovrin 10, and their derivatives, were heavily used in Chinese wheat crossing programs from the early 1970s to the late 1980s[1]. However, no information is available on the presence of the 1BL/1RS translocation in Chinese wheats and its effect on quality. In this study, the composition of HMW and LMW gluten subunits, and the presence of the 1BL/1RS translocation and its effect on dough properties were investigated in Chinese wheat cultivars and advanced lines.

2 MATERIALS AND METHODS

A total of 250 wheat cultivars and advanced lines obtained from various breeding programs located in autumn-sown wheat regions were sown both in Beijing and Anyang in the 2001-02 season. Wheat flour samples were evaluated for protein content using NIR analysis (Instalab 610, Newport Scientific Sales and Services Pty. Ltd), SDS sedimentation value.[3]

and mixographic characteristics (mixing time and mixing tolerance) according to AACC method 54-40 A. The 1B/1R translocation and Glu-1 and Glu-3 allelic variations were identified using SDS-PAGE electrophoresis.[4,5]

3 RESULTS AND DISCUSSION

As shown in Table 1, the frequency of subunit 1 was dominant at Glu-A1, followed by that of the null allele. Subunits 7+8 and 7+9 shared a high proportion of the variation at Glu-B1, and the frequency of 2+12 was greater than that of 5+10 at Glu-D1. At Glu-A3, the frequency of Glu-A3a was larger than that of Glu-A3d, followed by that of Glu-A3c, and Glu-A3e. At Glu-B3, the Glu-B3j allele showed the highest frequency while Glu-B3b was the least frequent allele. The frequencies of other GluB3 alleles, Glu-B3d and Glu-B3, were rather low (Table 1). Thus the frequency of the Glu-B3j allele showed that close to 50% of Chinese wheats carry the 1BL/1RS translocation. Significant variation in the frequency of the 1BL/1RS translocation was observed in wheats from different regions; the frequencies for the Northern China Plain (Zone I), Yellow and Huai Valley (Zone II), Middle and Lower Yangtze Valley (Zone III), and Southwestern Region (Zone IV) were 54.0%, 50.4%, 6.9%, and 34.6%, respectively.

Table 1 *Comparison of the effect of individual glutenin subunits in cultivars and advanced lines of wheat*

Locus	Subunit	Number	Frequency (%)	Protein content (14%mb)	SDS sed. vol. (ml)	Mixing time (min)	Mixing tolerance(min)
	2*	23	9.2	15.6a	16.0b	2.5b	2.4b
Glu-A1	1	129	51.4	15.8a	17.7a	2.8a	2.7a
	N	99	39.4	15.6a	16.1b	2.3b	2.4b
	7+8	96	42.9	15.7a	17.4ab	2.9a	3.0a
Glu-B1	7+9	113	50.4	15.7a	16.3b	2.4b	2.2b
	14+15	15	6.7	16.0a	18.8a	2.7b	2.8ab
	5+10	62	25.5	15.9a	18.0a	3.5a	3.5a
Glu-D1	2+12	150	61.7	15.7a	16.5b	2.3b	2.2b
	4+12	31	12.8	15.4a	17.0b	2.5b	2.5b
	GluA3d	78	31.7	15.9ab	17.5a	3.0a	2.9a
Glu-A3	GluA3c	51	20.7	15.5b	16.3ab	2.3bc	2.3bc
	GluA3a	93	37.8	15.6b	17.2ab	2.6b	2.6b
	GluA3e	24	9.8	16.1a	15.4b	2.1c	1.9c
	GluB3d	52	22.4	16.0a	19.2a	3.3a	3.4a
Glu-B3	GluB3b	29	12.5	16.0a	17.4b	2.6b	2.7b
	GluB3f	39	16.8	15.6a	17.7b	2.8ab	3.0ab
	GluB3j	112	48.2	15.6a	15.5c	2.2b	1.9c

Different letters indicate significant differences at the 5% probability level.

As presented in Table 2 from the data in Beijing, the analysis of variance showed that the effects of all *Glu-1* and *Glu-3* loci examined on SDS sedimentation values and mixing properties were significant at the 1% probability level, while *Glu-B3* and *Glu-D1* played a more significant role in determining mixing properties. For example, *Glu-B3* accounts for 19%, 22%, and 30% of the variation for SDS sedimentation values, mixing time, and mixing tolerance, respectively, while *Glu-D1* explains 31% and 23% of the variation for mixing time and mixing tolerance, respectively. Loci interaction, such as *Glu-A1* x *Glu-B1*, *Glu-A1* x *Glu-B3*, *Glu-B1* x *Glu-A3*, *Glu-D1* x *Glu-A3*, and *Glu-A1* x *Glu-D1* x *Glu-B3*, also played an important role in determining dough properties.

Table 2 ANOVA analysis of *Glu-1* and *Glu-3* loci effects on processing quality of cultivars and advanced lines of wheat

Locus	DF	Protein content (14% m.b.)		SDS sed. vol. (ml)		Mixing time (min)		Mixing tolerance (min)	
		MS	%	MS	%	MS	%	MS	%
Glu-A1	2	1.8	1	80.0**	6	7.4**	6	3.1**	2
Glu-B1	8	1.4	4	26.8**	7	2.5**	8	5.8**	14
Glu-D1	3	1.7	2	41.3**	4	25.0**	31	25.2**	23
Glu-A3	4	2.9*	5	26.3**	4	5.3**	9	5.7**	7
Glu-B3	7	1.7	5	78.5**	19	7.6**	22	14.0**	30
Glu-A1*Glu-B1	8	1.7	5	25.2**	7	0.8*	3	1.7**	4
Glu-A1*Glu-A3	6	2.4*	6	10.0	2	0.7	2	1.1*	2
Glu-A1*Glu-B3	9	1.5	5	1.6	1	0.8**	3	2.3**	6
Glu-B1*Glu-D1	13	1.2	6	15.8*	7	0.8*	4	0.7	3
Glu-B1*Glu-A3	14	0.9	5	7.8	4	1.5**	9	1.6**	7
Glu-B1*Glu-B3	16	0.5	3	4.8	3	0.9**	6	0.0	0
Glu-D1*Glu-A3	11	1.0	4	8.3	3	1.6**	7	2.7**	9
Glu-A3*Glu-B3	14	1.2	7	2.3	1	0.1**	0	0.2*	1
Glu-A1*Glu-B1*Glu-D1	3	0.0	0	18.6	2	0.0	0	3.2**	3
Glu-A1*Glu-B1*Glu-A3	6	1.0	2	2.8	1	0.6	2	2.0**	4
Glu-A1*Glu-D1*Glu-A3	6	1.8	4	22.6**	5	1.1**	3	3.0**	6
Glu-A1*Glu-D1*Glu-B3	7	0.7	2	6.6	2	0.4	1	2.1**	5
Glu-B1*Glu-A3*Glu-B3	4	4.2**	7	8.5	1	0.0	0	0.0	0
Error	106	0.9		7.3		0.3		0.4	

This is only one part of the ANOVA analysis table.
% Sum of squares as percentage of the total sum of squares, which can be interpreted as an indication of the relevance of various characteristics.
*and ** Significant at the 5% and 1% probability levels, respectively.

Based on the data from two locations, for SDS sedimentation values the gluten subunit loci could be ranked as: *Glu-B3* > *Glu-B1* > *Glu-A1* > *Glu-A1* x *Glu-D1* x *Glu-A3* > *Glu-D1* = *Glu-A3*; for mixing time: *Glu-D1* > *Glu-B3* > *Glu-B1* x *Glu-A3* > *Glu-B1* > *Glu-D1* x *Glu-A3* > *Glu-A1* = *Glu-B1* x *Glu-B3*; and for mixing tolerance: *Glu-B3* > *Glu-D1* > *Glu-B1* > *Glu-D1* x *Glu-A3* > *Glu-A3* = *Glu-B1* x *Glu-A3* > *Glu-A1* x *Glu-B3* = *Glu-A1* x *Glu-D1* x *Glu-A3* > *Glu-A1* x *Glu-D1* x *Glu-B3* > *Glu-A1* x *Glu-B1* x *Glu-A3*.

As shown in Table 1, the allelic effect on protein content in general was not significant. For *Glu-1* loci, subunits 1, 7+8, and 5+10 are significantly associated with strong dough properties in comparison with other alleles, indicating their crucial role in determining breadmaking quality. For *Glu-3* loci, *GluA3d* and *GluB3d* show significant positive contribution to dough properties. *Glub3j*, associated with the presence of the 1BL/1RS translocation, plays a strong negative role in bread making quality. The effect of the 1BL/1RS translocation is presented in Table 3.

Table 3 *Effect of the 1BL/1RS translocation on wheat processing quality*

Location	Type	Number	Protein content (14% m.b.)	SDS sed. vol. (ml)	Mixing time (min)	Mixing tolerance (min)
Beijing	Non-1RS	139	15.8A	18.0A	2.9A	3.0A
	1BL/1RS	112	15.6A	15.5B	2.2B	1.9B
Anyang	Non-1RS	139	15.7A	14.2A	2.2A	2.2A
	1BL/1RS	112	15.3B	11.7B	1.9B	1.6B
Mean	Non-1RS	139	15.8A	16.1A	2.6A	2.6A
	1BL/1RS	112	15.5B	13.6B	2.1B	1.8B

Different letters indicate significant differences at the 1% probability level.

3 CONCLUSIONS

The high frequency of undesirable HMW and LMW gluten subunits and the presence of the 1BL/1RS translocation are responsible for weak gluten properties in Chinese wheats. The reduction of the 1B/1R frequency and the integration of desirable subunits at *Glu-1* and *Glu-3* (such as 1, 7+8, 5+10, *Glu-A3d*, and *Glu-B3d*) could lead to the improvement of gluten quality in Chinese wheats.

References

1. He Z H, Rajaram S, Xin Z Y and Huang G Z. A History of Wheat Breeding in China, 2001, Mexico D.F., CIMMYT.
2. He Z H, Peña R J and Rajaram S. *Euphytica,* 1992, 64: 11-20.
3. Peña R J, Amaya A, Rajaram S and Mujeeb Kazi A. *J. Cereal Sci,* 1990, 12: 105-112.
4. Jackson E A, Morel M H, Sontag-Strohm T, Branlard G, Metakovsky E V and Redaelli R. *J. Genet & Breed,* 1996, 50: 321-326.
5. Gupta R B and Shephered K W. *Plant Breeding,* 1992, 109: 130-140.
6. He Z H, Lin Z J, Wang L J, Xiao Z M, Zhuang Q S and Wan F S. *J. of Chinese Agricultural Sciences,* 2002, 35 (4): 359-364.

Genetics and Quality 101

CHARACTERISATION OF LMW-GS GENES AND THE CORRESPONDING PROTEINS IN COMMON WHEATS

T.M. Ikeda[1], T. Nagamine[1,2] and H. Yano[1]

[1]Department of Crop Breeding, National Agricultural Research Center for Western Region, 6-12-1, Nishifukatsu, Fukuyama, JAPAN 721-8514
[2]Present address: Tochigi Branch, Tochigi Prefectural Agricultural Experiment Station, Otsuka, Tochigi, JAPAN 328-0007

1 INTRODUCTION

Wheat seed storage proteins are composed of two major fractions, gliadin and glutenin. Glutenin consists of high-molecular-weight (HMW) and low-molecular-weight (LMW) subunits. The HMW glutenin subunits (HMW-GSs) are encoded by *Glu-A1*, *Glu-B1*, and *Glu-D1* on the long arm of chromosomes 1A, 1B and 1D, respectively; the LMW glutenin subunits (LMW-GSs) are encoded by *Glu-A3*, *Glu-B3* and *Glu-D3* on the short arm of these chromosomes. It has been shown that allelic variations of HMW-GSs and LMW-GSs affect dough properties in various wheat cultivars. LMW-GSs were classified into three types, LMW-i, LMW-m and LMW-s, based on the amino acid at the N-terminal end, which corresponded to isoleucine, methionine and serine residues respectively[1,2]. Although LMW-m and LMW-s types were identified as protein products[2], the LMW-i type has not yet been identified as a protein product. In our previous study, we comprehensively analysed the composition of LMW-GS genes in a soft wheat cultivar, Norin 61, and classified these genes into 12 groups based on their deduced amino acid sequence identity. In this study, we identified the proteins corresponding to these groups by determining the N-terminal amino acid sequence of all LMW-GSs of Norin 61 separated by 2-D gel electrophoresis.

2 METHOD AND RESULTS

2.1 Plant materials and methods

We used Norin 61, which is a Japanese soft wheat cultivar with poor breadmaking properties, for this study. Two-D gel electrophoresis was carried out by PROTEAN IEF Cell (Bio-Rad Lab. Co. Ltd.) using a Immobiline DryStrip gel (13 cm) ranging from pH6-11 (Amersham Biosciences Co. Ltd.) as the first dimension and 10 % SDS-PAGE (14 cm x 14 cm) as the second dimension. The separated LMW-GS spots were blotted onto a PVDF membrane (SeqBlott, Bio-Rad Lab. Co. Ltd.) and their N-terminal amino acid sequences were determined by a PPSQ-21A protein sequencer (Shimadzu Co. Ltd.).

2.2 Characterisation of low-molecular-weight glutenin genes

In our previous study, we classified LMW-GS genes of a Japanese soft wheat, Norin 61, into 12 groups based on the deduced amino acid sequence identity in the N- and C-terminal conserved domains[3] (Table 1). Groups 1, 2, 5, 6, 7, 8, 9 and 10 shared similar N-terminal amino acid sequences to those of the LMW-m type. Among these LMW-GSs, groups 6, 7, 8, 9 and 10 have the first cysteine residue at position 5, while groups 1, 2 and 5 have the first cysteine residue in a repetitive domain. Group 3 has the N-terminal domain of the LMW-s type with asparagine at position 3. The N-terminal sequence of the mature protein of this group was expected to start from the serine residue at position 4^4. Group 4 also had an asparagine residue at position 3, but this group was distinguished from group 3 by the presence of an isoleucine residue at position 1 and a serine residue at position 11. Group 5 showed two additional repeats consisting of five hydrophobic amino acids (PIIIL and PVIIL) interrupting the N-terminal repetitive domain. Group 9 shared its amino acid sequence with group 8, except at the C-terminal end (Table 1). Groups 11 and 12 were characterised by the absence of cysteine residues in the N-terminal domain and the presence of all eight cysteine residues in the C-terminal conserved domain. The loci of individual groups were determined by group-specific PCR primer sets using nullisomic-tetrasomic lines of Chinese Spring lacking group 1 chromosomes (Ikeda et al. unpublished)[3,5]. These primer sets distinguished between these groups except between groups 7 and 8, and groups 11 and 12, which share almost identical amino acid sequences (Table 1). Groups of three, two and seven LMW-GS were encoded by the *Glu-A3*, *Glu-B3* and *Glu-D3* loci, respectively (Table 1).

Table 1. *Classification of LMW-GSs based on the predicted N- and C-terminal amino acid sequences of LMW-GSs and their loci*

Group	Predicted N-terminal amino acid sequence	Predicted C-terminal amino acid sequence	Locus
1	METSHIPGLEKPS	SVPFGVGTQVGAY	*Glu-D3*
2	METSHIPSLEKPL	IVPFGVGTRVGAY	*Glu-B3*
3	MENSHIPGLERPS	RVPFGVGTGVGGY	*Glu-B3*
4	IENSHIPGLEKPS	SVPFGVGAGVGAY	*Glu-D3*
5	METSRVPGLEKPW	IMPFSIGTGVGGY	*Glu-D3*
6	MDTSCIPGLERPW	SVPFGVGTGVGAY	*Glu-A3*
7	METSCISGLERPW	SVPFGVGTGVGAY	*Glu-D3*
8	METSCIPGLERPW	SAPLGVGSRVGAY	*Glu-D3*
9	METSCIPGLERPW	SVPFGVGTQVGAY	*Glu-D3*
10	METRCIPGLERPW	SVPFDVGTGVGAY	*Glu-D3*
11	ISQQQQPPLFSQQ	SVPLGVGIGVGVY	*Glu-A3*
12	ISQQQQQPPFSQQ	SVPLGIGIGVGVY	*Glu-A3*

Genetics and Quality

2.3 Identification of the corresponding LMW-GS proteins

To identify proteins corresponding to the LMW-GS groups, we separated a glutenin fraction of Norin 61 by 2-D gel electrophoresis. Eleven spots were LMW-GSs, while the remaining spots showing similar molecular weights were α/β and γ-gliadins. The N-terminal sequences of the 11 spots matched 10 out of 12 groups of LMW-GSs (Table 1). The two groups (groups 6 and 7) of LMW-GS genes encoded by *Glu-A3* and *Glu-D3* could not be detected. Since the genes encoding these groups did not have a nonsense mutation, silencing of these genes may have occurred[3]. Among the largest LMW-GSs, a major spot was the LMW-s type showing the N-terminal amino acid sequence starting with a serine residue (SHIPGLERPSQQ)(Table 2). However, a minor spot appearing slightly more acidic than the major spot was the LMW-m type showing a methionine residue at the N-terminal end (MENSHIPGLERP) (Table 2). Comparing their N-terminal amino acid sequences, both LMW-GSs matched group 3, encoded by the *Glu-B3* locus, with different processing sites. This is the first finding that a minor fraction of LMW-s type is processed at the same site as the LMW-m type. Another protein spot starting from a serine residue was also identified, and matched the N-terminal amino acid sequence of group 4 encoded by the *Glu-D3* locus (Table 2). This result showed that LMW-s type was encoded by at least two genes on different loci (*Glu-B3* and *Glu-D3*).

We also found three protein spots which have the N-terminal amino acid sequences corresponding to LMW-i type[1]. One spot matched group 11 and the other two spots matched group 12 (Table 2). These LMW-GSs, encoded by *Glu-A3*, have not previously been detected as protein products. Since spots 12 and 12* share the same amino acid sequence, these spots might be encoded by the same gene but have different modifications such as phosphorylation. We also detected N-terminal amino acid sequences matching the other groups, excluding groups 6 and 7 (ie. groups 1, 2, 5, 8, 9 and 10). From this result, we could assign all of the LMW-GSs to 10 groups of LMW-GSs in Norin 61.

Table 2. *N-terminal amino acid sequence of LMW-GSs of Norin 61 and their locus*

N-terminal amino acid sequence	Group*	Locus
METSHIPGLEKP	1	*Glu-D3*
METSHIPSLEKP	2	*Glu-B3*
SHIPGLERPSQQ	3	*Glu-B3*
MENSHIPGLERP	3	*Glu-B3*
SHIPGLEKPSQQ	4	*Glu-D3*
METSRVPGLEKP	5	*Glu-D3*
METSCIPGLERP	8&9	*Glu-D3*
METRQIPGLERP	10	*Glu-D3*
ISQQQQPPLFSQ	11	*Glu-A3*
ISQQQQQPPFSQ	12	*Glu-A3*

*: The N-terminal amino acid sequence corresponding to groups 6 and 7 was not found.

3 CONCLUSION

We have identified the N-terminal amino acid sequences of LMW-GSs in Norin 61 and their loci. There were ten groups of LMW-GS genes among 12 matched LMW-GS proteins. Three, three and four spots were encoded by *Glu-A3*, *Glu-B3* and *Glu-D3*, respectively. The LMW-GSs include LMW-i, LMW-m and LMW-s types. LMW-s type was encoded by two genes located at *Glu-B3* and *Glu-D3*. LMW-i type was encoded by two genes at the same *Glu-A3*. We found that a small fraction of the LMW-m type was encoded by the same gene as the LMW-s but was processed at a different site. Characterisation of individual LMW-GSs based on amino acid sequence data will be useful to identify LMW-GSs involved in gluten quality by gene-specific PCR. Further analysis is necessary to identify LMW-GS variation among wheat cultivars.

References

1. S. Cloutier, C. Rampitsch, G.A. Penner, O.M. Lukow, *J. Cereal Sci.*, 2001, **33**, 143.
2. E.J.-L. Lew, D.D. Kuzmicky, D.D. Kasarda, *Cereal Chem.*, 1992, **69**, 508.
3. T.M. Ikeda, T. Nagamine, H. Fukuoka, H. Yano, *Theor. Appl. Genet.*, 2002, **104**, 680.
4. S. Masci, R. D'Ovidio, D. Lafiandra, D.D. Kasarda, *Plant Physiol.*, 1998, **118**, 1147.
5. T.M. Ikeda, K. Nakamichi, T. Nagamine, H. Yano, A. Yanagisawa, *JARQ*, 2003, **37**, 99.

THE USE OF SE-HPLC FOR QUALITY PREDICTION IN TWO AFRICAN COUNTRIES

M.T. Labuschagne, E. Koen and T. Dessalegn

1 INTRODUCTION

SE-HPLC is a powerful tool to study native protein aggregates and physicochemical basis of baking strength, and has potential for rapid assessment of baking quality of bread wheat genotypes in breeding programmes.[1] SE-HPLC accurately separates the three main classes of wheat endosperm proteins, glutenin, gliadins and albumins-globulins.[2] The results obtained with this technique have been highly correlated with breadmaking quality.[3] Correlations were found between quantity of polymers and technological parameters linked with mixing, [4,5,6] gluten baking strength and loaf volume.[7] In a country like Ethiopia, almost no quality testing facilities are available, although HPLC machines are available at some stations. The aim of this study was to determine whether SE-HPLC could be used for quality prediction in Ethiopian as well as South African wheat breeding programmes.

2 MATERIAL AND METHODS

South Africa: Eighteen hard red wheat and facultative bread wheat cultivars were grown in 1999 at four different localities throughout the Free State province of South Africa under dryland conditions. Plots consisted of five rows (5m with 45 cm inter row spacing). A randomised block design with four replicates was used.

Ethiopia: Thirteen popularly grown Ethiopia cultivars and two South African cultivars of known quality were used. The trials were grown at two environments in Ethiopia in 2001: Adet Research Center, which is a higher protein potential area, and Motta, which is a low protein potential area. A RCB design with three replications was used. The plot size was 3 m^2 at both localities. After harvesting, the material was transported to South Africa.

Quality analysis: AACC procedures[8] were mainly used and all tests were done in triplicate: Hectoliter weight (kg hl^{-1}), breakflour yield (AACC 26-21A), flour yield (AACC 26-21A), flour colour (Kent Jones,C76), flour protein content (AACC 39-11), SDS-sedimentation test (AACC 56-70), vitreous kernels (kernels were sliced and counted), single kernel characteristic system (SKCS) - hardness index (AACC 53-31), SKCS-seed diameter (AACC 53-31), SKCS-seed weight (AACC 53-31), mixograph development time

(AACC 54-40A), farinograph water absorption (AACC 54-21), alveograph P/L ratio (AACC 54-30A), alveograph strength (AACC 54-30A) and loaf volume (AACC 10-09).

SE-HPLC: Proteins were extracted from the same wheat flour with a two-step extraction procedure developed by Gupta and colleagues.[9] The first step extracts the proteins soluble in dilute SDS, while the second extract contains proteins soluble only after sonication. In 1999, facilities for sonication were not available at the UFS yet, and only SDS extractable proteins were measured. Size exclusion HPLC analyses were performed on a Varian HPLC system using a BIOSEP SEC-4000 Phenomenex column. Separation was achieved in 30min by loading $20\mu l$ of sample into an eluant of 50% (v/v) acetonitrile and water containing 0.1% (v/v) trifluoroacetic acid at a flow rate of 0.2ml/min. Proteins were detected by UV absorbance at 210nm. Areas of the different peaks were calculated. The percentage of total unextractable polymeric protein in the total polymeric protein was calculated according to the method of Gupta.[9] The measured HPLC fractions were: (a) SDS-soluble (b) SDS-insoluble, where A = large polymeric proteins (LPP), B=smaller polymeric proteins (SPP), C= large monomeric proteins (LMP) mainly gliadins, D=smaller monomeric proteins (SMP) mainly albumins and globulins. All statistical analysis was done with Agrobase 2000 software.

3 RESULTS AND DISCUSSION

At Henneman, the LPP positively and LMP negatively correlated with loaf volume (only significant correlations are mentioned here). The other correlations were LPP and SPP (positively) and LMP (negatively) with flour yield (Table 1). At Wesselsbron1 LPP negatively correlated with flour yield. At Wesselsbron2 SPP correlated positively with flour protein, and at Senekal SPP correlated with mixing time, both important for baking quality. All the other significant correlations were with kernel characteristics. Across all four localities, with the exception of the correlation of SMP with farinograph absorption, all correlations were with kernel characteristics. At both Motta and Adet, SDS-insoluble LMP negatively correlated and at Motta SDS-insoluble SPP positively correlated with SDS-sedimentation. At Motta, SDS-insoluble SPP, and TUP correlated negatively and SDS-insoluble LMP positively with protein content. Both SDS-insoluble SPP and TUP correlated positively and SDS-insoluble LMP negatively with mixograph development time.

When the two data sets were combined (Table 1), for the SDS-soluble fractions, LPP was highly correlated with SDS-sedimentation, vitreous kernels and flour protein content. SPP were correlated with vitreous kernels and flour protein content. The LMP negatively correlated with SDS-sedimentation, vitreous kernels and flour protein content. SMP positively correlated with vitreous kernels and flour protein content. Concerning the SDS-insoluble protein fractions, the LPP was again highly correlated with SDS-sedimentation, vitreous kernels and flour protein content. The LMP was negatively correlated with vitreous kernels and flour protein content. TUP was positively correlated with SDS-sedimentation, vitreous kernels and flour protein content. LUP was positively correlated with all the same characteristics with the exception of vitreous kernels. SDS-sedimentation and flour protein content are generally seen as important indicators of bread making quality. Across Motta and Adet, both the SDS-soluble and SDS-insoluble LPP correlated with SDS-sedimentation and flour protein content.

Table I Significant correlations between specific protein fractions and quality characteristics in Ethiopia and South Africa

Location	Character1	Character2	Correlation	Location	Character 1	Character 2	Correlation
Henneman	LPP	Flour yield	0.516*	Adet	SDS-insol LMP	SDS-sedimentation	-0.577*
	SPP	Flour yield	0.503*		TUP	Hardness	0.562*
	LMP	Flour yield	-0.524*	Motta	SDS-insol SPP	SDS-sedimentation	0.653*
	LPP	Loaf volume	0.630**		SDS-insol SPP	Flour protein	-0.605*
	LMP	Loaf volume	-0.617**			Mixo development	0.589*
Wesselsbron1	LPP	Flour yield	-0.589**		SDS-insol LMP	SDS-sedimentation	-0.559*
Wesselsbron2	LMP	HLM	0.499*			Flour protein content	0.709**
	SMP	HLM	-0.516*			Mixo development	-0.566*
	SMP	Vitreous kernels	0.522*		TUP	Flour protein content	-0.707
	SPP	Flour protein	0.520*			Mixo development	0.642*
	SPP	SKCS hardness	-0.57**	Motta and Adet	SDS-sol LPP	SDS-sedimentation	0.446*
	LMP	SKSC hardness	0.507**			Vitreous kernels	0.513**
Senekal	LPP	Flour yield	0.603**			Flour protein content	0.649**
	SPP	Mixing time	0.580**		SDS-sol SPP	Vitreous kernels	0.420*
	LMP	Flour yield	-0.521**			Flour protein content	0.620**
	LPP	SKCS diameter	0.510**		SDS-sol LMP	SDS-sedimentation	-0.404*
	SPP	SKCS diameter	0.560**			Vitreous kernels	-0.502**
	LMP	SKCS diameter	-0.581**			Flour protein content	-0.702**
All localities	LMP	HLM	0.303**		SDS-sol SMP	Vitreous kernels	0.546**
	SMP	HLM	-0.356**			Flour protein content	0.675**
	LPP	Vitreous kernels	-0.303**		SDS-insol LPP	SDS-sedimentation	0.526**
	SPP	Vitreous kernels	-0.309**			Vitreous kernels	0.432*
	LMP	Vitreous kernels	0.324**			Flour protein content	0.555**
	SMP	Farino absorption	-0.335**		SDS-insol SPP	SDS-sedimentation	0.590**
	LPP	SKCS hardness	-0.257*		SDS-insol LMP	SDS-sedimentation	-0.742**
	SPP	SKCS diameter	0.263*			Vitreous kernels	-0.419*
	SPP	SKCS hardness	-0.358**			Flour protein content	-0.562**
	LMP	SKCS hardness	0.336**		SDS-insol SMP	SDS-sedimentation	-0.383*
					TUP	SDS-sedimentation	0.595**
						Vitreous kernels	0.415*
						Flour protein content	0.571**
					LUP	SDS-sedimentation	0.486**
						Flour protein content	0.451*

LPP = large polymeric proteins, SPP = small polymeric proteins, LMP = large monomeric proteins, SMP = small monomeric proteins, TUP = % total unextractable polymeric proteins, LUP = % large unextractable polymeric proteins. In South Africa only SDS-soluble proteins were tested, * p = 0.05, ** p = 0.01

The TUP and LUP showed the same relationship with these two characteristics. The SDS-soluble and -insoluble SPP also correlated with flour protein content and SDS-sedimentation, respectively. The LMP showed an inverse relationship with both protein content and SDS-sedimentation. Both the soluble- and insoluble- large polymeric proteins, which are mainly the high molecular weight glutenins, highly significantly influenced the most important quality characteristics. The smaller polymeric proteins, which are mainly the low molecular weight glutenins, also positively influenced these characteristics. Other authors,[4,5] investigating sonicated proteins, found that the first fraction (high molecular weight glutenins) can be used as a measure of breadmaking quality, and that the third fraction (gliadins) was negatively correlated with loaf volume. The correlation between quantity of polymers and technological parameters was also confirmed in other studies.[6] In Ethiopia no quality testing facilities are available at breeding stations, and so far, yield and disease resistance was used as criteria for cultivar release. The need for quality testing has become more urgent in the last few years. Setting up a quality laboratory is extremely expensive. If a HPLC machine is available, it can be used to assess quality, and in early generations, half kernels can be used to test quality potential.

4 CONCLUSIONS

Across the four South African environments, only the SDS-soluble proteins were used as criteria, but no significant relationship was present between protein fractions and any rheological or baking characteristics. If the SDS-insoluble proteins were included, the picture would probably have looked totally different. In Ethiopia, the two environments were very diverse. Baking tests were not done in this case, but very strong relationships were present between the LPP, TUP and LUP and SDS-sedimentation and flour protein content, both which are seen as good predictors of bread making quality. A strong negative relationship was present between the LMP and SDS-sedimentation and flour protein content. Therefore, in the Ethiopian environments, large SDS-soluble and -insoluble LPP were predictors of good quality, while high amounts of SDS-soluble and -insoluble LMP were an indicator of poor quality.

References

1. Dachkevitch, T. and Autran, J., *Cereal Chemistry,* 1989, **66,** 448.
2. Larroque, O.R., Gianibelli, M.C., Batey, I.L. and MacRitchie, F. *Electrophoresis,* 1997, **18,** 1064.
3. Batey, I.L., Gupta, R.B. and MacRitchie, F. *Cereal Chemistry* 1991, **68,** 207.
4. Singh, N.K., Donavan, G.R. and MacRitchie, F., *Cereal Chemistry,* 1990, **67,** 161.
5. Singh, N.K., Shepherd, K.W., Gupta, R.B., Moss, H.J. and MacRitchie, F. In 'Gluten proteins' (W. Bushuk and R. Tkachuk, eds.). Am. Assoc. Cereal Chem., USA, 1991, 129.
6. Gupta, R.B., Batey, I.L. and MacRitchie, F. *Cereal Chemistry,* 1992, **69,** 125.
7. Haddad, L.E., Aussenac, T., Fabre, J. and Sarraf, A. *Cereal Chemistry,*1995, **72,** 598.
8. AACC (American Association of Cereal Chemists). Approved methods of the American Association of Cereal Chemists Inc., St. Paul, 1995.
9. Gupta, R.B., Khan, K. and MacRitchie, F. *J. Cereal Sci.,* 1993, **18,** 23.
10. Johansson, E., Prieto-Linde, M and Jönsson, J.O. *Cereal Chemistry,* 2001, **78,** 19.

BIOCHEMICAL AND FUNCTIONAL STUDIES OF WHEAT ISOLINES CONTAINING SINGLE TYPE HIGH MOLECULAR WEIGHT GLUTENIN SUBUNITS (HMW-GS)

O.R. Larroque[1], B. Margiotta[4], M.C. Gianibelli[1], F. Bekes[1], P. Sharp[3] and D. Lafiandra[2]

[1] CSIRO Plant Industry, PO Box 1600, Black Mountain, Canberra, ACT 2601, Australia.
[2] Università degli Studi della Tuscia, Dipartimento di Agrobiologia e Agrochimica, Via S. Camillo de Lellis, 01100 Viterbo, Italy.
[3] The University of Sydney, Plant Breeding Institute, PMB 11, Camden NSW 2570, Australia.
[4] Istitituto di Genetica Vegetale – CNR, Via Amendola 165/A, Bari, Italy.

1 INTRODUCTION

Glutenin is the polymeric component of gluten. It consists of high (HMW-GS) and low molecular weight subunits (LMW-GS), based on their electrophoretic mobility in SDS-PAGE. In its native state, HMW-GS provide the polymer's backbone and LMW-GS branch the structure, in both cases through disulfide bonding. Gliadins, the monomeric component of gluten, interact with the polymer via non-covalent forces. HMW-GS are coded by genes located in the long arm of group 1 chromosomes (*Glu-A1, Glu-B1, Glu-D1*). Representing only about 10% of the total endosperm proteins, they are major determinants of end-use quality in bread wheat. Each locus includes two tightly linked genes encoding two types of HMW-GS, named x- and y-type subunits[1]. The x-type subunits have a slower electrophoretic mobility in SDS-PAGE and higher molecular weight than the y-type subunits. These two types also show differences in the length of the repetitive region and in the number of cysteines residues. Usually, at the *Glu-A1* locus only the x-type is expressed. Some alleles at the *Glu-B1* locus only expressed the x-type (i.e 7, 21) but generally they express both types (i.e. 17+ 18, 7+8, 7+9) while x- and y-type are always expressed at the *Glu-D1* locus (i.e. 2+12, 5+10). Therefore, it is difficult to assess the effect of single HMW glutenin subunits.
Near isogenic lines of bread wheat showing unusual combinations of HMW-GS have been recently developed and will be assessed in order to elucidate the effect of single type HMW-GS on polymer structures and dough functional properties[2].

2 MATERIALS AND METHODS

Plant material: Near isogenic lines of bread wheat differing in composition of HMW-GS produced at the Istituto di Genetica Vegetale, CNR, Bari (Italy) were used in this study[2]. Lines included single subunit Bx 7, single subunit By18 and single subunit Dy12*. Two sets of deletion lines produced at CSIRO Plant Industry were also used[3].
Protein extraction for Size Exclusion HPLC: Samples were extracted using 0.5% sodium dodecyl sulfate - phosphate buffer (pH 6.9) according to standard procedures[4,5].

For reversed-phase HPLC, reduced and alkylated glutenin extracts were prepared according to Larroque et al.[6]

HPLC system and columns: Protein extracts were subjected to HPLC using a Phenomenex BIOSEP-SEC 4000 (5 μm, 500 Å, 7.8 mm x 300 mm) column (Phenomenex, Torrance, CA, USA) was utilized for size exclusion (SE-HPLC) and a Vydac C18 (10 μm, 4.6 mm x 250 mm) column (Vydac, Hesperia, CA, USA) was used for reversed-phase (RP-HPLC).

Multi-angle laser light scattering (MALLS): Eluates from the SE-HPLC separations were detected by a multiangle laser light scattering detector (MiniDawn, Wyatt Technlology Corp. Santa Barbara, CA) with 3 detection angles at 45, 90 and 135 degrees.

Mixograph: A 2 g CSIRO prototype Mixograph was used for the assessment of dough quality properties[7].

3 RESULTS AND DISCUSSION

Percentage of unextractable polymeric protein (%UPP) was the parameter that varied the most from all the SE-based measurements performed (Figure 1). A sharp drop in its value was found in all the analysed single HMW-GS lines when compared with the line with 5 subunits of HMW glutenins (1, 17+18, 5+10). %UPP is an indicator of polymer size and therefore MALLS detection gave also a clear differentiation in molar mass between the single type HMW-GS and the + + + line (Figure 2), the latter being much larger in size.

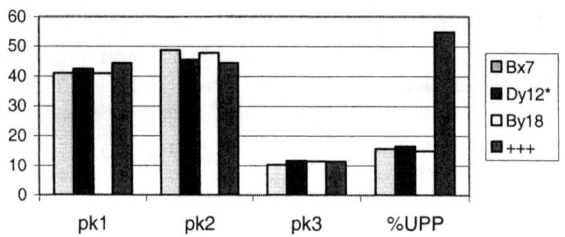

Figure 1 *Proportion of the main size classes of wheat endosperm protein (pk1: glutenins, pk2: gliadins, pk3: albumins+globulins) and %UPP determined by SE-HPLC in single type HMWS-GS (Bx7, Dy12*, By18) and control (+++: 1, 17+18, 5+10) isolines.*

When individual subunits where analysed by RP-HPLC, Dy12* percentages showed a slight increase when compared with the +++ isoline or with the average of Dy-types of 50 Australian cultivars (indicated as % of the subunit in the total reduced glutenin extract).[8] On the other hand, Bx7 and By18 showed levels two and three times larger, respectively (Table 1, Figure 3) indicating a certain compensation effect in the genetic lines with only one subunit. Bx7 showed similar percentages (when referred as % in total reduced glutenin) as in over-expressing Bx7 lines (i.e extra strong Canadian wheat cv. Glenlea). Nevertheless, gluten strength was quite weak. This is likely to be due to a much lower H/L ratio than in cv. Glenlea (results not shown).

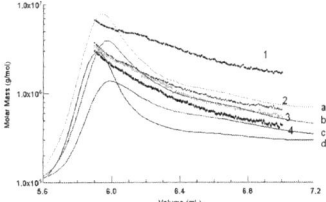

Figure 2 *Molar mass (g/mol) distribution of unreduced protein extracts from single type HMWS-GS (Bx7, Dy12*, By18) and control isolines subjected to SE-HPLC (Peak 1 sector of the chromatogram shown). Letters refer to chromatograms and numbers to molar mass distribution. 1&d: +++; 2&b: By18; 3&a: Bx7 and 4&c: Dy12**

Table 1 *HPLC-based results obtained from single type HMW-GS (Bx7, Dy12*, By18) and other deletion lines.*

	%HMW	%LMW	H/L	%prot	%Dy	%By	%Dx	%Bx	%Ax	%UPP
Bx7	17.00	83.00	0.20	13.01				17.00		15.57
Dy12*	4.26	95.74	0.04	14.22	4.26					16.46
By18	12.30	87.71	0.14	13.79		12.29				14.80
1 17+18 2+12	30.69	69.32	0.44	10.68	3.45	3.68	8.94	9.56	5.07	54.80
2* 17+18 2+12	29.62	70.38	0.42	11.68	2.57	4.41	7.25	10.59	4.81	49.72
2* 17+18	21.65	78.35	0.28	11.80			4.95	11.28	5.43	31.36
17+18 2+12	25.62	74.38	0.34	9.73	3.06	4.63	7.14	10.81		49.85
17+18	18.78	81.22	0.23	12.18		5.77		13.01		24.12

All lines with single-type subunit showed a much shorter mixing time when compared with the full dotation HMW-GS control (Figure 4).

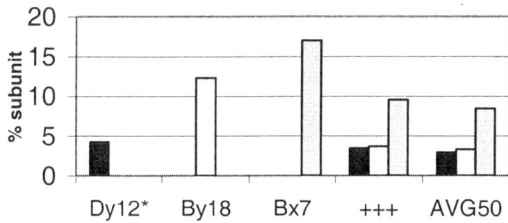

Figure 3 *Proportion of Dy, By, Bx individual subunits in total reduced glutenin extracts from single type HMWS-GS (Bx7, Dy12*, By18) and control (+++: 1, 17+18, 5+10) isolines subjected to RP-HPLC. Average values from a survey of 50 Australian cultivars is also shown (AVG50).*

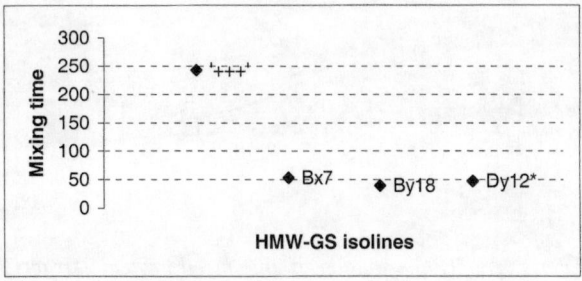

Figure 4 *Mixing behaviour (evaluated as mixing time using a 2g mixograph) of single type HMWS-GS (Bx7, Dy12*, By18) and control (+++: 1, 17+18, 5+10) isolines.*

4 CONCLUSION

Glutenin polymers made from single HMW-GS (Bx7, Dy12*, By18), either being x-type or y-type, do not show significant differences in size distribution or in the functional properties they confer to wheat doughs. When compared with the full HMW-GS dotation isoline (+ + +), all single HMW-GS lines showed significantly lower values for parameters associated with insoluble protein (%UPP) and a very poor mixing performance. A change in the balance between HMW-GS and LMW-GS, evaluated by the H/L ratio, is likely to be responsible for this behaviour.

References

1. P.I. Payne, K.G. Corfield and J.A. Blackman, *J. Sc. Food Agric.,* 1981, **32**, 51.
2. B. Margiotta, G. Colaprico, M. Aramini, S. Masci, and D. Lafiandra, *These proceedings*
3. G.J. Lawrence, F. MacRitchie. and C.W. Wrigley, *J. Cereal Sci.*, 1988, **7**, 109.
4. I.L Batey, R.B. Gupta, and F. MacRitchie, *Cereal Chem.*, 1991, **68**, 207.
5. R.B. Gupta, K. Khan, and F. MacRitchie, *J Cereal Sci.*, 1993, **18**, 3.
6. O.R. Larroque, F. Bekes, C.W. Wrigley and W.G. Rathmell, in: *Wheat Gluten,*
7. P.R. Shewry and A.S. Tatham eds., Royal Society of Chemistry special publications, Cambridge, 2000, p. 136.
8. F.Bekes, P.W. Gras, R.B.Gupta, D.R. Hickman and A.S. Tatham, *J. Cereal Sci.*, **19**, 3.
9. O.R. Larroque, M.B. Cuniberti, M.C. Gianibelli and F. Bekes, *Proceedings of the 52nd Australian Cereal Chemistry Conference*, Royal Australian Chemistry Institute, Melbourne, 2002, p. 214.

INFLUENCE OF GLUTENIN AND PUROINDOLINE COMPOSITION ON THE QUALITY OF BREAD WHEAT VARIETIES GROWN IN PORTUGAL

A.S. Bagulho[1], M.C. Muacho[1], J.M. Carrillo[2] and C. Brites[3]

[1] Estação Nacional de Melhoramento de Plantas, Elvas, Portugal
[2] Unidad de Genética, ETSIA-Agrónomos, Universidad Politécnica, Madrid, Spain
[3] Estação Agronómica Nacional, Oeiras, Portugal

1 INTRODUCTION

The end-use quality of bread wheat is related to a large extend, to the glutenin composition, particularly the high molecular weight glutenin subunits (HMW-GS)[1,2]. However, bread wheat dough is a multiphasic and multicomponent system and some other flour components also influence the technological quality of varieties.

In this study we analyse the variability of HMW glutenin and puroindoline composition in a collection of commercial bread wheat varieties, and its relationship with technological quality parameters. Puroindolines are two basic cysteine-rich proteins, with high affinities for binding lipids[3]. The genes coding for these proteins exhibit some mutations, which are associated with the expression of hard wheat texture[4], and affect milling and breadmaking quality traits[5].

Our main purposes are: to clarify the relationship between puroindoline mutations and dough functional properties; to compare the effects of HMW-glutenin and puroindoline allelic variants on wheat technological quality and to identify the level of significance of HMW glutenin allelic variants in separated soft and hard classes of bread wheat.

2 MATERIAL AND METHODS

2.1 Plant Material

A set of 49 commercial bread wheat varieties (*Triticum aestivum* L.) commonly grown in Portugal was sown in a complete randomised trial with three replicates in 1999/2000 and 2001/2002.

2.2 Methods

2.2.1 Glutenin and Puroindoline Composition. Glutenin allelic composition was characterised in a previous work[6]. Puroindoline allelic composition was obtained by DNA isolation[7] and PCR amplification of *pinA* and *pinB* coding regions with specific primers, that identify loss of puroindoline a (PinA) or point mutation in puroindoline b (PinB)[8,9,10].

2.2.2 *Quality Evaluation.* Hectolitre weight (Hect) (ISO 7971-2:1995) and thousand-kernel weight (TKW) (ISO 520:1977) were determined. Grain samples from each replicate were ground in a Cyclotec mill (0.5-mm screen) and used to determine SDS-sedimentation test (SDS)[11], kernel hardness (H) and protein content (Prot) by NIR, according to the AACC method 39-70A and the ICC Standard N° 159:1995.

Physical dough testing by micro-farinograph (ABS, DT, Stab, DS, QN) (ISO 5530-1:1997 adapted) and micro-alveograph (W, P, L, P/L) (ISO 5530-4:2000 adapted) were carried out on flour produced on Chopin CD1 mill. Bulked samples from the three replicates were used.

2.2.3 *Statistical analysis.* Variance analyse (General Linear Model Procedure, F-test on type III sum of squares, SAS, 1999) was first carried out with all cultivars (n=49) to study the effects of different *Glu-1* and *pin-D1* alleles on technological mean values. Subsequently, to remove the effect of the *Hardness* gene, we analysed hard (n=39) and soft (n=10) cultivars in separate classes. The differences between the pairs of alleles least-squares means were tested using the t-test (LSMEANS procedure).

3 RESULTS AND DISCUSSION

3.1 Allelic Variation at the *Glu-1, PinA-D1/PinB-D1 Loci*

The 49 cultivars exhibited some variability in HMW-GS, particularly in the subunits encoded at *Glu-B1*. At the *Glu-D1* locus, the subunits 5+10 (allele *d*) were present in the majority of lines (73 %).

Table 1 *Allelic variation at the puroindoline-encoding loci and mean hardness (H) of the varieties analysed*

Soft		Hard					
pinA-D1a pinB-D1a	H±SD †	pinA-D1b pinB-D1a	H±SD	pinA-D1a pinB-D1b	H±SD	pinA-D1a pinB-D1d	H±SD
Alva	32±8	Argueil	95±15	Abental	71±8	Armie	64±6
Amazonas	28±7	Búfalo	90±11	Almansor	102±11		
Arieiro	38±7	Cortex	73±6	Anza	76±8		
Centauro	30±8	Golia	96±13	Areal	71±12		
Dourado	34±9	Greina	88±7	Arpain	67±9		
Fiuza	38±7	Guadalupe	82±10	Arpège	78±13		
Libero	32±7	Lima	76±7	Atir	72±7		
Oderzo	31±10	Mercero	94±9	Bonpain	68±10		
Sever	28±5	Mira	96±8	Castan	77±10		
Sideral	28±7	Podenco	81±7	Côa	69±6		
		Roxo	91±3	Eufrates	68±11		
		Sorraia	91±4	Fitti	69±7		
		Tigre	91±6	Idra	57±7		
		Torero	84±7	Jordão	82±11		
		Trapio	90±10	Lancer	69±12		
		Trida	81±6	Orion	68±14		
		Tua	89±7	Pandas	75±11		
				Pinzon	73±9		
				Regain	67±12		
				Sagittario	77±8		
				Sarina	77±7		

† Mean and standard deviation were calculated with the values of the three replicates and two years (n=6).

The cultivars could be divided into four groups on the basis of PCR amplification of *pinA* and *pinB* coding regions (Table 1). Soft cultivars exhibited wild-type alleles (*pinA-D1a*, *pinB-D1a*) while the two prevalent mutations in hard cultivars were: absence of PinA (*pinA-D1b*) and the substitution Gly-46 to Ser-46 in PinB (*pinB-D1b*). Only one cultivar (Armie) had the Trp-44 to Arg-44 mutation in PinB (*pinB-D1d*).

3.2 Influence of HMW-GS and Puroindoline Mutations on Wheat Quality

3.3.1 Total Collection. The *pinA-D1/pinB-D1* alleles had significant effects on the grain hardness, water absorption, degree of softening, dough strength and tenacity results. Only the farinograph results were significantly affected by *Glu-A1* and *Glu-B1*. Contrarily to expectations[1], *Glu-D1* didn't significantly influence any quality parameter (Table 2).

Table 2 *Analysis of variance for hardness (H), alveograph (W, P) and farinograph (ABS, DT, Stab, DS, QN) values of the total collection (n= 49)*

		Mean Squares							
Source of Variation	df	H	W (10^{-4} J)	P (mm)	ABS (%)	DT (min)	Stab (min)	DS (FU)	QN
Glu-A1	2	16,8	3650	120	3,9	6,5	23,1	880*	5272*
Glu-B1	6	63,5	3096	391	10,2	28,6*	30,7**	482	2463
Glu-D1	1	2,7	80,9	1,4	2,2	13,2	0,9	0,3	21,5
pin-D1	3	4766**	11394*	3955**	92,2**	10,9	5,7	1418**	2814
Error	36	54,4	3238	683	8,3	10,3	8,5	269	1307

H=grain hardness, W=strength, P=tenacity, ABS=water absorption, DT=dough development time, Stab=stability, DS= degree of softening, QN=farinograph quality number.
*, ** significant at the 5% and 1% levels, respectively.

3.3.2 Hard Wheat Varieties. The hardness mutations only had significant effects on grain hardness and water absorption (Table 3). Grain hardness (H) was higher in varieties without PinA (Table 4) supporting previous studies[5,12]. The farinograph water absorption of these varieties was also higher than in varieties with the Gly-46 to Ser-46 mutation in PinB. Other studies[12] also found no effect on mixograph absorption.

The varieties with subunits 1 and 2* (*Glu-A1*) were associated with high farinograph quality number (QN) when compared with varieties with the null allele (Table 4). Varieties with subunit 2* also had lower values for degree of softening (DS). The effects of *Glu-B1* alleles were less significant and affected only the farinograph development time (DT). This result was supported by only one variety with subunits7+15.

Table 3 *Analysis of variance for hardness (H) and farinograph values (ABS, DT, DS, QN) of the hard varieties (n= 39) and for thousand-kernel weight (TKW), strength (W), water absorption values (ABS) of the soft varieties (n=10)*

		Hard Mean Squares						Soft Mean Squares		
Source of Variation	df	H	ABS (%)	DT (min)	DS (FU)	QN	df	TKW (g)	W (10^{-4} J)	ABS (%)
Glu-A1	2	4,2	3,8	9,1	899*	5351*	2	34,3**	20,1	2,2
Glu-B1	6	59,5	8,6	34,9*	238	2907	3	12,6*	2451**	13,6*
Glu-D1	1	5,8	2,7	26,5	155	257	1	48,0**	3946**	10,6
Pin-D1	2	515**	29,3*	3,1	144	1048	-	-	-	-
Error	27	65,7	8,8	10,9	252	1421	3	1,0	75,9	1,4

*, ** significant at the 5% and 1% levels, respectively

3.3.3 Soft Wheat Varieties. The influence of hardness mutations was eliminated. The three *Glu-1* loci significantly affected the thousand-kernel weight (TKW). *Glu-B1* and *Glu-D1* also had a significant effect on dough strength (Table 3). The varieties with subunits 7+9 (*Glu-B1*) and 5+10 (*Glu-D1*) have significantly higher strength than the other varieties (Table 5).

Table 4 *Significant difference of allele least squares means in hard wheat varieties*

Locus	Alleles compared	No. varieties	H	ABS (%)	DS (FU)	QN
Glu-A1	a vs c	8 vs 12	n.s.	n.s.	n.s.	48*
	b vs c	19 vs 12	n.s.	n.s.	-22*	54*
pin-D1	b† vs b	17 vs 21	13,5**	2,9*	n.s.	n.s.
	b† vs d	17 vs 1	20,7*	n.s.	n.s.	n.s.

Glu-A1: a=1, b=2*, c=N; *Glu-B1*: b=7+8, c=7+9, d= 6+8, i=17+18; *Glu-D1*: a=2+12, d=5+10.
† the first allele refers to *pinA-D1b*, and the second to *pinB-D1b* or *d*.
*, ** significant at the 5% and 1% levels, respectively, according the student's t-test; n.s. no significant.

Table 5 *Significant difference of allele least squares means in soft wheat varieties*

Locus	Alleles compared	No. varieties	TKW (g)	ABS (%)	W (10^{-4} J)
Glu-A1	b vs a	1 vs 5	9,4**	n.s.	n.s.
	b vs c	1 vs 4	8,6**	n.s.	n.s.
Glu-B1	c vs b	3 vs 5	n.s.	n.s.	55,8**
	c vs d	3 vs 1	-4,9*	n.s.	84,9**
	c vs i	3 vs 1	-4,7*	n.s.	82,1**
	d vs b	1 vs 5	5,4*	n.s.	n.s
	i vs b	1 vs 5	5,1*	6,7*	n.s.
Glu-D1	a vs d	1 vs 9	9,1**	n.s.	-81,9**

4 CONCLUSIONS

Puroindoline mutations explain variation in hardness values and are tightly related with water flour absorption. Gluten strength-related parameters are significantly influenced by puroindoline mutations and this effect seems to mask the glutenin effect.

The effect of *Glu-B1* and *Glu-D1* loci on rheological tests was only evident in soft wheat varieties, so the application of the *Glu-1* score in breeding selection for gluten strength seems to be more effective in soft bread wheat germplasm. In hard bread wheat breeding programs, puroindoline composition could be used in addition to the *Glu-1* score.

References

1. P.I. Payne, M.A. Nightingale, A.F. Krattiger and L.M. Holt,. *J. Sci. Food Agric.*, 1987, **40**, 51
2. C. Brites, A.S. Bagulho, M. Rodriguez-Quijano and J.M. Carrillo, in *7th Workshop Gluten 2000*, eds P.R. Shewry and A.S. Tatham, Royal Society of Chemistry, Cambridge, 2000, p.55
3. L. Dubreil, J. Compoint and D. Marion, *J. Agric.Food Chem.*, 1997, **45**, 108
4. C.F. Morris, M. Lillemo, M.C. Simeone, M.J. Giroux, S.L. Babb, K. Kidwell, *Crop Sci.*, 2001, **41**, 218
5. J .M. Martin, R.C. Frohberg, C.F. Morris, L.E. Talbert and M.J. Giroux, *Crop Sci.*, 2001, **41**, 228
6. G. Igrejas, H. Guedes-Pinto, V. Carnide and G. Branlard, *Plant Breed.*, 1999, **118**, 297
7. S.L. Dellaporta, J. Wood and J.B. Hicks, *Plant Mol. Biol. Rep.*, 1983, **1**, 19
8. M. Gautier, M. Aleman, A. Guirao, D. Marion and P. Joudrier, *Plant Mol. Biology*, 1994, **25**, 43
9. M.J. Giroux and C.F. Morris, *Theor. Appl. Genet.*, 1997, **95**, 857
10. V. Corona, L. Gazza, R. Zanier and N.E. Pogna, *J. Genet. & Breed.*, 2001, **55**, 187
11. J.W. Dick and J.S. Quick, *Cereal Chem.*, 1983, **60**, 315
12. M.J. Giroux, L. Talbert, D.K. Habernicht, S. Lanning, A. Hemphill and J.M. Martin, *Crop Sci.*, 2000, **40**, 370

Acknowledgements: Financial support for this research was provided by a grant (SFRH / BD / 1098 / 2000) from Fundação para a Ciência e a Tecnologia

VARIABILITY AND GENETIC DIVERSITY FOR ENDOSPERM STORAGE PROTEINS IN SPANISH SPELT WHEAT

L. Caballero, L.M. Martín, and J.B. Alvarez

Departamento de Genética, Escuela Técnica Superior de Ingenieros Agrónomos y de Montes, Universidad de Córdoba, Apdo. 3048, E-14080 Córdoba, Spain.

1 INTRODUCTION

The evolution of agriculture in the last fifty years has produced an erosion of the genetic base of wheats. The study of the storage proteins has shown that the frequency of some allelic variants is higher than others. This low variability is a direct consequence of plant breeding programs, in which a limit number of sources has been used, with strong pressure to replace traditional varieties, with new cultivars. Consequently, a search has been made for new alleles in local varieties in Germplasm Banks, including hulled wheats such as spelt (2n=6x=48; AABBDD; *Triticum aestivum* ssp. *spelta* L. em. Thell.). This wheat was extensively grown in Spain before the 20th century, but is now endangered, and only survives in marginal farming areas of Asturias (North of Spain), where traditional farming systems still survive.[1]

The increasing interest in low-input system, driven by ecological and economic factors, has led to a growing interest in increasing genetic variability. Sustainable agriculture and health food products have gained increasing popularity that has led to a renewed interest in hulled species such as spelt wheat.[2,3] Furthermore, spelt has been described as a crop that grows on poor soils in mountain regions, is tolerant to cold and excess humidity and resistant to several diseases.[4,5] In addition, this species could be a rich sources of useful genes, such as genes for endosperm storage proteins (glutenins and gliadins), which are directly related to breadmaking quality.

The aim of this work was to determine the genetic diversity of a wide collection of spelt wheat of Spanish origin using the polymorphism of endosperm storage proteins.

2 MATERIAL AND METHODS

Three hundred and thirty three spelt wheat accessions of Spanish origin were analysed. Geographic distribution data supplied by Swiss Federal Research Station for Agroecology and Agriculture allowed us to group these accessions in fifty locations or populations.

Gliadins were extracted with 1.5M dimethylformamide and fractionated by A-PAGE (T= 8.5; C= 2.67). Reduced and alkylated glutenin subunits were separated by electrophoresis in vertical SDS-PAGE slabs in a discontinuous Tris-HCl- SDS buffer system (pH: 6.8/8.8) and

(T=8 and 10, C=1.28) with and without 4M urea.

The frequencies of glutenin and gliadin patterns were calculated at the population and collection levels. Proportion of polymorphic loci (P), average number of alleles per locus (A) and effective number of alleles per locus (Ne), were used to evaluate the genetic diversity within populations. The gene diversity over all populations (Ht), together with the average allelic gene diversity within (Hs) and among (Dst) populations, was calculated. The relative magnitude of genetic differentiation among populations, Gst, was estimated as Dst/Ht.

3 RESULTS AND DISCUSSION

3.1 Variability in endosperm storage proteins

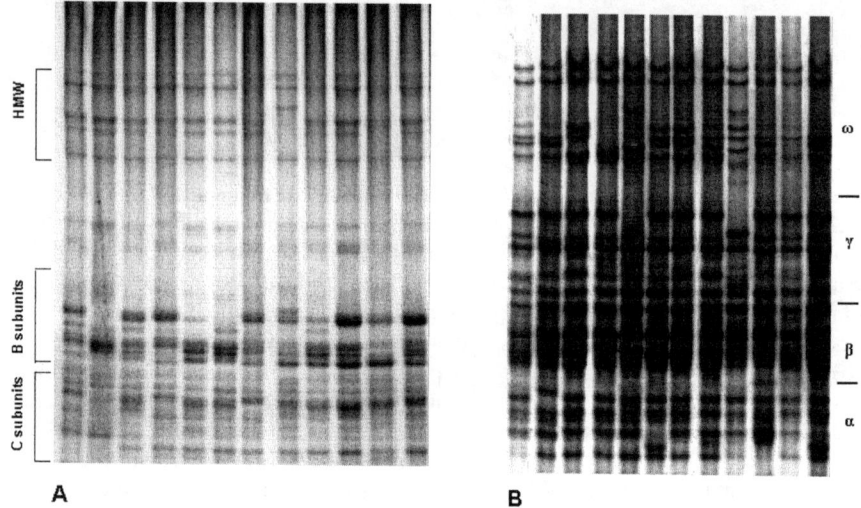

Figure 1. *Samples of variation in spelt.* **A**, *HMW and LMW glutenin subunits in SDS-PAGE, and* **B**, *gliadins in A-PAGE.*

A high level of variation was found for all loci evaluated. A total of 17 allelic variants were found for the HMW glutenin subunits, three variants for the *Glu-A1* loci, six for the *Glu-B1* loci and eight for the *Glu-D1* loci, of which six had not been described previously. With all these variants, twenty different combinations were detected, with the most frequent combination being 1, 13+16, 2+12, which was present in 223 of the accessions. In contrast, nineteen combinations appeared in only one accession.

For the LMW glutenin subunits, 50 different patterns were detected for the B-LMW group and 23 for the C-LMW group, with 77 different combinations. The combination of the patterns B1 and C2 was the most frequent appearing in 30 accessions, followed by the combination B2 and C3 that was detected in 14 accessions. Thirty-two combinations were present only in single accessions.

A higher level of polymorphism was found for the gliadins, with 53 different patterns for the ω-gliadins, 37 for the γ-gliadins and 17 and 13 for the β- and α-gliadins, respectively. The combination of these patterns formed 128 and 39 different combinations for *Gli-1* and *Gli-2* loci, respectively, with the most frequent combination being ω1-γ3 for the *Gli-1* loci appearing in 27 accessions. Seventy-four combinations were unique for one accession. The combination formed by the pattern β1 and α1 for the *Gli-2* loci was detected in 174 accessions while 13 combinations appeared in one accession.

In order to assess the distribution of variants in different populations, the criterion of Marshall and Brown was used,[6] who classified the variants as common when they have a frequency greater than 5% and rare with a frequency lower of 5%. Furthermore, these variants can be of wide distribution in many populations or local distribution occurring in one or a few populations. According to this classification, 38 of these variants were classified as common and wide distribution, 68 variants were rare with a wide distribution, and 84 were very rare (frequency lower than 1%), 54 of which had local distributions.

3.2 Genetic diversity

The mean value of A (number of alleles per locus) was 3.155 (range between 1.333 and 5.000), which was higher than that detected by Nevo and Beiles,[7] who studied 42 loci in *T. turgidum* ssp. *dicoccoides* from Israel and Turkey and also higher than the value found by Hegde et al.,[8] when they determined the genetic diversity for 10 isozymes in 35 populations of diploid goat grass species, which were 1.50 and 1.22, respectively. The same occurred with the rest of parameters that were higher than for the last species. These populations had high levels of polymorphism, with the mean value for the percentage of polymorphic loci (P) being 0.76 (range: 0.33-1.00). In the last species, this value was 0.22 and 0.17, respectively.

Table 1.- *Differentiation of seed storage protein diversity within and between 50 spelt populations.*

Locus	Allele	Ht	Hs	Dst	Gst
Glu-A1	3	0.237	0.166	0.071	0.300
Glu-B1	6	0.224	0.147	0.077	0.344
Glu-D1	8	0.200	0.166	0.034	0.170
B-LMW	50	0.955	0.748	0.207	0.217
C-LMW	23	0.829	0.646	0.183	0.221
ω-gliadins	52	0.959	0.755	0.204	0.213
γ-gliadins	37	0.938	0.745	0.193	0.206
β-gliadins	18	0.620	0.499	0.121	0.195
α-gliadins	12	0.294	0.230	0.064	0.218
Mean		0.584	0.455	0.128	0.231

Ht, Hs, Dst and Gst, see explain in the text.

Further characterisation of the diversity for all loci in spelt is given in Table 1. The distribution among and between populations for all variants was calculated. The average of total genetic diversity, Ht, across of all loci evaluated for the fifty populations was 0.584 (range from 0.200 for *Glu-D1* loci to 0.959 for ω-gliadins). This parameter can be divided into two components, the genetic diversity within and between populations, Hs and Dst, respectively. The average relative differentiation among populations was Gst = 0.231 (range from 0.170 for *Glu-D1* to 0.344 for *Glu-B1* loci). For all loci the greatest genetic diversity was present in populations with 76.9% of the total genetic diversity, the variation among populations being 23.1%. This could be related to fixation effects of some variants by genetic drift.

4 CONCLUSIONS

Data from the present work suggest that the seed-storage proteins of spelt wheat are not associated with adaptiveness. Their distribution could be understood in the cultural context of agricultural society, which could have selected materials empirically with compositions suitable for food uses. This is favoured for the fact that spelt wheat is an autogamous species that will easily fix one or more alleles for each locus.

Because of the antiquity of the analysed collection (1939) and the great homogeneity within them, the loss of variability is probably a consequence of the effects of genetic drift developed before it was collected. This loss is typical of neutral traits such as storage proteins of cereal crops, could have increased subsequently due to the decrease in the cultivated area of this crop in Spain since the end of 1960's. Consequently, the saving of these accessions, together with others stored in Germplasm Banks, is fundamental in maintaining diversity for the plant breeding of this crop, and in particular for quality improvement.

For this reason, these materials are being recovered and multiplied to maintain their variability. Analysis of the different agronomic characters, including their quality, must then be carried out. This variability could be used for the improvement of the spelt wheat *per se* or to increase the genetic base of the durum and bread wheats by introgression.

References

1   L. Peña-Chocarro and L. Zapata-Peña, in *Triticeae III* ed. A.A. Jaradat, Science Publisher Inc., 1998, p. 45
2   E.S.M. Abdel-Aal, P. Huci and F.W. Sosulski, *Cereal Chem.,* 1995, **72**, 621.
3   G.S. Ranhotra, J.A. Gelroth, B.K. Glaser and K.J. Lorenz, *Lebensm.-Wiss. Technol.*, 1995, **28**, 118.
4   A.B. Damania, L. Pecetti and S. Jnan, in *Wheat genetic resources: Meeting diversity needs* eds. J.P. Srivastava and A.B. Damania, Wiley, Chichester, 1990, p. 57.
5   G.H.J. Kema, *Euphytica*, 1992, **63**, 207.
6   D.R. Marshall and A.H.D. Brown, in *Crop genetic resources for today and tomorrow*, eds.O.H. Frankel and J.G. Hawkes, Cambridge University Press, Cambridge, 1975, p. 53.
7   E. Nevo, and A. Beiles, *Theor. Appl. Genet.*, 1989, **77**, 421.
8   S.G. Hegde, J. Valkoun and J.G. Waines, *Theor. Appl. Genet.*, 2000, **101**, 309.

Acknowledgements

This research was supported by grant AGL2001-2419-CO2-02 from the Spanish Ministry of Science and Technology and the European Regional Development Fund from the EU. The senior author is grateful to *Ramón y Cajal+ Programme of the Spanish Ministry of Science and Technology for the financial support.

BIOCHEMICAL AND TECHNOLOGICAL INDICATORS OF PASTA QUALITY

M. Carcea[1], N. Guerrieri[2], E. Marconi[3], S. Salvatorelli[1], E. Franchi[2] and M.C. Messia[3]

[1]Istituto Nazionale di Ricerca per gli Alimenti e la Nutrizione, Via Ardeatina 546, 00178 Roma, Italy, carcea@inran.it
[2]DISMA, Università di Milano, Via Celoria 2, 20133 Milano, Italy.
[3]DISTAAM, Università del Molise, Via De Sanctis, 86100 Campobasso, Italy

1 INTRODUCTION

All over the world, durum wheat is the preferred material for pasta making thanks to its milling performance, colour and gluten rheological properties. It is well known to cereal chemists and biochemists that the ability of a durum wheat cultivar to give a determined cooking quality could be dependent on one or more of its components.

However, the problem of correlating component composition and structure to pasta cooking performance is not easy due to a series of problems referring to the number of potentially involved components, their high molecular weight and limited solubility, the difficulty to separate or isolate components in a relatively pure state without altering them, the interactions between different components and the influence on the components themselves of the different technological steps involved in pasta manufacturing such as milling, dough mixing, extruding and drying. Proteins and gluten in particular have been the most studied amongst durum wheat components; however, regardless of the work done, it still remains to be elucidated what the structural features of gluten proteins responsible for pasta cooking quality are.

The present work is an attempt to contribute to clarifying the correlations between biochemical features of gluten proteins and cooking performance by applying new analytical techniques which involve a stepwise controlled reduction of the gluten complex by means of reducing agents such as DTT. The reduced samples are then analysed by SDS PAGE.

2 MATERIALS AND METHODS

Dry spaghetti was prepared in a pilot plant according to a low (maximum temperature 60°C) and a high (maximum temperature 90°C) drying diagram from semolinas belonging to 5 different cultivars which are amongst the most cultivated in Italy, namely Colosseo, Creso, Duilio, Ofanto and Simeto. Since results of previous studies showed that protein quality and quantity can act independently on cooking quality on the basis of the drying technology adopted, samples were selected with very similar gluten contents but different gluten quality (Tab.1). Raw materials were characterised by a number of quality tests and spaghetti cooking quality was determined according to a sensorial and a chemical test. The list of methods is as follows:

Gluten Quantity and Quality: ICC Method n° 158 [1].

Semolina Colour or Yellow Index : Minolta Chroma Meter-200. b* value used as yellow index.

Cooking quality was assessed by cooking 100g of spaghetti in 1L of unsalted boiling tap water. Optimum cooking time was indicated when the white core of the pasta disappeared when squeezed between two glass plates (Approved Method 66-50, AACC, 2000) [2].

Sensorial Cooking Quality: Trained panel of three assessors according to Cubadda, R, 1988 [3].

Total Organic Matter: ICC Method n°153 [1].

Furosine: HPLC procedure of Resmini et al. (1990) [4].

Gliadins in Semolina: determined after sequential extraction according to Capelli et al., 1998 [5].

Solubilised proteins: 20 mg of sample were solubilised in 1ml of 0.2M Tris buffer pH8.5 containing 8M urea, for 30 min at 25°C with magnetic stirring. The solubilised proteins were centrifuged at 8000 x g and quantified with the modified Bradford method (Eynard et al., 1994) [6].

Protein reduction: stepwise controlled reduction of the gluten complex was performed by means of reducing agents such as DTT. The reduced samples were then analysed by SDS PAGE.

3 RESULTS AND DISCUSSION

As regards quality parameters of semolina relative to proteins, Table 1 clearly shows that the selected cultivars possessed very similar dry gluten contents (except for Simeto) but different gluten strength with the cvs Colosseo and Simeto reaching the highest values (56 and 77 respectively). The yellow index didn't allow differentiation between the cvs. The studied cvs also possessed different amounts of gliadins in their gluten ranging from 9% for Colosseo to 2.6% for Ofanto. Creso, Duilio and Simeto showed more similar contents (4.6, 4.4 and 5.4, respectively).

Table 1 *Some quality parameters of durum wheat semolina*

Sample	Dry gluten (% d.m.)	Gluten Index	Yellow index b*
Creso	10.6	49	17
Colosseo	10.8	56	18
Duilio	10.7	45	19
Ofanto	10.8	43	20
Simeto	9.7	77	20

Cooking quality as judged by the sensorial method shows that, generally speaking, HT samples were more satisfactory than LT samples with HT Colosseo reaching the highest score. On the contrary, values of TOM were similar for HT and LT in most cases and were within the range indicated as being typical of good cooking quality (2.1 to 1.4 g/100g). However, HT Colosseo pasta, which obtained the highest sensorial judgement concomitantly exhibited the lowest TOM score (1.42 g starch/100 g).

Furosine content was strongly influenced by the heat intensity in the drying process. Furosine is an unnatural amino acid formed from the acid hydrolysis of Amadori compound (ε-deoxy-fructosyl-lysine) produced during the early stages of the Maillard reaction. High values were consistently found in all HT samples and much lower values in LT samples (Tab.2).

Table 2 *Cooking quality and furosine content of pasta samples manufactured at low (LT) and high (HT) temperature*

PASTA SAMPLES	SENSORIAL METHOD					TOM (g starch/ 100 g)	FUROSINE (mg/100 g protein)
	Stickiness	Bulkiness	Firmness	Total Score	Judgement		
Creso LT	58	40	60	40	unsatisfactory	1.46	126.5
Creso HT	60	53	48	54	sufficient	1.46	556.3
Colosseo LT	58	35	30	41	unsatisfactory	1.52	124.6
Colosseo HT	65	58	58	60	fair	1.42	582.4
Duilio LT	60	50	65	59	sufficient	1.49	118.7
Duilio HT	60	55	50	55	sufficient	1.49	563.7
Ofanto LT	46	33	34	38	poor	1.60	174.3
Ofanto HT	10	10	0	10	very poor	2.35	596.2
Simeto LT	50	45	60	52	sufficient	1.49	197.4
Simeto HT	60	43	63	55	sufficient	1.42	649.5

The protein solubility in pasta in the presence of SDS and 8M urea decreased from LT to HT samples indicating a polymerisation of the gluten proteins due to heat treatment. The LT process reduced protein solubility by about 20% and the HT process by 85% (Fig. 2). SDS-PAGE electrophoresis showed a broad band that cannot be reduced by the reducing agent in the HT samples (on top of the gel); this band was also present in the LT samples but was less intense.

The stepwise reduction of semolina, LT and HT pasta belonging to the five different cvs provided information on the disulphide bridge accessibility to the reducing agent (DTT). The cv Ofanto was the most accessible, semolina and LT pasta being completely reduced by 5mM DTT. For the other samples, reduction was more gradual. For the cv Duilio, the behaviour of LT pasta and semolina were similar. However, for the cvs Simeto, Colosseo and Creso, the LT samples reduction required less reducing agent. The HT samples cannot be reduced completely but the accessible disulphide bridges require only 2 or 5mM DTT. Only the HT Colosseo pasta, which reached the highest score of all in cooking quality (Tab. 2), showed a different behaviour and required 15mM DTT.

Figure 2 *Percent of solubilised proteins in pasta dried at low (LT) and high temperature (HT)*

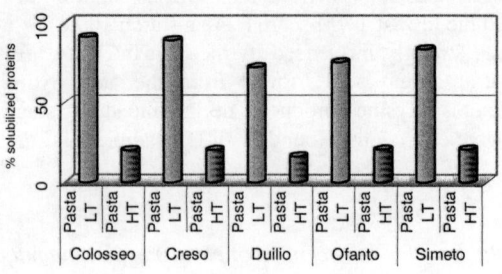

Figure 3 *Percent of protein reduction in pasta dried at low (LT) and high temperature (HT)*

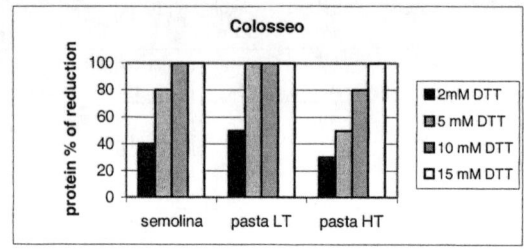

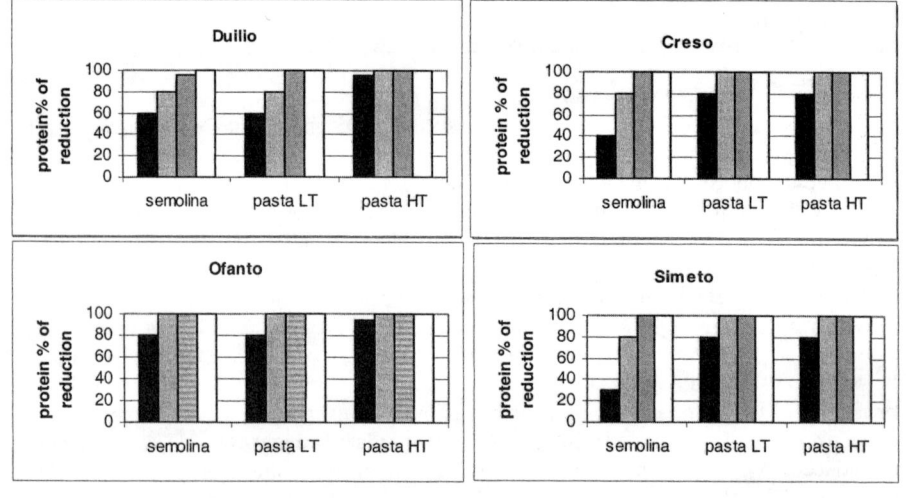

References

1 AACC, Approved Methods of the AACC, 10th ed., The Association, St.Paul, MN, USA.
2 ICC, Standard Methods of the ICC, 1995. The Association, Vienna.
3 R. Cubadda, In *Durum wheat: Chemistry and Technology*, G. Fabriani and C. Lintas, eds. Am. Assoc. Cereal Chem., St.Paul, MN., 1988, p. 217.
4 P. Resmini, L. Pellegrino, G. Battelli, *Ital. J. Food Sci.*, 1990, **2**, 173.
5 L. Capelli, F. Forlani, F. Perini, N. Guerrieri, *Electrophoresis*, 1998, **19**, 1.
6 L.Eynard, N. Guerrieri, P. Cerletti, *Cereal Chem.*, 1994, **71**, 434.

GENETIC VARIATION OF THE STORAGE PROTEINS IN THE STRUCTURAL MUTANT FORMS OF *T. AESTIVUM* L.

T. Dekova[1], S. Georgiev[1], K. Gecheff[2]

[1]Department of Genetics, Faculty of Biology, Sofia University, Bul. Dr. Tsankov 8, Sofia 1164, Bulgaria
[2]AgroBioInstitute, Bul. Dr. Tsankov 8, Sofia 1164, Bulgaria

1 INTRODUCTION

The majority of seed proteins are stored in the starchy endosperm in the form of prolamins; these are unique to cereal grains and account for over half of the total seed nitrogen. The prolamins of wheat are highly polymorphic polypeptide mixtures of >50 components with Mr. values ranging from 30 to 90 kDa (Payne, 1987). The wheat prolamins are divided into two groups on the basis of function: the glutenins and gliadins. These together confer the properties of elasticity (strength) and extensibility (viscosity). The gliadins are monomeric molecules (30 - 75 kDa) divided into several classes (α-, β- and ω-gliadins). In contrast, the glutenins form large polymeric structures as a result of intermolecular disulphide bonds. The glutenins are divided into a low-molecular-weight glutenin subunit (LMW) group and a high-molecular-weight glutenin subunit (HMW) group. The HMW (65 - 90 kDa), which represent approximately 0.5% of the total seed dry weight, have been studied extensively because of their effect on elasticity and hence the bread-making quality of wheat dough.

The HMW genes (Glu-1) are located on the long arms of homologous chromosomes 1A, 1B and 1D (Payne *et al.*, 1982). Genes encoding LMW (Gli-1), γ-gliadins and ω-gliadins are grouped at loci on the short arms of chromosomes 1A, 1B and 1D. The genes for α-gliadins (Gli-2) are found on the short arms of chromosomes 6A, 6B and 6D. Based on their electrophoretic mobility and isoelectric points, the HMW proteins are divided into high Mr x-type and low Mr y-type subunits. According to D`Òvidio *et al.* (1995), six HMW genes are present in hexaploid wheat but, because of gene silencing, only three of these (coding for subunits 1Bx, 1Dx and 1Dy) are always expressed in all cultivars The 1Ax and 1By are present in some cultivars but not always expressed and no cultivar contains 1Ay because the gene is always silent. According to Harberd *et al.* (1987), the gene silencing of 1Ay could be caused by the presence of a transposon-like insertion in the coding region.

2 METHOD AND RESULTS

2.1 Genotypes and 1D SDS-PAGE

The main aim of the present investigation was to establish the genetic variation of high- and low-molecular proteins by 1D SDS-PAGE of 27 sphaerococcum mutant forms obtained by Georgiev (1982) and their control forms. We also investigated the nulli-tetrasomic lines of *T. aestivum* cv. Chinese Spring. Extraction of total proteins and 1D SDS-PAGE in 10% polyacrylamide were performed according to Laemlli (1970), modified by Payne *et al.* (1980). Protein extraction was carried out according to Privalsky *et al.* (1983) with a few modifications (Dekova *et al.*, 1992) which refer mainly to the separate components of the buffer system. Gels were stained in a mixture of methanol : acetic acid : water (5:1:5) containing 0.25 % Coomassie blue R250 and destained in methanol : acetic acid : water (5:1:5) overnight. High molecular weight standard proteins (Pharmacia) (200 to 14 kDa) were used. HMW glutenins were designated according to Payne and Lawrence (1983) and LMW glutenins according to Nieto-Taladriz *et al.* (1997). The flour protein percentage was determined using a standard micro-Kjeldahl method.

2.2 Assessment the variation of the glutenin subunits

The HMW glutenin phenotypes of the mutant and control forms of *T. aestivum* are shown in Fig. 1, 2, 3 and 4. Electrophoretical analyses by 1D SDS-PAGE of wheat mutant forms revealed the presence of 30-34 bands (Fig. 4). The results illustrated by the protein profiles of different mutant forms of wheat (Fig. 1 and 2) indicate that the lower MW region (40 to 30 kDa) was the least variable, whilst the high and medium MW region (90 and above to 60 kDa) had the greatest protein subunit variability. Differences were observed more frequently in the 60 kDa and higher zone than in the 90 kDa zone (Fig. 1 and 2) for the sphaerococcum mutant forms 49Al sph, 49Al aest., 169/613, 6512 sph, 6512c and 613 sph. The mutant forms 49Al sph, 49Al aest. and 169/613 possessed only one additional band in the 100 kDa zone (Fig. 1, lanes 2, 4 and 5 and Fig. 2, lanes 2, 3 and 6). Eight to fifteen HMW glutenin subunits of different mobility were identified (Fig. 1, 2 and 4). The Glu-B1 locus showed much more allelic variation than the other loci. At the Glu-A1 locus the subunit 1, which according Piergovanni and Blanco (1999) is related to the good quality of bread wheat, was observed only in sphaerococcum mutant forms 49Al, 169/613 (Fig. 1).

It must also be noted that SDS-PAGE of total protein probably masks part of the genetic variation in mutant forms, probably because protein subunits of different pH and amino acid composition but with the same MW would migrate to the same place in the gel. The tested mutant forms were also analysed for protein content. A higher protein content characterised the mutant forms 613sph and 6512 (mean value: 19-24.9%). It should be noted that the sphaerococcum mutant forms MT47 of hexaploid Triticale (Georgiev, 1982) and 613 sph, 6512 sph, 49Al sph and 169/613 possessed an inserted Ac/Ds-like controlling element (Georgiev *et al.*, 2000).

Fig. 4 represents the electrophoretical patterns of mutant forms MT47, 613sph, 169/613sph, 6512sph and 6512/126sph after 1D SDS-PAGE (Panel A) and 2M urea treatment (Panel B). Better separation after preliminary 2M urea incubation was observed in the 70 kDa zone, whilst some of the SDS-PAGE established bands were missing in the low- and high-molecular weight zones of 30-40 kDa and >100 kDa. Since the bands appear only after reduction with ME, it is likely that they occur as disulphide-linked aggregates in the endosperm of mutant and control forms.

The results of nullisomic-tetrasomic lines of "Chinese Spring" by 1D SDS-PAGE revealed presence of a new band in the 60 kDa zone (Fig. 3). This component is always present in the compensating nullisomic-tetrasomics of the homoeologous group 4 (4A and 4D) which is in accordance with previous report of Gupta et al. (1991). For example, the band between 60/70 kDa was absent from CS N4DT4B (Fig. 3, lane 9) and showed an increased staining intensity in CS N4AT4D (lane 8) stock carrying four doses of chromosome 4D as expected. Thus, it is concluded that this band is controlled by genes on chromosome 4D. The most interesting findings of the present study are the lack of a clearly expressed protein band in the 65/67 kDa zone in CSN4DT4B and the presence of a new one on chromosome 4B. As shown in Fig. 3, considerable differences in staining intensity of the 65/67 kDa bands were observed. Analysis of the stocks carrying four doses of chromosome 4B (CSN4AT4B and N4DT4B, lanes 7 and 9), showed increased staining intensity in comparison with the stocks carrying four doses of chromosome 4D(CSN4AT4D) with a faint staining intensity (lane 8). It should be noted that the same band of 65/67 kDa with a faint staining intensity was assigned in CSN5AT5D and CSN5DT5B (pattern not shown). Taking into account the established differences in staining intensity of the 65/67 kDa band in compensating nulli-tetrasomic lines, it may be concluded that this protein is controlled by gene(s) on the 4B chromosome. This is most probably a protein band corresponding to β-amylase (high-molecular-weight albumins) about which Gupta et al. (1991) also found out is controlled by genes localized on the 4DL, 4AL and 5AL chromosome arms. The band between 60/70 kDa is more intense in lines carrying four doses of chromosome 3B (CSN3AT3B, CSN3DT3B) (Fig. 3, lanes 1 and 6) than CS N3BT3A, CS N3BT3D and CS N3DT3A (lanes 3, 4 and 5). Thus, these bands are either controlled by chromosome 3B, or chromosome 3B showed dose effects.

3 CONCLUSION

The genetic variation of the HMW proteins in sphaerococcum mutant forms is of importance to elucidate the nature of these structural mutations, as well as the fundamental role of these proteins for the improvement of the grain quality of cereals. Analyses of aneuploid stocks of "Chinese spring" wheat showed that some of the seed protein bands separated by SDS-PAGE (Mr 65-70kDa) are controlled by genes on chromosomes 4D, 4A, 5A, 5D and a new chromosome 4B.

References

1. T. Dekova, T. Dimitrov, S. Georgiev, V. Petkova, *Genet. Breed.*, 1992, **25**, 306
2. R. D'Ovidio, S. Masci, E. Porcedu, *Theor. Appl. Genet.*, 1995, **91**, 189
3. S. Georgiev, WIS, 1982, **55**, 32
4. S. Georgiev, T. Dekova, I. Atanassov, Z. Angelova, V. Mirkova, A. Dimitrova, L. Stoilov, *Biotechnol. & Biotechn.*, EQ, 2000, **1**, 25
5. R. B. Gupta, K. W. Shepherd, F. MacRitchie, *J. Cereal Sci.*, 1991, **13**, 221
6. N. P. Harberd, R. B. Flavell, R. D. Thompson, *Mol. Gen. Genet.*, 1987, **209**, 326
7. U. K. Laemlli, *Nature*, 1970, **227**, 680
8. M. T. Nieto-Taladriz, M. Ruiz, M. C. Martinez, J. F. Vazquez, J. M. Carrillo, *Theor. Appl. Genet.*, 1997, **95**, 1155
9. P. I. Payne, C. N. Law, E. E. Mudd, ***Theor. Appl. Genet.***, 1980, **58**, 113
10. P. I. Payne, L. M. Holt, A. J. Worland, C. N. Law, *Theor. Appl. Genet.*, 1982, **63**, 129
11. P. I. Payne, G. J. Lawrence, *Cereal Res. Commun.*, 1983, **11**, 29
12. P. I. Payne, *Ann. Rev. Plant Physiol.* 1987, **38**, 141

13. A. R. Piergiovanni, A. Blanco, *Cereal Res. Commun.*, 1999, **27**, 205
14. M. L. Privalsky, L. Sealy, J. M. Bishop, J. P. McGrath, A. D. Levinson, *Cell*, 1983, **32**, 1257

GENETIC VARIATION OF THE STORAGE PROTEINS IN THE SPHAEROCOCCUM MUTANT FORMS OF HEXAPLOID WHEAT

T. Dekova, Y. Yordanov, S. Georgiev

Department of Genetics, Faculty of Biology, Sofia University, Bul. Dr. Tsankov 8, Sofia 1164, Bulgaria

1 INTRODUCTION

The study of the main seed storage proteins (glutenins and gliadins) is essential because of their importance in bread-making quality (Payne *et al.*, 1983) and their use in varietal identification (Payne *et al.*, 1981). The storage proteins of the wheat endosperm are usually classified into two major groups: gliadin, consisting of a complex mixture of single polypeptides, and glutenin, consisting of protein aggregates up to a molecular weight of 20×10^6 formed by polypeptides that are held together through disulphide bonds.

The glutenins are divided into a low-molecular-weight glutenin subunit (LMW) group and a high-molecular-weight glutenin subunit (HMW) group. Based on their mobilities on SDS-PAGE, HMW are divided into x-(83 – 88 kDa) and y-type (67 – 74 kDa) subunits, whereas three groups of LMW can be identified: B (42 – 51 kDa), C (30 – 40 kDa) and D (55 – 70 kDa) subunits. High-molecular-weight (HMW) glutenin subunits are encoded by genes located in Glu-1 loci on the long arms of group 1 chromosomes.

Genes encoding LMW and gliadins are grouped at loci on the short arms of chromosomes 1A, 1B and 1D and the short arms of chromosomes 6A, 6B and 6D. The B-LMW subunits coded by genes at the Glu-3 loci have been linked only with ω and γ-gliadins coded at the Gli-1 loci on the short arms of chromosomes 1A and 1B (Payne *et al.*, 1984; Pogna *et al.*, 1990). According to D'Ovidio *et al.* (1995), six HMW genes are present in hexaploid wheat but, because of gene silencing, only three of these (coding for subunits 1Bx, 1Dx and 1Dy) are always expressed in all cultivars. The 1Ax and 1By are present in some cultivars but not always expressed, and no cultivar contains 1Ay because the gene is always silent.

Recently, different transposable elements, retrotransposons and CACTA elements, have been shown to be involved in specific interactions with certain glutenin and gliadin genes in *Triticeae* genomes (Wicker *et al.*, 2003).

2 METHOD AND RESULTS

2.1 SDS-PAGE analysis of HMW subunits in sphaerococcum mutant forms 613 and 6512 and their derivatives

The aim of the present study was to estimate the variation of the HMW glutenin subunits in *sphaerococcum* mutant forms 613 and 6512 and their derivatives. Among the progeny of these mutant forms, some plants showed typical *sphaerococcum* phenotype with dense ears, compact and loose ears (*spelta* type) as well as mutants bearing only the *sphaerococcum* symptoms (Fig. 4 and 5). Several types of *sphaerococcum* mutant forms were used in our experiments: a) typical *sphaerococcum* mutant forms 613 and 6512, b) derivatives of mutant forms 613 and 6512 with reverted phenotype – type aestivum; 2k x 613 with dense ear; 613 *subcompactum*; 613 semi bearded and 613 type spelta and c) *sphaerococcum* mutant forms with *compactum* ears, dense ears and spelta type (Fig. 4 and 5). Variegation of the *sphaerococcum* phenotype could be detected in ear glume and kernel. These kernels were used for further experiments – SDS-PAGE. Mutant forms 613 and 6512 have been obtained after EMS treatment of *T. aestivum* cv. S. ranozreika 2 (Georgiev, 1976). In these mutant forms, Georgiev *et al.* (2000) established the presence of controlling elements such as Ac/Ds. Mutant forms 2k x 613 with dense ear type aestivum as well as 2k x 613 spelta type with bearded ear and 49L/13 were obtained after crossing the mutant forms 613 and 6512 with the control form of *T. aestivum* cv. S. ranozreika 2. Extraction of total proteins and 1D SDS-PAGE in 10% polyacrylamide were performed according to Laemlli (1970), modified by Payne *et al.* (1981). Protein extraction was carried out according to Privalsky *et al.* (1983) with a few modifications (Dekova *et al.*, 1992) which refer mainly to the separate components of the buffer system. Gels were stained in a mixture of methanol : acetic acid : water (5:1:5) containing 0.25% of Coomassie blue R250 and destained in methanol : acetic acid : water (5:1:5) overnight. The flour protein percentage was determined using a standard micro-Kjeldahl method.

2.2 Variability in HMW-glutenin subunits of sphaerococcum mutant forms

The SDS-PAGE electrophoretic patterns of HMW-glutenin of endosperm proteins from individual grains of sphaerococcum mutant and control forms are shown in Fig. 1, 2 and 3. The electrophoretic patterns of sphaerococcum mutant forms 613 and 6512 are quite distinctive in comparison with the control form. The slow-moving group corresponds to the HMW glutenin subunits, whereas those with medium mobility appear to be albumins. Most of the mutant forms with HMW glutenin subunits are derivatives of the mutant forms 613 and 6512. Each accession synthesises from two to six subunits (Fig. 1-3). Significant variation was also found between different derivatives of 613 and 6512. A clear-cut variation was observed in mutant forms (Fig. 1, lanes 2, 4, 5, 7 and 8). Mutant forms 2, 4 and 5 are derivatives of the 613 sphaerococcum form and possessed the aestivum phenotype. The heterogeneity of the HMW subunits was also observed in the electrophoretical patterns of the mutant forms of 6512 (Fig. 2, 3). Among these mutant forms (Fig. 2, lanes 4, 5, 6 and 7 and Fig. 3, lanes 3, 4 and 5), variation in the banding pattern of HMW glutenin subunits was shown; these are also derivatives of the 6512 *sphaerococcum* mutant form and possessed a different phenotype – type *aestivum*, spelta, compactum and mutant form with dense ears (Fig. 5). Thus, the mutant forms 613 and 6512 and their derivatives clearly differed in their genetic makeup. However, the variation among the mutant forms and their derivatives is still difficult to interpret, but it might be possible to explain this phenotype variation with the presence of controlling elements such as Ac/Ds (Georgiev et al., 2000). This "transposon hypothesis" could explain several properties. Firstly, that the presence of an active transposon could explain why drastic phenotype variation coexists in *sphaerococcum* mutant forms; secondly, the transposable elements are probably in close association with *sphaerococcum* genes and possibly contribute to regulatory functions that may alter gene expression of the *sphaerococcum* mutant forms. At present, this is still an open question and further understanding of transposons might be important for future analysis of cereal genomes. The fast-moving group consists mainly of LMW glutenin subunits (corresponding to the B and C LMW glutenin subunits of Jackson et al., 1983) although some albumin and globulin bands may also be present in this region of the gel. The survey of *sphaerococcum* mutant forms showed extensive variation in both slow- and fast-moving bands. However, even more variation was exhibited among the *sphaerococcum* mutant forms and the control form. Thus, there is much scope for further genetic analysis of this variation including inheritance of the total protein content. A sample of the variation observed among *sphaerococcum* mutant forms is shown in Fig. 1, 2 and 3. Each mutant form possesses approximately 13 bands in the region of fast-moving bands and most variation occurred among the slower-moving bands in this region, probably corresponding to the B subunits of LMW glutenins (Fig. 1 lanes 4, 5, 6, 7 and 8 and Fig. 2, lanes 3, 4 and 5, Fig. 3, lanes 3, 4, 5, 6 and 7).

3 CONCLUSION

The genetic variation of the HMW proteins in *sphaerococcum* mutant forms is important in elucidating the nature of these mutations as well as the fundamental role of these proteins for the improvement of the grain quality of cereals.

References

1. T. Dekova, T. Dimitrov, S. Georgiev, V. Petkova, *Genet. Breed.*, 1992, **25**, 306
2. R. D'Ovidio, S. Masci, E. Porceddu, *Theor. Appl. Genet.*, 1995, **91**, 189
3. S. Georgiev, *Genet. Breed.*, 1976, **3**, 218
4. S. Georgiev, T. Dekova, I. Atanassov, Z. Angelova, V. Mirkova, A. Dimitrova, L. Stoilov, *Biotechnol. & Biotechn.*, EQ, 2000, **1**, 25
5. E. A. Jackson, L. M. Holt, P. I. Payne, *Theor Appl Genet.*, 1983, **66**, 29
6. U. K. Laemlli, *Nature*, 1970, **227**, 680
7. P. I. Payne, K. G. Corfield, L. M. Holt, J. A. Bleckman, *J Sci Food Agric.*, 1981, **32**, 51
8. P. I. Payne, G. J. Lawrence, *Cereal Res. Commun.*, 1983, **11**, 29
9. P. I. Payne, E. A. Jackson, L. M. Holt, *J. Cereal Sci.*, 1984, **2**, 73
10. N. E. Pogna, J. C. Autran, C. Mellini, D. Lafiandra, P. Feillet, *J. Cereal Sci.*, 1990, **11**; 15
11. M. L. Privalsky, L. Sealy, J. M. Bishop, J. P. Grath, A. D. Levinson, *Cell*, 1983, **32**, 1257
12. T. Wicker, R. Guyot, N. Yahiaoui, B. Keller, *Plant Physiol.*, 2003, **132**, 52

DURUM WHEAT DOUGH STRENGTH RELATIONSHIP TO POLYMERIC PROTEIN QUANTITY AND COMPOSITION

N.M. Edwards[1], M.C. Gianibelli[2], N.P. Ames[3], J.M. Clarke[4], J.E. Dexter[1], O.R. Larroque[2], and T.N. McCaig[4]

[1] Canadian Grain Commission, Winnipeg, Canada
[2] CSIRO Plant Industry, Canberra, Australia
[3] Cereal Research Centre, Agriculture and Agri-Food Canada, Winnipeg, Canada
[4] Semiarid Prairie Research Centre, Agriculture and Agri-Food Canada, Swift Current, Canada

1 INTRODUCTION

Quantity and composition of polymeric storage proteins are generally considered to be the primary contributors to variations in functional properties among wheat cultivars. Small scale measurement of polymeric protein may provide a useful means of predicting dough strength characteristics of durum wheat.

2 MATERIALS AND METHODS

Durum wheat cultivars and breeding lines (n=96) originating from twelve countries were grown out in each of two locations in western Canada. Harvested wheat was assessed for test weight and was milled to semolina.[1]

Protein content (%N x 5.7) of both wheat and semolina was determined by combustible nitrogen analysis (LECO model FP-428 Duman CNA Analyzer, St. Joseph, MI, USA). Traditional measurements of durum dough strength included alveograph,[2] 2g direct drive mixograph[3] performed at fixed water absorption of 50%, and gluten index.[4]

Semolina samples were extracted for SDS-soluble (without sonication) polymeric protein and SDS-soluble (with sonication) polymeric protein according to the procedure of Gupta et al[5], with modifications[6]. A Phenomenex BIOSEP-SEC 4000 (5 μm, 500 Å, 7.8 mm x 300 mm) column (Phenomenex, Torrance, CA, USA) connected to a Beckman System Gold HPLC was used for size exclusion (SE-HPLC) separations. The proportion of protein that was extractable only upon sonication is reported as % unextractable polymeric protein (UPP). Extraction and SE-HPLC of total protein from semolina was used to calculate glutenin to gliadin ratio.[7] Electrophoresis of high M_r glutenin subunits was conducted using SDS-PAGE,[8] and the glutenin alleles were classified according to Payne and Lawrence.[9]

3 RESULTS AND DISCUSSION

Durum wheat test weight and protein content (Table 1) varied among cultivars and lines and between growing locations. Similarly, there was a range in gluten strength as measure by both traditional means and by proportion of UPP (Table 1).

Table 1 *Durum wheat and semolina quality characteristics.*

	Location 1			Location 2		
	Mean	Minimum	Maximum	Mean	Minimum	Maximum
Wheat Properties						
Test Weight, kg/hL	80.9	75.7	84.1	83.4	79.0	85.9
Wheat Protein, %	13.8	11.8	16.6	12.6	9.5	16.6
Semolina Yield, %	67.0	62.1	69.3	67.8	63.4	71.2
Semolina Properties						
Semolina Protein, %	13.0	11.0	15.7	11.5	8.6	15.5
Alveograph P, mm	75	21	140	72	20	131
Alveo.L, mm	81	6	152	63	5	115
Alveo. W, x 10^{-4} joules	185	7	336	141	6	317
Gluten Index, %	40	0	89	41	1	93
Mixograph Mixing Time, min	3.9	1.8	4.8	3.6	1.5	5.6
Glutenin / Gliadin	0.95	0.61	1.31	1.09	0.71	1.60
% UPP	39.9	15.1	52.3	41.0	18.4	58.3

3.1 Wheat Properties

A wide range in protein content was evident among cultivars and lines, and between growing locations (Table 1). Location 1 reported higher average proteins than did location 2. Although the average test weight differed by 2.5 kg/hL, the average semolina milling yield was comparable.

3.2 Semolina Characteristics

Large differences in dough strength were observed among durum cultivars and lines, as evident from the range of values for alveograph parameters, gluten indices and mixograph mixing times (Table 1). The greater average alveograph L value for location 1 corresponded with higher protein content, an observation previously made for Canadian durum wheats.[10] As a result of the higher L values, average alveograph W values were also higher for location 1. Gluten indices differed little between growing locations, suggesting that is not influences by protein content. The range in mixing strength was easily seen by 2g mixograph testing (Fig. 1).

3.3 Polymeric Protein

The range in %UPP among durum cultivars and lines was wide and correlated well with traditional dough strength measurements (mixograph MT r=0.70, bandwidth at peak r=0.65, time to maximum bandwidth r=0.78, see Figs. 1 and 2; alveograph P r=0.61 and W r=0.72; GI r=0.77, see Fig 3). It was interesting to note that at %UPP less than approximately 30%, the GI values were less than 5. Past 30% UPP there was a strong positive relationship. Glutenin to gliadin ratio showed little relationship to dough strength parameters (GI r=0.32; alveograph parameters r=0.18 or less; mixograph parameters r=0.36 or less; %UPP r=0.26). Presence of HMW glutenin subunit 20 consistently corresponded with poor gluten strength (Fig. 3), in agreement with the work of Margiotta et al.[11]

4 CONCLUSIONS

Strong relationships were found between traditional durum dough strength measurements and %UPP, making it a useful predictor of dough strength. Glutenin to gliadin ratio does not appear to exert much influence over dough strength parameters. Durum wheat possessing HMW glutenin subunit 20 is of generally poorer quality than those durum wheats possessing 6+8, 7+8 or 14+15.

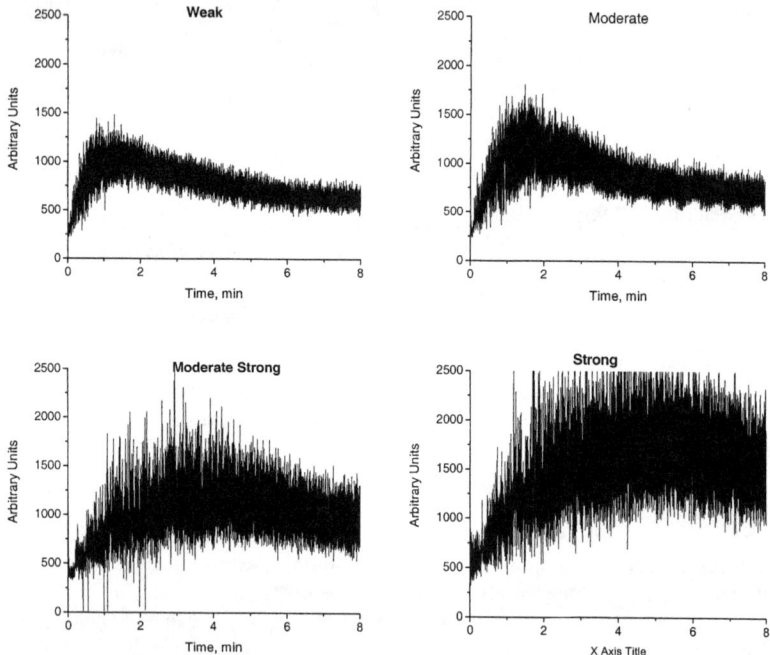

Figure 1 *Mixograph curves representing the range in mixing strength found within the sample population.*

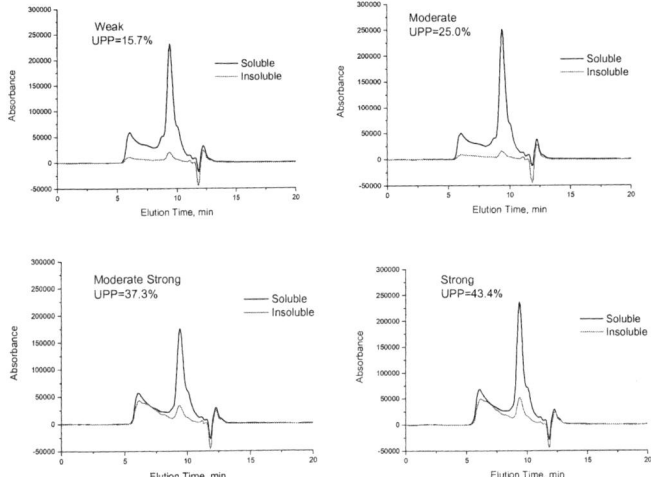

Figure 2 *Percentage of soluble and insoluble polymeric protein from the samples represented in Fig. 1. Note that the proportion of insoluble polymeric protein, and therefore the calculated %UPP, increased with increasing mixing strength.*

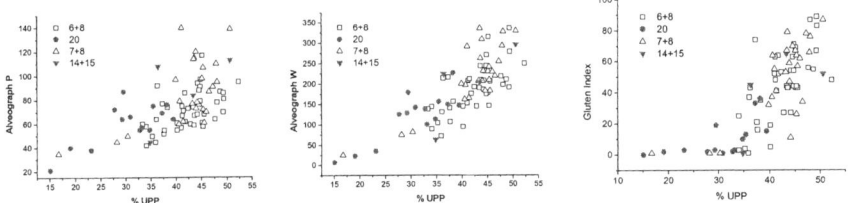

Figure 3 *Alveograph P, and W, and GI versus %UPP*

References

1. J.E. Dexter., R.R. Matsuo, J.E. Kruger, *Cereal Chem.*,1990, **67**, 405.
2. ICC 1980. Standard Methods of the International Association for Cereal Science and Technology. The Association, Vienna, Austria.
3. C.R. Rath, P.W. Gras, C.W. Wrigley, C.E.Walker, *Cereal Foods World*, 1990, **35**, 572.
4. American Association of Cereal Chemists. Approved methods of the AACC, 10th edn. The Association, St. Paul, MN 2000.
5. R.B. Gupta and F. MacRitchie, *J. Cereal Sci.* 1993, **18**, 23.
6. O.R. Larroque and F. Bekes, *Cereal Chem.* 2000, **77**, 451.
7. Batey, R.B. Gupta, F. MacRitchie, *Cereal Chem.*, 1991, **68**, 207.
8. R.B. Gupta and F. MacRitchie, *J. Cereal Sci.*, 1991, **14**, 105.
9. P.I. Payne and G.J. Lawrence, *Cereal Res. Commun.*, 1984, **11**, 29.
10. J.E. Dexter, K.R. Preston, D.G. Martin, E.J. Gander, *J. Cereal Sci.*, 1994, **20**, 139.
11. B. Margiotta, G. Colaprico, R. D'Ovidio, and D. Lafiandra, J. Cereal Sci., 1993, **17**, 221

LMW-i TYPE SUBUNITS ARE EXPRESSED IN THE WHEAT ENDOSPERM AND BELONG TO THE GLUTENIN FRACTION

P. Ferrante, C. Patacchini, S. Masci, R. D'Ovidio, D. Lafiandra

Dipartimento di Agrobiologia e Agrochimica, Università della Tuscia, Via S.C. de Lellis, 01100 Viterbo, Italy

1 INTRODUCTION

Gluten is the main functional component of wheat and is the main source of the visco-elastic properties of a dough.

Gluten proteins are classified into two main classes based on their solubility in alcohol water solutions: the gliadins and the glutenins.

The gliadins are monomeric proteins that can have intramolecular disulphide bonds and are soluble in alcohol-water solutions, while glutenins are polymeric proteins joined together by intermolecular disulfide bonds and are insoluble in alcohol-water solutions. Glutenin subunits are classified into two main groups on the basis of their mobility in SDS-PAGE under reducing conditions: high molecular weight glutenin subunits (HMW-GS) and low molecular weight glutenin subunits (LMW-GS).

While the structures of HMW-GS and their encoding genes have been well characterized in different cultivars and functional roles have been established for many subunits, there is a lack of information about LMW-GS due to their large number and similarity. LMW-GS are encoded by genes at the orthologous *Glu-3* loci and are represented by several different components. The presence of particular allelic forms has been correlated with differences in the gluten properties, in particular in durum wheat[1].

On the basis of the their N-terminal amino acid sequences, LMW-GS fall into three broad classes: those starting with either a methionine (LMW-m) or serine (LMW-s) are commonly found[2], whereas LMW-GS starting with an isoleucine (LMW-i) were not found by N-terminal sequencing, although corresponding nucleotide sequences have been reported. Even if there were indications of their presence[3,4], there was no direct proof of their expression in wheat endosperm. Thus, the aim of this study is to demonstrate that LMW-i types are actual component of wheat glutenin polymers by means of cloning and expression of a LMW-i type gene (LMW1A1) from *Triticum durum* cultivar Langdon using an *Escherichia coli* expression system.

2 MATERIALS AND METHODS

The LMW1A1 gene is present on chromosome 1A and contains an open reading frame of 1161 bp, including the signal peptide. It codes for a protein of 387 amino acids with a theoretical molecular weigh of 42501. This gene does not present the typical structure of LMW-GS genes, because it is missing the short N-terminal region and, consequently, the repetitive domain starts soon after the signal peptide.

2.1 Heterologous Expression of LMW1A1 and Purification of the LMW-i Type Polypeptide

PCR was performed on genomic DNA from *Triticum durum* cultivar Langdon with a pair of primers specific for the LMW1A1 gene (primers a and b[5]).

The PCR product so obtained was cloned by means of the Gateway expression system (Invitrogen). *Escherichia coli* BL21Star(DE3) pLysS cells were transformed with the expression vector containing the coding region of LMW1A1 gene without the sequence specific for the signal peptide. Bacterial cell, the culture conditions and purification procedure were as reported[6]. The identity of the heterologous protein was confirmed by western blot analysis performed with a polyclonal antibody anti-LMW-GS[6].

2.2 Two-Dimensional Electrophoresis (IEF X SDS-PAGE)

Glutenin subunits were prepared from the durum wheat cultivar Langdon (LND) and its D-genome-chromosome intervarietal substitution lines LND 1D(1A) and LND 1D(1B). They were separated by two-dimensional electrophoresis (IEF X SDS-PAGE), whose first dimension was performed by means of the IPG-Phore system (Amersham), by using 7 cm strips (pH 3-10). The second dimension was performed on ready-made Bis-Tris Novex gels (Invitrogen), with a 4-12% acrylamide gradient. Gels were stained according to Neuhoff et al[7].

2.3 Electroblotting and N-Terminal Amino Acid Sequencing

Western blotting was performed on two-dimensional gels by means of a Mini Trans-Blot Electrophoretic Transfer Cell (Bio-Rad). The protein spot (from cv Langdon) with the same electrophoretic mobility as the heterologous LMW-GS was submitted to N-terminal amino acid sequencing. This latter was perfomed by the Molecular Structure Facility at University of California, Davis by means of an Applied Biosystems Procise 494 sequencer.

3 RESULTS

The nucleotide sequence of the LMW1A1 gene is deposited in the EMBL Nucleotide Sequence Database (Accession nr. AJ293097).

The heterologous LMW1A1 polypeptide was expressed in significant amounts at three hours after induction (Figure 1A) and, although it is not expressed at high level, the isolation of inclusion bodies, where the heterologous protein is accumulated, allows a significant concentration of this protein (Figure 1A and 1B).

In order to determine if the LMW1A1 gene is expressed *in vivo* and if it is part of the monomeric or polymeric fraction of gluten, we have compared the two dimensional electrophoretic patterns of the heterologously expressed protein with that of a glutenin preparation of the durum wheat cultivar Langdon, from which the LMW1A1 gene was isolated. As confirmation, we made a comparison with the two-dimensional patterns of its D-genome-chromosome intervarietal substitution lines LND 1D(1A) and LND 1D(1B), since the LMW1A1 gene is present on chromosome 1A (Figure 2).

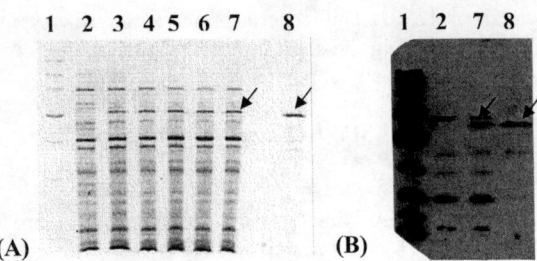

Figure 1 *(A) SDS-PAGE and (B) western Blotting analysis to verify the expression and the purification of LMW glutenin subunit 1A1. Lanes 1, MW marker, lanes 2, total cell proteins before induction of heterologous expression; lanes 3-6, total cell proteins after 1, 2, 3, 4 hours induction, respectively; lanes 7, total cell proteins after 3 hours induction; lanes 8, purified LMW glutenin subunit 1A1. Arrows indicate the LMW 1A1 subunit.*

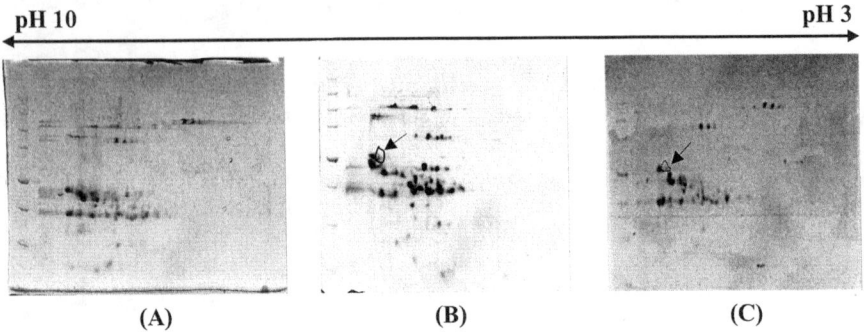

Figure 2 *Comparison of the 2D patterns (IEF X SDS-PAGE) of glutenin preparations of cv. Langdon (B) and its Chinese Spring 1D substitution lines (A) LND 1D(1A) and (C) LND 1D(1B). The circled spot indicated by an arrow is the LMW 1A1 glutenin subunit, that is not present in LND 1D(1A), as expected.*

Superimposition of the patterns of the heterologous LMW1A1 protein and glutenin proteins from the cultivar Langdon allowed the identification of a weak component in this latter (data not shown). This protein spot was still present in the LDN 1D(1B) substitution line, but absent in the LDN 1D(1A) substitution line, indicating that it is a chromosome 1A encoded protein.

This protein spot was electroblotted and submitted to N-terminal amino acid sequencing: IXQQQQP- was the resulting sequence. Although the second amino acid could not be identified, the remaining sequence shows a perfect homology with the reported LMW-i type genes, thus demonstrating that LMW-i type glutenin subunits are actually expressed in wheat endosperm and that they are part of the glutenin polymers.

4 CONCLUSION

Although there were indirect indications of the presence of LMW-i type subunits in the wheat endosperm, they had never been found by direct N-terminal sequencing, and consequently there was no decisive prove of their presence in wheat endosperm. Here we demonstrate that LMW-i types are present in wheat endosperm and correspond to polymeric proteins, opening the possibility to study their role in polymer formation and gluten quality

References

1. N.E. Pogna, D. Lafiandra, P. Feillet, J.-C. Autran, *J. Cereal Sci,.* 1988, **7**, 211
2. E.J.L. Lew, D.D. Kuzmicky, D.D. Kasarda, *Cereal Chem.*, 1992, **69**, 508
3. S. Cloutier, C. Rampitsch, G.A.,Penner and O. M. Lukow, *J. Cereal Sci.*, 2001, **33**, 143
4. T.M. Ikeda, T. Nagamine, H. Fukuoka and H. Yano, *Theor Appl Genet*, 2002, 104, 680-687
5. R. D'Ovidio, M. Simeone, S. Masci and E. Porceddu. *Theor. Appl. Genet.*, 1997, **95**, 1119
6. C. Patacchini, S. Masci, R D'Ovidio and D. Lafiandra. *J. Chromat. B*, 2003, **786**, 215
7. V. Neuhoff, N. Arold, D. Taube and W. Ehrhardt. *Electrophoresis*, 1988, **9**, 255

Acknowledgments

This work was supported by the Italian Minister for University and Research (MIUR), projects "Aspetti biochimici, genetici e molecolari delle proteine della cariosside dei frumenti in relazione alle caratteristiche nutrizionali e tecnologiche dei prodotti derivati" (PRIN 2002) and "Espressione genica ed accumulo di proteine d'interesse agronomico nella cellula vegetale: meccanismi trascrizionali e post-trascrizionali" (FIRB RBNE01TYZF)

QUALITATIVE AND QUANTITATIVE ANALYSIS OF GLIADIN COMPOSITION IN AN OLD HUNGARIAN WHEAT POPULATION

A. Juhász[1,2], R. Haraszi[1], O. Larroque[1], M.C. Gianibelli[1], F. Békés[1], Z. Bedő[2]

1 CSIRO Plant Industry, GPO Box 1600, Canberra, ACT 2601 Australia, angela.juhasz@csiro.au
2 Agricultural Research Institute of the Hungarian Academy of Sciences, Brunszvik 2, Martonvásár, 2462 Hungary

1 INTRODUCTION

The old Hungarian wheat varieties bred in the first half of the last century, were renowned for their good bread-making quality. Dough prepared from their flour could be used for producing highly extensible filo pastry. One of these old Hungarian cultivars, Bánkúti 1201 has been the focus of research attention in the last five years. Based on earlier results, Bánkúti 1201 is a wheat population, containing lines with different storage protein composition resulting in different dough properties. Most of the lines can be characterised as showing a special allelic composition at the *Glu-3/Gli-1* complex loci and high protein, gluten and gliadin content[1]. The aim of this study was to analyse the gliadin component to determine their relation with rheological properties.

2 MATERIALS AND METHODS

Fifteen lines of the Bánkúti 1201 population grown in the experimental field of the ARI HAS, in Hungary were involved in this study. Gliadin composition was determined by A-PAGE (pH3.1) and A-PAGE x SDS-PAGE[2]. Individual gliadin bands were numbered based on their relative mobility. Quantitative analysis of gliadin composition was determined by RP-HPLC[3]. Gliadins were integrated in blocks defined by retention time. ω_1: 5-14 min; ω_2: 14-17 min; ω_3: 17-23 min, α_1: 23-29 min; α_2: 29-36 min, γ_1: 36-41 min; γ_2: 41-50 min. Rheological properties were evaluated using a 2g-Mixograph^{TM} [4] and Micro Extension Tester[5]. Statistical analyses were carried out using GENSTAT 4.1 (Lawes Agricultural Trust, Rothamsted).

Table 1. *Relationship between gliadin composition and amounts of gliadin fractions and sub-fractions (ANOVA)*

N=30	ω%	α%	γ%	ω_1%	ω_2%	ω_3%	α_1%	α_2%	γ_1%	γ_2%
F value	1.89	25.39***	2.69	9.61***	11.77***	3.53*	1.82	19.12***	0.94	1.45
Pattern 1	14.98 a	53.63 b	32.72 b	6.42 c	2.12 a	6.44 a	10.31 a	43.32 b	19.03 b	13.69 b
Pattern 2	16.36 b	54.91 c	29.66 a	5.70 b	2.38 b	8.28 c	11.76 a	43.14 b	17.89 a	11.77 a
Pattern 3	14.98 a	46.65 a	32.11 b	5.07 a	2.70 c	7.21 b	12.56 b	34.08 a	18.55 a	13.56 b

3 RESULTS AND DISCUSSION

3.1 Gliadin composition and expressed amounts of gliadin fractions

The Bánkúti 1201 lines were divided in three groups based on their gliadin composition determined using one and two-dimensional PAGE (Figure 1). 67% show pattern 1, identified as *mma* on *Gli-1* loci, 13% pattern 2 (*mga*) and 20% possess pattern 3 (*abg*). The three gliadin patterns combined with differences in expression pattern, resulting in significantly different amounts of ω -, α- and γ -gliadins (Figure 1 and Table 1). The highest total ω gliadin content was measured in lines with pattern 2. Lines possessing pattern 1 produced the highest amount of ω_1 subfraction, pattern 3 expressed the most ω_2 gliadins and the amount of ω_3 subfraction was the highest in pattern 2. A more simple expression profile was detected in γ-gliadins. Both of total γ-gliadins and γ_1, γ_2 subfractions expressed in highest amount in lines which have gliadin pattern 1 and lines belong to pattern 2 had the poorest γ-gliadin expression.

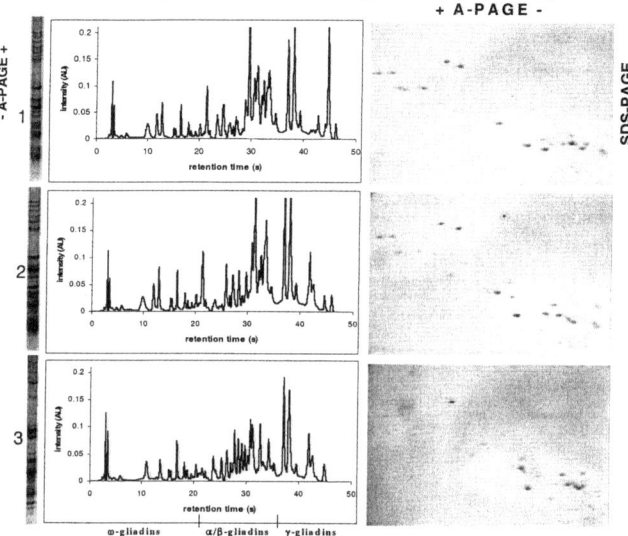

Figure 1. *Characteristic gliadin composition and expression pattern of gliadin fractions of Bánkúti 1201 lines B82, B1 and B44 representing the three groups detected in the Bánkúti 1201 population*

3.2 Relationship of qualitative gliadin composition on dough properties

The presence or absence of individual band patterns had significant effect on dough properties. Only the results of mixing time (MT), peak dough resistance (PR), resistance at breakdown (RBD), extensibility (Ext) and maximum resistance by extension (Rmax) are presented. In this sample set gliadin patterns have significant effect on all of the dough parameters evaluated, however in generally *Glu-1* and *Glu-3* loci affected stronger than *Gli-1 and Gli-2* loci (results not shown).

Table 2. *Unbalanced one-way analysis of variance of effects of gliadin pattern on dough properties*

N=16	MT (sec)	PR (AU)	RBD (AU)	Ext (mm)	Rmax (AU)
F value	38.19***	11.62**	16.92***	13.32***	125.93***
Pattern 1	175.2 a	409 b	20.50 b	1116 b	349 b
Pattern 2	174.0 a	356 a	21.00 b	1086 b	255 a
Pattern 3	266.5 b	360 a	11.50 a	687 a	730 c

To exclude the effects of high and low-molecular weight glutenin subunits a smaller sample set with same HMW and LMW-background ($2*^B$ 7+9 2+12, f i c) was chosen (Table 2). Regardless the linkage between *Glu-3* and *Gli-1* loci all three gliadin patterns are to be found in this narrowed sample set. Due to the small sample number the results of statistical analysis are orientating.

Based on our results gliadin pattern 3 has positive influence on dough strength and stability and affect negatively extensibility. Lines with patterns 1 and 2 produced more extensible but weaker dough. The presence of gliadin proteins in pattern 1 resulted in higher peak dough resistance. Using the relative mobility values of bands the effect of individual gliadin bands were evaluated. Those bands present in all three gliadin patterns were excluded. As a result of this analysis some groups of bands were identified (highlighted in black in Figure 2) showing positive effect on dough strength, peak resistance and extensibility, respectively. A small group of bands were identified not affecting RBD and Rmax but having decreasing effect on MT (signed with a star in Figure 2). Some bands which are present in 80% of the lines (Pattern 1 and 2) show positive effect on extensibility.

Figure 2. *Individual gliadin bands showing affecting significantly dough strength, peak resistance and extensibility. (+ effect: black, - effect: white)*

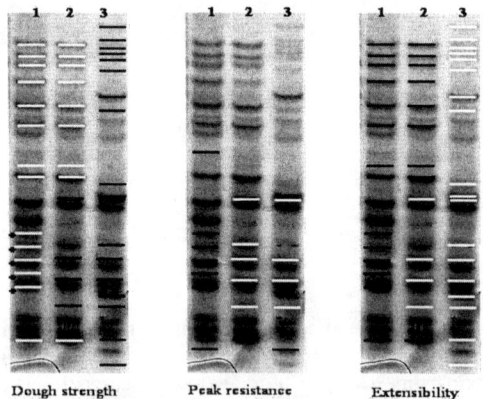

3.2 Relationship of quantitative gliadin composition on dough properties

Multiple regression analysis was carried out involving all 30 samples to describe relationship between gliadin fractions and dough properties (Table 3 and 4). Relatively high regression coefficient (R^2) was found by MT affected negatively by amounts of all three gliadin fractions. Similar negative effects were found by RBD and Rmax. PR decreased by increasing of ω% and increased if the amount of α-gliadins was higher. Based on our results extensibility is less describable by amounts of gliadin fractions, however expression level of γ gliadins seems to have positive effect on this parameter. Using the amounts of gliadin subfractions to the regression model some interesting conclusions can be drawn (Table 4).

Table 3. *Multiple regression analysis of effect of gliadin fractions on dough properties*

N=30	R^2	ω-gli %	α-gli %	γ-gli %
		Beta value		
MT (sec)	0.7258	-0.56***	-0.65***	-0.29**
PR (AU)	0.5865	-0.60***	0.38**	-0.03
RBD (AU)	0.5918	0.43***	0.64**	0.12
Ext (mm)	0.3170	-0.04	0.30	0.53**
Rmax (AU)	0.6102	-0.53***	-0.59***	-0.21

Table 4. *Multiple regression analysis of effect of gliadin subfractions on dough properties*

	R^2	ω_1%	ω_2%	ω_3%	α_1%	α_2%	γ_1%	γ_2%
N=30		Beta value						
MT (sec)	0.7637	-0.19	0.27	-0.57***	-0.26	-0.64***	-0.14	-0.09
PR (AU)	0.6057	-0.20	0.04	-0.73***	0.10	0.42	-0.12	0.01
RBD (AU)	0.6766	0.10	-0.26	0.62***	0.47**	0.69***	0.29*	-0.16
Ext (mm)	0.6227	0.14	-0.003	-0.30	-0.28	0.20	0.07	0.45*
Rmax (AU)	0.6913	-0.14	0.30	-0.65***	-0.24	-0.58**	-0.30*	0.08

The amounts of ω subfractions do not bear evenly on rheological parameters. Only subfraction ω_3, the less hydrophobic ω fraction has significant negative effect on dough strength. Similarly, mainly α_2 subfraction shows significant effect on the measured parameters.

4 CONCLUSION

A set of Bánkúti 1201 lines were analysed to evaluate the impact of qualitative and quantitative gliadin composition on rheological properties. Three different gliadin compositions were identified differing in both gliadin pattern and expression profile. Patterns 1 and 2 more common in the population have positive correlation with dough extensibility, and some less frequent bands show dough strengthening effect. Based on our HPLC results only the amount of some subfractions has significant effect on dough properties. Mainly amounts of ω_3 and α_2 subfractions have dough weakening effect and only γ_2 subfraction had medium level effect on extensibility. Based on our results extensibility is more influenced by allelic composition of gliadins than their expressed amounts.

References

1. A. Juhász, O.R. Larroque, L. Tamás, S.L.K. Hsam, F.J. Zeller, F. Békés and Z. Bedő, *Theor and Appl Genet*, 2003, 107:697-704
2. E.A. Jackson, M.H. Morel, T. Sontag-Strohm, G. Branlard, E.V. Metakovsky and R. Redaelli *J. Genet & Breed.*, 1996, 50, 321.
3. O.R. Larroque, F. Bekes, C.W. Wrigley and W.G. Rathmell, in *Wheat gluten* eds.P.R. Shewry and A.S. Tatham, Royal society of Chemistry special publications, Cambridge,2000, p. 136.
4. AACC 54-40A, Approved method of the AACC, approved 1961, revised 1988, The Association, St. Paul, MN
5. C.R. Rath, P.W. Gras, Z. Zhen, R. Appels, F. Békés and C.W. Wrigley In: *Proceedings of the 44th Australian Cereal Chemistry Conference*, Royal Australian Chemistry Institute, Melbourne, 1994, p. 122

ADDITIVE AND EPISTATIC EFFECTS OF *GLU-1*, *GLU-3* AND *GLI-1* ALLELES ON CHARACTERISTICS OF BAKING QUALITY AND AGRONOMIC PERFORMANCE IN FOUR DOUBLED HAPLOID WHEAT POPULATIONS

B. Killermann[1], G. Zimmermann[1] and W. Friedt[2]

[1] Bavarian State Research Center for Agriculture, Institute for Crop Production and Plant Breeding, Lange Point 6/II, D-85354 Freising, Germany
[2] Justus-Liebig-University Giessen, Institute of Agronomy & Plant Breeding I, Heinrich-Buff-Ring 26-32, D-35392 Giessen, Germany

1 INTRODUCTION

Differences in baking quality are mainly based on the qualitative and quantitative composition of the *high* and *low molecular weight glutenin subunits* (HMW-GS, LMW-GS) and gliadins encoded by gene loci *Glu-1*, *Glu-3* and *Gli-1*, respectively. In Germany the main characteristic for quality classification of wheat cultivars is the bread volume determined in the Rapid Mix Test (RMT). Varying correlations are found with other attributes used for quality characterization and selection, depending on the genetic background of the material investigated. The aim of this work was to determine the allelic variation of the glutenins and gliadins in four doubled haploid (DH) wheat populations and to investigate the individual and combined effects of alleles in a common background. The DH populations represent a wide range of genetic background of the Central and West European gene pool. The parents used differed with regard to *Glu-1* alleles and, to some extent, to the *Glu-3* and *Gli-1* loci, and they were selected to cover a wide scope of different quality features. Primary emphasis was placed on the HMW-GS, because there was little variability of the LMW-GS and gliadins in the DH populations. Therefore, a greater part of baking quality variability than possible up to now should be explained by protein markers. In this presentation we show the results for baking volume (RMT) in detail.

2 MATERIALS AND METHODS

In our investigation we used four DH populations comprising 495 lines derived from anther culture (Table 1). The HMW-GS, LMW-GS and gliadin alleles were determined by means of SDS-PAGE and A-PAGE[1,2] and classified according the numbering system of Payne & Lawrence[3] and Jackson et al.[1]. The whole spectrum of quality features used in breeding and flour processing in Germany was analysed, i.e. Rapid Mix Test bread volume, kernel hardness, protein content, Zeleny sedimentation value and rheological characteristics from Brabender Extensograph and Farinograph[4]. The most important agronomic characteristics were established in two years of field trials (yield performance and structure, straw length, grain properties and resistance characteristics). The effects of allelic variation of glutenins and gliadins on quality characteristics were studied using the

Genetics and Quality 145

GLM procedure of the SAS statistical package[5]. All main and interaction effects were entered into the models and least squares means of the effects were estimated.

Table 1 *Allelic combinations of the parental cultivars of four doubled haploid wheat populations*

Pop N	parental cultivars	Glu-1 (HMW-GS)			Glu-3 (LMW-GS)			Gli-1 (gliadins)		
		Glu-A1	Glu-B1	Glu-D1	Glu-A3	Glu-B3	Glu-D3	Gli-A1	Gli-B1	Gli-D1
Pop1 142	Atlantis/ Bovictus	a, 1 c, N	d, 6+8 c, 7+9	a, 2+12 d, 5+10	a e	j^1 j^1	c c	b b	l^1 l^1	b b
Pop2 103	Flair/ CWW 92-6	c, N a, 1	d, 6+8 i, 17+18	d, 5+10 a, 2+12	e e	g g	c c	m b	f f	b b
Pop3 150	Atlantis/ Lindos	a, 1 c, N	d, 6+8 c, 7+9	a, 2+12 d, 5+10	a d	j^1 g	c c	b o	l^1 f	b b
Pop4 100	W 84332/ Bussard	c, N a, 1	d, 6+8 c, 7+9	a, 2+12 d, 5+10	e a	g g	c c	f f	c c	b b

[1] *Glu-B3j*, *Gli-B1l* = rye alleles (cultivars containing the 1BL/1RS wheat-rye-translocation)

3 RESULTS AND DISCUSSION

For the four most important quality traits, i.e. baking volume (RMT), Zeleny sedimentation value (SED), dough softening degree in the farinogram (FARTER10) and the energy in the extensogram (EXFL), analyses of variance with pure *Glu-1* models (only main effects, main and interaction effects) were calculated within populations (Table 2).

Table 2 *R-squares (R^2) of the Glu-1 models (main effects, main and interaction effects) for the four most important quality traits RMT, SED, FARTER10 and EXFL*

	Pop1	Pop2	Pop3	Pop4
***Glu-1* main effects**				
RMT	0,47	0,41	0,07	0,28
SED	0,46	0,15	0,15	0,23
FARTER10	0,52	0,26	0,14	0,37
EXFL	0,64	0,40	0,66	0,64
***Glu-1* main effects and interaction effects**				
RMT	0,51	0,49	0,21	0,35
SED	0,50	0,20	0,21	0,36
FARTER10	0,56	0,33	0,22	0,37
EXFL	0,68	0,43	0,66	0,66

The *Glu-1* main effects in Pop1 and Pop2 explained 47 and 41 % of the variability in baking volume, respectively. The R-squares of Pop3 and Pop4 were much lower. By inclusion of the interaction effects into the models, i.e. epistatic effects between the *Glu-1* loci, 21 to 51 % of the variability could be attributed to the *Glu-1* alleles depending on the genetic background of the individual DH population. Similar results were found for the sedimentation value. Considerably higher R-squares were found for the dough-rheological traits, especially for the energy in the extensogram. The inclusion of the *Glu-3* and *Gli-1* alleles into the models led to a slight increase of the values, from 21 to 30 % and 51 to 56 %, respectively. In Pop3, 82 % of the variation of EXFL could be explained by variation of HMW-GS and LMW-GS. It can be assumed that the glutenin alleles have only partially additive effects in the analysed DH lines. This is a confirmation that the scoring system according to Payne et al.[6] cannot be maintained in its original form. Epistatic effects were established most clearly for Pop3, but they can also be found in other populations. The combination of the three favourable alleles *acd* (1, 7+9, 5+10) is not worth striving for in this population (Figure 1).

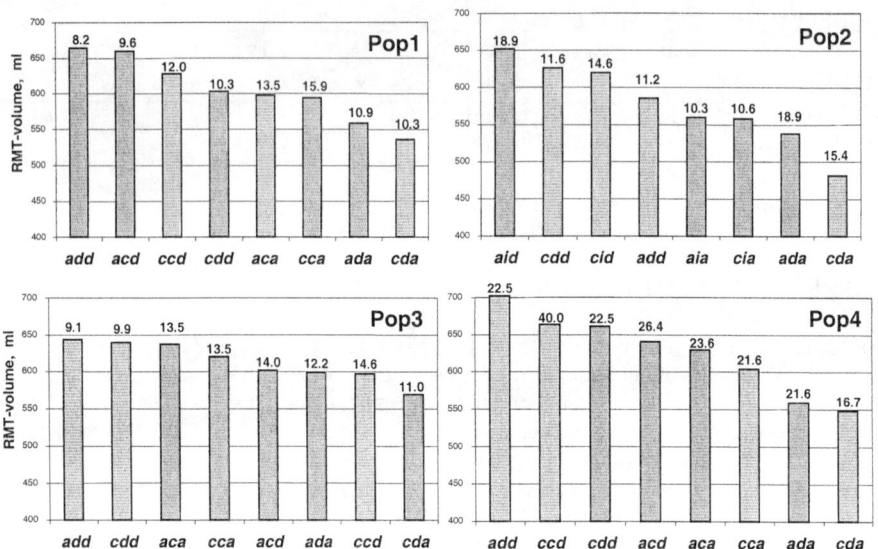

Figure 1 *Average RMT-volume and standard error of mean values of Glu-1 allele combinations in the lines of four DH populations in descending order*

It became obvious that the estimation of baking volume of the DH lines from the combination of the *Glu-1* alleles cannot be performed with a unique formula for the four DH populations. The genetic background of the parents is decisive for the weighting with which the main effects of the individual *Glu-1* gene loci and their interactions enter the multiple regression equations (Table 3). In particular, the dough characteristics of the parents play a major role in this respect. Either differing significant effects, or the same effects with varying weighting are included in the equations. Considerable differences were found among the populations in the phenotypic variances explained by the marker, ranging from 20 % in Pop3 up to 49 % in Pop2. *Glu-D1* showed without exception the greatest influence on baking volume. Pop1 is in accordance with the additive scoring system developed by Payne et al.[6]. The relationship between gluten protein alleles and agronomic traits was only weakly expressed and varied among the populations. Coefficients of determination that are worth mentioning were found in Pop2 for infestation by *Septoria tritici* and *Fusarium* at $R^2 = 0{,}25$ and 0,28, respectively (data not shown).

Table 3 *Multiple regression equations and R-squares (R^2) for the RMT-volume of four DH populations*

Pop1 (*Atlantis/Bovictus*)		
RMT-volume (ml/100 g flour) = 538	+ 37 *Glu-A1a* + + 21 *Glu-B1c* + + 74 *Glu-D1d*	$R^2 = 0{,}47$
Pop2 (*Flair/CWW 92-6*)		
RMT-volume (ml/100 g flour) = 482	+ 56 *Glu-A1a* + + 73 *Glu-B1i* + + 155 *Glu-D1d* - - 54 *Glu-A1a*Glu-B1i* - - 96 *Glu-A1a*Glu-D1d* - - 82 *Glu-B1i*Glu-D1d* + + 127 *Glu-A1a*Glu-B1i*Glu-D1d*	$R^2 = 0{,}49$
Pop3 (*Atlantis/Lindos*)		
RMT-volume (ml/100 g flour) = 577	+ 13 *Glu-A1a* + + 45 *Glu-B1c* + + 58 *Glu-D1d* - - 86 *Glu-B1c*Glu-D1d*	$R^2 = 0{,}20$
Pop4 (*Bussard/W 84332*)		
RMT-volume (ml/100 g flour) = 552	+ 64 *Glu-B1c* + + 129 *Glu-D1d* - - 89 *Glu-B1c*Glu-D1d*	$R^2 = 0{,}33$

Glu-A1a, *Glu-B1c*, *Glu-B1i* and *Glu-D1d* have values 1 and 0 for present and absent respectively.

4 CONCLUSIONS

Genetic background plays a major role in selection towards high bread volume by means of protein markers, so use of markers must be assessed within specific populations.

References

1 E.A. Jackson, M.-H. Morel, T. Sontag-Strohm, G. Branlard, E.V. Metakovsky and R. Redaelli, *J. Genet. & Breed.*, 1996, **50**, 321.
2 N.K. Singh, K.W. Shepherd and G.B. Cornish, *J. Cereal Sci.*, 1991, **14**, 203.
3 P.I. Payne, G.J. Lawrence, *Cereal Res. Commun.*, 1983, **11**, 29.
4 ICC-Standards, *Standard-Methoden der Internationalen Gesellschaft für Getreidechemie (ICC)*, Verlag Moritz Schäfer, Detmold Germany, 1986, ISBN 3-87696-010-X
5 *SAS Procedure Guide*, 3rd Edn., SAS Institute Inc., 1990a, Vers. 6, Cary, NC
6 P.I. Payne, M.A. Nightingale, A.F. Krattiger, L.M. Holt, *J. Sci. Food & Agric.*, 1987, **40**, 51-65

Acknowledgements

Thanks are due to Dr. G. Daniel (Bavarian State Research Center for Agriculture, Institute for Crop Production and Plant Breeding) for producing the DH populations. This research was funded by the GFP (Gemeinschaft zur Förderung der privaten deutschen Pflanzenzüchtung e.V., project number G 76/97 HS).

QTL ASSOCIATION WITH MEASURES OF GLUTEN STRENGTH ACROSS ENVIRONMENTS IN DURUM WHEAT

R.E. Knox[1], S. Houshmand[1], F.R Clarke[1], J.M.Clarke[1], and N.A. Ames[2]

[1] Agriculture and Agri-Food Canada, Semiarid Prairie Agricultural Research Centre, P.O. Box 1030, Swift Current, Saskatchewan, Canada S9H 3X2
[2] Cereal Research Centre, 195 Dafoe Rd., Winnipeg, Manitoba, Canada R3T 2M9

1 INTRODUCTION

Gluten strength is an important quality trait of durum wheat (*Triticum turgidum* L. var *durum*). The SDS-sedimentation test and gluten index are used as measures of gluten strength. SDS-sedimentation volume is correlated (r=0.78 to 0.85) with mixograph mixing time[1,2] and correlated (r = 0.70 to 0.77) with mixograph peak height (r = 0.70 to 0.77)[1]. Heritability for SDS-sedimentation volume ranges from 0.53 to 0.68.[3,4,5]

Gluten index has also been shown to predict gluten strength.[6] The heritability of gluten index was similar to SDS-sedimentation in preliminary studies by Clarke *et al.*[7] Blanco *et al.*[8] found seven quantitative trait loci (QTL) loci and Elouafi *et al.*[9] found five QTLs associated with SDS-sedimentation volume indicating SDS-sedimentation volume is complexly inherited.

The objective of this research was to discover QTLs associated with gluten strength measured by SDS-sedimentation and Gluten index in durum wheat genotypes used in a Canadian wheat breeding program and to determine the effect of environment on those QTLs.

2 MATERIALS AND METHODS

From the F_1 of the cross of a (Kyle*2/Biodur) F_9 inbred selection by Kofa a doubled haploid (DH) durum population of 155 lines was developed using the maize pollen method.[10] The (Kyle*2/Biodur) F_9 sel. was developed in the Swift Current breeding program, with quality attributes similar to Kyle, the predominant Canadian cultivar at the time. Biodur is a semidwarf cultivar from France and Kofa is a U.S. semidwarf cultivar. The (Kyle*2/Biodur) F_9 sel. had greater protein concentration than Kofa, but lower gluten strength. One hundred and fifty-one of the DH lines with 17 checks that included parents, were grown in 2000 in a randomized complete block design with two replications at Swift Current, Saskatchewan. In 2001 the trial was grown at Swift Current, Indian Head and Regina with 155 DH lines and 15 checks arranged in an alpha lattice of 22 blocks and two replications.

Gluten strength of each line and check was estimated using 3% sodium dodecyl sulfate (SDS) sedimentation values.[2] Gluten index of whole meal samples was measured using AACC Method 38-12[11]. Data were analyzed using a mixed-model ANOVA to assess differences among genotypes.

DNA was extracted from leaves of two-week-old seedlings of lines and parents. Parental DNA was tested with microsatellite primer pairs. Primers that generated polymorphism between the parents were tested on the entire population. MAPMAKER/EXP[12] was used to determine linkage and order of markers. Least square means of SDS sedimentation and gluten index data were used in genetic analysis.[13] The map data and gluten strength results were analysed using MQTL.[14] A marker was considered significant if at least one environment was P<0.001 or more than one environment P< 0.01.[15] Percent of genetic variance explained by the marker association was calculated based on Knapp.[16]

3 RESULTS AND DISCUSSION

Kofa had a significantly higher gluten index (P < 0.01) and SDS-sedimentation volume (P < 0.01) than the other parent, (Kyle*2/Biodur) F_9 sel. in all environments (Table 1). As expected from the large difference in the parents for measures of gluten strength, a wide range of variation was observed in the progeny for gluten index and SDS-sedimentation volume. In all environments and for all measures, the range in the progeny exceeded the parents.

Table 1 *SDS Sedimentation volume and gluten index means of parents and ranges in population (Kyle*2/Biodur) F_9 sel.//Kofa grown at Swift Current, Regina and Indian Head in 2000-2001.*

Year	Line Location	Gluten Index				SDS Sedimentation volume			
		Kofa Mean	(Kyle*2/ Biodur) F_9 sel. Mean	(Kyle*2/ Biodur)sel. //Kofa Range	SED	Kofa Mean	(Kyle*2/ Biodur) F_9 sel. Mean	(Kyle*2/ Biodur)sel. //Kofa Range	SED
2000	Swift Current	94.0	28.0	11.5-97.0	9.0	40.0	24	20.0-52.0	2.9
2001	Swift Current	89.5	22.1	14.3-97.0	5.6	52.5	35	32.5-60.0	2.7
	Regina	81.6	40.5	16.2-82.1	5.4	50.2	28.5	26.0-52.4	2.2
	Indian Head	65.5	37.5	16.3-85.0	4.3	52.0	29.5	25.5-55	2.5

SED = Standard Error of the Difference between the line means

Forty seven polymorphic loci were found. The loci included all chromosomes except 6A and 7A. There was one to as many as seven markers per chromosome.

QTL analysis showed the 1B locus to be important in gluten strength across environments (Table 2). This locus showed up in each environment for both measures of gluten strength. Other loci appeared to affect gluten strength under certain environments and not others. These loci border on statistically significant and are not consistent across measures of gluten strength. Not all loci involved in gluten strength are expected to be represented in these results because of the low number of markers on the map. At best two loci are represented by the present data. This compares to five loci reported by Elouafi *et al.*[9] and seven loci by Blanco *et al.*[8] in a multiple environment study.

The 1B locus was significant in our study as well as those of Elouafi *et al.*[9] and Blanco *et al.*[8] Clarke *et al.*[17] estimated the number of effective factors at nine for SDS-sedimentation volume and 12 for gluten index. The continuous distribution of the lines in their study supported such complex inheritance. However a possible bimodal appearance to the data reported by Clarke et al.[17] would indicate one of the loci had a major effect. It remains to be seen if the 1B locus is associated with modality of the distribution.

Blanco et al.[9] and Elouafi et al.[9] found an association between gluten strength and makers on chromosomes 1A, 3A, 3B, 4B, 5A, 6A, 6B, and 7B. We found a weak association with chromosome 2A.

SDS-sedimentation volume and gluten index can be highly heritable,[17] however a positive correlation of SDS-sedimentation volume and grain protein concentration was shown by Kovacs et al.[18] and Carrillo et al.[19] Variation from locus to locus over environments may be complicated by environmental and genotypic variation in grain protein concentration.

Table 2 *Simple interval mapping (SIM) test statistic and type I error rates for SDS Sedimentation value and gluten index in 2000-2001 at Swift Current, Regina and Indian Head, Canada.*

Trait	Chromosome	/Marker	Simple Interval Mapping (SIM) Test Statistic and Probability							
			2000		2001					
			Swift Current		Swift Current		Regina		Indian Head	
			TS^z	P	TS	P	TS	P	TS	P
GI^y	1B	WMC49	48.7	0.0001	68.4	0.0001	50.7	0.0001	38.1	0.0001
	2A	Xgwm339/ WMC114^w	NS	NS	NS	NS	15.5	0.0070	NS	NS
SV^x	1B	WMC49	61.9	0.0001	69.3	0.0001	82.8	0.0001	74.5	0.0001
	2A	Xgwm339/ WMC114	11.6	0.03	NS	NS	14.5	0.001	15.8	0.008

^{z} Test Statistic
^{y} Gluten Index
^{x} SDS-Sedimentation Volume
^{w} Linked markers with Xgwm339 located to chromosome 2A

We found that WMC49 explained 38 to 54% of the gluten index genetic variation depending on the environment, whereas Xgwm339 explained only about 4%. Similarily WMC49 explained from 53 to 61% of the genetic variation in sedimentation volume and Xgwm339 explained between 4 and 14% of the variation depending on environment.

4 CONCLUSIONS

Adequate gluten strength is critical to the manufacture of high quality pasta that will produce good cooking quality. It is therefore important to select durum for gluten strength during cultivar development. DNA markers are useful tools for assessing genetic loci involved in complex traits and should assist our understanding of gluten strength more effective select ion for this trait. In a sparse map, QTL loci were associated with markers on chromosome 1B and possibly 2A. The 1B locus was highly significant in all locations across both years for both SDS-sedimentation and gluten index. The 2A locus, in contrast, was significant for sedimentation value for two locations in only 2001 and only one of these locations appeared as significant for gluten index. Apparently some loci that contribute to gluten strength are more consistently expressed, while other loci are more ephemeral.

References

1. M. Ruiz and J.M. Carrillo, *Plant Breeding*, 1995, **114**, 40-44.
2. J. E. Dexter R. R. Matsuo F. G. Kosmolak D. Leisle and B. A. Marchylo, *Can. J. Plant Sci.*, 1980, **60**, 25-29.
3. M.O. Braaten, K.L. Lebsock and L.D. Sibbitt, *Crop Sci.*, 1962, **2**, 277-281.
4. J.M. Clarke, N.K. Howes, J.G. McLeod and R.M. DePauw, *Crop Sci.*, 1993, **33**, 956-958.
5. A. Sarrafi, R. Ecochard and P. Grignac, *Plant Var. Seeds*, 1989, **2**, 163-169.
6. R. Cubadda, M. Carcea and L.A. Pasqui, *Cereal Foods World* 1992, **37**, 866-869.
7. J.M. Clarke, F.R. Clarke, N.P. Ames, T.N. McCaig, and R.E. Knox, in *Durum Wheat Improvement in the Mediterranean Region: New Challenges,* eds. C. Royo, M.M. Nachit, N. DiFonzo and J.L. Araus, International Centre for Advanced Mediterranean Agronomic Studies (CIHEAM), Zaragoza, 2000, pp. 439-446.
8. A. Blanco, M.P. Bellomo, C. Lotti, T. Maniglio, A. Pasqualone, R. Simeone, A. Troccoli and N. DiFonzo, *Plant Breeding* 1998, **117**, 413-417.
9. I. Elouafi, M.M. Nachit, A. Elsaleh, A. Asbati and D.E. Mather, in *Durum Wheat Improvement in the Mediterranean Region:New Challenges,* eds. C. Royo, M.M. Nachit, N. DiFonzo and J.L. Araus, International Centre for Advanced Mediterranean Agronomic Studies (CIHEAM), Zaragoza, 2000, pp. 505-509.
10. R.E. Knox, J.M. Clarke and R.M. DePauw, *Plant Breeding*, 2000, **119**, 289-298.
11. American Association of Cereal Chemists. *Approved Methods of the AACC,* 10th ed. Method 38-12, approved November 1995, The Association, St. Paul, MN, 2000.
12. E.S. Lander, P. Green, J. Abrahamson, A. Barlow, M.J. Daly, S.E. Lincoln and L. Newburg, *Genomics*, 1987, **1**, 174-181.
13. A. Paterson, *Molecular dissection of complex traits*. CRC press, New York, 1998.
14. N.A. Tinker and D.E. Mather, *Journal of Agricultural Genomics*, 1995, **1** (http://www.ncgr.org/research/jag/papers95/paper295/indexp295.html)
15. E. Lander and L. Kruglyak, Nature Genet., 1995, **11**, 241-247.
16. S.J Knapp, in *Mapping quantitative trait loci. In: DNA-Based Markers in Plants*, eds. R.I. Phillips and I.K. Vasil, Kluwer Academic Publishers, Netherlands, 2001, pp. 59-99.
17. F.R. Clarke, J.M. Clarke, N.P. Ames, T.N. McCaig and R.E. Knox, in *Proc. 2nd Int'l. Workshop Durum Wheat and Pasta Quality: Recent Achievements and New Trends*, ed. M.G. Egidio, Industria Grafica Failli Fausto s.n.c., Rome, Italy, 2002, pp. 281-284.
18. M.I.P. Kovacs, N.K. Howes, J.M. Clarke, and D.Leisle, *J. Cereal Sci.* 1997, **27**, 47-51.
19. J.M. Carrillo, J.F. Vasquez and J. Orellana, *Plant Breeding,* 1990, **104**, 325-333.

Acknowledgements

We gratefully acknowledge the financial support of the Matching Investment Initiative of Agriculture and Agri-Food Canada and the Wheat Producer Check-Off administered by the Western Grains Research Foundation and the technical assistance of Jay Ross, Shawn Yates, Brad Meyer and Isabelle Piche. Marker and map information provided by Dr. Daryl Somers is very much appreciated.

DEVELOPMENT OF NILS OF BREAD WHEAT DIFFERING IN HIGH MOLECULAR WEIGHT GLUTENIN SUBUNITS AND THEIR USE IN QUALITY RELATED STUDIES

B. Margiotta[1], G. Colaprico[1], M. Aramini[2], S. Masci[2], C. Patacchini[2], D. Lafiandra[2]

[1] Plant Genetics Institute, CNR, Bari, Italy
[2] Department of Agrobiology and Agrochemistry, University of Tuscia, Viterbo, Italy

1 INTRODUCTION

The combined use of proper genetic material and analytical tools have been critical in establishing the role of the high molecular weight glutenin subunits (HMW-GS) in affecting flour bread-making properties.

The use of aneuploid lines and electrophoretic techniques has enabled location of HMW-GS genes. They are encoded by pair of tightly linked genes (encoding one high Mr x-type subunit and one low Mr y-type subunit) at the *Glu-1* loci present on the long arms of chromosome 1A, 1B and 1D of bread wheat. However, bread wheat cultivars contain from three to five high molecular weight glutenin subunits due to the silencing of some genes. It has been suggested that increasing the number of expressed HMW-GS genes above the current maximum of five could lead to improved bread-making quality as a result of a higher amount of large glutenin polymers, but also the value of lines with a different number of subunits has been stressed in relation to different end-uses. For instance, Payne and Seekings[1] have produced wheat lines with a single HMW-GS which have proved to be very extensible and suitable for biscuit production.

In order to understand the correlation between number and type of HMW-GS and quality properties of bread wheat doughs and gain information on glutenin polymer composition, we are currently developing new bread wheat lines with an altered number and composition of HMW-GS, based on the triple HMW-GS null previously developed by Lawrence et al.[2]. This have been developed thanks to the extensive analysis of bread wheat landraces which have permitted to identify rare alleles at the different *Glu-1* loci. Null forms lacking x- and/or y-type subunits have also been detected. Combination of the different null *Glu-1* alleles has resulted in the production of further genotypes with unusual HMW-GS composition.

2 METHODS AND RESULTS

The bread wheat lines have been analysed by SDS-PAGE of fully reduced samples. The relative electrophoretic patterns of some of them are reported in Figure 1.

Partial reduction of the glutenin polymer has been used to release oligomers in order to gain information on subunit and disulfide arrangement of native glutenin[3,4]. Association between HMW-GS has been determined by two-dimensional electrophoresis (SDS-PAGE of partially reduced samples vs. fully reduced samples). The DTT concentrations used for partial and progressive reduction of first dimension samples ranged from 0.05 mM up to 60 mM. The best DTT concentration to detect HMW-GS dimers was 5mM. An example of the two-dimensional pattern of the line with the HMW-GS composition 1, 5+10, null, is reported in Figure 2.

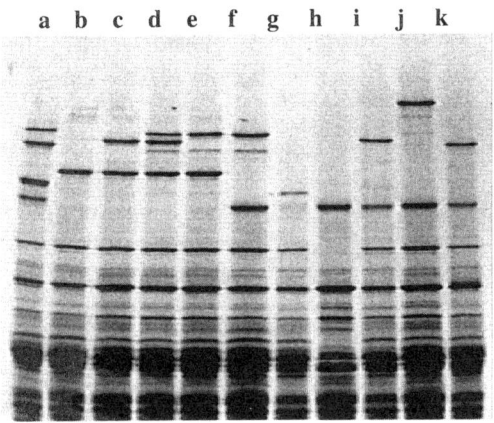

Figure 1: *SDS-PAGE of the near isogenic bread wheat lines with different HMW-GS composition. (a) reference line with the full HMW-GS complement, 1, 17+18, 5+10; (b) null, 7, null; (c) null, 7, 2; (d) 1, 7, 2; (e) 1, 7; null; (f) 1, null, 12; (g) null, 8, null; (h) null, null, 12; (i) null, null, 2+12; (l) 2.2*, null, 12; (m) null, null, 5*+12.*

The results obtained from the analyses of different lines have provided information on covalent linkages between different HMW-GS. In particular, the following associations have been detected: Ax1-By18; Ax1-Dy10; Dx5-Dy10; Dx5-By18; Bx17-By18; Bx17-Dy10. The formation of x-x dimers[3] or y-y[5] has not been observed, although it is not possible to exclude their presence. These observations might be in agreement with the hypothesis of a preferential linkage between x- and y-subunits[6,7], even if the finding of prevalent x-y linkages might be due to a preferential release in the conditions here used rather than to a preferential linkage between these subunit types. However, the coexistence of x- and y- subunits in a genotype might be correlated to the formation of higher size polymers, since we have not observed any release of oligomers from the residue (aggregated fraction) of a line containing only the 1Ax subunit, whereas they were released when the soluble fraction (composed mainly by monomers and oligomers) was not removed (data not shown).

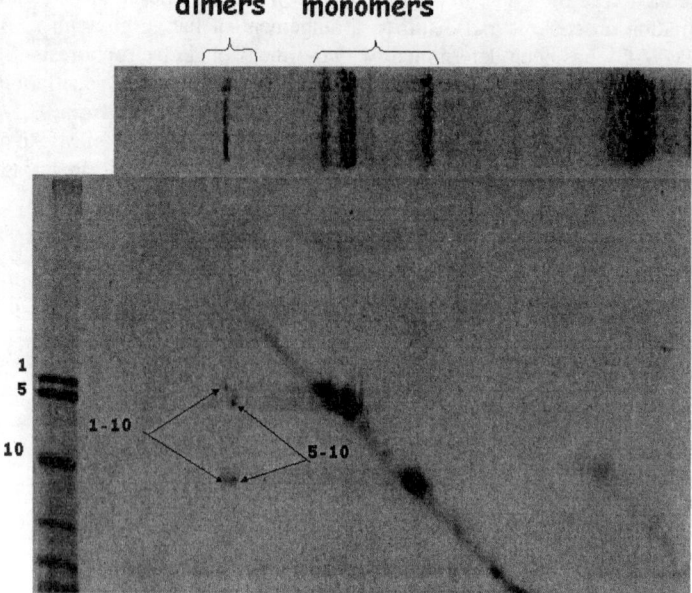

Figure 2: *Two dimensional SDS-PAGE (partially reduced sample vs. fully reduced sample) of the bread wheat line containing HMW-GS Ax1 and Dx5+Dy10. Dimers composed of subunits 1 and 10, and 5 and 10 are shown.*

3 CONCLUSIONS

This preliminary work has shown the potentiality of the materials developed in order to understand HMW-GS organization inside the glutenin polymer. Quality data on some of the lines produced are described in the paper by Larroque et al. (these proceedings). The material is further being increased in order to perform complete qualitative tests and evaluation of the size and composition of the glutenin polymers.

References

1. P.I. Payne and J.A. Seekings, *Proc. 6th Int. Gluten Workshop*, 1996, C.W Wrigley ed., 14
2. G.J. Lawrence and F. MacRitchie, C.W. Wrigley, *J. Cereal Sci.*, 1988, **7**, 109
3. W.E. Werner, E.A. Adalsteins and D.D. Kasarda, *Cereal Chem.*, 1992, **69**, 535
4. M.P. Lindsay, J.H. Skerritt, *J. Agric. Food Chem.*, 1998, **46**, 3447
5. P. Köhler, H.-D. Belitz, and H. Wieser, *Z. Lebensm. Unters. Forsch.*, 1993, **196**, 239
6. H.P. Tao, E.A. Adalsteins and D.D. Kasarda, *Biochim. Biophys. Acta*, 1992, **1159**, 13
7. Y. Shimoni, A.E. Blechl, O.D. Anderson and G. Galili, *J. Biol. Chem.*, 1997, **272**, 15488

Acknowledgments

Research supported by the Italian Ministry for University and Research (MIUR), project "Aspetti biochimici, genetici e molecolari delle proteine della cariosside dei frumenti in relazione alle caratteristiche nutrizionali e tecnologiche dei prodotti derivati" (PRIN 2002)

RELATIONSHIP BETWEEN Glu-D1/Glu-B3 ALLELIC COMBINATIONS AND BREAD-MAKING QUALITY-RELATED PARAMETERS COMMONLY USED IN WHEAT BREEDING

R. J. Peña, H. Gonzalez-Santoyo and F. Cervantes

International Maize and Wheat Improvement Center, (CIMMYT) Apdo. Postal 6-641, Mexico, D.F. 06600, MEXICO

1 INTRODUCTION

Gluten strength and extensibility, the main characteristics defining the bread making quality of bread wheat (*Triticum aestivum* L.), are determined mainly by the gliadin and glutenin subunits present in a given genotype. Within the high- (HMW-GS) and the low- (LMW-GS) molecular weight glutenins, those controlled by genes located at the *Glu-D1* and *Glu-B3* loci, respectively, are the ones that contribute most to variations in gluten properties.[1-3]

In the wheat breeding program of the International Maize and Wheat Improvement Center (CIMMYT), the main quality objective is to improve gluten strength and extensibility to obtain wheat germplasm suitable for the manufacture of diverse baking and non baking products (pan-type bread, flat breads, noodles, etc.). This has been addressed partly by reducing as much as possible the frequency in new crosses of common HMW-GS alleles of the *Glu-A1* (null allele), *Glu-B1* (7, 6+8, 20, 20+20, 13+19), and *Glu-D1* (3+12, 4+12) loci, known to be associated with inferior dough mixing properties, strength, and extensibility. Although the frequency of genotypes possessing weak and inextensible gluten has been substantially reduced, many new lines still possess undesirable quality attributes, particularly limited gluten extensibility. Hence, correcting HMW-GS composition has not been sufficient to improve the gluten properties of CIMMYT wheat germplasm. Recently, we started to determine glutenin subunit composition using an even shorter version of the rapid SDS-PAGE procedure of Singh *et al*[4]. This procedure gives good electrophoretic separations of HMW-GS and LMW-GS and, therefore, permits rapid determination of HMW-GS and the LMW-GS compositions for breeding purposes.

The aim of this work was to compare the quality characteristics of wheat genotypic groups representing various *Glu-D1/Glu-B3* glutenin subunit combinations and to determine the association of quality parameters commonly used in wheat breeding (SDS-sedimentation, mixograph dough mixing time, alveograph strength and extensibility, and bread loaf volume) to HMW-GS and LMW-GS, known to play a major role in bread making quality.

2 MATERIALS AND METHODS

Spring bread wheat cultivars (363) of diverse origin where grown under irrigation in Sonora, Mexico, during the 2001-02 crop cycle. The quality parameters measured were SDS-sedimentation (SDS-S) as described by Peña et al.[5]; mixograph mixing time (MT); alveograph dough strength and extensibility (W, P/L, using variable water absorption); and bread loaf volume (LV, using the straight dough baking method), according to AACC official methods.[6]

Glutenin and gliadin protein fractions were obtained according to Singh et al.[4] with the following modifications:

1. Gliadins were extracted in a single step. Whole meal (20 mg) was extracted with 1.5 ml of 50% propanol (v/v) for 5 min using continuous vortex mixing, followed by incubation (20 min at 65^0 C), vortexing (5 min), and centrifugation (5 min at 6,600 g). The supernatant and the residue so obtained were taken as the final gliadin and glutenin fractions, respectively. These two fractions were separately prepared according to Singh et al.[4] for SDS-PAGE analysis.
2. The protein separating gel had a single concentration of acrylamide (13.5% T) and was prepared using 1M Tris buffer, with a pH of 8.5 instead of the conventional pH 8.8. Gels were run at 15 mA for 20-21 h.

HMW-GS were classified using the nomenclature of Payne and Lawrence[7], and the LMW-GS and gliadins according to the nomenclature of Singh et al.[4] and Jackson et al.[8]

The various genotypic groups examined were unbalanced and therefore compared for quality characteristics by ANOVA using a GLM procedure.[9]

3 RESULTS AND DISCUSSION

Figure 1 shows SDS-PAGE patterns of glutenin and gliadin allelic variants. Our simplified gliadin extraction procedure still leaves some contaminant gliadin in the glutenin residue (Figure 1A). However, the faint omega-gliadin bands helped to confirm the identification of corresponding linked *Glu-B3* alleles. The use of Tris buffer at pH 8.5 in the running gel improved resolution of glutenin bands as compared to results at pH8.8, especially for B-type LMW-GS. Figure 1B shows the SDS-PAGE separation of gliadins, which helped us identify LMW-GS of the B and D groups.

While our primary aim was to associate quality parameters with glutenin subunits coded by *Glu-D1* and *Glu-B3* alleles, we also observed other allelic variation among *Glu-1* and *Glu-3* glutenins. Three allelic *Glu-A1* variants (0, 1, 2*) occurred in the population. When *Glu-D1* 2+12 was present, no significant differences among *Glu-A1* alleles were observed for all quality parameters. However, when *Glu-D1* 5+10 was present, the *Glu-A1* null allele (0) was slightly inferior in quality to subunit 1 for all parameters evaluated. Four allelic variants at *Glu-B1* occurred (7*+8, 7+8, 7+9, 17+18). In most cases, quality of 7+9 lines was inferior to that of at least one (7*+8) other allele, partly because of the high frequency of 7+9 in lines having the 1B/1R translocation, which has negative effects on bread making properties.

Mean values for quality parameters of the *Glu-D1* genotypic groups 5+10 and 2+12 are shown in Table 1. Group 5+10 had significantly larger values for MT, W, and LV than did group 2+12, but there were no significant differences in SDS-S and P/L between these groups. These results confirm that variation among *Glu-D1* HMW-GS is a major contributor to gluten strength.

IMPORTANCE OF HMW AND LMW GLUTENIN SUBUNITS AND THEIR INTERACTIONS ON BREAD-MAKING QUALITY

L.A. Pflüger[1], D. Lafiandra[2], S. Benedettelli[3]

[1] Instituto de Recursos Biológicos, CIRN, INTA, Las Cabañas y Los Reseros s/n, (1712) Castelar, República Argentina.
[2] Dipartimento di Agrobiologia ed Agrochimica, Università degli Studi della Tuscia, Via San Camilo de Lellis s/n, 01100 Viterbo, Italy.
[3] Department of Agronomy Science and Land Management, University of Florence, Piazzale delle Cascine 18, Florence, Italy.

1 INTRODUCTION

A random population of recombinant inbred lines (RILs), derived from the cross between 'Synthetic' (*T. turgidum* ssp. *durum* x *Ae. tauschii*) and *T. aestivum* cv. 'Opata 85', developed with the major objective of creating a molecular-marker linkage map of hexaploid wheat, was used to explore the effects of individual HMW and LMW glutenin alleles, and their interactions, on gluten strength. Such lines permit estimation of the genetic effects of particular loci in many related lines. The contribution of the storage proteins from durum wheat and *Ae. tauschii* of synthetic hexaploid wheat is of particular interest due to effects on characteristics that may arise from allelic variation at the *Glu-B3* locus of durum wheat and at *Glu-D1* and *Glu-D3* loci from *Ae. tauschii*.

2 MATERIAL AND METHODS

F_8 seeds of 87 RILs derived from the cross above described were sown in Viterbo. The material was electrophoretically analysed and advanced by single plant propagation to F_{10}.

Gliadins were separated by A-PAGE (aluminium lactate buffer, pH 3.1)[1]. Glutenins were analysed by SDS-PAGE in 8% polyacrylamide gel and in 7.5-13% linear acrylamide gradient gel with 1.28% cross-linker concentration. Glutenins were also analysed by one-dimensional A-PAGE and two-dimensional A-PAGE x SDS-PAGE[2].

The HMW-GS were designated based on the numbering system for wheat[3] and *Ae. tauschii*[4].

Each of the RILs was classified according to its allelic composition at the *Glu-1* and *Glu-3* loci. Eighty-seven lines were used for a field experiment. Fifty-two out of 64 (2^6) possible combinations were present in the 87 selected RILs. These lines were sown as single rows (plot) in a randomised block design with three replications. The SDS-sedimentation volume test (SDSS) was used as measure of flour gluten strength. SDSS was performed using 1 g of whole-grain meal[5].

The SDSS data for each allelic combination were analysed statistically applying the random effect model of the ANOVA, to compute the expected value of the variance of each source of variance (main effect of each loci and first order interactions).

Genetics and Quality 159

3 RESULTS

3.1 Protein Characterisation of RILs

The parental genotypes showed the following composition in HMW-GS: 'Opata 85' (2*, 13+16, 2+12), 'Synthetic' (null, 7+8, 1.5+T2). All eight possible combinations between these loci were represented in the F_8 lines.

For the identification of gliadin blocks, coded by each group 1 chromosome, 'Synthetic', 'Altar 84', 'Opata 85' and the RILs were analysed by A-PAGE. The same material was analysed by SDS-PAGE and one-dimensional A-PAGE to identify different alleles present at the *Glu-3* loci. Based on the gliadin allelic composition of the genotypes analysed, it was possible to attribute some LMW-GS in the A-PAGE and SDS-PAGE patterns to each of the three *Glu-3* loci on chromosomes 1A, 1B and 1D. For the statistical analysis HMW-GS and LMW-GS alleles of 'Opata' and 'Synthetic', have been indicated as "1" and "2", respectively (Fig. 1).

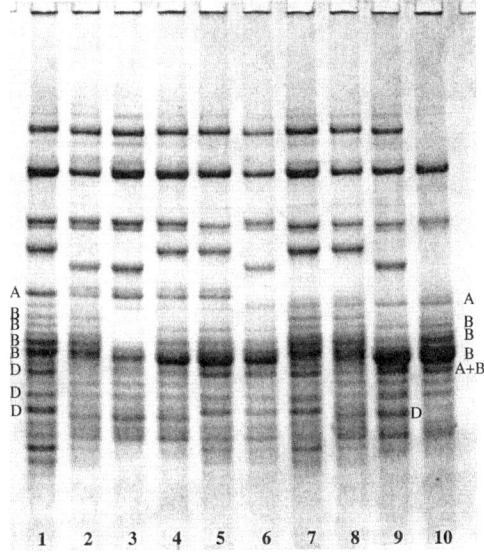

Figure 1: *One-dimensional A-PAGE patterns for LMW-GS. Letters A, B and D indicate subunits encoded by genes at the Glu-A3, Glu-B3 and Glu-D3 loci, respectively. Lanes: 1) Opata (LMW-GS code 1 1 1); 9): Synthetic (2 2 2); 10): Altar 84 (2 2 -). Lanes 2 to 8: RILs with diverse combinations of LMW-GS; 2): 1 1 1; 3): 1 0 2; 4): 1 2 2; 5): 1 2 1; 6): 2 2 1; 7): 2 1 1; 8): 2 1 2.*

3.2 HMW and LMW and First-order Interaction Effects

The effect of each allelic variant at the *Glu-1* and *Glu-3* loci is shown in Figure 2. Lines with the allelic composition 1-2-2-1-2-1 positively influenced the sedimentation values, whereas lines with the opposite composition had a negative effect.

For the *Glu-1* loci, only *Glu-A1* has a significant variance for grain quality. At this locus, lines with the 'Opata' allele (subunit 2*) produced a positive effect on sedimentation volume compared to the negative contribution of the 'Synthetic' allele (null). Allelic variation at both the *Glu-B1* and *Glu-D1* loci had no significant influence on sedimentation volumes. No differences were found among lines possessing bands 13+16 ("Opata type") or 7+8 ("Synthetic type") at the *Glu-B1* locus. The mean sedimentation values of lines with the *Glu-D1* pair of glutenin subunits 2+12 was similar to those carrying the Glu-D^t1 allelic variant 1.5+T2.

ANOVA results for LMW-GS indicated that SDS-sedimentation volume was significantly affected by the allelic variation at the *Glu-A3* and *Glu-B3* loci, whereas variation at *Glu-D3* locus had no significant influence. The allelic variations observed at *Glu-A3* and at *Glu-B3* explained about 12.83% and 12.79% of the total variance, respectively.

Allelic variation at the *Glu-3* loci, including their interactions explained 32.1% of total variation (Fig.2).

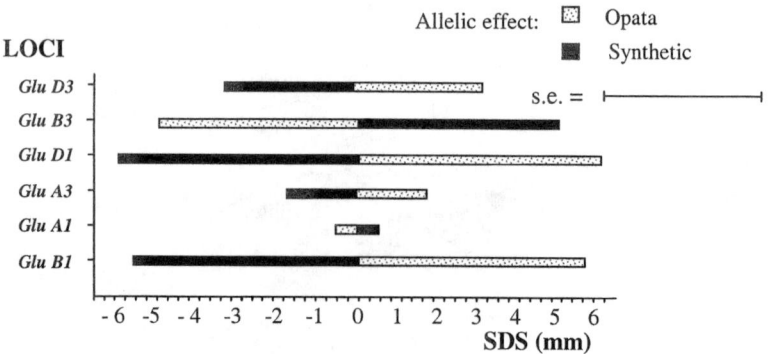

Figure 2: *Effect of the allelic HMW-GS and LMW-GS loci on the range of sedimentation volume. Total and Relative contribution to the total variance (partials H^2): Total H^2 = 0.5671, H^2 HMW= 0.139, H^2 LMW= 0.256, H^2 LMW + interaction within LMW = 0.321, H^2 HMW + LMW = 0.395, H^2 interaction between HMW + LMW = 0.106876.*

Considering a model with only additive effects of HMW-GS and LMW-GS, 39.05% of total variance was explained, while a model taking into account additive and epistatic effects explained 56.71% of total variance. Five out of fifteen first-order interactions were significant.

4 DISCUSSION

Generally, it is considered that HMW-GS are the major determinant of bread-making quality and the LMW-GS are relegated to a minor role. Our results show clearly that both LMW-GS and HMW-GS have significant effects on SDS-sedimentation volume. The parental genotypes have a limited variability of HMW-GS on bread-making quality (H^2 =

13.9%). This allowed the influences of LMW-GS on sedimentation volume to be more easily observed.

The results show that while there was no significant effect of individual allelic variation at the *Glu-B1* and *Glu-D1* loci, there was a significant effect of the *Glu-A1* locus. Lines with subunit 2* showed positive effect on sedimentation volume, in agreement with the expectation that the null allele is normally associated with weaker dough.

The significant effect of the *Glu-A3* locus on sedimentation volume observed in our study agrees with other results[6,7].

The positive effect on sedimentation volume observed for *Glu-B3* locus of 'Synthetic', suggests that there is a scope for improving gluten quality of bread wheat by utilising this allele, normally present in durum wheat varieties and associated with superior gluten properties.

Our results show that HMW-GS and LMW-GS account for only a part of the variation on sedimentation volume (13.9% and 25.6%, respectively). Both additive and epistatic effects between high- and low-molecular weight glutenin subunit loci had a significant effect on sedimentation volume of the RILs analysed. Manipulation of the LMW-GS composition could modify the bread-making quality of lines with specific HMW-GS composition. The relative importance of interactions indicated that they should be taken into consideration in quality improvement schemes.

References

1 K. Khan, A.S. Hamada and J. Patek, *Cereal Chem.*, 1985, **62**, 310.
2 M.H. Morel, *Cereal Chem.*, 1994, **71**, 238.
3 P.I. Payne and G.J. Lawrence, *Cereal Res Comm*, 1983, **11**, 29.
4 R.J. Peña, J. Zarco Hernández and A. Mujeeb-Kazi, *J. Cereal Sci.*, 1995, **21**, 15.
5 J. Dick and J.S. Quick, *Cereal Chem.*, 1983, **60**, 315.
6 R.B. Gupta and K.W. Shepherd, in *Proc. 7th Wheat Genet. Symp.*, eds. T.E. Moller and R.M.D. Koebner, Bath Press, Bath, 1988, p. 943.
7 R.B. Gupta, N.K. Singh and K.W. Shepherd, *Theor. Appl. Genet.*, 1989, **77**, 57.

UNDERSTANDING THE FUNCTIONALITY OF WHEAT HIGH MOLECULAR WEIGHT GLUTENIN SUBUNITS (HMW-GS) IN CHAPATI MAKING QUALITY

G. Sreeramulu*, R. Banerjee, A. Bharadwaj, P.P.Vaishnav

Popular Foods Division, Hindustan Lever Research Centre, 64, Main Road, Whitefield, Bangalore-560 066, India (* correspondence author).

1 ABSTRACT

Forty five commercially released Indian wheat varieties were analysed for qualitative differences in high molecular weight glutenin subunits using SDS-PAGE. These varieties have also been characterized for dough and chapati quality attributes. Correlation among the differences in protein composition and dough, chapati (Indian unleavened flat bread) quality characteristics revealed that the wheat varieties possessing HMW-GS, 1B20 produced chapatis with higher consumer acceptance.

2 INTRODUCTION

In India over 200 wheat varieties have been released for commercial cultivation during last five decades. Out of the 76 million tons of wheat produced annually, 60-70% is being consumed in the form of wheat flour (Atta) for chapati (Indian unleavened flat bread). Chapati is the staple diet of >2/3 population in the Indian subcontinent. The majority of the wheat varieties released for commercial cultivation were bred for resistance to abiotic and biotic stresses and also for improving yield potential. These varieties are known to yield moderate to very good chapatis.[1] However, the biochemical basis in wheat leading to good chapati quality characteristics is largely unknown. Hence, a systematic study was carried out to identify the biochemical markers responsible for chapati quality. Because wheat glutenin proteins control the visco-elastic properties of dough and chapati, attempts were made to understand their role if any, in chapati pliability, the major consumer preference attribute.

3 MATERIALS AND METHODS

3.1 Materials

Wheat varieties used in this investigations were: BBY 2, C 306, DEWA 9107, DL 7882, DL 8033, GW 1, GW 173, GW 190, GW 273, GW 496, GW 5, GW 1, GW 173, GW 5, HALNA, HD 2198, HD 2329, HD 2338, HDR 199, HI 1077, HI 1418, HI 1454, HI 977,

HI 1555, HI 1456, HI 1555, HI 2033, HI 9498, HI 838, Hindi-62, HJ 1011, HW 2004, HW 2033, HW 542, HYB 65, JWS 17, K 65, Karachia 65, LOK 1, MACS 2496, HI-8498, MP 147, Narmada 4, PBW 343, PBW 524, PUSA, RAJ 1482, RAJ 2329, RAJ 3077, RR 21, Sharbati, Sonalika, Sujata, WH 147 and WH 542. Seeds of these varieties were procured from different breeding stations and Agricultural Universities in India. In some cases the same variety was procured from different breeding stations. All chemicals used for protein extraction and electrophoresis were of analytical grade.

3.2 Methods

3.2.1 Total Seed Proteins Extraction

Total seed proteins from endosperm halves of single wheat kernel were extracted in a sample buffer (60 mM Tris pH 6.8, 2% SDS, 15% Glycerol, 1% DTT) according to the method described elsewhere.[2] Twenty random seeds from each variety were analysed for HMW-GS composition by SDS-PAGE as mentioned below.

3.2.2 SDS-PAGE Analysis for HMW-GS

SDS-PAGE was performed on Hoefer vertical electrophoresis system.[3] The stacking gel had 3% acrylamide, 0.08% bisacrylamide, 0.1% SDS, 125 mM Tris-Cl (pH 6.8). The separating gel contained 0.15 (w/v) SDS and 375 mM Tris-Cl (pH 8.9). A 10% acrylamide gel concentration was used for the analysis of glutenins and 8% acrylamide gel concentration was used for separation of 2* from the subunit 2. Electrophoresis was performed at constant current of 40 mA per gel. After electrophoresis, proteins were visualized by staining overnight with 0.025% Coomassie Brilliant Blue-R250 containing 18% methanol, 7% acetic acid and 5% trichloro acetic acid. Destaining of the gels was performed by placing them into 3% NaCl solution for 6 hours.[4]

3.2.3 Chapati Preparation and Sensory Evaluation

Chapati were prepared with 25 g dough by sheeting to 150 mm diameter with 1.5 mm thickness. Freshly prepared chapati were served to a trained sensory panel of 14 members for pliability evaluation on a 0-10 scale.

4 RESULTS AND DISCUSSION

4.1 Relationship Between the HMW-GS Variation and Chapati-Making Quality

Out of the 45 varieties, 15 variations were identified for high molecular weight glutenin subunits (HMW-GS). An example of representative variations for HMW-GS for Indian wheat varieties is shown (Figure 1). Two alleles were found for *Glu-D1*, three alleles for *Glu-A1* and six alleles were identified for the *Glu-B3* locus (Figure 1). In the literature, it is clearly established that the qualitative variations amongst HMW-GS in wheat varieties showed significant influence on bread-making quality.[5] Such knowledge has been lacking for chapati-making quality. The analysis of the 45 wheat varieties for various attributes of chapati quality revealed a range of pliability, which was proven to be an important attribute for chapati-making quality. The influence of qualitative variations in HMW-GS for different wheat varieties on pliability variation for chapati is shown in Table 1.

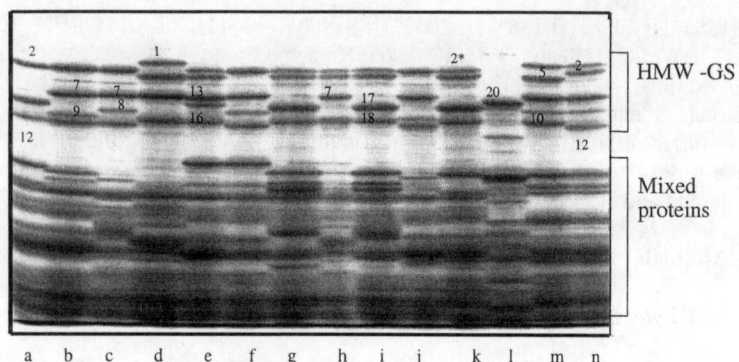

Figure 1 SDS-PAGE analysis of total protein extracts from representative variants amongst high molecular weight glutenin subunits (HMW-GS) in Indian wheat varieties. Lane a: KW-11, b: Raj-2329, c: WH-417, d: PB-343, e: PUSA, f: GW-173, g: K-9107, h: WH-542, i: GW-273, j: HD 2329, k: HI-977, l: HI-8498, m: BBY-2, n: BBY-2 from different source

Wheat varieties can be categorized into three groups based on the relationship between the pliability scores and HMW-GS composition (Table 1). The *Glu-B1x20* together with *Glu-A1null* alleles contributed significantly to the pliability of chapati. Similar observations

Table 1. Relationship between HMW-GS allelic variation and chapati-making quality (pliability score).

Variety	QDA Score	Wheat Genome			Bread
	Pliability	D1	A1	B1	Score
BBY-2	6.07	2+12	1	17+18	2+ 3+3 =8
JWS 17	6.29	5+10	1	7+8	4+ 3+3 =10
DI 7882	6.38	2+12	1	17+18	2+ 3+3 =8
DL 8033	6.44	5+10	1	7+9	4+3+2 =9
KHARACHIA	6.57	2+12	2*	17+18	2+3+3 =8
HYB-65	6.58	2+12	2*	13+16	2+3+2 =7
HI 1418	6.75	2+12	2*	7+8	2+3+3 =8
GW 173	6.95	2+12	2*	7+8	2+3+3 =8
HI 977	7	5+10	2*	17+18	4+3+3 =10
GW 273	7.13	5+10	2*	17+18	4+3+3 =10
LOK1	7.5	2+12	2*	17+18	2+3+3 =8
HI 8498	7.8	DURUM	N	20	1+2 =3
HW 2004	7.9	2+12	N	20	2+1+2 =5
SONALIKA	8.17	2+12	2*	7+9	2+3+2 =7
C-306	8.19	2+12	N	20	2+1+2 =5

have been made recently by other workers[6]. HMW-GS *1B20* is reported to have only two cysteine residues one in the N-terminal region and the other in the C-terminal domain[7]. The number of cystenine residues in other HMW-GS varies from 3-5[8]. We can speculate from these preliminary observations that the two cysteine residues present in HMW-GS *1B20* might be responsible for the specific gluten network, resulting in more pliable chapatis. In contrast, Sonalika, possessing *Glu-A1 2** and *Glu-B1 7+9* alleles, had pliability of 8.17. However, it is important to analyse a large number of wheat varieties in order to establish a clear relationship between HMW-GS composition and pliability of chapati. The information from the study of HMW-GS in wheat varieties could be utilized to improve the bread or chapati making quality[9].

5 CONCLUSION

Wheat varieties possessing HMW-GS *1B20* together with *1Anull* were found to have higher pliability scores for chapati. However, this finding must be further authenticated by analysing a large number of wheat varieties grown under similar agro-climatic conditions.

References

1. B.K. Misra, *Wheat Research Needs Beyond 2000 AD* (Eds. S. Nagarajan, G. Singh and B.S. Tyagi), Narosa Publishing House, New Delhi, Madras, Mumbai, Calcutta, London. pp 313.
2. G.J. Lawrence, Aust. J. Agric. Res., 1986, **37**, 12.
3. N.K. Singh, K.W. Shepherd and A. Cornish, *J. Cereal Sci.*, 1991, **14**, 203.
4. G. Sreeramulu and N.K. Singh, *Electrophoresis*, 1995, **16**, 362.
5. P.I. Payne, *Ann.l Rev. Plant Phys.*, 1987, **38**, 141.
6. F.M. Anjum, G.L. Lookhart and C.E. Walker, *J. Sci. Food Agric.*, 2000, **80**, 219.
7. F. Buonocore, C. Caporale and D. Lafiandra, *J. Cereal Sci.*, 1996, **23**, 195.
8. P.R. Shewry, N.G. Halford and A.S. Tatham, *J. Cereal Sci.*, 1992, **15**, 105.
9. R. Anwar, S Masood, M.A. Khan and S. Nasim. *Pak. J. Bot.*, 2003, **35**, 61.

Acknowledgments

Authors thank, Dr. S. Nagarajan, Dr. B.K. Misra, Dr. R.K. Gupta, Dr. Hanchinal, and Dr. H.N. Pandey for providing wheat varieties and Dr. A.V. Sawant for helpful suggestions.

Environmental Effects

PROTEOMIC ANALYSIS OF WHEAT ENDOSPERM PROTEINS: CHANGES IN RESPONSE TO DEVELOPMENT AND HIGH TEMPERATURE

W.J. Hurkman[1], C.K. Tanaka[1], W.H. Vensel[1], J.H. Wong[2], Y.Balmer[2], N. Cai[2], and B.B. Buchanan[2]

[1]US Department of Agriculture, Agricultural Research Service, Western Regional Research Center, 800 Buchanan St., Albany, CA 94710 and [2]Department of Plant and Microbial Biology, University of California, 111 Koshland Hall, Berkeley, CA 94720

1 INTRODUCTION

Environmental interactions during grain-fill alter the time course for grain development and influence final grain weight, protein content, and starch content.[1] Identification of molecular events influenced by environment will provide new insights into mechanisms that determine grain yield and quality. In this paper, we used a combined 2-DE, mass spectrometry proteomics approach to identify gliadins and glutenins, metabolic proteins and thioredoxin target proteins in wheat endosperm. Protein profiles were established to determine the effect of high temperature on accumulation patterns during grain development. The gluten proteins have been extensively characterized, because of their importance to the value of the wheat crop for baked products. Little is known about the far less abundant metabolic proteins that are involved in biosynthetic and regulatory processes central to grain development. Even less is known of mechanisms for regulating the accumulation and activity of the metabolic proteins.

2 METHOD AND RESULTS

2.1 Fractionation of Wheat Endosperm Proteins

Wheat (*Triticum aestivum* L. cv. Butte 86) plants were grown in a climate-controlled greenhouse[1] with an average maximum daytime temperature of 24°C and nighttime temperature of 17°C. At 10 days post anthesis (dpa) plants were transferred to a second greenhouse with identical growing conditions, but with respective daytime and nighttime temperatures of 37 and 28°C. Harvest maturity was reduced from 44 dpa for the 24/17°C regimen to 26 dpa for the 37/28°C regimen. Grain was harvested throughout development; endosperm was collected, frozen in liquid nitrogen, and stored at -80°C. Endosperm was fractionated sequentially[2] to obtain a KCl-insoluble fraction containing the gliadins and glutenins and a KCl-soluble/methanol-insoluble fraction containing mostly enzymes and other metabolic proteins (henceforth called "metabolic proteins"). High resolution 2-DE[2] revealed striking qualitative differences between these two fractions (Figure 1).

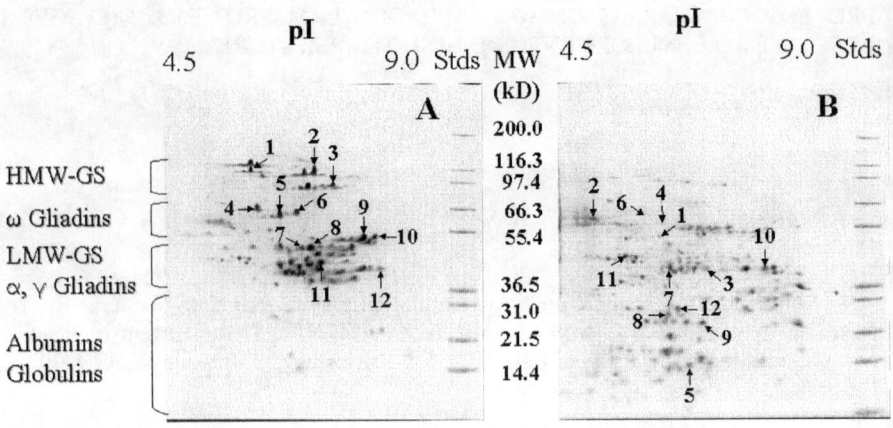

Figure 1 2-DE of proteins in the gliadin/glutenin (A) and metabolic protein (B) fractions from wheat flour. Protein classes according to Payne et al. [3] Arrows in A and B refer to accumulation profiles shown in Figures 2 and 3, respectively. A: 1, Ax2*; 2, Bx7; 3, 1By9; 4-6, 1B ω-gliadins; 7-9, γ-gliadins; 10-12, LMW-GS. B: 1, alanine aminotransferase; 2, protein disulfide isomerase; 3, 3-glyceraldehyde phosphate dehydrogenase; 4, ADP-glucose pyrophosphorylase; 5, α-amylase inhibitor 0.19; 6, pyrophosphate-fructose-6-phosphate phosphotransferase; 7, 3- glyceraldehyde 3-phosphate dehydrogenase; 8, peroxiredoxin; 9, α-amylase/subtillisin inhibitor; 10, peroxidase; 11, serpin; 12, avenin.

2.2 Accumulation Profiles of Gliadins and Glutenins

The identities of the HMW-GS, LMW-GS, ω-gliadins, and γ-gliadins indicated in Figure 1A were obtained using mass spectrometry. Protein accumulation profiles based on spot volumes were created using computer image analysis (Progenesis Workstation 2002.1, Nonlinear Dynamics, Newcastle upon Tyne, UK). The gluten proteins accumulated coordinately under the two temperature regimens (Figure 2) and the ratios of the different proteins essentially remained constant irrespective of temperature and time to harvest maturity. Under the 24/17°C regimen, proteins began to accumulate by 10 dpa and reached maximum levels by 22-28 dpa with little change at 40 dpa. The HMW-GS, γ-gliadins, and LMW-GS remained at high levels until grain maturity (not shown), whereas the ω-gliadins peaked at 16 dpa and decreased thereafter. Profiles were similar for the 37/28°C regimen, except that maximal accumulation of the proteins occurred at 18-22 dpa.

2.3 Identification of Metabolic Proteins and Thioredoxin Target Proteins

Mass spectrometry was used to identify nearly 200 metabolic proteins in the endosperm. Because they were not coordinately regulated and, thus, were not all present at a given time point, it was necessary to use early (10 dpa) and late (36 dpa) developmental stages to obtain the full protein complement. Identification revealed that the proteins function in a variety of cell processes: ATP-linked, carbohydrate metabolism, cell division, nitrogen metabolism, protein degradation, protein synthesis, storage, stress, and vitamin biosynthesis. As would be expected in the endosperm, the majority of proteins function in protein synthesis, starch metabolism, and defense. To gain insight into the regulation of the metabolic proteins, potential thioredoxin target proteins were identified (Figure 1B) using a fluorescent thiol-specific probe, monobromobimane.[4, 5] Twenty-three proteins were identified and found to function in metabolism, protein synthesis, protein degradation, storage, and stress response.

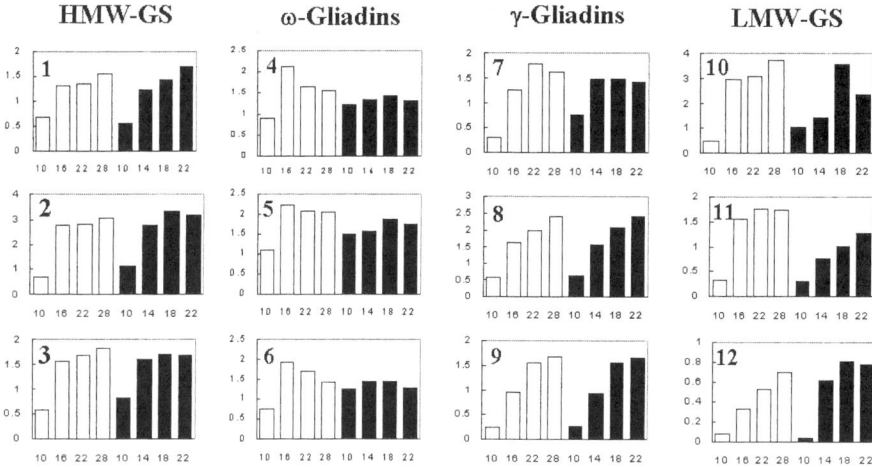

Figure 2 *Accumulation profiles of gliadins and glutenins during grain development and in response to high temperature.* □, $24/17°C$; ■, $37/28°C$. *Relative spot volume (y axis) graphed as a function of dpa (x axis). Proteins 1-12 identified in Figure 1A.*

2.4 Accumulation Profiles of the Metabolic Proteins

The accumulation profiles for the metabolic proteins were much more complex than for the gliadins and glutenins. In general, the acidic to neutral pI proteins were predominant at 10 dpa and the low- to mid-MW proteins were predominant at 36 dpa (not shown). Profiles of the metabolic proteins that are thioredoxin targets (Figure 3) could be separated into those that were most abundant during early, middle, middle-late, or late stages of development and, thus, provide candidates for markers of endosperm age. Unlike the storage proteins, the thioredoxin targets were not coordinately regulated, probably because they have specific functions at different stages of development. Like the storage proteins, many of the metabolic proteins peaked earlier during development under the high temperature regimen, consistent with the shortened time to harvest maturity.

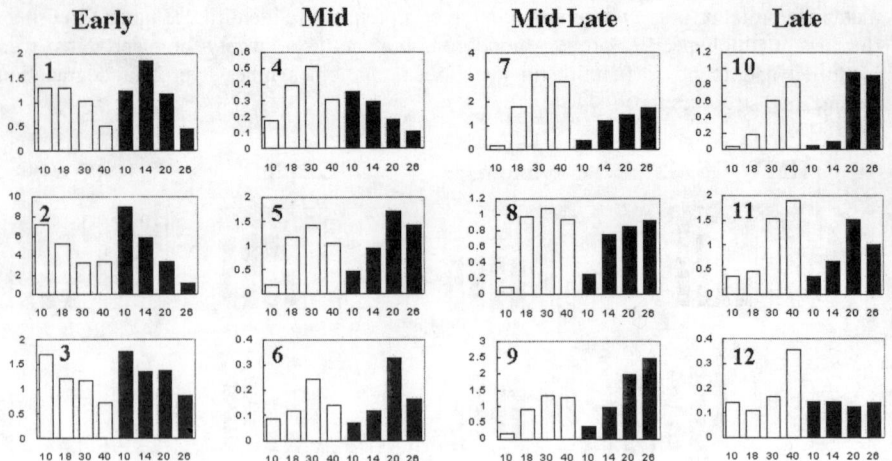

Figure 3 *Accumulation profiles of thioredoxin target proteins during grain development and in response to high temperature.* □, *24/17°C;* ■, *37/28°C. Early, Mid, Mid-Late, and Late refer to stages where proteins were maximal at 10, 30, 30-40, and 40 dpa, respectively, under the 24/17°C regimen. Relative spot volume (y axis) graphed as a function of dpa (x axis). Proteins 1-12 identified in Figure 1B.*

3 CONCLUSION

Proteomics demonstrated the coordinate regulation of the gliadins and glutenins, the timing of metabolic processes during endosperm development, and their response to high temperatures during grain-fill. The identification of thioredoxin target proteins opens a new door for elucidation of mechanisms regulating metabolic processes in wheat endosperm, such as those based on phosphorylation and thiol redox status. These approaches will provide new insight into factors responsible for grain yield and quality.

4 REFERENCES

1 S.B. Altenbach, F.M. DuPont, K.M. Kothari, R. Chan, E.L. Johnson and D. Lieu, *J. Cereal Sci.*, 2003, 37, 9.
2 W.H. Vensel, L. Harden, C.K. Tanaka, W.J. Hurkman and W.F. Haddon, *J. Biomolec. Tech.*, 2002, 13, 95.
3 P.I. Payne, L.M. Holt, J.G. Jarvis and E.A. Jackson, *Cereal Chem.*, 1985, 62, 319.
4 J.H. Wong, Y. Balmer, N. Cai, C.K. Tanaka, W.H. Vensel, W.J. Hurkman and B.B. Buchanan, *FEBS Letters*, 2003, 547, 151.
5 H. Yano, J.H. Wong, Y.M Lee, M.-J. Cho and B.B. Buchanan, *Proc. Acad. Natl. Sci. USA*, 2001, 98, 4799.

ANALYSIS OF THE EFFECT OF HEAT SHOCK ON WHEAT STORAGE PROTEINS

T. Majoul[1], E. Bancel[2], E. Triboi[3], J. Ben Hamida[1], G. Branlard[2]

[1] Unité de Biochimie, Faculté des Sciences de Tunis, Campus Universitaire 1060, Tunisia
[2] INRA ASP-UBP, 234 av. du Brezet 63039 Clermont Ferrand Cedex 02, France
[3] INRA APAC, 234 av. du Brezet 63039 Clermont Ferrand Cedex 02 France

1 INTRODUCTION

Elevated temperatures during grain filling have been identified as a major source of variation in dough properties and quality characteristics of different wheat crops in the world. The protein content that constitutes the prime measure of wheat quality has been extensively studied in relation to heat stress. Many studies reported that the increase in temperature throughout grain filling exerts a significant influence on the quantity of proteins that accumulate in the grain, as well as the ratios between storage protein fractions.[1] In previous works we analysed the effect of heat treatment during grain filling on the polypeptide patterns of bread wheat grain.[2,3] Using proteomic tools we showed that protein synthesis was severely affected by heat stress. The present study extends this previous work and describes the effects of a heat shock applied in a few days during grain filling on the protein composition of bread wheat grain. We were particularly interested in storage proteins composed of gliadins, high molecular weight glutenins (HMW) and low molecular weight glutenins (LMW).

2 MATERIALS AND METHODS

2.1 Plant Material

Wheat plants (*Triticum aestivum* cv. Récital) were grown under field conditions, in 2 m^2 containers that were transferred to climate tunnels soon after flowering. They were subjected to two thermal regimes, 18°C/10°C (day/night) for control and treated plants apart from the heat shock. Stressed plants were subjected to a heat shock of 4 hours at 38°C with a cooler temperature of 20°C for 16 hours from the 18 day after anthesis (DAA) to the 21 DAA. Four samples were harvested during the post-anthesis period, here we discuss only the sample harvested just after the application of heat shock and corresponding to 25 DAA and 22 DAA for control and treated plants, respectively. After harvest, grains were lyophilized and stored at 4°C before analysis.

2.3 Extraction and Separation of Total Proteins

Total proteins were extracted from the whole wheat grain and separated by 2-DE as described,[3] with a modification in the immobiline pH gradient (IPG) step that was performed with a total run of 60 kVh (0.3 kVh for the first step, 2.9 kVh for the second step and 56.8 kVh for the last step). Six to seven replicates of 2-DE were performed for each grain sample. Stained gels were scanned and analyzed using Melanie-3 software (GeneBio, Geneva Switzerland). Variance analysis of spot volume and comparisons of mean values of treated versus control samples were performed using SAS statistical software and a statistical significance level of $p < 0.05$.

2.2 Extraction and Separation of Storage Proteins

Storage proteins were extracted from the whole wheat grain as described.[4] They were separated by two-dimensional electrophoresis (2-DE), stained and scanned as performed for total proteins.

3 RESULTS

Statistical analyses that were performed on total proteins revealed significant changes in the amounts of some spots that were also detected in the storage proteins patterns. As shown in Figure 1, two of these spots were identified as gliadins (gli 1, gli2) whereas five others to three LMW glutenin subunits (LMW1, LMW 2, LMW3) and to HMW glutenin subunits 5 and 6 (sub5 and sub6). Table 1 illustrates a comparison of the amounts, expressed as the volume of each spot, under control and stressed conditions.

Protein name	Control volume	Stress volume	X fold
LMW1	2.83	4.07	1.43
LMW2	0.96	1.13	1.17
LMW3	0.5	0.73	1.45
Gli1	1.2	1.6	1.34
Gli2	1.37	2.05	1.48
Sub5	0.75	0.92	1.2
Sub6	1.4	1.75	1.25

Table 1 V*olume of storage proteins that were differentially expressed under heat shock; the third column shows the ratio of the volume of the spot under stressed conditions to that under control conditions.*

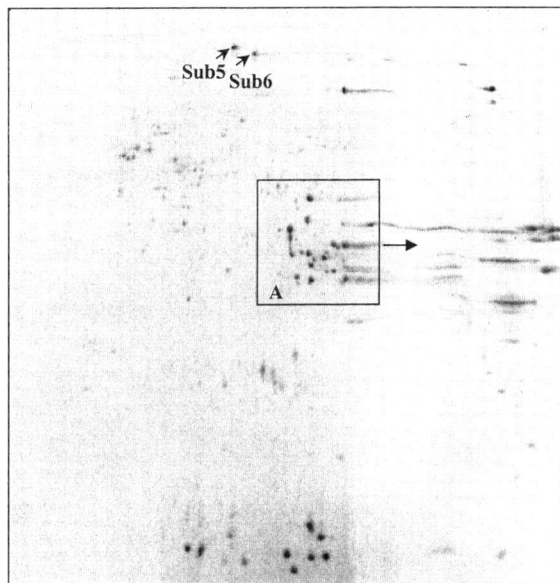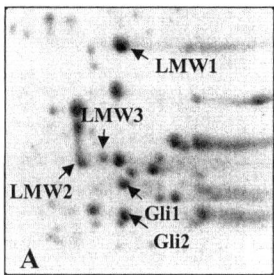

Figure 1 *A two-dimensional gel of total proteins extracted from the wholewheat grain in control conditions at 25 DAA. Sub5 and sub6 are the two HMW glutenins that were significantly changed under heat shock. Region A shows the gliadins and LMW glutenins that were differentially expressed under heat shock.*

4 CONCLUSION

These results confirm the involvement of storage proteins in the heat stress response of bread wheat grain. In effect, we observed that heat shock induced increases in the amounts of seven spots identified as storage proteins. Taking account of the fact that LMW subunits are very similar structurally to the gliadins, we suggest that gliadins are more affected by the stress then glutenins. It has been reported that the activation of gliadin synthesis is due to the presence of heat stress elements (HSE) in the upstream regions of some gliadin genes.[5]

However, we noticed that some gliadins were not affected by the stress. This may be explained by the existence of a mechanism of gliadin transport that is different from that functioning through the Golgi complex. In effect, under stress conditions, gliadins were reported to be transported from the endoplasmic reticulum directly to the vacuole bypassing the Golgi complex.[5]

References

1 E. Triboï and A.M. Triboï-Blondel, *Aspect Applied Biology*, 2001, **64**, 91.
2 T. Majoul, E. Bancel, E. Triboï, J. Ben Hamida, G. Branlard, *Proteomics,* 2003, **3**, 175.
3 T. Majoul, E. Bancel, E. Triboï, J. Ben Hamida, G. Branlard, *Proteomics* (in press).
4 J. Dumur et al, *J. Plant. Physiol*, submitted.
5 G. Galili, Y. Shimoni, S. Gioini-Silfen, H. Levanony, Y. Altsshuler and N. Shani, *Plant Physiol. Biochem*, 1996, **34**, 245.

GLUTENIN PARTICLES ARE AFFECTED BY GROWING CONDITIONS.

C. Don[1], G. Lookhart[2], H. Naeem[3], F. MacRitchie[3] and R.J. Hamer[1]

[1]CPT-TNO-Wageningen University, the Netherlands. [2]USDA-ARS GMPRC, Manhattan, KS. [3]Grain Science and Industry Department, Kansas State University, Manhattan, KS, USA.

1 INTRODUCTION

Wheat quality is governed by a combination of genetic and environmental factors. Growing conditions can seriously affect the quality of any given variety. The changes in processing requirements resulting from this are not always understood. Also, some varieties seem to be more susceptible than others. The quantity of unextractable glutenin protein aggregates (like GMP, Glutenin Macropolymer, and %UPP (% Unextractable Polymeric Protein), is thought to play an important role in dough mixing properties. Recently it was discovered that GMP consists of glutenin particles[1]. Size measurements revealed a clear relation between glutenin particle size and dough mixing requirements[2]. We therefore studied the effect of growing conditions, in particular heat stress on the quantity and properties of glutenin particles.

2 METHOD AND RESULTS

2.1 Flour lines and analyses

Four near-isogenic lines were used, Lance C (HMWGS 2*, 17+18, 5+10) + Lance A (HMWGS 2*, 17+18, 2+12) and Warigal A (null, 7+8, 5+10) + Warigal B (null, 7+8, 2+12). The wheat lines were grown to maturity under controlled conditions, using various temperature regimes –t°C day/ t°C night- to simulate six different stress levels. Wheat kernels were milled on a Quadrumat JR. Mixing properties were determined with a National Mixograph. GMP was isolated by dispersing flour in 1.5% SDS followed by ultracentrifugation. The GMP gel-layer rigidity was measured with a Bohlin VOR; particle properties were determined viscometrically, with a Coulter Laser particle sizer and observed with confocal scanning laser microscopy (CSLM).

2.2 SDS extractable proteins, GMP and UPP

Protein data and mixogram data provide insight on how the stress regimes have affected wheat flour quality parameters. Table I summarizes the results.

Table I *Flour proteins, GMP, UPP and MT for Lance and Warigal samples vs. treatment*

Treatment day/night, timing	Lance 2+12				Lance 5+10				Warigal 2+12				Warigal 5+10			
	P mg	G%	U%	MT'	P mg	G%	U%	MT'	P mg	G%	U%	MT'	P mg	G%	U%	MT'
1. 20°C/16°C. till maturity	127	20.2	40.2	3.6	154	34.6	50.6	5.3	135	21.1	37.3	2.7	147	27.2	41.4	4.1
2. 30°C/18°C, 72h 16DAA then 1	132	18.6		3.0	164	34.7		4.6	144	22.0		2.4	142	28.6		4.4
3. 35°C/20°C, 72h 16DAA then 1	119	19.1	40.5	2.8	167	29.3	45.3	3.3	142	23.3	41.4	2.8	141	30.0	48.4	4.5
4. 35°C/20°C, 16 DAA till maturity	132	15.1		3.1	154	23.3		3.2	145	16.0		2.9	144	22.9		4.2
5. 40°C/25°C, 16 DAA, till maturity	117	9.0		5.8	142	19.5		3.6	136	9.7		5.4	133	13.3		5.3
6. 40°C/25°C, 25 DAA, till maturity	127	16.4	33.2	3.5	145	30.2	46.2	3.8	122	15.6	30.4	3.0	136	20.2	33.2	3.2

DAA=Days after anthesis, **P** = SDS extractable + SDS unextractable protein (mg from 1g flour)
G = % of GMP protein in **P**, **U** = % UPP determined by Naeem et al.[3], **MT**= mixograph peak time

Heat stress is more severe going from treatment 1 to 5. At treatment 6, the heat stress is started later (25 DAA instead of 16 DAA), so this treatment is considered less severe than treatment 5. Protein content ranged between 12 – 17%. Overall, Lance 2+12 showed the lowest protein and GMP content; Lance 5+10 had the highest protein content; Warigal 5+10 and 2+12 showed protein and GMP values in between that of Lance 2+12 and 5+10. With increasing heat stress both % Protein and % GMP are affected, although for all lines % GMP is more sensitive than % Protein. Treatment 5 resulted in minimal GMP levels for all varieties. GMP levels are generally higher for 5+10 varieties, than for 2+12 varieties. In general Treatment 5 resulted in a minimal amount of GMP, indicating that heat stress strongly affects GMP. %UPP and GMP values correspond well ($R^2=0.86$, n=12), indicating that UPP and GMP refer to a similar fraction of glutenin. Heat treatment is reported to significantly affected dough mixing time values. With the 2+12 varieties dough mixing time increased from 3-5.8 min with the severity of temperature treatment. With the 5+10 varieties effects on % GMP are similar, but effects on mixing time could not directly be related to heat stress conditions. The quantity of insoluble glutenins like GMP or UPP is regarded to be a good predictor of dough development time (MT)[4]. However, with this sample set the correlation between %GMP and MT was poor (figure 1 $R^2<0.1$).

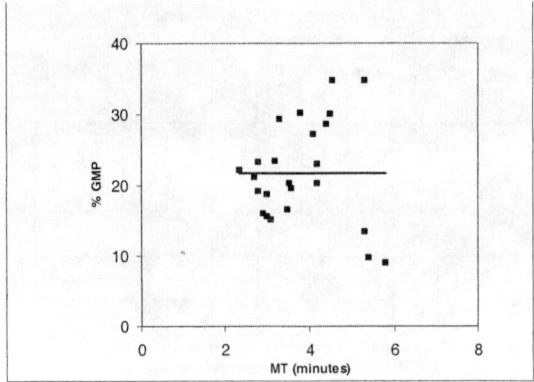

Figure 1 *MT vs. % GMP*

2.3 Glutenin particle size analysis

In a recent paper we have reported a clear relation between glutenin particle size and dough mixing requirements[2]. This study also indicated that GMP from flour contains very large particles, their size is much larger (>>20μm) than reported for wheat starch.
As a control starch granule size distributions were measured, they did not show large differences in relation to stress (data not shown). Addition of excess DTT to the GMP dispersion, while running the particle size measurement system, results in a strong decrease of particle sizes, especially in the > 20μm region. However even after reduction, protein aggregates appeared to be present, this is thought to be a result of the poor water-solubility of the HMW and LMW glutenins. GMP-gel can be solubilized combining 6% SDS with reducing agent (DTT), washing and centrifuging GMP-gel in this solvent leaves a very small quantity of white material. Collecting this remnant material for particle size analysis showed that this material has an average size of just 5μm. This demonstrates that the GMP dispersion consists of protein, but a little remnant starch can be present, this is apparently B-starch. The glutenin particle size measurements showed that severely heat stressed samples have very large glutenin particles. In figure 2 an example is given for Lance 5+10.

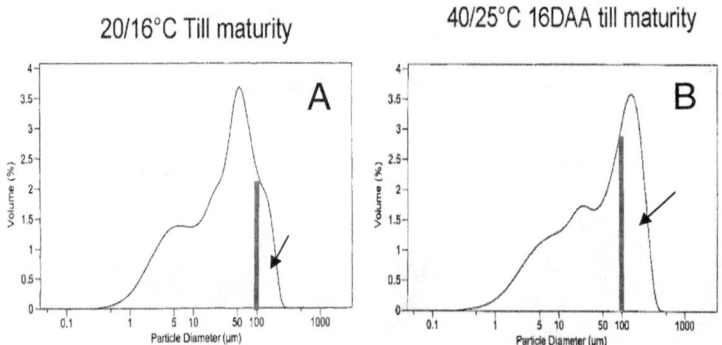

Figure 2 *Example of how stress conditions affect GMP particle size distribution*

In the Coulter Laser examples given in fig. 2 the protein concentration during the measurement is very similar. Figure 2 B indicates a more prominent presence of large particles (> 100mu) in heat stressed samples than in samples grown under mild conditions (2 A). Figure 3 clearly shows that the average particle voluminosity correlates better with MT than quantity. Glutenin particle size has been shown to play a role in the physical breakdown of glutenin particles (R2=0.94)[4]. We observed that in severely stressed samples more SDS soluble proteins were present in relation to SDS insoluble proteins. This suggests that effects of chemical assembly and/or breakdown mechanisms should also be considered.

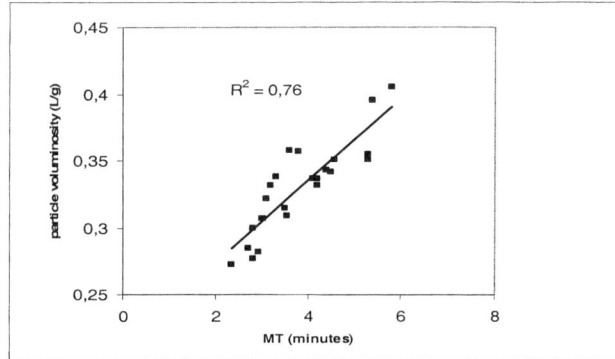

Figure 2 *MT vs. GMP particle size*

3 CONCLUSION

Within this set of samples it appears that glutenin particle size provides a new key to better understand how heat stress affects changes in dough MT. Further unravelling of the mechanism behind these tentative findings requires the linkage of GMP particle size and mixing properties with the HMWGS composition. So far the results show that the effects of heat stress are similar on 5+10 lines and 2+12 lines.

References

1 Don J.A.C, Lichtendonk W.J., Plijter J.J. and Hamer R.J. *J. Cereal Sci,* **37,** 1
2 Don J.A.C, Lichtendonk W.J., Plijter J.J. and Hamer R.J. *J Cereal Sci,* **38,** 157
3 Naeem H.A., Lookhart G.L. and MacRitchie F., unpublished results
4 Gupta R.B., Batey L., and MacRitchie F., *Cereal Chem.,* **69,** 125

GRAIN PROTEIN POLYMER FORMATION: INFLUENCES OF CULTIVAR, ENVIRONMENT AND DOUGH TREATMENT.

E. Johansson[1], M.L. Prieto-Linde[1], R. Kuktaite[1], A. Andersson[1], H. Larsson[2]

[1]Department of Crop Science, The Swedish University of Agricultural Sciences, Box 44, SE-230 53 Alnarp, Sweden
[2]Department of Food Technology, Lund University, PO Box 124, S-221 00 Lund, Sweden

1 INTRODUCTION

It is important to understand the reasons for the variation in breadmaking quality between wheat flours and cultivars, and research within the area has existed in many countries for a long period. Correlations have been established between particular proteins and protein subunits and different breadmaking quality parameters[1-6]. Another crucial parameter determining breadmaking quality is the amount and size distribution of polymeric proteins in cultivars and during bread dough formation[7-9].

The amount and size distribution of polymeric proteins in a flour is determined by the cultivar[8], and by environmental influences mainly during the grain-filling period[10-13]. During the dough formation process, the amount and size distribution of polymeric proteins are changed, which is related to the background of the flour and to the breadmaking process applied[9].

The aim of the present study was to investigate the formation of and changes in wheat grain polymeric proteins during grain development and dough mixing. The aim was to find cultivar and environmental reasons in grain polymer formation that could explain variation in breadmaking quality.

2 MATERIALS AND METHODS

2.1 Materials

The plant material comprised spring and winter wheat cultivars and breeding lines grown in Sweden during the period 1975 to 2003, in different locations (mainly in the south of Sweden but up to Central Sweden), and with different nitrogen fertilizer rates.

The plant material also comprised wheat cultivars and breeding lines of Baltic origin, grown in the Baltic countries[14]. Furthermore, spring wheat cultivars grown in climate chambers in nutrient solution culture[15] and in the glasshouse[13] were included.

For the dough studies, commercially available Swedish flour mixtures of varying gluten strength were used.

2.2. Methods

The cultivars were tested for quality at Svalöf Weibull AB, Svalöv, Sweden according to Johansson and Svensson[2].

For investigations of protein compositions, the proteins were extracted from white flour and separated on polyacrylamide gels in the presence of SDS according to Johansson[1].

For investigations of the amount and size distribution of protein groups and protein polymers and monomers, proteins were extracted from white flour and separated by RP- and SE-HPLC by the methods developed by Wieser and Seilmeier[16] and Gupta et al.[7] as described in Johansson et al.[8].

Statistical analyses, Spearman rank correlations, analyses of variance (ANOVA), principal component and principal factor analyses, were carried out using SAS[17].

3 RESULTS AND DISCUSSION

3.1 Variation in amount and composition of proteins

Similarly to earlier studies[16], protein concentration was significantly positively correlated with the total amounts of glutenins and gliadins and also with the amounts of most mono- and polymeric proteins.

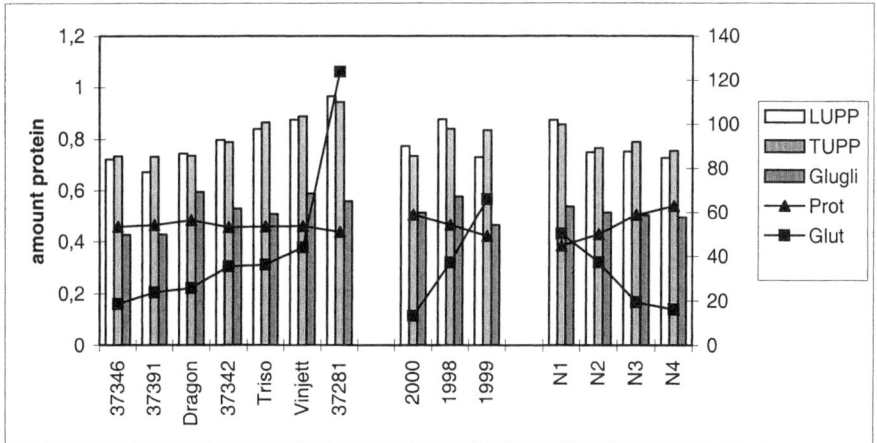

Figure 1. *Variation in the percentage of large unextractable polymeric protein in the total large polymeric protein (LUPP), total unextractable polymeric protein in the total polymeric protein (TUPP), glutenin/gliadin ratio (Glugli), glutograph dough development time (Glut) and 5 times protein concentration (Prot) in different wheat lines and years, and with different fertilizer rates.*

Cultivar influences giving rise to variation in gluten strength were found to influence the relation between SDS-soluble and –insoluble polymeric proteins, leading to a significant positive correlation between the gluten strength and the percentage of total unextractable polymeric protein in the total polymeric protein (TUPP) and large unextractable polymeric protein in the total large polymeric protein (LUPP; Fig 1). Environmental variation in gluten strength was found to be significantly positively correlated with SDS-insoluble proteins and negatively correlated to SDS-soluble proteins. This also led to a significant positive correlation with the percentage of LUPP and/or TUPP (Fig 1). Timing of nitrogen application was related to amount and size distribution of polymeric proteins. Late nitrogen application was negatively correlated with TUPP and LUPP.

3.2 Stability in breadmaking quality

Differences in stability of breadmaking quality parameters over environments was found between cultivars[8,18]. For variation over years, generally wheat cultivars with weaker gluten, containing HMW glutenin subunits 2+12, showed higher stability compared WITH stronger cultivars, containing HMW glutenin subunits 5+10[18]. The reason for this might be that quality cultivars for bread-making can be very strong under certain conditions, but weak under other. Cultivars with weak gluten cannot become much weaker during environmental conditions generally leading to weak gluten characteristics. As cultivar variation exists in breadmaking quality stability, the character is most likely to be genetically determined, although the reason might not be the protein composition.

3.3 Relationships between protein composition and the amount and size distribution of polymeric proteins

Specific differences in protein composition are related to differences in gluten strength between cultivars[1-6]. Specific protein composition partly influenced the amount and size distribution of polymeric protein in the grain. This was particularly valid for HMW glutenin subunits 2+12 and 5+10. Subunits 2+12 were correlated with a lower percentage of LUPP and TUPP than subunits 5+10. Cultivar differences in gluten strength were also related to differences in extractability of proteins, indicating differences in protein polymer conformation and/or disulphide bonds.

3.4 Protein accumulation during grain development

Protein concentration and influx time of nitrogen vary in the spikelet of wheat[15]. Thus, differences in amount and size distribution of polymeric proteins might exist within the spikelets of a spike.

In the beginning of the grain-filling period, high values of SDS-extractable monomeric proteins were found in the grains (most likely different types of enzymatic proteins). During grain maturation the SDS-extractable monomeric proteins decreased and SDS- unextractable polymeric proteins accumulated.

3.5 Influence of dough mixing time on the amount and size distribution of polymeric proteins

Dough treatment influenced the amount and size distribution of polymeric proteins. The highest percentages of TUPP and LUPP were present at about the optimal dough mixing time for the specific flour. For strong flour, the highest LUPP values were found during long mixing times, while for weak flour, the highest LUPP values were found during shorter mixing times[19].

4. CONCLUSIONS

The amount and size distribution of polymeric proteins is thus of high importance for breeding cultivars of good and even quality between years. A better understanding of the processes leading to built-up and break-down of protein polymers are of importance. This is also important when trying to explain the bread dough formation process.

References

1. E. Johansson, *Plant Breed.*, 1996, **115**, 57.
2. E. Johansson and G. Svensson, *Cereal Chem.*, 1995, **72**, 287.
3. E. Johansson, P. Henriksson, G. Svensson and W. K. Heneen, *J Cereal Sci.* 1993, **17**, 237.
4. G. J. Lawrence, H. J. Moss, K. W. Shepherd and C. W. Wrigley, *J. Cereal Sci.* 1987, **6**, 99.
5. P. I. Payne, M. A. Nightingale, A. F. Krattiger and L. M. Holt, *J. Sci Food Agric.* 1987, **40**, 51.
6. A. K. Uhlen, *Norwegian J. Agric. Sci.,* 1990, **4**, 1.
7. R. B. Gupta, K. Khan and F. MacRitchie, *J. Cereal Sci.*, 1993, **18**, 23.
8. E. Johansson, M.-L. Prieto-Linde and J. Ö. Jönsson, *Cereal Chem.*, 2001, **78**, 19-25.
9. R. Kuktaite, M.-L. Prieto-Linde, H. Larsson and E. Johansson, 2003a (manuscript).
10. E. Johansson and G. Svensson, *J. Sci. Food Agric.* 1998, **78**, 109.
11. E. Johansson, H. Nilsson, H. Mazhar, J. Skerritt, F. MacRitchie and G. Svensson, *J. Sci. Food Agric.* 2002, **82**, 1305.
12. E. Johansson, M.-L. Prieto-Linde, G. Svensson and J. Ö. Jönsson, *J. Agric. Sci.* 2003a, **140**, 275.
13. E. Johansson, M.-L. Prieto-Linde and A. Andersson, 2003b (manuscript)
14. E. Johansson, R. Kuktaite, M.-L. Prieto-Linde, R. Koppel, V. Ruzgas, A. Leistrumaite and V. Strazdina, 2003c (accepted).
15. A. Andersson, E. Johansson and P. Oscarson, 2003 (submitted).
16. H. Wieser and W. Seilmeier, *J. Sci. Food Agric.*, 1998, **76**, 49.
17. SAS. *User's Guide: Statistics.* Version 5 Edition. Cary, NC, USA: SAS Institute Inc. 1985.
18. E. Johansson, G. Svensson and S. Tsegaye, *Acta Agric. Scand.* 2000, **49**, 225.
19. R. Kuktaite, H. Larsson, S. Marttila, K. Brismar, M. Prieto-Linde and E. Johansson, 2003b In *Gluten workshop 2003. (these Proceedings)*

Acknowledgments

This work was supported by The Cerealia Foundation R&D, The Swedish Farmers Foundation, The Swedish Research Council for Environment, Agricultural Sciences and Spatial Planning, The VL-stiftelsen, The SL-stiftelsen and Svalöf Weibull AB.

RELATIONSHIP BETWEEN SOME PROLAMIN VARIANTS AND QUALITY PARAMETERS AT DIFFERENT LEVELS OF NITROGENOUS FERTILIZER IN DURUM WHEAT

J.M. Carrillo, M. Ruiz, M.C. Martínez, M. Rodríguez-Quijano and J.F. Vázquez

Unidad de Genética, E.T.S.I. Agrónomos. Universidad Politécnica de Madrid, Spain.

1 INTRODUCTION

In durum wheat the composition of endosperm prolamins plays an important role in the determination of semolina quality.

The nitrogen fertilization has long been recognized as a critical nutrient for wheat. The nitrogen input highly influences the grain yield, protein content and quality. Sustainable systems for reducing chemical inputs are becoming more important and they are required in many countries.

The integration of wheat quality and sustainability goals, eliminating negative impacts on the environment and assuring suitable levels of wheat quality, is a real objective in the activity of many wheat researchers.

The aim of this work is to determine in durum wheat the relationship between quality parameters and gluten proteins (prolamins) at low and high nitrogen fertiliser rates.

2 MATERIAL AND METHODS

A field experiment was established involving the parents and 37 F_3 lines of the progeny of the durum wheat cross 'Antón' x 'Blanco de Nolas'. The quality parameters analysed were vitreousness (V), protein content (P), SDS-sedimentation volume (SDSS), mixograph parameters: mixing time (MT), peak height (PH), height three minutes after the peak of the curve (H3) and breakdown resistance (BDR), and alveograph gluten strength parameter (W). All the quality characteristics were recorded in a trial with three levels of nitrogen applications, low ($N_{0.5}$= 40 u. /Ha), medium (N_1= 80 u. /Ha) and high ($N_{1.5}$= 120 u. /Ha), with two replications at each of the three levels. The medium nitrogen level corresponded to that usually applied by the farmers. The alveograph test was performed with mixed grain of the 32 lines of the two replications while the rest of the traits were carried out with the 37 lines of the two blocks (a total of 74 lines).

The prolamin allelic variants studied in F_3 lines were at the loci *Glu-A1*, *Glu-B1*, *Gli-A1/Glu-A3* and *Gli-B1/Glu-B3* (Table 1 and Figure 1). Prolamins were extracted[1] and fractionated in SDS-PAGE[2] and A-PAGE[3].

Table 1 *Allelic prolamin variants analysed for quality and the location of genes controlling them*

Locus	Antón	Blanco de Nolas
Glu-A1	Null	HMW 1
Glu-B1	HMW 7+8	HMW 6+8
Glu-A3/Gli-A1	LMW 5/ω-20+ω-24	LMW 6/Null
Glu-B3/Gli-B1	LMW 8+9+13+16/γ-42	LMW 1+3+13*+16/γ-47

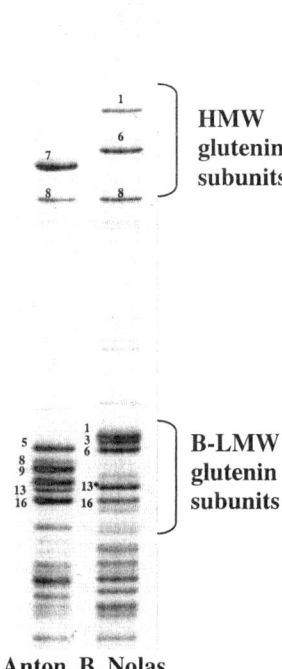

Figure 1 *SDS-PAGE electrophoregram of HMW and LMW glutenin subunits*

3 RESULTS AND DISCUSSION

3.1 Influence of N fertilisation on quality tests

The mean values of the quality tests for the three nitrogen fertilisation rates ($N_{0.5}$, N_1 and $N_{1.5}$) are shown in Table 2. Comparing the values obtained for the $N_{0.5}$ level (low nitrogen fertilisation) with the N_1 level (usual application) the values of the parameters V, P, SDSS, MH, H3 were significantly lower with $N_{0.5}$ than with N_1 fertilization rates. In contrast, the two levels of N showed no differential effects on MT, BDR and W. Correlations between protein content and the quality tests were determined. The traits V, MH and H3 were significantly correlated with P at $N_{0.5}$ and N_1 rates, while SDSS was only correlated at the $N_{0.5}$ level. There appeared to be a positive effect of protein content on V, MH and H3 in agreement with other[4,5]. So, the better values obtained with N_1 than with $N_{0.5}$ could be due to the greater protein content found with increasing N fertiliser rates. In contrast, the influence of N on SDSS could not be very important at normal and high N levels according to other studies[4,6]. Some studies, however, found a positive effect of P with SDSS values[7,8]. The other quality parameters showed different correlations with P: significant for MT, significant only at N_1 for BDR and not significant at any rate for W. Conversely other studies[9,10,11] with bread wheat found that increased P resulted in a higher alveogram index.

Table 2 *Mean values of the quality tests for each level of nitrogen fertilisation*

Quality test	$N_{0.5}$	N_1	$N_{1.5}$
V (%)	97.46 b*	98.62 a	98.54 a
P (%)	16.57 c	17.58 b	18.28 a
SDSS (mm)	30.49 b	34.65 a	32.89 a
MIXOGRAPH			
MT (s)	89.58 a	89.21 a	88.90 a
MH (mm)	86.07 b	90.43 a	89.67 a
H3 (mm)	62.78 b	66.57 a	64.11 b
BDR (%)	27.05 b	26.37 b	28.46 a
ALVEOGRAPH			
W (10^{-4} J)	71.50 b	75.45 ab	81.68 a

*Means with diverse letters are significantly different with the Duncan test

3.2 Effects of allelic variation in prolamins on quality

Table 3 *Percentage of variance explained by the nitrogen fertilisation effect (N) and by the nitrogen fertilisation plus prolamin locus effects.*

	V	P	SDSS	MT	MH	H3	BDR	W
N	17	34	9	0.1	18	6	11	6
N + *Glu-A1*	-	40	22	22	40	14	-	-
N + *Glu-B1*	-	-	15	-	-	-		
N + *Glu-B3*	-	-	24	4	-	-	17	21
N + *Glu-A3*	-	-	-	-	-	-	18	12

Table 3 shows the variation explained by the fertilisation effect and fertilisation plus prolamin locus effects for each quality test. The results indicated that fertilisation influenced V, P and BDR more than prolamin variants. The parameter MH was also highly affected by fertilisation. For the rest of the quality measurements, SDSS, MT, H3 and W, the prolamin effect was more important than N, particularly for MT and W where the fertilisation influence was no significant. All the glutenin loci showed significant effects on quality, especially *Glu-A1* on MT, MH and H3, and *Glu-B3* on BDR and W. The SDSS values were markedly affected by *Glu-A1* and *Glu-B3*. Moreover, interactions between N and prolamins were not significant. Considering the effect of prolamins on quality at the $N_{0.5}$ level (low Nitrogen fertilisation) the results indicated that prolamins at the locus *Glu-*

A1 improved SDSS, MT, MH, H3 and prolamins at *Glu-B3* increased SDSS values. In the case of BDR and W prolamin effects were only significant for fertilisation rates higher than $N_{0.5}$. Also, no correlation was found with P at $N_{0.5}$. However, W showed significant correlation with SDSS, MT, MH and H3 at $N_{0.5}$. Therefore, increasing these quality properties could result in an improving in W. High correlations between *Glu-A1* and *Glu-B3* allelic variants and SDSS, mixograph and alveograph parameters have been found in other studies of durum wheat with standard rates of N fertilisation[4,12,13,14].

4 CONCLUSION

Durum wheat quality decreased under low fertiliser conditions due to the positive effect of nitrogen on quality. Based on these results, gluten strength could be improved in low N fertilisation conditions by selecting lines with the best alleles at *Glu-A1* and *Glu-B3* (independent loci) and with high protein content.

References

1   N.K. Singh, K.W. Shepherd and G.B. Cornish, *J. Cereal Sci.*, 1991, **14**, 203.
2   P.I. Payne, C.N. Law and E.E. Mudd, *Theor. Appl. Genet.*, 1980, **58**, 113.
3   D. Lafiandra and D.D. Kasarda, *Cereal Chem.*, 1985, **62**, 314.
4   M. Ruiz and J.M. Carrillo, *Plant Breed.*, **114**,40.
5   C. Luo, G. Branlard G, W.B. Griffin and D.L. McNeil, *J. Cereal* Sci., 2000, **31**, 185. 6
6   J.C. Autran and G. Galterio, *J. Cereal Sci.*, 1989, **9**, 195.
7   R. Cubbada, N. Carcea and L.A. Pasqui, *Cereal Foods World Res.*, 1992, **37**, 866.
8   M.I.P. Kovacs, N.K. Howes, D. Leisle and J.H. Skerritt, *J. Cereal Sci.*, 1993, **18**, 43.
9   R. Ortiz-MOnasterio, J.I. Peña, R.J. Sayre and S. Rajaram, *Crop Sci.*, 1997, **37**, 892.
10 L. López-Bellido, M. Fuentes, J.E. Castillo and F.J. López-Garrido, *Fiel Crop Research*, 1998, **57**, 265.
11 J.A. Lloveras, A. López, J. Ferrán, S. Espachs and J. Solsona, *Agron.*, **93**, 1183.
12 D.L. du Cros, *J. Cereal Sci.*, 1987, **5**, 3.
13 E. Porceddu, T. Turchetta, S. Masci, R. D´Ovidio, D. Lafiandra, D.D. Kasarda, A.Impiglia and M.M. Nachi, Euphytica, 1998, **100**, 97.
14 C. Brites and J.M. Carrillo, *Cereal Chem*, 2001, 78, 59.

Acknowledgements

This work was supported by the grant AGL-2000-1280 from Comisión Interministerial de Ciencia y Tecnología (CICYT) of Spain.

INFLUENCE OF SULPHUR FERTILISATION ON THE QUANTITATIVE COMPOSITION OF GLUTEN PROTEIN TYPES IN WHEAT FLOUR

P. Koehler,[1] H. Wieser,[1] R. Gutser,[2] and S. von Tucher[2]

[1]German Research Centre of Food Chemistry, Lichtenbergstraße 4, D-85748 Garching, Germany
[2]Technical University of Munich, WZW Centre of Life Sciences, Department of Plant Sciences, Chair of Plant Nutrition, Am Hochanger 2, D-85350 Freising, Germany

1 INTRODUCTION

Dough properties and baking performance of wheat flours are decisively determined by the qualitative and quantitative composition of the gluten proteins[1]. This composition is strongly influenced by the cultivar (cv.) and the growing conditions. Whereas the cultivar determines the structure and the amount of the gluten proteins, the growing conditions only have a quantitative effect. Previous studies have shown that nitrogen (N) fertilisation significantly changes the amount and the ratio of specific gluten protein components[2]. The supply of sulphur (S) to the plant is also important for the quantitative composition of gluten proteins and, therefore, for the technological properties of wheat flour[3,4]. Qualitative studies have shown that gluten proteins are affected by S-fertilisation[5-8]. The aim of the present study was, therefore, to investigate the influence of S-fertilisation on the amount and the proportions of all gluten protein types present in wheat flour by means of a combined extraction/high-performance liquid chromatography (HPLC) method.

2 METHOD AND RESULTS

2.1 Wheat samples

The German spring wheat cv. Star was grown in containers on two different soils, B1 and B2 (7 kg of soil; B1: pH 5.6, sulphate-S 7mg/kg; B2: pH 6.1, sulphate-S 2 mg/kg) and fertilised with different S-levels (S0=0 mg, S1=40 mg, S2=80 mg, S4=160 mg per container, applied as calcium sulphate). N, phosphorous (P), potassium (K), and magnesium (Mg) supply were normal. Mature kernels were milled into wholemeal flours.

2.2 Analytical methods

The N-content was determined by the method of Dumas, the S-content by ICP-AES (Inductively Coupled Plasma-Atomic Emission Spectrometry) after microwave digestion. For the quantification of gluten protein types, 100 mg of flour were subsequently extracted with salt solution, 60 % (v/v) aqueous ethanol, glutenin extraction buffer and separated by

HPLC on C_8 silica gel[9]. The amounts of gliadins and glutenin subunits were calculated from the absorption areas of the peaks.

2.3 Basic determinations

Spring wheat cv. Star was grown in containers on two different soils (B1, B2). The plants were sufficiently supplied with N, P, K, and Mg. S was applied in four levels (S0, S1, S2, S4) and caused substantial differences in the kernel yield (Table 1). Wholemeal flours were prepared and analysed for their N- and S-content (Table 1). The N-content of the flours was not affected by S-fertilisation. Higher S-levels increased the S-content of the kernels. An S-content of 0.12 % can be considered as the lower limit for sufficient S-supply. Therefore, the kernels of the levels S0 and S1 were S-deficient and those of levels S2 and S4 were sufficiently supplied with S (Table 1).

Table 1 *Characterisation of wheat samples.*

S-level	mg S/ container	Kernel yield (g^a/container)		N-content (%)^a		S-content (%)^a	
		B1	B2	B1	B2	B1	B2
S4	160	65.1	32.8	2.73	2.97	0.14	0.17
S2	80	61.3	29.7	2.75	3.01	0.12	0.16
S1	40	55.3	26.3	2.66	2.90	0.09	0.13
S0	0	40.8	17.2	2.71	2.94	0.08	0.11

^{a} in dry mass.

2.4 Gluten proteins

Gluten protein types were analysed using an extraction/HPLC procedure[9]. The absorbance areas at 210 nm, which are highly correlated with the amount of protein[9], are shown in Table 2. The different S-levels did not affect the total amount of gluten proteins and the crude protein content. Only samples without S-fertilisation (S0) or fertilisation with low S-levels (S1) had significantly lower amounts of gluten proteins than the other samples.

Table 2 *Crude protein content and HPLC absorbance areas of gluten protein types^a*

| Wheat sample | CP (%) | Gluten | HPLC absorbance area (area/mg of flour) | | | | | | | | |
| | | | Gliadins | | | | | Glutenin subunits | | | |
			Total	ω5	ω1,2	α	γ	Total	ωb	HMW	LMW
B1-S4	13,4	1348	889	53	107	374	355	459	10	144	305
B1-S2	13,5	1345	889	72	154	389	274	456	12	164	280
B1-S1	13,0	1271	864	103	226	370	165	407	17	186	204
B1-S0	13,3	1135	791	115	251	305	120	344	19	171	154
B2-S4	14,6	1460	877	54	122	381	320	583	12	141	430
B2-S2	14,8	1468	878	61	141	387	289	590	14	160	416
B2-S1	14,2	1467	924	87	187	416	234	543	15	182	346
B2-S0	14,4	1388	912	121	272	373	146	476	22	201	253
±V (%)	0,7	1,4	2,3	2,5	2,6	1,8	4,7	2,1	7,2	1,4	2,5

^{a} CP = crude protein (N x 5,7), V = average coefficient of variation for n=2

For glutenin subunits the amounts of total glutenin (S4/S0: –25 and –18 %, respectively) and LMW subunits of glutenin (S4/S0: –50 and –41 %, respectively) clearly decreased in the kernels grown on both soils. The amounts of high-molecular-weight (HMW) subunits of glutenin (S4/S0: +19 and +43 %, respectively) and of glutenin-bound ω-gliadins (ω; S4/S0: +90 and +83 %, respectively) increased (Table 2). The influence of S-fertilisation on the total gliadins was different on both soils (B1: S4/S0: +4 %; B2: (S4/S0: –11 %). Within the gliadins types the γ-gliadins were mostly affected by S-deficiency as their amount was clearly reduced (S4/S0: –66 and –54 %, respectively). An opposite effect was found for ω-gliadins. In S-deficient plants the amount of the ω5 (S4/S0: +116 and +124 %, respectively) and also the ω1,2-type (S4/S0: +135 and –54 %, respectively) increased dramatically. The amount of α-gliadins was only weakly affected by S-fertilisation (Table 2).

2.5 Discussion

Table 3 shows that there is a close relationship between the proportions of the S-containing amino acids Met and Cys in gluten protein types and the influence of S-fertilisation on their amount. From the S4 to the S0-level the amount of the S-free ω-gliadins increased by 117 to 141 %, and the amount of HMW subunits of glutenin (Met + Cys ≈ 1.2 mol-%) increased by 20 to 43 %. The amount of S-rich protein types decreased. γ-gliadins (Met + Cys ≈ 3.9 mol-%) and LMW subunits of glutenin (Met + Cys ≈ 3.7 mol-%) were reduced by 54 to 66 % and 41 to 49 %, respectively. α-gliadins were only slightly affected (decrease 2 to 18 %) because of their medium Met and Cys content (≈ 2.7 mol-%). These figures confirm the qualitative results obtained in previous electrophoretic studies[5-8] and show the close relationship between the S-supply of the plant, the S-content of the gluten protein types, and their amount.

Table 3 *Met- and Cys-content (mol%) of gluten protein types (cv. Rektor[10]) and changes in the proportions of (S4/S0, cv. Star) caused by S-deficiency*

	Gliadins				Glutenin subunits	
	ω5	ω1,2	α	γ	HMW	LMW
Met	0.0	0.0–0.3	0.4–0.9	1.2–1.6	0.1–0.3	1.2–1.6
Cys	0.0	0.0	1.9–2.2	2.2–2.8	0.6–1.3	1.9–2.6
S4/S0 (B1)	+117	+135	-18	-66	+20	-49
S4/S0 (B2)	+141	+123	-2	-54	+43	-41

The changes of the percentages of gliadins and glutenins and of their protein types as affected by S-fertilisation are summarised in Table 4. The influence of the soil is primarily focused on the amount of total gliadin, which is generally higher for samples grown on B1 (66 to 70 %) than for samples grown on B2 (60 to 66 %). Therefore, the ratio of gliadin and glutenin is significantly different (B1: 1.9 to 2.3; B2: 1.5 to 1.9). However, for both soils the proportion of gliadin increases and that of glutenin decreases in S-deficient plants (S0, S1) leading to an increase of the gliadin/glutenin when less S is provided. The protein composition of samples grown with the highest levels of S (S2, S4) can be considered as normal[11]: proteins of the LMW group (α-, γ-gliadins, LMW subunits of glutenin) are major (20 to 29 %), ω-gliadins and HMW subunits of glutenin are minor components (4 to 11 %). All protein types except α-gliadins are heavily affected by S-deficiency. The proportion of ω-gliadins increases and the amount of γ-gliadins decreases, so that for the S0 level the percentages of the ω1,2 gliadins are increased 2-fold and those of the ω5-gliadins are equal to the γ-gliadins. Concerning low levels of S-supply, the amounts of LMW- and HMW subunits of gluten become similar.

Table 4 *Influence of S-fertilisation on the proportions of gluten protein fractions, individual protein types, and the gliadin/glutenin ratio*

	Gliadins					Glutenin subunits				GLI/
	Total	ω5	ω1,2	α	γ	Total	ωb	HMW	LMW	GLU
B1–S4	66	4	8	28	26	34	1	11	22	1,9
B1–S2	66	5	11	29	21	34	1	12	21	2,0
B1–S1	68	8	18	29	13	32	1	12	16	2,1
B1–S0	70	11	22	27	10	30	2	15	13	2,3
B2–S4	60	4	8	26	22	40	1	10	29	1,5
B2–S2	60	4	10	26	20	40	1	11	28	1,5
B2–S1	63	6	13	28	16	37	1	12	24	1,7
B2–S0	66	9	20	27	10	34	2	14	18	1,9

3 CONCLUSION

S-deficiency leads to a loss of kernel yield, not to a loss of total protein, to decreases of the amounts of S-rich γ-gliadins and LMW-subunits of glutenin, and to increases in the amounts of S-poor ω-gliadins and HMW-subunits of glutenin. The protein proportions are shifted to an increasing gliadin/glutenin ratio. ω-gliadins become major and γ-gliadins minor components, whereas the ratio of HMW and LMW subunits is well-balanced.

References

1   C.W. Wrigley and J.A. Bietz, 'Amino Acids and Proteins' in *Wheat - Chemistry and Technology*, ed. Y. Pomeranz, AACC, St. Paul, Minnesota, 1988, Vol I, pp. 159-275.
2   H. Wieser, and W. Seilmeier, *J. Sci. Food. Agric.*, 1998, **76**, 49.
3   P.J. Randall and C.W. Wrigley, 'Effects of the sulfur supply on the yield, composition, and quality of grain from cereals, oilseeds, and legumes' in *Advances in Cereal Science and Technology*, ed. Y. Pomeranz, AACC, St. Paul, Minnesota, 1986, Vol VIII, pp. 171-187.
4   F.J. Zhao, M.J. Hawkesford and S.P. McGrath, *J. Cereal Sci.*, 1999, **30**, 1.
5   C.W. Wrigley, D.L. Du Cros, M. Archer, P.G. Downie and C.M. Roxburg, *Aust. J. Plant Physiol.*, 1980, **7**, 755.
6   M.H. Moss, C.W. Wrigley, F. MacRitchie and P.J. Randall, *Aust. J. Agric. Research*, 1981, **32**, 213.
7   C.W. Wrigley, D.L. Du Cros, J.G. Fullington and D.D. Kasarda, *J. Cereal Sci.*, 1984, **2**, 15-24.
8   J.G. Fullington, D.M. Miskelly, C.W. Wrigley and D.D. Kasarda, *J. Cereal Sci.*, 1987, **5**, 233.
9   H. Wieser, S. Antes and W. Seilmeier, *Cereal Chem.*, 1998, **75**, 644.
10 H. Wieser, W. Seilmeier and H.-D. Belitz, *Getreide Mehl Brot*, 1991, **45**, 35.
11 H. Wieser and R. Kieffer, *J. Cereal Sci.*, 2001, **34**, 19.

ENVIRONMENTAL EFFECTS ON MEASUREMENT OF GLUTEN INDEX AND SDS-SEDIMENTATION VOLUME IN DURUM WHEAT

F.R Clarke[1], J.M.Clarke[1], N.A. Ames[2], and R.E. Knox[1]

Agriculture and Agri-Food Canada, [1]Semiarid Prairie Agricultural Research Centre, P.O. Box 1030, Swift Current, Saskatchewan, Canada S9H 3X2; [2] Cereal Research Centre, 195 Dafoe Rd., Winnipeg, Manitoba, Canada R3T 2M9 (clarkef@agr.gc.ca)

1 INTRODUCTION

Protein quality, particularly gluten strength, is an important factor in pasta manufacture and cooking quality, and thus an important selection criterion in durum wheat *(Triticum turgidum* L. var *durum)* cultivar development. Durum breeders have used the sodium dodecyl sulfate (SDS) sedimentation test extensively because of the small sample size required and of the rapidity of measurement.[1,2] More recently, the gluten index[3] has been applied to early generation selection as well as to gluten strength measurement in pasta industry laboratories.

Relatively little detailed information is available on the utility of the gluten index for selection in a breeding program. Ames et al.[4] found no genotype-environmental interaction in a small set of inbred genotypes. However, the effects of environmental conditions, particularly those that influence grain protein concentration, on gluten index have not been studied in detail. Numerous reports show low to moderate positive correlation between SDS-sedimentation volume and protein concentration.[1,5,6,7] There are a few studies of the association of gluten index with grain protein concentration, showing low[8] or zero[7,9] correlation between the traits.

Our objective was to compare the efficacy of SDS-sedimentation and gluten index for selection for gluten strength, and to assess response to environment.

2 MATERIALS AND METHODS

Two populations were developed for the study. A doubled haploid durum population of 155 lines was developed from the cross W9262-260D3/Kofa using the maize pollen method.[10] W9262-260D3 is an F_9 inbred line from the cross Kyle*2/Biodur and has intermediate gluten strength. Kofa has very strong gluten properties. Field experiments were grown in 2000, 2001 and 2002 comprising the DH population and 13 checks, including the parents, in an alpha lattice design with two replications, at three locations (Swift Current, Regina, and Indian Head, Saskatchewan) typical of the Canadian durum production area.

The second population consisted of the cross 'Fanfarron'/DT 369. Fanfarron (PI221411) is of Spanish origin and has weak gluten, and DT 369 (PI546362)[11] is a semidwarf line with strong gluten from our breeding program. Random inbred lines of this cross were developed from material grown as natural selection bulks from F_2 through F_7. Single F_8 seeds were planted for multiplication. Thirty-six F_8 derived lines were grown with the parents and seven other durum checks in an alpha lattice design with two replications. The trial was grown at Swift Current (1997, 1998,1999) and Regina (1998, 1999) Saskatchewan.

Grain protein concentration, gluten index and SDS-sedimentation volume were measured on grain harvested from all plots and locations. Whole meal was prepared on 25 g samples using a Udy mill. Protein concentration was estimated by NIR spectroscopy. SDS-sedimentation values were determined using 3% SDS as described by Dexter et al[1]. AACC Method 38-12 was followed for the determination of gluten index.[12]

Data were analyzed using mixed-model ANOVA to assess differences among genotypes, and to look at all interactions. Micro-environmental trends were assessed within trials using spatial analysis procedures as described below. SAS[13] (version 8.2) was used for all analyses

The Papadakis[14] method (described by Bartlett[15]) was performed on plot observed gluten index and SDS-sedimentation volume to remove environmental gradients. This adjusts observed values with an environmental gradient index used as a covariate. The gradient index for each plot is the mean of the residuals, calculated from the observed values minus the corresponding treatment mean. The gradient index for each plot was calculated from plots located immediately before and after that plot.[16] The moving mean trend index for each plot was the average of three neighbouring plots on each side (trend adjusted value = observed value - trend index).[17] Heritabilities were estimated using variance components[18] along with their confidence intervals.[19]

3 RESULTS AND DISCUSSION

The ANOVA showed significant (P<0.05) main effects and interactions for both populations. Within trial micro-environmental trends as detected by the moving mean and Papadakis methods were variable, and tended to be more uniform and consistent for SDS-sedimentation volume than for gluten index. Adjustment for trends with either moving mean or Papadakis tended to improve correlation among environments, particularly for SDS-sedimentation (not shown). However, heritability calculated from variance components was not improved following adjustment by the Papadakis method.

Table 1 *Heritability of gluten index and SDS-sedimentation volume of two durum wheat populations.*

		Heritability	95% confidence interval
W9262-260D3/Kofa	gluten index	0.95	0.94 – 0.96
	SDS-sed.	0.97	0.96 – 0.98
Fanfarron/DT 369	gluten index	0.84	0.70 – 0.90
	SDS-sed.	0.95	0.90 – 0.97

Both gluten index and SDS-sedimentation volume were highly heritable (Table 1) when measured in these inbred lines with multiple years, locations and replications. This suggests that either trait could be used for selection in a breeding program.

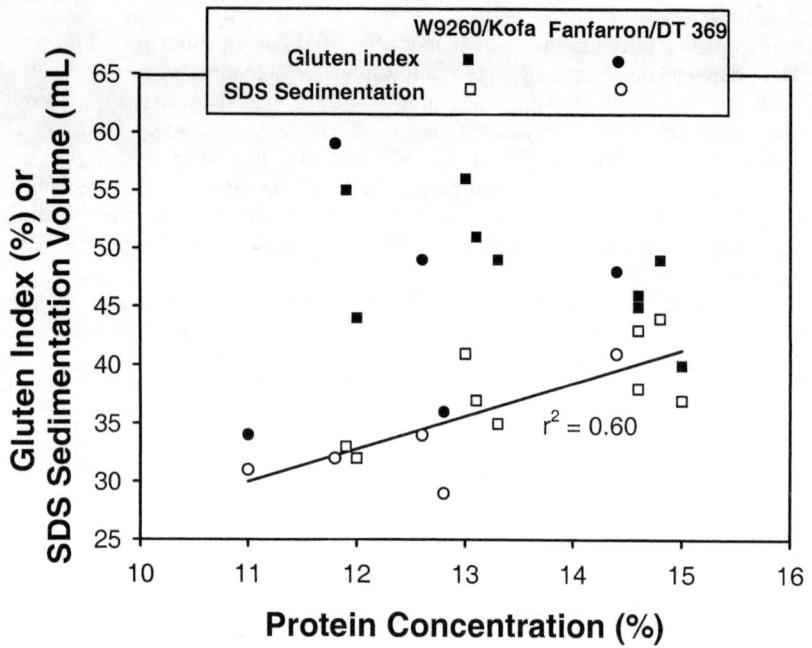

Figure 1 *Relationship of environment mean gluten index and SDS-sedimentation volume with grain protein concentration, and regression of SDS-sedimentation on protein for two durum wheat populations. Gluten index was not correlated with protein.*

The SDS-sedimentation volume was positively associated with grain protein concentration (Figure 1), but gluten index was not correlated with protein. These results agree with other published reports.[7,9] There was considerable variation in mean gluten index over environments. The range among environments was 40 to 56 for W9262/Kofa, and 34 to 59 for Fanfarron/DT 369. These results suggest that SDS-sedimentation and gluten index are partially independent measures of gluten strength, although the two parameters were highly correlated (r=0.67 to 0.90).

In summary, heritability of gluten index and SDS-sedimentation was similar. Sample size, labour and equipment requirements are less for SDS-sedimentation than for gluten index. However, we have concerns about the efficacy of SDS-sedimentation volume for selection purposes due to the association with protein concentration. We have observed large protein trends within trials.[20] It is possible to compensate for these trends using the moving mean in replicated or un-replicated trials, or the Papadakis method in replicated trials, if one determines gluten strength on all plots. However in early generation trials we use a step-wise selection process, discarding first on agronomic traits, then on protein concentration and yellow pigment so that perhaps only 15 to 25% of the population is screened for gluten strength. Consequently too few observations remain for determination of environmental trends. The gluten index appears to be relatively independent of protein concentration, and has the added advantage of being a gluten strength measurement used in the pasta industry. Further research is required to explain the environmental variation in gluten index.

References

1. J. E.Dexter, R. R.Matsuo, F. G Kosmolak, D. Leisle, and B. A. Marchylo, *Can. J. Plant Sci.* 1980, **60**, 25.
2. J.W. Dick and J.S Quick, *Cereal Chem.*,1983, **60**, 315.
3. R. Cubadda, M. Carcea, and L.A. Pasqui, *Cereal Foods World,* 1992, **37**, 866.
4. N.P. Ames, J.M. Clarke, B.A. Marchylo, D.E. Dexter and S.M. Woods, *Cereal Chem.* 1999, **76**, 582.
5. J.M. Carrillo, J.F. Vasquez and J. Orellana, *Plant Breed.*, 1990, **104**, 325.
6. M.I.P. Kovacs, N.K. Howes, D. Leisle and J. Zawistowski, *Cereal. Chem.*, 1995, **72**, 85.
7. N.P. Ames, J.M. Clarke, D.E. Dexter, S.M. Woods, F. Selles, and B. Marchylo, *Cereal Chem.* 2003, **80**, 203.
8. R.J. Pena in *Durum Wheat Improvement in the Mediterranean Region:New Challenges,* ed. C. Royo, M.M. Nachit, N. DiFonzo and J.L Araus, International Centre for Advanced Mediterranean Agronomic Studies (CIHEAM), Zaragoza, 2000, p. 423.
9. C. Brites and J.M. Carrillo, *Cereal Chem,* 2001, **78,** 59.
10. R.E. Knox, J. M. Clarke and R. M. DePauw, *Plant Breeding,* 2000, **119**, 289.
11. J.G. McLeod, T.F. Townley-Smith, R.M. DePauw, J.M. Clarke, C.W.B. Lendrum and G.E. McCrystal, *Crop Sci.* 1991, **31**, 1717.
12. American Association of Cereal Chemists. 2000. Approved methods of the AACC, 10th ed. Method 08-01, approved April 1961; Method 14-30, approved October 1974; Method 38-12, approved November 1995; Method 44-15A, approved October 1975. The Association: St. Paul, MN.
13. *SAS System for Mixed Models*, ed. R.C. Littell, G.A. Milliken, W.W. Stroup and R.D. Wolfinger, SAS Institute Inc., Cary, N.C., 1996, 633 p.
14. Papadakis,JS (1937). Méthode statistique pour des expériences sur champ. In: Bulletin de l'Institut d'Amélioration des Plantes B Salonique, No.23
15. M.S Bartlett, *J. Agric. Sci,* 1938, **28**, 418.
16. C. Brownie, D.T.Bowman, and J.W.Burton, *Agron. J.* 1993, **85**, 1244.
17. F.R. Clarke, R.J. Baker and R.M. DePauw, *Crop Sci.,* 1994, **34**, 1479.
18. Baker, R.J. 1986. Selection Indices in Plant Breeding. CRC Press, Boca Raton, FL. 218 pages.
19. S.J. Knapp, W.W. Stroup and W.M. Ross, *Crop Sci.,*1985, **25**, 192.
20. F.R.Clarke, J.M.Clarke, and R.M. DePauw, *Proceedings of the Ninth International Wheat Genetics Symposium,* ed. A.E. Slinkard, University Extension Press, University of Saskatchewan, Saskatoon, 1998, vol. 4, p. 142.

Acknowledgments

We gratefully acknowledge the financial support of the Matching Investment Initiative of Agriculture and Agri-Food Canada and the Wheat Producer Check-Off administered by the Western Grains Research Foundation and the technical assistance of Jay Ross, Marlin Olfert...

A DYNAMIC MODEL OF THE EFFECTS OF NITROGEN FERTILIZATION, WATER DEFICIT, AND TEMPERATURE ON GRAIN PROTEIN LEVEL AND COMPOSITION FOR BREAD WHEAT (*TRITICUM AESTIVUM* L.)

P. Martre,[1] J.R. Porter,[2] P.D. Jamieson,[3] S.M. Henton,[3] and E. Triboï[1]

[1] Unité d'Agronomie, Institut National de la Recherche Agronomique, F-63 039 Clermont-Ferrand, Cedex 2, France. E-mail: pmartre@clermont.inra.fr
[2] Department of Agricultural Sciences, Royal Veterinary and Agricultural University, 2630 Taastrup, Denmark
[3] New Zealand Institute for Crop & Food Research Ltd., Private Bag 4704, Christchurch, New Zealand

1 INTRODUCTION

Protein level and composition are key components of the end-use value for wheat grain.[1,2] Although the qualitative composition of the grain is genetically determined, the quantitative composition (i.e., the ratio between the different fractions) is significantly modified by the growing conditions, and there are significant genotype × environment interactions.[3,4]

In wheat, grain proteins can be separated on the bases of there solubility as albumin-globulin, amphiphilic, gliadin, and glutenin protein fractions. The first two fractions are mainly structural and metabolic proteins, whereas the last two are storage proteins. To date, no attempt has been made to model the partitioning of grain N to different protein fractions. We believe that such a model will help us to understand the physiological and genetic mechanisms of the reported genotype × environment interactions for wheat grain composition and end use values.

In this paper, we present the first model of the response of wheat grain protein composition to environmental variations. This grain model has been coupled with the crop simulation model Sirius,[5,6] allowing us to analyse the interaction between the vegetative sources and the reproductive sinks at the crop level. The aims of this model are (1) to analyze the mechanisms of the responses of grain protein level and composition to environmental variations; (2) and to analyze the genetic determinism of grain protein level and composition.

2 MODEL DESCRIPTION

Based on previous works,[7-8] the main hypotheses incorporated in this model are: (1) the accumulation of structural C is sink-driven and is a function of temperature; (2) the accumulation of structural N is driven by the accumulation of structural C, such as the ratio of these two fractions is constant during grain growth; (3) the accumulation of storage proteins and starch are independent and are both supply limited; (4) the allocation of structural N between albumin-globulin and amphiphilic protein fractions and the allocation of storage N between gliadin and glutenin fractions during grain growth is constant; (5) the duration in days of the accumulation of structural C and N and the onset of deposition of storage proteins and starch are determined by the end of DNA endoreduplication and cell division in the endosperm, respectively.

This grain model has been coupled with the crop simulation model Sirius V99.[5,6] Phenological development was not part of this study. Thus, the phyllochron in Sirius was adjusted so that the simulated and observed anthesis dates matched. A phyllochron value of 93 °Cdays was used for the controlled environment chamber experiments and 112 °Cdays was used for the field experiments. Where appropriate, others genetic parameters in Sirius were set as for cultivar Claire.

3 EXPERIMENTAL TREATMENTS

We have used three sets of data from experiments in Clermont-Ferrand, France. The data are all independent of the model and were not used in its development, except the treatments 0 and L0 of the controlled environment chambers and the field experiments, respectively (see below), which where used to parameterise the grain model, and to calibrate the soil content of organic N and the soil mineralization constant of the Sirius model by matching simulated values of crop N at anthesis with the observed values.

Details of the experimental treatments and procedures are given elsewhere.[8] Briefly, the first set was from a field experiment where the effect of N availability at anthesis in relation to the level of N nutrition before anthesis was studied. Three levels of N supply before anthesis were applied. Low N treatments (treatments termed L) were established on plots that had not received N since 1948; moderate N nutrition treatments (M) received 50 Kg N ha^{-1} at beginning of tillering; and high N nutrition treatments (H) were on plots where leaves of sugar beet from a previous cultivation and a cut of alfalfa had been buried. At anthesis, each plot was split into three sub-plots to which 0 Kg N ha^{-1} (L0, M0, and H0), 30 Kg N ha^{-1} (L30, M30 and H30) or 150 Kg N ha^{-1} (L150, M150, and H150) were applied. The second set was from an experiment where the effect of post-anthesis temperature on grain filling was analysed on crops grown outside under natural light in 2 m^2 containers. From 5 days after anthesis to grain maturity the containers were covered with a clear-top controlled environment chamber. Four air temperatures, relative to outside air temperature, were then applied: -5°C (treatment termed -5); 0°C (0); +5°C (+5); +5°C until 300°Cday, base 0°C, after anthesis then +10°C until harvest maturity (+5/+10); and +10°C until 300°Cday after anthesis then +5°C until harvest maturity (+10/+5). The third set was an experiment where the interaction between post-anthesis temperature and water deficit was studied. As for the former experiment crops where grown outside in 2 m^2 containers with two post-anthesis air temperatures (-5 °C and +5 °C) crossed with two post-anthesis watering rates (100%, treatment termed -5W and +5W, and 5 to 15%, treatment -5D and +5D, of daily crop evapotranspiration). In all three sets, plants were sampled from anthesis to grain maturity at 75 to 100 °Cdays intervals. Grains were ground to wholemeal flour and the albumin-globulin, amphiphilic, gliadin, and glutenin protein fractions were sequentially extracted.[8] Total N content for the wholemeal flour and for the protein fractions were determined using the Kjeldhal method.

4 MODEL EVALUATION

At the canopy level, *SiriusQuality* provided close simulations of the accumulation of grain dry mass and total protein for all three set of data (data not shown). Hence, the model gave estimate of grain N content and protein concentration close to observed values for over all

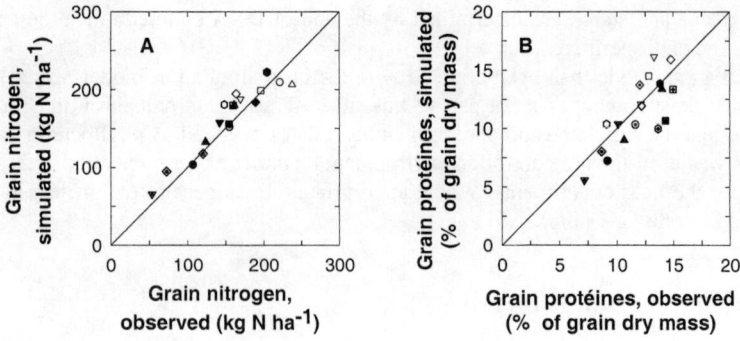

Figure 1 Simulated versus observed grain N content (A) and protein concentration (B) at harvest maturity for crops of wheat (Triticum aestivum L.). Crops were grown either in the field with different rates and timings of N fertilisation (y, treatment L0; ε, L30; κ, L150; d, M0; p, M30; j, M150; v, H0; в, H30; н, H150) or in control environment chambers with different post-anthesis temperature (c, -5; o, 0; i, +5; u, +5/+10; a, +10/+5) and watering rates (g, -5W; g, -5D; s, +5W; m, +5D). Lines are Y = X.

N, temperature and water treatments (Fig. 1). For grain N content, simulations and observations were well correlated ($r^2 = 0.89$, 16 d.f.) and the mean error of prediction was small (2 kg N ha^{-1} over a range of 54 to 238 kg N ha^{-1}; Fig. 1A). Grain protein concentration simulations were also well correlated with observations ($r^2 = 0.63$, 16 d.f.), and the mean error of prediction was 1.8% over a range of 7.2 to 15.0% (Fig. 1B).

At the grain level, as illustrated in Figure 2, the model gave accurate simulations of the timing of accumulation of the different protein fractions (Fig. 2), even for conditions of non-limiting soil N supply, such as the treatment H150 (Fig. 2D). Similar agreement was observed for the 14 other treatments of Figure 1 (data not shown). At maturity, simulated and observed quantity of the albumin-globulin ($r^2 < 0.01$, 16 d.f.), and amphiphilic ($r^2 = 0.12$, 16 d.f.) fractions were poorly correlated (Fig. 3AB). The inability of SiriusQuality to simulate the effect of N fertilisation and post-anthesis water deficit on the accumulation of structural protein fractions may be due to the effect of N availability on grain demand for structural C and N, which was not modelled. Agreement between simulated and observed

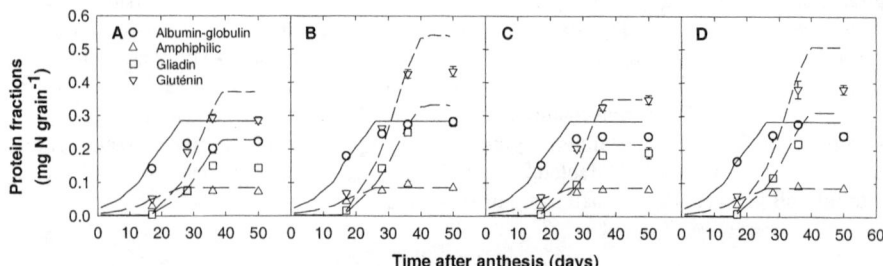

Figure 2 Observed (symbols) and simulated (lines) quantity of albumin-globulin, amphiphilic, gliadin, and glutenin protein fractions versus the number of days after anthesis. A, treatment L30; B, L150; C, H30; D, H150. Experimental data are mean ± 1 SE (n = 3 replicates each of 10 plants).

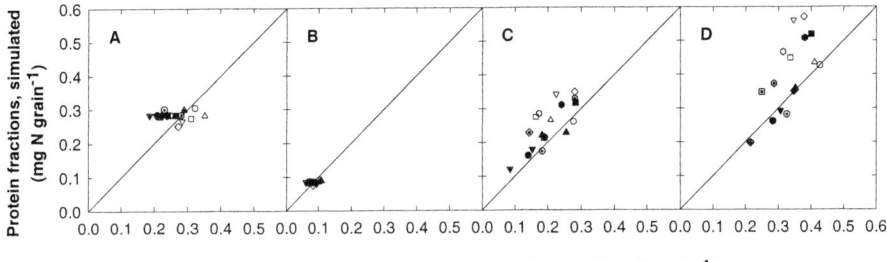

Figure 3 *Observed versus simulated quantity of albumin-globulin (A), amphiphilic (B), gliadin (C), and glutenin (D) proteins at harvest maturity for grain of wheat. Crops were grown either in the field with different rates and timings of N fertilisation or in control environment chambers with different post-anthesis temperature and watering rates. Symbols are as for Figure 1. Lines are Y = X*

quantities of gliadins ($r^2 = 0.59$, 16 d.f.) and glutenins ($r^2 = 0.42$, 16 d.f.) was better than for structural proteins, though their was a tendency to overestimate the final quantity of these two fractions (Fig. 3CD). The mean error of prediction was 0.062 mg N grain^{-1} over a range of 0.084 to 0.289 for the gliadins, and 0.101 mg N grain^{-1} over a range of 0.215 to 0.432 mg N grain^{-1} for the glutenins.

5 CONCLUSIONS

The close simulations of total N and storage proteins accumulation provided by *SiriusQuality* confirm that accumulation of grain N is source-regulated rather than sink-regulated.[6] *SiriusQuality* also gives a simple mechanistic framework that explains environmental effects on grain protein content and composition. Our assumptions that grain protein composition is a direct function of the total quantity of N per grain, and that N partitioning is not affected by the growing conditions appeared to hold over a significant range of N fertilisation and post-anthesis temperature and watering conditions. An important point for the use of this model to analyze the genetic determinism of grain protein composition is the low number of parameters needed (only seven) to model the accumulation of the different protein fractions.

References

1 P.R. Shewry and N.G. Halford, *J. Exp. Bot.*, 2002, **53**, 947.
2 P.L. Weegels, R.J. Hamer and J.D. Schofield, *J. Cereal Sci.*, 1996, **23**, 1.
3 F.R. Huebner, T.C. Nelsen, O.K. Chung and J.A. Bietz, *Cereal Chem.*, 1997, **74**, 123.
4 J. Zhu and K. Khan, *Cereal Chem.*, 2001, **78**, 125.
5 P.D. Jamieson, M.A. Semenov, I.R. Brooking and G.S. Francis, *Eur. J. Agron.*, 1998, **8**, 161.
6 P.D. Jamieson and M.A. Semenov, *Field Crops Res.*, 2000, **68**, 21.
7 P.J. Stone and M.E. Nicolas, *Aust. J. Plant Physiol.*, 1996, **23**, 727.
8 E. Triboï, P. Martre and A.-M. Triboï-Blondel, *J. Exp. Bot.*, 2003, **54**, 1731.

Gluten Rheology and Functionality

WHAT MAKES A GOOD THEORY OF GLUTEN VISCOELASTICITY?

P.S. Belton

School of Chemical Sciences and Pharmacy, University of East Anglia, Norwich, NR4 7TJ, UK

1. INTRODUCTION

It has long been recognised that the viscoelastic properties of gluten are unique and a major determinant of the quality of bread made from wheat flour. Understanding the origins of these properties is major intellectual challenge, as well as of considerable in importance in developing the science and technology of bread making. This paper attempts to analyse the criteria which any theory that attempts to explain the origins of the viscoelasticity of gluten must meet.

The first point to note is that a theory of gluten viscoelasticity is not theory of bread quality. Appropriate gluten rheological properties are a necessary, but not sufficient condition, for making bread of good quality. Any comprehensive theory of bread quality must therefore involve a theory of gluten viscoelasticity but a partial theory, concerned with the post-mixing phase, might merely assume the viscoelastic properties of gluten and concentrate on other effects that occur during rising and baking.

A similar issue arise when considering the construction of a theory of gluten behaviour, how much can be assumed as part of the axiom system and how much must be explained? The answer may in part be answered by considering the uses to which the theory will be put. An analogy from engineering and materials science is useful. In considering the construction of a building from concrete beams an engineer would only be concerned with the bulk properties of each beam, the theory of construction would then be based on the individual properties of beams, such as their size, weight and mechanical properties. Someone concerned with fabricating beams would be interested in the bulk properties of different sorts of concrete, with the properties of reinforcing materials and the results of combining them. A materials chemist would be concerned to know how to make different sorts of construction materials based on an understanding of the chemical properties of the constituent molecules.

In like manner theories of gluten behaviour may be constructed at various phenomenological levels. Practical engineering may wish to take bulk parameters for gluten and use them to predict behaviour in mixing plants etc. From the plant breeding point of view the parameters that can be manipulated are gene products, either directly or indirectly. This suggests that the starting point for a theory, which would be helpful, would be with proteins. The history of this meeting is a testimony to the fact that this view is commonly held. However the idea has been raised that the functioning unit in rheology is not at the molecular level but at a particulate level [1]. This is analogous to the ideas about the rheology of ungelled corn flour where the starch particles in a fluid water continuum determine the properties. This is opposed to the gelled state where the polymeric nature of the starch and the specific nature of the interactions of these polymers is important. The point here is not to comment on the validity of this idea but to demonstrate that a theory may start from a different place to where it might conventionally be considered to do so.

2. THE CRITERIA FOR A GOOD THEORY

There are three sets of criteria which must be met if a theory is to be considered viable: It must be:

- Consistent with facts
- Based on sound scientific principles
- A source of testable predictions.

Consistency with the facts implies some sort of discrimination between relevant and irrelevant observations and in any real, as opposed to ideal theory, will require some degree of simplification. However key facts that must be respected and incorporated or explained in the theory are:

1. The known properties of prolamins and the critical role of High Molecular Weight subunits,
2. The nature of dough and or gluten rheology and the factors which affect it,
3. Other relevant observations such as reconstitution experiments, microscopy etc.

The first and second criteria have been discussed in detail elsewhere[2], however any theory must explain key phenomena such as the observed maximum in resistance during mixing and must explain the critical role of the high molecular weight subunits. It goes without saying that there must also be credible mechanism of elasticity. This is not always the case, as discussed further in the next section. The problem with the third criterion is that there is a judgement has to be made about what is relevant and what constitutes a fact. An example is the gluten macropolymer usually referred to as GMP[3]. This material is extracted from dough or gluten by treatment with SDS and centrifugation. There is no doubt that the nature of this material changes throughout the dough mixing process and there is a correlation between its properties and dough behaviour. The problem is to what extent the structure of this material represents the structure that exists in the dough and what extent the observed behaviour is only a secondary indication of the structure in the dough. An analogy might be the extraction of myosin from muscle tissue by sodium chloride and water. A myosin gel results but the gel clearly does not reflect the high level of order in the muscle nor the separation of the myosin filaments by actin filaments. In area such as these the value of theories based on the gluten macropolymer will depend on the extent to which convincing evidence for the structure observed in the extracted material genuinely reflecting that in the dough.

The second criterion for a good theory is that it is based on sound scientific principles, among these requirements are consistency with the laws of thermodynamics, with accepted models of rheology and polymer behaviour.

As is discussed in more detail elsewhere[2] some authors[4] have confused free energies of mixing with those of solution and thus have used an inappropriate equation from Flory

Huggins theory to describe the factors affecting the solubility of prolamins. This has lead to a misunderstanding of the nature of the factors affecting prolamin solubility.

There often seems to be a view in the literature that any network will be intrinsically elastic. This is not the case, the observation of elastic behaviour requires a mechanism to explain it. A network of disulphide-bonded polymers is not necessarily intrinsically elastic, some means of storing energy is required. This may be purely entropic, but in general would not be expected to be so since interactions by hydrogen bonding are known to be important in gluten, most biopolymer gel systems are not entropic[5] and results on model systems clearly show that the elasticity of high molecular weight subunits has a very significant energetic contribution[6].

Any theory must make testable predictions. The actual nature of the predictions will depend on the starting point of the theory. However a general theory, as argued above, will take as its starting point the nature of the proteins involved in the gluten and build up a model of the system from there. Even if, as in the hyperaggregation model[1] the effective rheological unit is a particle rather than individual molecules there is still a need to explain why particles are formed and have their particular properties.

The problem for the theorist is therefore to start from the properties of the proteins and explain how they, in contact with water and under the influence of a variety of other chemical species, exhibit the rheological properties that they do. The theory therefore must involve some postulated mechanism and this mechanism must be able to predict behaviours. Assertions are not mechanisms. For example in the hyperaggregation model, it is asserted that hard fat will affect level III aggregation, i.e. formation of very large aggregates, however no mechanism is postulated so there so there is no reason to suppose that this is or is not the case, nor is it possible, on the basis of the theory, to predict what other entities might have an effect on this level of aggregation. In the model it is assumes that level III aggregation results in stronger, elastic gels and that these are favoured by a variety of conditions. However the logical path appears to be as follows:

(1) Level III aggregation results in strong, elastic doughs - ASSERTION
(2) Certain other materials added to doughs favour the formation of strong elastic doughs - OBSERVATION
(3) Therefore these materials encourage hyperaggregation. - CONCLUSION

The problems with this approach are that there needs to be a mechanism which shows that level three aggregation will result in strong elastic doughs and that the observation concerning the effect of other materials is a logical consequence of this. In the absence of a connecting mechanism, even if both 1 and 2 are true they might be operating through different mechanisms and 3 may not be valid.

The best test of a theory is that it can predict previously unobserved phenomena. This ensures that the theory is more than a good fit to existing on observations and that it is generating new knowledge.

1. CONCLUSIONS

It would be invidious to try and put existing ideas[1,5,8] about the rheology of gluten into some kind of league table. All of them have strengths and weaknesses. There is currently no adequate theory of gluten and dough rheology. Some of the problem areas that remain are listed below; these are a challenge to both theorists and experimentalists:

The early stages of mixing-
There is no clear picture of the hydration process or how individual flour particles cohere to form a hydrated continuum.

The complexity of the protein system-
Macropolymer[1] based models model explicitly involve HMW and LMW proteins. Other models such as the Loop and Train model[8] assume a simple idealised protein structure.

The relative roles of hydrogen and disulphide bonds roles-
One disulphide bond has an energy of about 220 kJmol^{-1}, this is about 10 hydrogen bonds, given the ratio of hydrogen bonding amino acids to cysteines in the prolamins it would be expected that hydrogen were more important than disulphide bonds.

How important are dityrosine cross links?-
Are dityrosine links important? Are they just another covalent cross link? Do they have some special role?

To what extent does the macropolymer reflect the actual structure within the gluten mass?
Are there methods by which the relationship can be examined?

References

1. R.J. Hamer and T Van Vliet, in *Wheat Gluten,* ed P.R.Shewry and A.S. Tatham, Royal Society of Chemistry, UK, 2000, p125
3. P.R. Shewry, N.G. Halford, A.S. Tatham, Y. Popineau, D. Lafiandra and P.S. Belton, *Advances in Food and Nutritional Science,* 2003, **45**, 219
2. C.Don, W.J. Lichtendonk, J.J. Plijter and R.J. Hamer, *J. Cereal Sci.* 2003, **38**, 157
3. H. Singh and F. Mac Ritchie, *J. Cereal Sci.* 2001, **33**, 231
4. S.B. Ross-Murphy, in *Biophysical Methods in Food Research,* ed H.W-S Chan, Blackwell, UK, 1984
5. A.S. Tatham, L.Hayes, P.R. Shewry and D.W.Urry, *Biochim. Biophys. Acta,* 2001, **1548**,187
6. P.S. Belton, *J. Cereal Sci.,* 1999, **29**, 103

Acknowledgement

The Biotechnology and Biological Sciences Research Council of the United Kingdom funded this work under research Grant No D14544.

VISCOELASTIC AND FLOW BEHAVIOUR OF DOUGHS FROM TRANSGENIC WHEAT LINES DIFFERING IN HMW GLUTENIN SUBUNITS.

J. Lefebvre, C. Rousseau and Y. Popineau

INRA, Centre de recherches de Nantes, B.P. 71627, 44316 NANTES cedex 3 (France)

1 INTRODUCTION

It is an established fact that i) high molecular weight glutenin subunits (HMW-GSU) play a major rôle in breadmaking quality of wheat cultivars, and ii) that the different subunits have not the same importance in this respect. For instance, among the HMW-GSU coded by the D genome, subunit pair 5+10 is associated with higher quality score than subunit pair 2+12. Since on the other hand the rheological properties of dough (which depend primarily upon those of gluten) have been for long recognised to be involved in the behaviour of dough all along the technological process and in the quality of the final products, there should be some links between the HMW-GSU composition of wheat and the rheological properties of gluten and dough.

Comparing near-isogenic wheat lines differing only by their HMW-GSU pattern is an elegant way to explore these functional links. We followed this approach in the case of the linear viscoelastic behaviour gluten from Sicco and from OlympicXGabo near-isogenic lines.[1,2] A complementary approach would be to consider transgenic lines obtained by reintroducing specific genes into the same genetical background. In this paper, we compare the rheological properties of the doughs from the control OlympicXGabo line (1/17+18/5+10), the double deleted line -/17+18/--, and the two transgenic lines 1/17+18/-- and -/17+18/5- obtained from the double deleted line by reintroducing the genes 1Ax1 and 1Dx5, respectively. The linearity limit of dough rheological behaviour is very low, contrary to that of gluten, when dough is submitted to large deformation during processing. We shall therefore address, besides the linear viscoelastic behaviour of dough in the linear regime, which is anyway a prerequisite for studies beyond the linearity limit, non-linear dough rheological properties. Although extensional deformation modes are also involved in the baking process, this work is restricted to shear deformation.

2 MATERIAL AND METHODS

The flour samples, the glutenin composition of which is given in Table 1, were kindly provided by Long Ashton plant breeding station. Doughs (3 g of flour, 2.2 g added water) were mixed to the consistency peak in a 2 grams Mixograph; an aliquot was immediately transferred to a Rheometric SR 2000 constant stress rheometer equipped with a cone-and-plate (cone: 2.5 cm, 0.1 radian) device. The samples were then covered with oil to prevent evaporation and left 1 h at rest before the measurements for stress dissipation.

Sequences of a frequency sweep and of a retardation test were applied to study dough rheology. Dynamic measurements were performed over the frequency range 0.06-100 rad/s under strain amplitude of 0.10 %, a value for which the samples could be still considered to behave linearly. Retardation test comprised 3h creep followed by 12h creep recovery; the creep stress values varied in the range 2-150 Pa. A fresh sample was prepared for each measuring sequence.

Table 1 *Composition of the flours of the control, double null near-isogenic, and transgenic wheat lines.*[3]

Wheat line	5.7xN % d.m.	Glutenin % protein	HMW-Glutenin subunit % Total HMW-GSU				
			1Ax1	1Bx17	1By18	1Dx5	1Dy10
886	13.7	44	16	33	16	26	10
8831	12.7	33	0	75	25	0	0
"1Ax1"	12.7	37	50	38	12	0	0
"1Dx5"	12.6	44	0	21	8	71	0

3 RESULTS AND DISCUSSION

3.1 Linear and Non-linear Behaviour in Steady-state Flow

The standard treatment of retardation curves readily gives the main rheological quantities characterising the viscoelastic behaviour of a material in the steady-state flow regime, namely the steady-state viscosity η, and the limiting compliance J_e^o.[4] J_e^o is a measure of the total elastic deformation of the material under steady flow. The longest retardation time is moreover obtained as $\tau_m = \eta J_e^o$. In the linear domain of the response, i. e. at low enough creep stress values σ, these quantities can be extracted equivalently from the creep or from the recovery terminal parts of the test. This is no longer the case beyond the linearity limit. We have in all cases calculated η from the creep compliance data and J_e^o from the recoverable compliance data obtained in the recovery part of the test.

The four lines display similar variations of η, J_e^o and τ_m with σ (Figs. 1 & 2). These quantities remain constant (with values η_o, $(J_e^o)_o$ and $(\tau_m)_o$, respectively, given in Table 2) up to $\sigma \sim 10$-20 Pa, stress value which can be therefore taken as the linearity limit for the steady-state behaviour. Above this limit, η and τ_m decrease steeply as σ increases, whereas J_e^o increases, indicating a breakdown of the global level of viscoelasticity in the non-linear domain. The lines range in the order 886>"1Dx5">"1Ax1">8831 for steady-state viscosity and global elasticity ($1/J_e^o$) levels; τ_m values range in the same order in the non-linear domain, but are rather close each other in the linear domain.

Qualitative identity of the steady flow viscous behaviour of the four lines is illustrated by the collapse of viscosity-shear rate data onto a single master curve when η/η_o is plotted

Gluten Rheology and Functionality

against the shear rate $\dot{\gamma}$ (Fig. 3). Cross equation with exponent m=1 fits accurately the master curve (Fig. 3):

$$\frac{\eta}{\eta_o} = \frac{1}{(1+(\dot{\gamma}/\dot{\gamma}_o)^m)} \qquad (1)$$

This exponent value corresponds to extreme shear-thinning behaviour; we observed the same type of behaviour for gluten.[5] On the contrary, no master curve could be found for $J_e^0(\sigma)$ data.

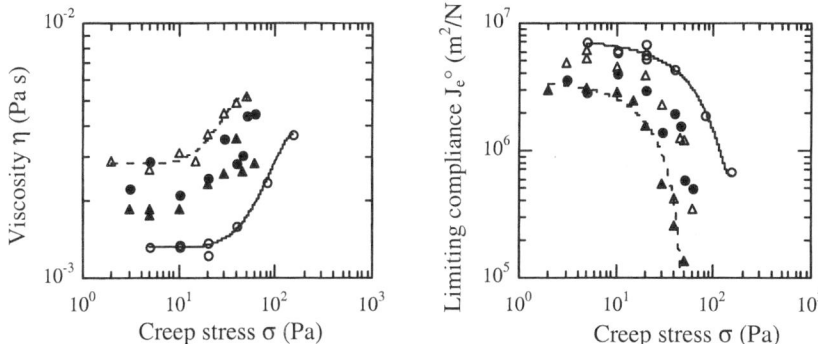

Figure 1 *Variations with creep stress of the steady-shear viscosity (left) and of the limiting compliance (right) of the doughs from the 886 (empty circles and continuous line), 8831 (filled triangles and interrupted line), "1Ax1" (filled circles), and "1Dx5" (empty triangles) wheat samples (the lines are guides to the eyes).*

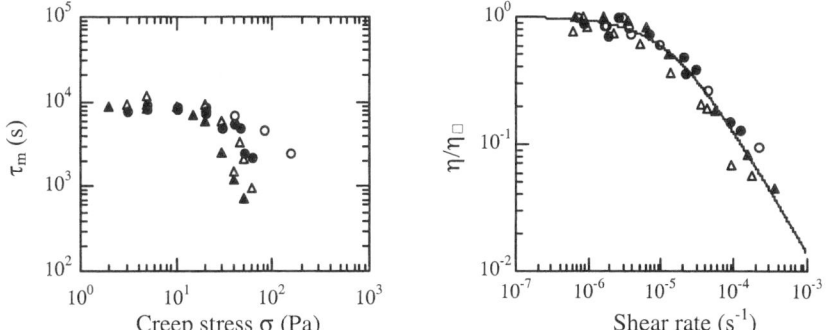

Figure 2 (left) *Dependence of the longest retardation time τ_m on the creep stress value. Same symbols as in Fig. 1.*
Figure 3 (right) *Master flow curve (see text). Same symbols as in Fig. 1. The continuous line represents the fit of equation (1) with m=1 ($\dot{\gamma}_o = (1.4 \pm 0.1)10^{-5}$ s^{-1}, $r^2=0.94$).*

3.2 The Linear Viscoelastic Response

As in the case of gluten,[1,2,5] the dynamic measurements frequency window frames a section of the viscoelastic plateau of doughs. These data have been analysed by fitting Cole-Cole functions to the components $J'(\omega)$ and $J''(\omega)$ of the complex compliance as described previously for gluten,[1,2,5] yielding the plateau compliance J_N^0, and the central frequency ω_o and the spread parameter n of the loss compliance peak which marks out the upper frequency limit of the viscoelastic plateau (Table 2).

Creep recovery data have been converted from the time to the frequency domain through the determination of the discrete retardation spectrum, following the same method as we described for gluten.[5] In this way, the frequency window can be extended down into the terminal region of the mechanical spectra. The mechanical spectra of the doughs obtained by combining dynamic and retardation data had similar shapes for all the wheat lines (results not shown). They ranged on the storage and loss moduli scale in the order 886>"1Dx5">"1Ax1">8831. As indicated by the values of $(\tau_m)_o$ and ω_o in Table 2, the composite spectra of 886 and "1Dx5" doughs were shifted to slightly higher frequencies as compared to those of 8831 and "1Ax1" doughs.

Table 2 *Rheological characteristics of the doughs in the linear viscoelastic domain*

Line	$10^3 (J_e^0)_o$ (m²/N)	$10^{-6} \eta_o$ (Pa s)	$10^{-3} (\tau_m)_o$ (s)	$10^4 J_N^0$ (a,b) (m²/N)	ω_0 (a,b) (rad/s)	n (a)
886	1.3	7.1	9.4	4.3± 0.1	0.42±0.03	0.359
8831	2.7	3.1	8.2	8.8±0.2	0.13±0.04	0.353
"1Ax1"	2.1	4.0	8.6	7.8±0.4	0.16±0.01	0.343
"1Dx5"	1.9	6.2	11.6	5.4±0.2	0.7±0.1	0.335

(a) mean and (b) standard error values of 9 (886), 8 (8831), 9 ("1Ax1), and 10 ("1Dx5") samples

3 CONCLUSION

Reintroduction of the corresponding genes in the double null line resulted in the expression of 1Ax1 and 1Dx5 HMW-GSU which proved rheologically functional. All four lines showed qualitatively similar linear and non-linear viscoelastic behaviours. As expected, 1Dx5 HMW-GSU was more "rheologically effective" than 1Ax1 HMW-GSU and its reintroduction lead to results very close to those obtained for the normal line. However, quantitative differences in rheological characteristics between the four lines were limited.

References

1 Y. Popineau, M. Cornec, J. Lefebvre and B. Marchylo, *J. Cereal Sci.*, 1994, **19**, 231.
2 J. Lefebvre, Y. Popineau, G. Deshayes and L. Lavenant, *Cereal Chem.*, 2000, **77**, 193.
3 Y. Popineau, G. Deshayes, J. Lefebvre, R. Fido, A. S. Tatham and P. R. Shewry, *J. Agric. Food Chem.*, 2001, **49**, 395.
4 J. D. Ferry, *Viscoelastic properties of polymers*, 3d ed., John Wiley and Sons, Inc., New York, 1980.
5 J. Lefebvre, A. Pruska-Kedzior, Z. Kedzior and L. Lavenant, *J. Cereal Sci.*, (in press).

LARGE-DEFORMATION PROPERTIES OF WHEAT FLOUR AND GLUTEN DOUGH IN UNI- AND BIAXIAL DEFORMATION

E.L. Sliwinski[1,2], P. Kolster[2] and T. van Vliet[1]

1 Wageningen University and Research Centre, Department of Agrotechnology and Food Sciences, Food Physics Group, P.O. Box 8129, 6700 EV Wageningen, The Netherlands
2 Agrotechnological Research Institute, Industrial Protein Group, P.O. Box 17, 6700 AA Wageningen, The Netherlands

1 INTRODUCTION

During all stages of the baking process (i.e. mixing, proofing, moulding and baking) bread dough is subjected to large deformations including fracture[1]. During fermentation the expanding gas cells deform the dough essentially in biaxial extension at relatively small deformation rates up to strains of 300%[1,2]. During the mixing process dough is deformed in a combination of shear and uniaxial elongation at deformation rates some orders of magnitude higher than those applied during fermentation[2,3]. Therefore to be able to predict dough behaviour during processing it is of key importance to have information about the large deformation properties of flour dough in both uniaxial and biaxial extension.

2 EXPERIMENTAL

2.1 Materials

Three European wheats, Vivant, A6/13 and Hereward differing in bread making performance were selected. Samples were milled to a milling extraction of approximately 70%. Flours were stored at –20°C. Data of parental flours and glutens are presented in table 1.

Table 1 *Analytical and baking data for wheat flours and glutens isolated thereof.*

Cultivar	Protein Content[1] (% dry basis)		Glutenin Content[1] (% total protein)		Bread making Performance[2]	
	Flour	Gluten	Flour	Gluten	Mixing time (min)	Volume (cm³/gflour)
Vivant	10.3	86	39	44	5.5	6.6
A6/13	12.0	85	39	44	7.5	7.5
Hereward	11.0	83	42	49	10.0	7.2

[1] Data from Sliwinski *et al.*, 2003a[4].
[2] Data are from baking tests in which a large mixer was used[5].

Glutens were isolated from flour in a batter process as described previously[5]. After mixing and centrifugation of the batter the gluten layer was separated from the other phases and further washed by hand. The wet gluten was frozen using liquid nitrogen, freeze-dried and milled in a Retsch Grinder using a 150 µm sieve. Glutens were stored at –20°C.

2.2 Dough preparation and rheological tests

Flour and gluten doughs were prepared by mixing in a 10 g National Mixograph. For both dough types water contents should be high enough to assure complete hydration of the gluten, while mixing times were defined as those giving the highest stress at a chosen strain at a constant strain rate on the condition that doughs were still manageable with respect to stickiness[6,7]. Water contents and mixing times used are given in table 2.
Biaxial extension tests were performed by lubricated uniaxial compression in a Zwick mechanical testing apparatus LBR 2000 equipped with a 50 N load cell as described previously[8]. Uniaxial extension tests were also performed using the same testing apparatus equipped with a 50 N load cell and a Kieffer Extensibility Rig as described previously[9].

3 RESULTS AND DISCUSSION

3.1 Results on flour dough

Force-displacement curves were obtained in lubricated uniaxial compression for flour-water-doughs of Vivant, A6/13 and Hereward. From these curves stress-strain curves were calculated at a constant strain rate. Typical examples for optimally mixed Vivant, A6/13 and Hereward flour dough at a constant strain rate of 0.01 s^{-1} are shown in figure 1.

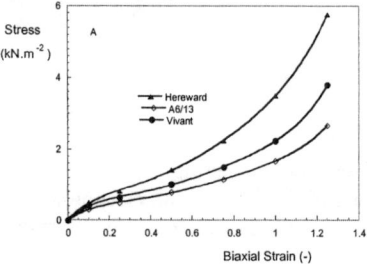 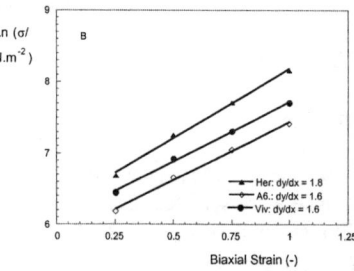

Figure 1 *Biaxial stress as a function of strain at a constant strain rate of 0.01 s^{-1} for A6/13, Vivant and Hereward flour doughs. Stress on normal scale (1A) and on log scale (1B). T = 25°C.*

The stress level increased going from A6/13, to Vivant and Hereward dough. For the latter the biaxial stress is two times higher than for A6/13 dough. Strain hardening was calculated from the slopes of the curves in figure 1B. It was highest for Hereward dough. Force-displacement curves obtained in uniaxial extension were recalculated into stress-strain curves at a constant strain rate of 0.01 s^{-1}[7,9] (see figure 2). For all cultivars a similar pattern was observed: an increase of stress with increasing strain. Similar trends were reported by others[10,11]. At larger strains the stress increases going from Vivant to A6/13 and Hereward. In figure 2B the stress is given on a logarithmic scale. It shows that the value for strain hardening increases in the order Vivant < A6/13 < Hereward.

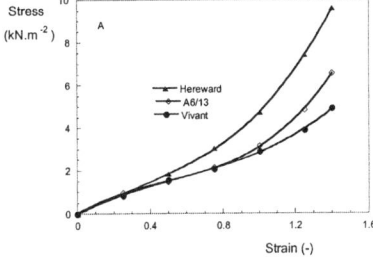 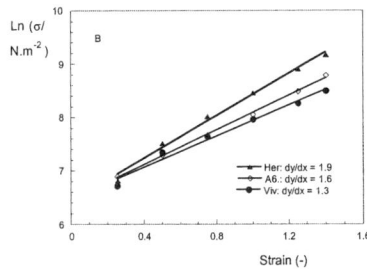

Figure 2 *Uniaxial stress as a function of strain at a constant strain rate of $0.01\ s^{-1}$ for Vivant, A6/13 and Hereward flour doughs. Stress on normal scale (2A) and on log scale (2B). T = 25°C.*

3.2 Comparison of uni- and biaxial extension for flour and gluten doughs

In figure 5 stress-strain curves are shown for Hereward flour and gluten dough in uniaxial and biaxial extension both calculated at a strain rate of $0.01\ s^{-1}$. For both dough types the stress is higher in uniaxial than in biaxial extension, the larger the strain the larger the difference. For Vivant and A6/13 similar differences were observed.

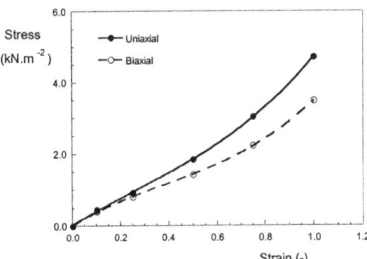 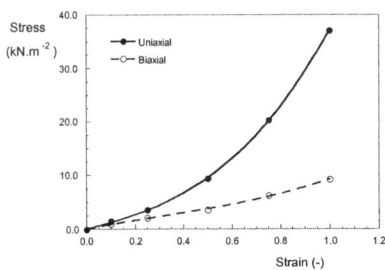

Figure 5 *Stress as a function of strain in uniaxial and biaxial extension at a constant strain rate of $0.01\ s^{-1}$ for Hereward flour dough (5A) and gluten dough (5B). T = 25°C.*

Some key characteristics calculated from the stress-strain curves in uniaxial and biaxial extension for Vivant, A6/13 and Hereward flour and gluten doughs are presented in table 2. Comparison of the mechanical properties of flour and gluten dough shows (*i*) both dough types show strain hardening in uniaxial and biaxial extension, (*ii*) stress and strain hardening coefficients are higher in uniaxial than in biaxial extension, (*iii*) in uniaxial extension values for the stress and strain hardening coefficients are much higher for gluten than for flour dough. The effect of type of deformation on the stress and strain hardening is much higher for gluten than for flour dough. The relative differences in the stress between gluten and flour dough differ between the cultivars. These results show that in elongational flow orientation of structure elements in flour and gluten dough plays an important role, which is in line with the fact that gluten shows birefringence when stretched uniaxially[12].

Table 2 *Applied water contents and mixing times and stress and strain hardening coefficients in uniaxial and biaxial extension for flour and gluten doughs. T = 25°C.*

Cultivar	Dough Preparation		Biaxial Extension		Uniaxial Extension	
	Water Content[1]	Mixing Time[2]	σ_B (kN.m^{-2})	dlnσ_B/dε_B (-)	σ (kN.m^{-2})	dlnσ/dε (-)
	(%)	(min)	$\varepsilon=0.75$; $\dot{\varepsilon}=0.01$ s^{-1}	$\dot{\varepsilon}=0.01$ s^{-1}	$\varepsilon=0.75$, $\dot{\varepsilon}=0.01$ s^{-1}	$\dot{\varepsilon}=0.01$ s^{-1}
Flour dough						
Vivant	39	7	1.4	1.6	2.0	1.3
A6/13	40	11	1.1	1.6	2.2	1.6
Hereward	39	14	2.0	1.8	3.0	1.9
Gluten dough						
Vivant	64	10	1.3	1.8	5.0	2.5
A6/13	62	10	2.1	1.8	5.3	2.5
Hereward	62	14	6.3	2.0	20.0	2.5

[1] Percentage of added water on total dough weight.
[2] In 10 g Mixograph.

4 CONCLUSIONS

Both flour and gluten dough become stronger when they are stretched and especially in the direction in which they are stretched. This conclusion is based on two observations: (i) both in uniaxial and biaxial extension stress increases more than proportionally with the strain and (ii) values for stress and for strain hardening are higher in uniaxial than in biaxial extension. Likely orientation effects play an important role in these observations. Finally, the clear similarity between the results for gluten and flour dough indicate that the large-deformation properties of the latter are largely determined by the gluten fraction.

References

1. Bloksma A.H., *Cereal Foods World*, 1990, **35**, 237
2. Van Vliet T., Janssen A.M., Bloksma A.H. and Walstra P., *J. Texture Stud.*, 1992, **23**, 439
3. Gras P.W., Carpenter H.C., Anderssen R.S., *J. Cereal Sci.*, 2000, **31**, 1
4. Sliwinski E.L., Kolster P., Prins A. and van Vliet T., *J. Cereal Sci.*, 2003a, in press
5. Sliwinski E.L., Kolster P. and van Vliet T., *J. Cereal Sci.*, 2003b, in press
6. Sliwinski E.L., van der Hoef F., Kolster P. and van Vliet T. 2003c, *Rheologica Acta*, in press
7. Sliwinski E.L., Kolster P. and van Vliet T., *Rheologica Acta*, 2003d in press
8. Kokelaar J.J., van Vliet T.and Prins A., *J. Cereal Sci.*, 1996, **24**, 199.
9. Dunnewind B., Grolle K., Sliwinski E.L. and van Vliet T., *J. Texture Stud.*, 2003, in press.
10. Tschoegl N.W., Rinde J.A. and Smith T.L., *Rheologica Acta*. 1970, **9**, 223.
11. Meissner J., in 'Proc. 1st International Symposium on Food Rheology and Structure', eds. J. Windhab and B. Wolf. ETH, Zurich, 1997, p 27.
12. Slade L., Levine H., Finley J.W., in 'Protein Quality and the Effects of Processing, eds. R.D. Philips and J.W. Finley, Marcel Dekker, New York, 1989, p 9.

EXTENSIONAL RHEOLOGY MEASUREMENTS AS PREDICTORS OF WHEAT QUALITY

G. Mann, F. Békés and M.K. Morell

CSIRO Plant Industry, GPO BOX 1600, Canberra, ACT 2601, Australia.

1 INTRODUCTION

Improving understanding about the rheological behaviour of wheat flour doughs is an important goal from a cereal science perspective, since it will assist with improving the milling, baking, food processing, and wheat breeding programs.

In a breeding program, with only limited amounts of test material available, the success of early generation wheat quality screening relies on application of small scale methods in measuring the rheological properties of the dough[1]. In that respect, dough extensional properties are of special interest to the breeder and end-user. In particular, improved methods for extension testing will greatly assist in identifying the relationships between the HMW/LMW glutenin and gliadin composition of the wheat and the dough rheological properties underpinning different end uses. In addition, we are interested in defining the molecular dynamics occurring during its extension, in order to better understand how the rheological properties are controlled at the molecular level.

The current choice of small-scale micro-extension testers is between the TA-XT2i Texture Analyzer with the Kieffer dough & gluten extensibility rig and the micro extension tester (CSIRO prototype)[2]. We have re-evaluated these instruments for their ability to discriminate between the uniaxial extensional properties of wheat flour doughs. A key element controlling the uniaxial extensional behaviour of doughs using each instrument was found to be the geometry of the dough piece and dough hook. Because the diameters of the hooks which extend the dough samples and the overall geometry of the individual dough piece are different, the stress-response with which a dough piece resists during uniaxial extension therefore will be different. This has led to the formulation of improved small scale extension methodology using two platforms, the CSIRO prototype instrument and modified Kieffer apparatus.

The results have shown that this methodology provides better reproducibility and differentiation between wheats with different genetic backgrounds, reflecting the molecular mechanisms occurring during uniaxial extension testing.

2 MATERIALS AND METHODS

2.1 Materials

Germplasm: Eight flour samples covering a wide range of dough strength and extensibility were selected for this study from a single cross CD87 x Katepwa double haploid population of 182 progeny[3]. Based on their HMW glutenin subunit allelic composition, the samples can be sub-divided into 4 groups, each containing two samples with the (2*, 7+9, 2+12), (2*, 7+9, 5+10), (2*, 7+8, 2+12) and (2*, 7+8, 5+10) alleles of the Glu1A, Glu1B and Glu1D loci, respectively.

2.2 Methods

2.2.1 *The Mixing of the Doughs*: The mixing properties of the eight lines have been determined on a 2g Mixograph^{TM} (TMCO, Lincoln, NE, USA). Prior to all extension testing, doughs with a final mass of 3.5 g were mixed to peak dough development.

2.2.2 *CSIRO method: Prototype Micro-Extension Tester (MET)*: Dough samples for extension testing (~1.7 g / test) were moulded into cylinders approximately 6 mm diameter with a prototype moulder Extension was carried out in duplicate on a micro-extension tester with a 19 mm gap and 6 mm hook operating at 10 mm/s.

2.2.3 *Kieffer method: Texture Analyser with Kieffer Dough & Gluten Extensibility Rig*: Dough samples for extension testing (~1.0 g / test) were moulded with a Kieffer moulder Extension was carried out in triplicates on Kieffer dough & gluten extensibility rig with a 13 mm gap and 1.5 mm hook diameter operating at 10 mm/s (The original Kieffer method uses speed of 3.3 mm/s)[4, 5].

2.2.4 *The prototype Micro-Extension Tester (MET) with Kieffer Moulder*
In this method, the Kieffer moulder was used to prepare the dough strips (as discussed in Section 2.2.3) which were then extended in the Micro-Extension Tester (MET) (as discussed in Section 2.2.2) was used to characterise the extensional properties of each dough strip.

2.2.5 *Improved Kieffer method: Texture Analyser with Kieffer Dough & Gluten Extensibility Rig*: An improved version of the commercially available method (Stable Microsystems) to measure wheat flour dough uniaxial extension properties was evaluated using the Kieffer dough and gluten extensibility rig. Extension was carried out in triplicates on a modified Kieffer dough & gluten extensibility rig with a 19 mm gap and 6 mm hook diameter operating at 10 mm/s.

3. RESULTS AND DISCUSSION

3.1 Comparison of the Extensional Measurements

An illustration of the differences between the Extensograms recorded for the eight flours with four different methodologies compared are shown in Figures 1 to 4. The four figures clearly indicate, the mechanism of extension differs depending on the sample size and the diameter of the hook performing the extension. Comparison of Figure 1 to Figures 2, 3 and 4 show that the differences in R_{max} and Extensibility (Ext_{Rmax}) are due to the sample size, since the CSIRO prototype MET method uses dough pieces with larger diameter. This would affect the R_{max} as thicker samples resist the extension in proportion to the ratio of cross-sectional area.

Gluten Rheology and Functionality

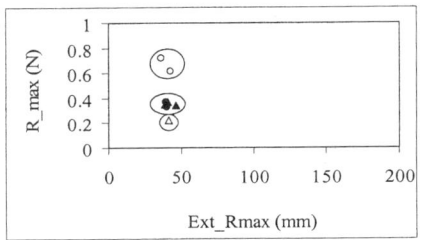

Figure 1. *Grouping of the 8 DH lines by CSIRO prototype MET method.*

Figure 2. *Grouping of the 8 DH lines by Kieffer extension method.*

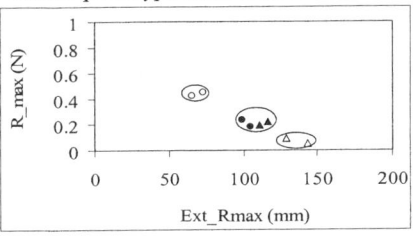
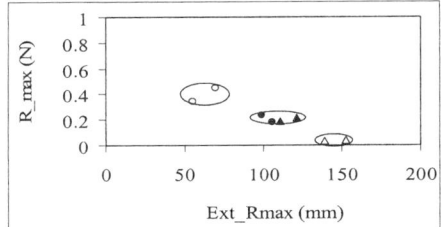

Figure 3. *Grouping of the 8 DH lines by CSIRO Hybrid method.*

Figure 4. *Grouping of the 8 DH lines by Modified Kieffer method.*

Key to the symbols used in figures 1 to 4: HMW-GS allelic composition of flour samples used. (O: 2*, 7+8, 5+10); (●: 2*, 7+9, 5+10); (▲: 2*, 7+8, 2+12); (△: 2*, 7+9, 2+12).

The extensions of the smaller dough piece with the smaller diameter hook resulted in more rapid rupture of the dough, therefore giving rise to shorter extensions, but stronger elastic response (Figure 2). In comparison, Figures 3 and 4 illustrate that the extension of the smaller dough pieces with the larger diameter hook gave more gradual ruptures with lower R_{max}, longer Ext_{Rmax} and relatively weaker initial elastic responses. The diameter of the hook, which extends the dough piece, is different for the two small extension testers. The ratio of the hook diameter of MET to the hook diameter of Kieffer rig (Texture Analyser) is 4:1. Therefore, for a defined dough mass and geometry, the stress experienced during the extension at a given speed, with the smaller hook diameter, is greater than for the larger hook diameter. Furthermore it is evident from Figure 1 and Table 1 that the smaller stretching hook diameter caused premature rupturing of the individual dough strips during extension with the Kieffer method, which gave rise to less extensibility of the doughs.

It is evident from Table 1 that all of the methods evaluated in this study have good reproducibility for the R_{max} as the low SE and CV% values suggest. However the reproducibility of the CSIRO hybrid and Modified Kieffer methods were better for Ext_{Rmax} values than the other two methodologies and this is reflected on the lower SE and CV% values. Considerably low Ext_{Rmax} values of MET method may be due to the sample preparation phase since all of the other three methodologies use Kieffer moulder which reduces the sample drying during resting period in the humidifier.

Table 1. *Summary of the overall mean, standard error of the mean (SE) and coefficient of variation (CV%) of the R_{max} and Ext_{Rmax} obtained from four different methodologies.*

	MET (Micro extension tester)		Kieffer method		CSIRO Hybrid method		Modified Kieffer method	
	R_{max} (N)	Ext_{Rmax}(mm)	R_{max}(N)	Ext_{Rmax}(mm)	R_{max} (N)	Ext_{Rmax} (mm)	R_{max} (N)	Ext_{Rmax} (mm)
Overall Mean	0.41	40.45	0.16	70.20	0.23	105.20	0.21	106.76
SE	0.02	2.07	0.01	2.71	0.01	2.98	0.01	2.46
CV%	9.76	12.27	8.27	13.80	7.98	7.70	9.20	7.07

CONCLUSION

From our results it can be concluded that all of the methodologies evaluated here gave good reproducibility however, CSIRO hybrid and modified Kieffer micro extension methodologies developed in our laboratory using modified hook geometries and the longer dough piece length divided the eight flour samples into three distinct classes, which corresponded to their HMW-GS composition. Furthermore, the stress-response of dough during uniaxial extension, at a given speed depends on the geometry of the hook that extends it as well as the geometry of the dough pieces.

References

1. F. Békés, O.M. Lukow, S. Uthayakumaran, and G. Mann, *Wheat gluten protein analysis* (Eds. P.R. Shewry and G. Lookhart) 2003. Chapter **9**.
2. C.R. Rath, P.W. Gras, Z. Zhen, R. Appels, F. Békés, and C.W. Wrigley, *Proceedings of the Australian Cereal Chemistry Conference*, 1994., **122**
3. S. J. Kammholz, A. W. Campbel, M. W. Sutherland, G. J Hollamby, P. J. Martin, R. F Eastwood, I. Barclay., R. E. Wilson, P. S. Brennan, and J. A. Sheppard, *Aust. J. Agric. Res.*, 2001, **52**, 1079.
4. R. Kieffer, F. Garnreiter, and H.D. Belitz, *Zeitschrift für Lebens. Forschung*, 1981, **172**, 193.
5. J. Smewing, *TA.TX2 texture analyser handbook* 1995.

DOUGH MIXING STUDIES ON THE MICRO Z-ARM MIXER

R. Haraszi[1], F. Bekes[1] M.L. Bason[2], J.M.C. Dang[2], J.L Blakeney[2]

[1]CSIRO Plant Industry, GPO Box 1600, Canberra ACT 2601, Australia
[2]Newport Scientific Pty. Ltd., Unit 1, 2 Apollo Street, Warriewood NSW 2102, Australia

1 INTRODUCTION

The small-scale emulation of dough rheology in traditional technological processes is of interest to wheat researchers and breeders. The latest member of the small-scale testing family is the micro Z-arm mixer. The prototype small-scale sigma-arm dough mixer was developed by CSIRO, Australia and BUTE, Hungary. It tests 4 g of flour and measures water absorption and dough mixing parameters similar to the Brabender Farinograph.[1,2] The objectives of this study were to give an overview of recent research related to the comparison and evaluation of results from the micro Z-arm instrument to those of the Farinograph^{TM} using various bowl and blade configurations, the determination of the effect of temperature on water absorption and dough-mixing properties, and the indication of some differences between mixing actions of a 2g-Mixograph and a micro Z-arm mixer.

2 METHODS, RESULTS AND DISCUSSION

2.1 Dough Mixing with a Micro Z-Arm Mixer

Wheat flour samples used in this study were provided by BRI, Australia. Mixing experiments with a micro Z-arm mixer were conducted using 4g of test flour per mix.[2,3] The reproducibility of DDT using the micro Z-arm mixer is 6.01%, while in a Farinograph^{TM} it is 9%.[4] Standard errors for stability (17.84%) and breakdown (12.32%) obtained from a micro Z-arm mixer are large, while these values in a Farinograph^{TM} are 11%[5] and 7%[4], respectively. Considering that the smaller size mixers generally have larger variance in mixing parameters,[4,5] the standard errors of parameters obtained from a micro Z-arm mixer are suitable for screening type of applications. The bandwidth-type parameters (BW@PDD, BW@BD, TMBW and MBW) are not standard parameters measured on a traditional Farinograph^{TM}, but those of a mixograph mixing curve have proven to be important as some of them can be related to dough strength.[6] There was only one significant relationship found among bandwidth type parameters and other mixing properties obtained with a micro Z-arm mixer; stability and BW@PDD (n=36, F=0.325, P=0.05). They do not seem to be important in mixing studies; however they may correlate with extensional, alveograph or fundamental rheological properties.

Table 1 *Comparison of mixing properties and water absorption obtained with a micro Z-arm mixer using different bowl and blade configurations.*

Config.	peak (VU)		WA (%)		DDT (min)		ST (s)		BD (AD*s)	
	mean± std	cv%	mean± std	cv%	mean± std	cv%	mean± std	cv%	mean± std	cv%
A	500±4.8	0.97	64.1±0.1	0.23	5.9±1.7	28.56	131.6±37.3	28.35	21332.5±6263.4	29.36
B	507±4.7	0.93	64.3±0.1	0.22	6.7±0.7	10.09	115.5±5.9	5.07	14352.9±3471.7	24.19
C	480.2±11.1	2.32	63.5±0.3	0.51	5.9±0.8	13.12	146.3±76.5	52.26	12393.6±2031.7	16.39
D	503.4±12.8	2.54	64.2±0.4	0.59	5.1±0.4	8.4	159.6±8.9	5.55	14925.7±794.8	5.33
E	533.6±5.3	1	65±0.2	0.24	5.1±0.5	9.39	159.6±33.6	21.03	20586.4±1759.7	8.55

A = bowl1 - blades1, B = bowl1 - blades2, C = bowl2 - blades2, D = bowl2 - blades1, E = bowl1- blades3

2.2 Comparison between Farinograph and a Micro Z-Arm Mixer Using Different Bowl and Blade Configurations

Comparison of mixing properties and water absorption obtained with a micro Z-arm mixer and a conventional Farinograph™ was carried out recently.[2,7] It is important to investigate the effect of any alteration of the most important instrument parts, such as the bowl and blades, on the measured parameters. For comparison between bowls and blades, five commercial wheat flour samples were tested, in duplicate, using two bowls and three pairs of blades and some of their combinations. Samples of known Farinograph™ water absorption were tested by the same operator on the micro Z-arm mixer using the standard method and control software.[3] Mixing parameters and water absorption results obtained with a micro Z-arm mixer using various configurations are shown in Table 1. The relationship (regression equations) between WA prediction by each configuration on the micro Z-arm and the Farinogoraph™ is shown in Table 2. The bowl1–blades1 configuration on the micro Z-arm mixer produced excellent results compared to the Farinograph™ with mean peak torque of 500.0 VU and water absorption of 64.1% (Table 1). The other configurations resulted in good comparability; however, a new validation process of the micro instrument with each configuration can significantly reduce the observed errors in peak torque values. Results show good repeatability for water absorption and peak torque measurements (Table 1). The relatively high cv% values for dough development time, stability and breakdown are due to strong flours used giving indistinct (flat) peaks. A further set of seven samples were used to establish the predictability (water absorption calculation at any added water level) of water absorption measurements. The standard error of prediction from a single measurement was 1.65% compared to that of a multiple (n=5) determination (1.15%). The observed coefficients of regression (Table 2) indicated that for the configurations used, the linear equations accounted for at least 95% of the

Table 2 *Regression equations for water absorption prediction from results of Farinograph™ and micro Z-arm tests. R^2 is the estimate of the variability accounted for by the regression; RMS is the root mean square of residuals of the fit*

Configuration	Regression	R^2	RMS
A	z-arm WA = 0.907 FWA + 5.94	0.95	0.61
B	z-arm WA = 0.972 FWA + 1.99	0.96	0.54
C	z-arm WA = 0.854 FWA + 8.79	0.96	0.54
D	z-arm WA = 1.085 FWA − 5.32	0.98	0.48
E	z-arm WA = 0.859 FWA + 10.01	0.96	0.52

total variability between Farinograph™ and micro Z-arm mixer for water absorption, which shows that the micro Z-arm mixer was capable of predicting water absorption values that were comparable to the Farinograph™ results.

2.3 Effect of Temperature on Dough Mixing Properties Using a Micro Z-arm Mixer

To study the effect of temperature, one flour sample was tested, in duplicate, at 25, 30, 35, and 50°C, and the dough properties compared. Temperature had a significant effect on peak torque and predicted WA, with both parameters decreasing non-linearly with increasing temperature (Figure 1). These results are in agreement with a previous study[8] which found that peak time and WA decreased logarithmically with increasing temperature of the circulating water. WA prediction is based on peak torque, which in turn depends on temperature. Temperature also had a non-linear effect on stability, dough development time and breakdown (Figure 1). It is therefore essential to control temperature to obtain accurate estimates of water absorption and mixing properties.

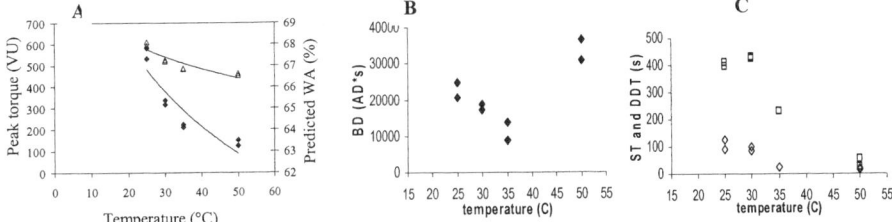

Figure 1 *Effect of temperature on (A) peak torque (△) and predicted WA (υ) (B) stability (◊) and dough development time (□) and (C) breakdown of a wheat flour sample, obtained from a micro Z-arm mixer*

2.4 Comparison between Dough Mixing Properties Obtained from Small-scale Z-arm and Pin Mixers

A 2g-Mixograph[9] representing pin mixing action and a micro Z-arm mixer representing Z-arm type mixing action were used for the investigation of relationship among mixing parameters of 18 bread wheat flour samples. Although the two mixing curves obtained with pin- and Z-arm -type mixing action show very similar performance, the derivable parameters have large differences in their functions. For example, the height of a Z-arm curve is well known to be proportional to dough consistency and can be used for water absorption prediction, while the whole shape of a mixograph curve can be related to this feature.[10, 11] It was suggested earlier[12] that instead of mixing time, the number of revolutions to reach peak dough development in a mixograph better represents the mixing requirement, because the mixing process is rate independent. Mixing was found to be rate-independent in both small- scale pin- (results not shown) and Z-arm (F=0.79 at P=0.54, Figure 2B) -type mixers, which confirmed previous observations with traditional dough mixing equipment.[13] The number of revolutions is a good indication of mixing requirement of both types of mixing. However, the number of revolutions of pin- and Z-arm mixings show significant relationship (r=0.811, P<0.001), there was no other significant correlation found between the corresponding mixing properties obtained with different mixing actions. One of the most relevant differences is that dough stability is related to different mechanism when different mixing actions are used. These are some indications of the different dough mixing performance.

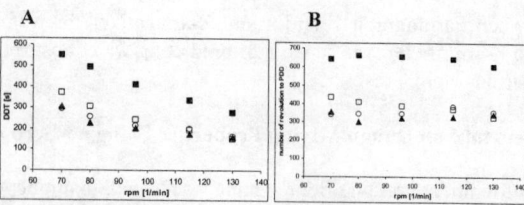

Figure 2 *Relationship of mixing speed (rpm) to (A) dough development time (DDT) and (B) the number of revolutions (nRev) of mixing with a micro Z-arm mixer. Samples: (■) Chara, (□) Janz, (△) Meering, (O) Whistler*

3 CONCLUSIONS

Reproducibility of mixing parameters obtained with a micro Z-arm mixer was found to be satisfactory, thus a micro Z-arm mixer is suitable for screening type of applications. The mixer with various bowl-blades configuration showed good comparability. The observed differences in peak torque values can be reduced significantly with a validation process of the micro instrument with each configuration. Linear equations for various bowl-blades configurations accounted for at least 95% of total variability between Farinograph^{TM} and micro Z-arm mixer for water absorption, which shows that the micro Z-arm mixer was capable of predicting water absorption values that were comparable to the Farinograph^{TM} results. Mixing properties and water absorption decreased non-linearly with increasing temperature, indicating the importance of temperature control in dough-mixing tests. Dough mixing process in a 2g-Mixograph and a micro Z-arm mixer showed rate-independence and confirmed the usefulness of the number of revolutions instead of dough development time as an indicator of mixing requirement. Corresponding mixing properties obtained with a 2g-Mixograph and a micro Z-arm mixer did not show significant correlations, except for the number of revolutions.

References

1. P.W. Gras, J. Varga, C.R. Rath, S. Tömösközi, D. Fodor, A. Salgó and F. Békés in *Proceeding of 11th International Cereal and Bread Congress*, Australia, 2000
2. P.W. Gras, R. Haraszi, M.L. Bason, C.R. Rath, J. Varga, S. Tömösközi, A. Salgo and F. Békés, *Cereal Foods World*, 2003, submitted
3. R. Haraszi, P.W. Gras, S. Tömösközi, A. Salgó, M. Morell and F. Békés, *Cereal Chem.*, 2003, accepted
4. International Standard, ISO 5530-1 1988, Part 1
5. B.L. D'Appolonia and W.H. Kunerth in *The Farinograph handbook*, AACC, USA, ISBN 0-913250-37-6
6. F. Békés, P.W. Gras, R.S. Anderssen, Appels, R., *Austr. J.of Agr. Res.*, 2001, **52**, 1325-1338
7. M.L. Bason, J.M.C. Dang, J. Blakeney, R. Haraszi and F. Bekes, *RACI 2003*, Australia, 2003
8. E.G. Bayfield, and C.D. Stone, *Cereal Chem.*, 1960, **37**, 233.
9. AACC Method 54-40A, 1988
10. J.L. Hazelton, O.K. Chung, J.J. Eastman, C.E. Lang, P.J. McCluskey, R.A. Miller, M.A. Shipman and C.E. Walker, *Cereal Chem.*, 1997, **74**, 400-402
11. M.E. Ingelin and O.M. Lukow, *Cereal Chem.*, 1999, **7**, 9-15
12. R.S. Anderssen, P.W. Gras and F. MacRitchie, *J. of Cereal Sci.*, 1998, **27**, 167-177
13. S. Zounis and K.J. Quail, *J. of Cereal Sci.*, 1997, **25**, 185-196

THE EFFECTS OF DOUGH MIXING ON GMP RE-AGGREGATION AND DOUGH ELASTICITY DURING DOUGH REST

C. Don[1], W.J. Lichtendonk[1], J.J. Plijter[1] and R.J. Hamer[1,2]

[1]TNO Nutrition and Food Research, P.O. Box 360, 3700 AJ Zeist, The Netherlands
[2]Centre for Protein Technology TNO-WUR, P.O. Box 8129, 6700 EV, The Netherlands

1 INTRODUCTION

Glutenins play a pivotal role in wheat flour quality. The quantity of glutenin macropolymer or GMP, a highly aggregated fraction of glutenin oligomers consisting of HMWGS and LMWGS, has been shown to parallel flour quality[1]. Recently it was shown that GMP from flour contains spherical glutenin particles and that dough contains disrupted GMP particles [2,3]. The initial glutenin particle size in flour affects flour mixing requirements. In this paper, we present new results that indicate that physical properties of the disrupted particles play a key role in the increase of dough elasticity during resting.

2 METHODS AND RESULTS

2.1 Flour materials and measurements

Varieties Baldus, Soissons and Estica with protein contents of 12.5, 10.6 and 12.4 %, respectively, were used in mixing and dough resting experiments. GMP was prepared from freeze-dried dough by dispersion in 1.5% SDS followed by ultracentrifugation in a Kontron Centrifuge. Mixing was done in a Brabender plastograph; GMP dispersion viscometry was done with a Ubbelohde. Flow relaxation half-time was determined with a Bohlin VOR. HMW/LMW ratio was determined with RP-HPLC.

2.2 Results with mixing and resting experiments

Baldus, Soissons and Estica had time to peak (TTP) values of 27, 10 and 5 minutes, respectively. At peak resistance, only a small quantity of GMP (<0.1g) can be isolated from the dough sample. Clearly, the particles are disrupted during mixing[3], the decreased particle size renders them soluble in SDS. For all three varieties, the average glutenin particle size at TTP is similar. Viscometric analysis of GMP dispersions (Figure 1a) showed that the Huggins constant, K', increases with mixing energy and follows the order Baldus>Soissons>Estica. The K' reflects the tendency of GMP particles to interact with each other. SDS soluble protein components typically have a much lower K' than GMP (< ca 0,5). The 5+10 variety Baldus shows the most pronounced increase in K'. Voluminosity of insoluble GMP particles increases during dough rest for Baldus, Soissons and Estica that were previously mixed to their respective TTP (Figure 1b). The voluminosity of the SDS soluble proteins decrease (att. reversed scale). The nodes indicate the dough rest-time.

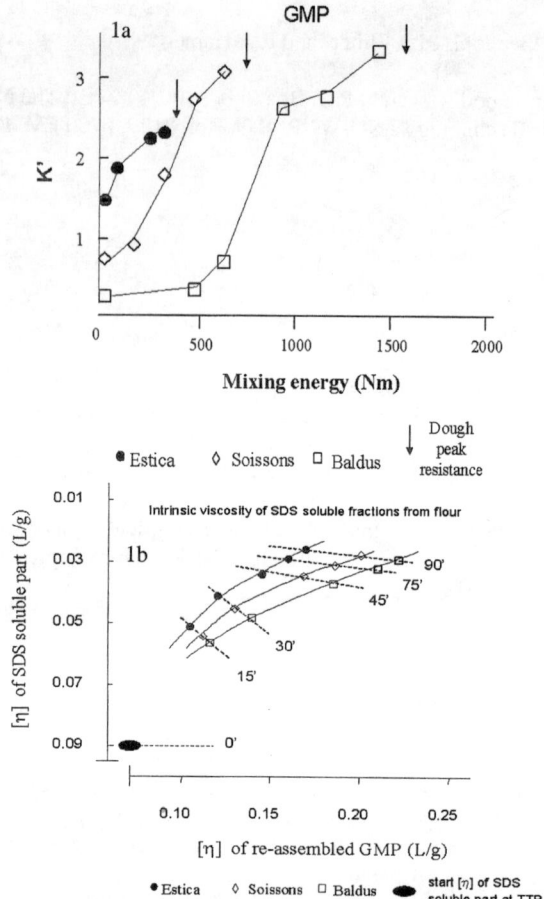

Figures 1a & b: *K' vs, mixing and SDS insoluble aggregate formation vs. dough rest*

When mixed to TTP, the GMP particles start at the same origin of 0.09 L/g. During dough rest, the SDS insoluble aggregates increase in the following order Baldus>Soissons>Estica. The formation of large SDS insoluble aggregates parallels the K' increase for the three flour varieties, suggesting that K' of the disrupted particles provides an indication of their re-aggregation behaviour. The rate of decrease of [η] of the SDS soluble portion during rest follows the order: Estica>Soissons>Baldus. This order would suggest that Estica has the highest initial rate of GMP formation. This corresponds to an increase in GMP quantity, but is not paralleled by an increase in size. Apparently, it is easier to form larger particles with Baldus than with Estica. It has been known that during dough rest, SDS soluble glutenins re-assemble to form SDS insoluble GMP. In this paper, we studied how glutenin particle voluminosity and flow relaxation half-times are affected. Figure 2a-c gives an overview of GMP quantity, voluminosity and t1/2 vs. dough rest time for the three varieties tested.

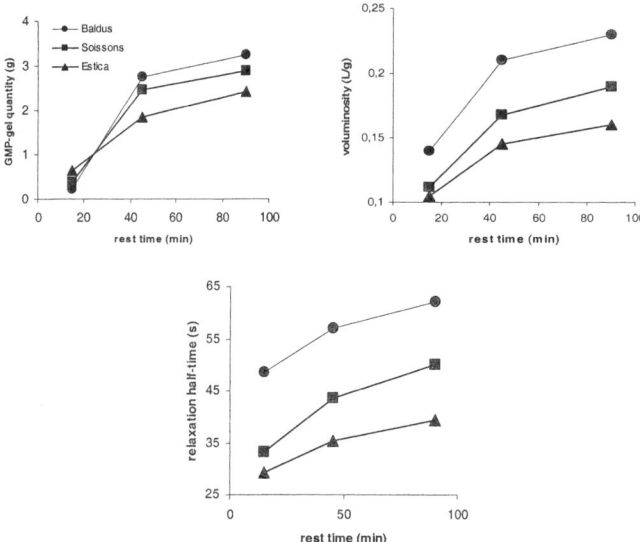

Figure 2a-c: *GMP quantity vs. rest, GMP particle voluminosity vs. rest and $t_{1/2}$ vs. rest*

Figure 2a&b illustrate how GMP quantity and glutenin particle voluminosity increase during rest for dough mixed to TTP. Figure 2c shows how dough relaxation half-time increases with dough rest time. For visco-elastic materials like dough, the relaxation-half-time parallels the elasticity. In this study, Baldus dough is more elastic than Estica dough. Glutenins are known to contribute to the elastic properties of dough; gliadins contribute to the viscous properties of dough. Therefore glutenin quantity could be a key in understanding differences in dough elastic properties. However, in this study, the R^2 between the GMP quantity in dough and the dough $t_{1/2}$ is just 0.42 (n=9), indicating that GMP quantity in dough is not sufficient to understand the link between GMP and dough rheological properties. The quantity of GMP in flour represents all glutenins in the dough system. Both SDS soluble and insoluble proteins correlate well (0.70) and confirm earlier findings [1]. When particle voluminosities are correlated against $t_{1/2}$, an R^2 of 0.85 (n=9) is found. Clearly, it is not only glutenin quantity that affects dough elasticity, but the interplay between quantity and size that determines dough elastic properties. When total glutenin quantity and SDS insoluble particle sizes are combined, a correlation R^2 of 0.96 is found.

3 CONCLUSIONS

Both the quantity and nature of the glutenins strongly affect GMP re-aggregation and dough rheological properties. In this paper, we demonstrated that the size of glutenin particles affects dough elasticity. However, factors determining the GMP size distribution after resting are still in question Differences in K' -a factor that affects the association of particles- demonstrate that this coefficient is dependent on flour variety. It has been suggested that LMWGS can act as end-blockers thus limiting glutenin oligomer size. Baldus, having the largest particles has the highest HMW/LMW ratio, whereas Estica has the lowest ratio (0.83, 0.75 and 0.62 for Baldus, Soissons and Estica respectively). The aspect of the HMW-GS composition of the glutenin particles and its effect on their properties deserves further attention.

References

1 Weegels P.L., van de Pijpekamp A.M., Graveland A., Hamer R.J. and Schofield J.D. *J. Cereal Sci,* **23,** 103
2 Don J.A.C, Lichtendonk W.J., Plijter J.J. and Hamer R.J. *J Cereal Sci,* **37,** 1
3 Don J.A.C, Lichtendonk W.J., Plijter J.J. and Hamer R.J. *J Cereal Sci,* **38,** 157

POLYMER CONCEPTS APPLIED TO GLUTEN BEHAVIOR IN DOUGH

F. MacRitchie and H. Singh

Department of Grain Science and Industry, Kansas State University, Manhattan, KS 66506, U.S.A.

1 INTRODUCTION

When energy is imparted to flour and water in the right proportions, the mixture of a solid (flour) and a liquid (water) is transformed into a dough which has both liquid and solid properties; i.e., it is said to be viscoelastic. As the dough is subjected to tensile and shear stresses during the mixing process, photomicrographs in which the starch and gluten protein have been differentially stained, show that the gluten, which initially occurs as discrete lumps, changes into a continuous network of protein strands[1]. In any multi-phase system, the properties are determined mainly by the continuous phase. Thus, the properties of dough are essentially those of hydrated gluten, modified by starch which acts as a filler. In order to understand properties of dough, it is preferable to approach the topic from the most general position. Many polymers exhibit the property of viscoelasticity. Therefore, we can apply concepts that have emerged from the extensive studies of polymers to try to understand what might be happening at a molecular level in dough.

2 METHOD AND RESULTS

2.1 Simple observations relevant to dough behavior

Gluten proteins consist of two main groups – gliadins and glutenins, in roughly equal proportions. Gliadins are single-chain proteins with molecular weights in the range 30,000 to 70,000. Glutenins are multiple-chain proteins with molecular weights above 100,000 and thought to extend to tens of millions[2,3]. When gliadin is mixed with starch and water, a purely viscous material is formed and there is no development stage as in a normal dough. In contrast, glutenin, provided it is above its glass transition temperature, forms a rubbery material with low extensibility,. The elastic properties that appear in a dough during mixing and are reflected in the development stage measured by a recording dough mixer, are due to glutenin. It is only when gliadin and glutenin are present together that optimum properties suitable for leavened products is obtained.

2.2 Molecular basis of viscoelastic behavior

What then is happening at a molecular level? For a material to exhibit purely viscous properties, requires that there must be a net movement of molecules relative to one another. To illustrate purely elastic behavior, we can use vulcanized rubber as an example, a polymeric material in which long chains are held together by cross links. In the rested state, molecular segments between cross links are in a relaxed conformation resembling a random coil. On applying a stress, these segments will be stretched, giving rise to an elastic restoring force[4]. The cross links prevent any net movement of molecules so that, on releasing the stress, the molecules return to their initial conformation, that of lowest free energy.

Dough combines viscous and elastic properties. Thus, when it is stretched and the stress removed, there is a partial recovery of the initial dimensions due to elastic recoil but the recovery is not complete because there has been relative molecular movement. Unlike rubber, gluten is not a cross-linked system in polymer terminology. However, glutenin comprises very large molecules which, like rubber, are extended by shear and tensile stresses imparted during dough mixing, giving rise to elastic restoring forces. The larger the molecular size of glutenins, the greater will be the strain rate required to extend the molecules and, providing the strain rate (i.e., mixing intensity) is sufficiently high, also the greater will be the elastic contribution to the dough properties. A critical mixing intensity is required to develop a dough and its value increases with increasing strength of the flour[5]; in other words, increases as the molecular weight distribution of the glutenin shifts to higher values. A phenomenon that is relevant to this discussion is that which has been termed "unmixing".[6] If a dough is developed by mixing and then mixed at an intensity (speed) below its critical mixing intensity, this lower mixing intensity is insufficient to maintain the extended configurations of the glutenins and it reverts to properties of an undeveloped dough. On a molecular scale, the extended glutenin molecules evidently return to their coiled state and the dough loses its elasticity. The motion of the dough facilitates retraction of the extended molecules more than in a rested dough.

2.3 Entanglements in polymers

We need to explain how a developed dough gives high resistance to deformation (or stiffness) when there are no cross-links to retard relative molecular motion. Another general concept that has emerged from polymer studies is that of entanglements. As the molecular weight of a given polymer is increased, rheological properties such as viscosity increase. This is because, for a molecule to move relative to neighboring molecules, it has to overcome frictional forces and these increase with the degree of interaction, which depends on molecular surface area and therefore size. However, at a certain molecular weight, which is characteristic for a given polymer, the slope of the viscosity-molecular weight relationship dramatically becomes steeper. Another factor besides the normal friction is acting. This has been related to widely separated points on the large molecules and has been termed entanglements[7]. The spacing between entanglements is a characteristic of a given polymer and is increased in the presence of diluents and plasticizers. In a dough system, water and other small molecules, including gliadins, act as plasticizers.

2.4 Entanglements in glutenin

Do entanglements play a role in dough behavior? If they do, then it would be predicted that there is a critical molecular weight for glutenin above which entanglements contribute to increased dough strength. Let us consider whether there is any evidence for this.

In a study of 74 recombinant inbred wheat lines[8], the correlation between the total polymeric protein and extensigraph maximum resistance (R_{max}) was quite low ($r^2 = 0.18$) whereas the correlation between Unextractable Polymeric Protein (UPP) and R_{max} was high ($r^2 = 0.86$). UPP is essentially a measurement related to the molecular weight distribution of glutenin, the main component of polymeric protein; a high UPP value signifies a greater proportion of glutenin that is insoluble and therefore of highest molecular weight. The correlations can be interpreted to mean that not all the glutenin contributes to R_{max} but only that portion of highest molecular weight. In order to pursue this further, a study was carried out on flours from 150 wheat samples comprising 30 lines each grown at five locations[9]. The chromatograms of the protein from all samples were run on size-exclusion HPLC (SE-HPLC) and the first peak, in which polymeric proteins elute, was divided into 0.4 minute intervals. The cumulative area of the chromatogram up to each elution time was then correlated with R_{max} for flours from each of the 150 lines. As elution time increased, correlation coefficients increased, reached a maximum value and thereafter decreased. The maximum value occurred at an elution time at which about 60% of the polymeric protein had eluted. This comprises the glutenin of highest molecular weight. The elution time for the highest correlation coefficient corresponded to a molecular weight of 250,000 based on a calibration with standard proteins. This can only be considered as a rough estimate of the molecular weight of glutenin. Nevertheless, the result does support the idea that there is a critical molecular weight above which glutenin contributes to dough strength.

2.5 Cross-linking of gluten proteins

Gluten is a non-crosslinked system, unlike elastomers such as rubber. Although glutenin molecules can be very large, they are discrete molecules that can be solubilized. However, the disulfide bonds and sulfhydryl groups in gluten give the potential for cross-linking under certain conditions, as has been reported when gluten is heated at high temperature[10,11]. Generally, crosslinking of glutenin is observed at temperatures below 100°C but gliadins do not appear to react until temperatures near or above 100°C depending on other conditions such as pH. When gluten was heated in an autoclave at a temperature of 120°C, SE-HPLC measurements showed a progressive decrease in the area corresponding to the gliadin peak. At short times of heating, a corresponding increase in the area of the polymeric peak occurred, consistent with a transfer of gliadins into this peak. At longer times, the gliadin peak continued to decrease but the total area of the chromatogram also began to decrease. This was evidently due to a decrease in solubility of the polymeric protein accompanying the formation of very large polymers. The gliadins appeared to cross-link with glutenins as no absorption was observed in the chromatograms at elution times expected for gliadin oligomers. Omega-gliadins, which have no cysteine residues, did not participate in the changes, supporting the view that disulfide bonding was responsible for the polymerization.[10,11] Additional evidence for disulfide bonding was that the process could be reversed by using a reducing agent such as mercaptoethanol.

3 CONCLUSION

Dough development involves extension of large glutenin molecules, creating elastic restoring forces. There is evidence that molecular entanglements of glutenins contribute to dough strength. Only glutenin molecules above a critical size appear to contribute. This size may correspond to the critical molecular weight for entanglements. Cross-linking of gluten proteins through disulfide bonds can be induced at high temperatures (> 100°C).

References

1. Moss, R.. *Cereal Sci. Today*, 1974, **19**, 557.
2. Stevenson, S.G., and Preston, K.R. *J. Cereal Sci.* 1996, **23**, 121.
3. Wahlund, K.-G., Gustavsson, M., MacRitchie, F., Nylander, T., and Wannerberger, L.. *J. Cereal Sci.* 1996, **23**, 113.
4. Treloar, L.R.G. The Physics of Rubber Elasticity, Clarendon Press, Oxford, UK, 1975.
5. Kilborn, R.H. and Tipples, K.H. *Cereal Chem.*, 1972, **49**, 34.
6. Tipples, K.H. and Kilborn, R.H. *Cereal Chem.* 1975, **52**, 248.
7. Doi, M. and Edwards, S.F. The Theory of Polymer Dynamics. Clarendon Press, Oxford, UK, 1986.
8. Gupta, R.B., Khan, K. and MacRitchie, F. *J. Cereal Sci.* 1993, **18**, 23.
9. Bangur, R., Batey, I.L., McKenzie, E. and MacRitchie, F., *J. Cereal Sci.* 1997, **25**, 237.
10. Schofield, J.D., Bottomley, R.C., Timms, M.F. and Booth, M.R., *J. Cereal Sci.* 1983, **1**, 241.
11. Weegels, P.L., Verhoek, J.A., de Groot, A.M.G. and Hamer, R.J., *J. Cereal Sci.*, 1994, **19**, 31.

RHEOLOGICAL MECHANISMS OF STABILITY OF BUBBLE EXPANSION IN BREADMAKING DOUGHS

B.J. Dobraszczyk[1], J.D. Schofield[1], J. Smewing[2], M. Albertini[3], and G. Maesmans[4]

[1] School of Food Biosciences, The University of Reading, Reading, RG6 6AP, U.K.
[2] Stable Micro Systems Ltd., Lammas Road, Godalming, Surrey GU7 1YL, U.K.
[3] Weston Research Laboratories Ltd., Vanwall Road, Maidenhead, SL6 4UF, U.K.
[4] Tate & Lyle, Amylum Europe N.V., Burchtstraat 10, B-9300 Aalst, Belgium

1 INTRODUCTION

Baking quality is strongly governed by the growth and stability of bubbles. The rheological properties of bubble cell walls in bread doughs are considered to be important in relation to their stability and gas retention during proof and baking. The limit of expansion during baking is related directly to their stability, due to coalescence and the eventual loss in gas retention on bubble rupture. The glutenin polymers representing the very high molecular weight (HMW) end of the gluten molecular weight distribution (MWD) are known to be largely responsible for variations in breadmaking performance amongst different cultivars[1]. Molecular size and structure of the gluten polymers that make up the major structural components of wheat are related to their rheological properties via modern polymer rheology concepts for HMW polymer melts and concentrated solutions. It is increasingly being recognized that the key mechanisms determining the rheology of HMW polymer melts and concentrated solutions arise from physical structural interactions between polymer molecules[2,3], and that much of the rheological behaviour of these polymers is independent of their chemistry. Interactions between polymer chain entanglements and branching are seen to be the key mechanisms determining the rheology of HMW polymers. Recent work confirms the observation that dynamic shear plateau modulus is essentially independent of variations in MW amongst wheat varieties of varying baking performance and is not related to variations in baking performance, and that it is not the size or chemistry of the individual soluble glutenin subunits, but the secondary structural interactions between these polymers that are responsible for the rheological properties of the insoluble polymer network and are mainly responsible for variations in baking performance[4]. Molecular entanglements of HMW polymers under large deformation are related to changes in the relaxation spectrum and strain hardening, which reflect the expected qualitative differences in the underlying MWD of glutenin polymers, and are being investigated as rapid methods of discriminating variations in MWD between cultivars which vary in baking quality.

2. METHODS AND RESULTS

Biaxial Extensional Rheology of Doughs At Elevated Temperatures

The rheological properties of doughs in biaxial extension were obtained during dough bubble inflation. Doughs were inflated on a Dough Inflation System attachment mounted on a TAXTPlus texture analyser (Stable Micro Systems Ltd., Godalming, U.K.) at various temperatures (25 – 60°C) and at a target constant strain rate of $0.1s^{-1}$. A constant strain rate of $0.1s^{-1}$ was used to inflate dough bubbles up to failure, and stress, Hencky strain and failure strain were recorded during bubble inflation. Strain hardening was calculated by fitting an exponential curve to the stress-Hencky strain curve. Tests were performed at various temperatures from 25°C to 60°C by placing the entire texture analyser in a heated cabinet.

Six different flour varieties from the 1999 harvest were used to provide a wide spread of baking performance: five U.K. flours (cvs.Charger, Hereward, Rialto, Riband and Soissons) and a commercial US flour provided by Pillsbury. Doughs were mixed on a Brabender Farinograph to optimum consistency of 600BU using a simple recipe of 300g flour, 6g salt and water. Samples for bubble inflation were prepared according to the methods described by Dobraszczyk[5,6]. The samples were placed in the thermal cabinet at the desired temperature and allowed to equilibrate for 30 minutes. To prevent sample drying the surface of each sample was coated using paraffin oil and covered with clingfilm. Samples were then inflated at the set strain rate using the heated moist air circulating in the cabinet. Test baking was carried out on the six flour samples using two methods: a Chorleywood Bread Process (CBP) baking procedure at Weston Research Laboratories Ltd., U.K., equivalent to that used in commercial practice in the UK, and a lower energy long fermentation procedure using a Mixograph to mix the dough at Amylum, Belgium. Loaf volume was measured by seed displacement.

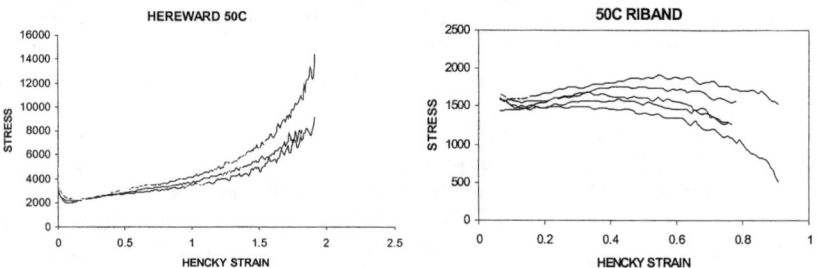

Figure 1. *Stress-Hencky strain curves for dough bubble walls at 50°C*

Figure 1 shows stress-strain curves measured at a constant strain rate ($0.1s^{-1}$) at 50°C for 2 flours: Hereward (a standard UK breadmaking variety) and Riband (a standard UK biscuit flour). Hereward maintains good strain hardening properties until 50°C, beyond which the bubbles become unstable, showing a decrease in strain hardening and failing at lower strains. Riband shows no strain hardening at 50°C, indicating that bubbles in Riband will have little resistance to failure and will coalesce rapidly at higher temperatures, whereas bubble walls in Hereward remain stable until beyond 50°C. Figure 2 shows changes in strain hardening against temperature for the six wheat varieties. Strain hardening decreased with increasing temperature for most flours, with the exception of Pillsbury and Soissons flours which showed a maximum at around 50°C. Riband started at a relatively low strain hardening value just above 1.0 and then decreased rapidly beyond 40°C. It is predicted that below a strain hardening value of around 1.0 bubble stability is no longer maintained and that this level marks an effective failure point below which bubbles will fail and coalesce on further expansion, as predicted by the Considere failure criterion[6,7]. Riband, Rialto and Charger all fall below the n=1 line at temperatures between 45-50°C, whereas Hereward, Soissons and Pillsbury do not fall below this line up to 60°C. This effectively discriminates two groups of flours, those where the bubble walls remain stable until starch gelatinisation occurs (at around 65-70°C), and those where the bubble walls become unstable at lower temperatures at around 45-50°C (Figure 3). This shows that bubble wall instability (indicated by a strain hardening value of 1), is increased to progressively higher temperatures with increasing baking volume, allowing the bubbles to resist coalescence and retain gas for much longer. Bubble wall instability in poorer breadmaking varieties occurs at much lower temperatures, giving earlier bubble coalescence and release of gas, resulting in lower loaf volumes and poorer texture.

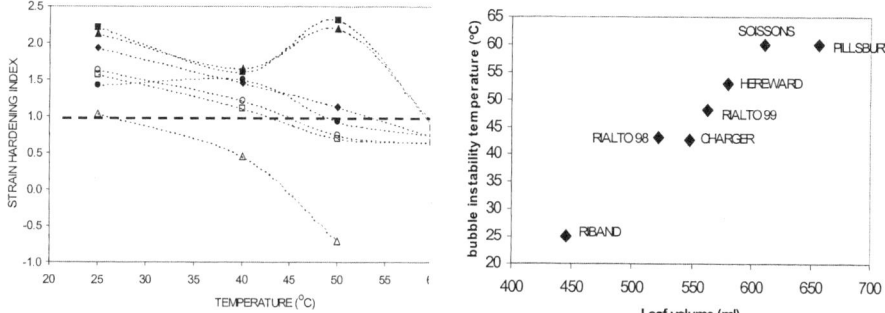

Figure 2. *Bubble wall strain hardening values for a number of wheat varieties inflated at constant strain rate ($0.1s^{-1}$) at various temperatures on the SMS Dough Inflation System.* ■ = *Pillsbury* ▲ = *Soissons* ● = *Rialto 1999* ◆ = *Hereward* ○ = *Rialto 1998* □ = *Charger* △ = *Riband*

Figure 3. *Bubble wall instability temperature vs loaf volume*

The Considere instability criterion predicts that above a strain hardening value of 1 (n=1) the bubble walls should be stable and that the failure strain of the bubble walls should be directly proportional to their strain hardening properties[6,7]. Below n=1 the bubble walls should be unstable and it is expected that their failure strain should decrease rapidly. Figure 4 shows bubble wall failure strain versus strain hardening index for all flour varieties used at various temperatures. This shows two distinct regions: one above a strain hardening value of 1 where failure strain is linearly related to strain hardening, and one below n=1, where the failure strain of the bubble walls decreases rapidly with decreasing strain hardening, which are mainly samples inflated at higher temperatures. This confirms the validity of the Considere instability criterion, and indicates that a strain hardening value of 1 defines a region below which the bubble walls are unstable and fail rapidly. Once the strain hardening of bubble walls falls below n=1, the bubble walls are no longer stable and will coalesce rapidly, resulting in lower baking volume and poorer texture. A value for strain hardening of 1 effectively acts as an indicator of baking performance: if the dough properties remain above this value during the expansion phase (proof and the early stages of baking) encompassing a temperature range from 30C up to starch gelatinisation (approximately 65°C), then the bubble walls remain stable and steady expansion will occur. If the strain hardening properties fall below 1 over this range, then bubble wall instability occurs giving rise to coalescence and release of gas and the cessation of expansion (Figure 5). This clearly shows that a strain hardening value of around 1 discriminates well between commercial flour blends of poor to moderate quality and those considered to be good and excellent.

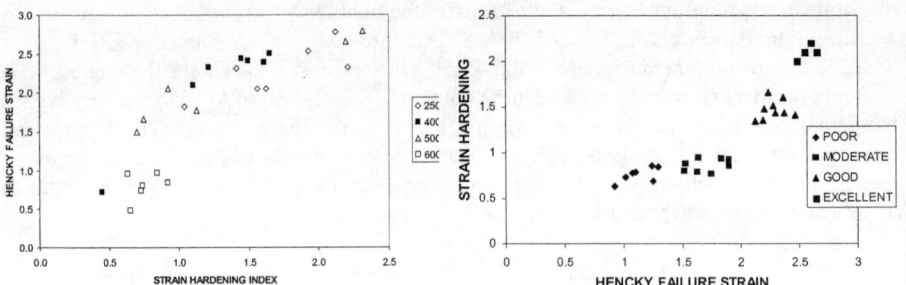

Figure 4. *Bubble wall failure strain vs. strain hardening index at various temperatures.*

Figure 5. *Discrimination of baking performance for a number of commercial flours inflated at $50^{\circ}C$.*

3. CONCLUSIONS

As predicted by the Considere failure criterion, when strain hardening falls below a value of around 1, bubble walls are no longer stable and coalesce rapidly, resulting in loss of gas retention and lower volume and texture. Bubble walls with good strain hardening properties remain stable for longer during baking, allowing the bubbles to resist coalescence and retain gas for much longer. Strain hardening in poorer breadmaking varieties starts to decrease at much lower temperatures, giving earlier bubble coalescence and release of gas, resulting in lower loaf volumes and poorer texture.

References

1. P.L. Weegels, R.J. Hamer and J.D. Schofield, *J. Cereal Sci.*, 1996, **23**, 231
2. M. Doi and S.F. Edwards, The Theory Of Polymer Dynamics, Oxford University Press, Oxford, 1986.
3. J.D. Ferry, Viscoelastic Properties of Polymers, John Wiley and Sons, 1980
4. B.J. Dobraszczyk and M.P.Morgenstern, *J.Cereal Sci.* (in press).
5. B.J. Dobraszczyk, *Cereal Foods World*, 1997, **42**, 516
6. B.J. Dobraszczyk, J. Smewing, M. Albertini, G. Maesmans and J.D. Schofield, *Cereal Chem.*, 2003, **80**, 218
7. B.J. Dobraszczyk and C.A. Roberts, *J. Cereal Sci.*, 1994, **20**, 265

EFFECT OF HIGH-PRESSURE AND TEMPERATURE ON THE FUNCTIONAL AND CHEMICAL PROPERTIES OF GLUTEN

R. Kieffer and H. Wieser

German Research Centre of Food Chemistry, Lichtenbergstraße 4, D-85748 Garching, Germany

1 INTRODUCTION

The physicochemical properties of gluten, such as elasticity, viscosity, adhesivity, cohesivity, solubility or film and foam forming capability, make it particularly interesting for many food and non-food uses. This renewable biopolymer is available in large amounts at a reasonable cost. It will be even more interesting if its properties can be adapted to a particular purpose, e.g. by reactions under high pressure. The goal of the present research was to study the effects of high-pressure treatment, with and without combined heating, on gluten rheological properties and gluten protein extractability.

2 MATERIAL AND METHODS

Gluten from the wheat cultivar Contra was isolated by pasting the flour and washing the dough with 0.4 mol/L of NaCl solution. The isolated gluten was freeze-dried before the extraction of gliadin with 60% ethanol.The extract was diluted with tenfold the amount of ice water to precipitate gliadin and the precipitate lyophilised. For high pressure treatment, gluten and gliadin were rehydrated and centrifuged at 2000 x g for 10 min.

Micro extension tests were done with the texture analyser TA.XTplus (Stable Microsystems, GB) fitted with the SMS/Kieffer-Rig[1]. Gliadin was washed 3x with 60% ethanol and the residue was dissolved in 50% 1-propanol with 2 mol/L urea and 1% dithioerythreitol (DTE) in 0.05 mol/L TRIS-HCl buffer, pH 7.5.

HPLC separation and quantification on Nucleosil C8 was carried out at 50 °C [2]. For SDS electrophoresis on NuPAGE 10%Bis-Tris gel (Invitrogen, NL), samples were put in Tris-buffer containing 2% SDS for 12 h and reduced with 50 mmol/L DTE by heating at 70 °C for 10 min.

For kinetic studies, gluten and isolated gliadin were subjected to high pressure (0.1 to 800 Mpa) for 5 to 20 min at temperatures ranging from 30 to 80 °C. Chaotropic salts were used to modify water structure and, thereby, influence gluten hydrophobic and hydrophilic interactions. No additives that could react with proteins were used.

3 RESULTS AND DISCUSSION

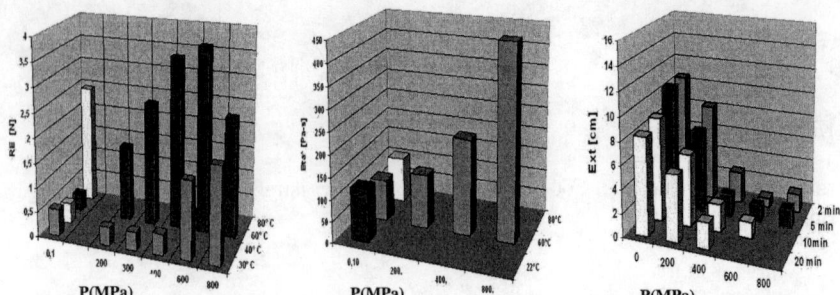

Figure 1 *Resistance of gluten (RE) as a function of pressure and temperature treatment for 10 min.*

Figure 2 *Dynamic viscosity (Eta*) of gliadin as a function of pressure and temperature treatment for 10 min.*

Figure 3 *Extensibility of gluten (Ext) as a function of pressure and time of treatment at 60 °C.*

The effect of pressure on gluten strongly depended on temperature (Figure 1). At 30 °C, a significant increase in firmness (RE) was detectable at pressures above 400 Mpa. At 60 °C, this threshold was lowered to about 200 MPa. However, at pressures exceeding 500 MPa and 60 °C, firmness decreased again as a result of the loss in gluten extensibility and cohesiveness. In this case, dried gluten no longer became a cohesive mass on rehydration. Without pressure treatment, firmness increased at temperatures higher than 60 °C. The irreversible changes at 60 °C resembled those observed with pressure-induced changes.

Gliadin, on the other hand, behaved differently at higher temperatures (Figure 2). At 80 °C without pressure, its viscosity did not change. When pressure was applied, viscosity increased dramatically, but cohesiveness, stickiness and solubility in 60% ethanol remained unchanged. SDS electrophoresis revealed no change in band intensity or band patterns before or after reduction with dithiothreitritol (Figure 6). In contrast, electrophoresis of gluten showed that the unreduced SDS soluble proteins, gliadins included, largely disappeared with pressure at 60°C. On reduction, no differences were observed between the pressurised and untreated material.

Prolonged pressure and high temperature treatment intensified gluten modifications. The limit, beyond which cohesiveness and gluten firmness decreased, shifted with time to lower pressure and temperature. Extension tests were not possible after treatment with 600 MPa for 20 min or 800 MPa for 10 min. Extensibility was much more sensitive to pressure than firmness and decreased significantly at 200 MPa and 60 °C (Figure3).

Important chemical changes accompanied these high-pressure induced rheological changes. Increasing intensity of treatment lowered the solubility of gluten proteins in solvents such as 60% ethanol, SDS, AUC or ethylenediamine(the latter two normally can dissolve up to 95% of native gluten). HPLC showed that extractability of gliadin in 60% ethanol decreased (Figure 4). After reduction, almost all the HMW and LMW fractions could be regained. The HPLC separations differentiated the behaviour among the gliadin types. The ω-gliadins, which have no cysteine, did not change, while α- and γ-gliadins rapidly disappeared. This result shows that disulphide linking is important in pressure-induced gluten hardening.

Similar hardening of gluten beyond 60 °C without added pressure (Figure 1) was not accompanied by decrease in the solubilities of α- and γ-gliadins (Figure 5). It may be concluded that only covalent bonding happened at high-pressure conditions.

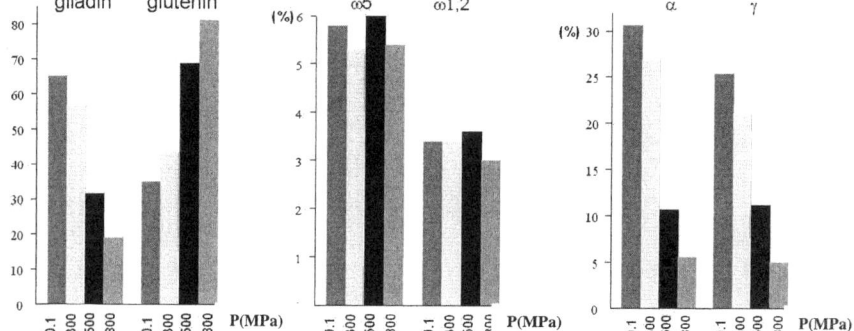

Figure 4 *Gluten treated at 60 °C for 10 min at different pressures. Quantification of the gliadin and glutenin fractions and of ω-, α- and γ-gliadins by RP-HPLC.*

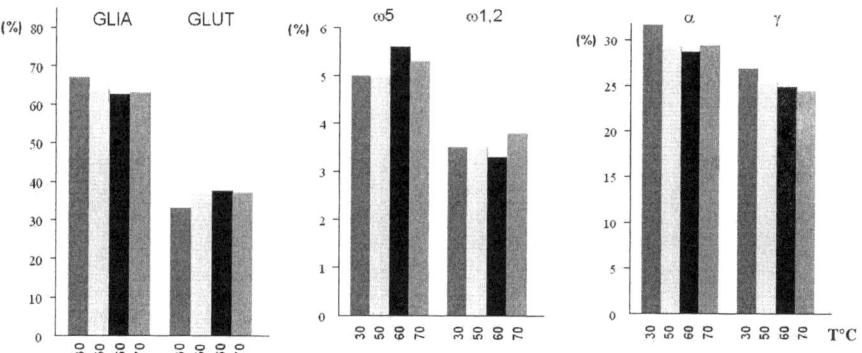

Figure 5 *Gluten treated at 0.1MPa for 10 min at different temperatures. Quantification of the gliadin and glutenin fractions and of ω-, α- and γ-gliadins by RP-HPLC.*

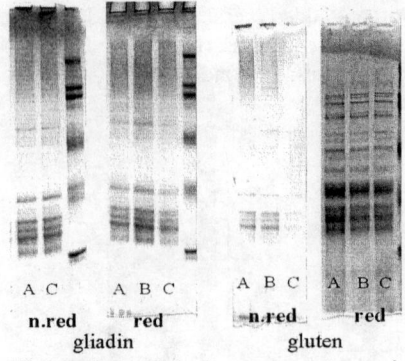
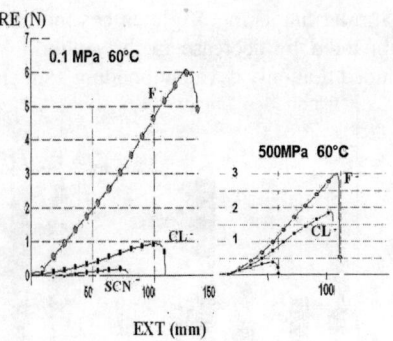

Figure 6 SDS-electrophoresis of not reduced (n.red) and reduced (red) gliadin and gluten: A=0,1Mpa/30°C; B=0,1Mpa/70°C; C=500Mpa/70°C

Figure 7 Resistance of gluten treated for 10 min in the presence of anions of the Hofmeister series (2 mol/L)

Changes in gluten firmness are dependent on water structure, which can be influenced by the Hofmeister series salts (Figure 7). Ions with a high salting out potential increase gluten resistance and reduce the relative changes induced by pressure and temperature treatment. For example, the relative strengthening of gluten was less with chloride than with rhodanide. With fluoride, gluten got soft and remained extensible.

3 CONCLUSION

High-pressure treatment decreases gluten protein solubility, increases firmness, decreases extensibility, and if treatment is too intensive, destroys cohesivity. Additional heating enhances these effects. Chaotropic salts can control the degree of gluten modification. The specific decrease in solubility of the cysteine-containing α- and γ-gliadins give strong evidence that intermolecular disulfide bonds are formed under high-pressure.

References

1   R. Kieffer, H. Wieser, MH. Henderson, A. Graveland, *J. Cereal Sci.*, 1998, **27**, 53
2   H. Wieser, S. Antes and W. Seilmeier W, *Cereal Chem.*, 1998, **75**, 644.

Acknowledgements

This research project was supported by FEI (Forschungskreis der Ernährungsindustrie, e.V.,Bonn) the AIF and the Ministry of Economics. Project No. 13178N.

CARBON ATOM AND THE THERMAL STABILITY OF WHEAT GLUTEN PROTEINS: EFFECT ON DOUGH PROPERTIES AND NOODLE TEXTURE

M.I.P. Kovacs[1], B.X. Fu[2], S.M. Woods[1], and K. Khan[3]

[1] Agriculture and Agri-Food Canada, Cereal Research Centre, 195 Dafoe Road, Winnipeg, Manitoba
[2] Canada, R3T 2M9, Canadian International Grains Institute, 1000-303 Main Street, Winnipeg, Manitoba, Canada, R3C 3G7
[3] North Dakota State University, Department of Cereal and Food Sciences, Harris Hall, Fargo, North Dakota, 58105 USA

ABSTRACT

There is a need to develop more sensitive and reliable tests to help breeders select wheat lines of appropriate quality. Gluten thermostability, measured by the viscoelasticity of heated gluten, was assessed for its usefulness in evaluating quality of wheats in breeding programs. A set of 20 cultivars and/or breeders' lines with diverse dough strengths and allelic variations of high M_r glutenin subunits coded at the *Glu-A1*, *Glu-B1* and *Glu-D1* loci (N = 20) were used in this study. Thermostability of the isolated wet gluten was determined by measuring its viscoelastic properties, and was related to noodle texture, flour protein content, protein composition, dough physical properties and other quality prediction tests.

1 INTRODUCTION

Selection of early generation lines for end-use quality is an important objective of wheat breeding programs. Pasta and Asian noodles are generally made from durum wheat (tetraploid) and bread wheat (hexaploid), respectively. Many tests have been used to predict pasta cooking quality of durum wheat, such as protein and gluten content, protein composition, mixograph mixing characteristics, alveograph characteristics, sedimentation volume (SV), swelling index of glutenin (SIG), pigment, gluten thermal stability [thereafter, cooked gluten viscoelasticity (CGV)], pasta disc viscoelasticity (PDV) and sensory tests.

There is no standard method to evaluate quality of noodle made from bread wheat. This is partly because of the large variation of noodle types and flour used. The quality of udon noodle, made from low protein soft wheats with strong gluten, appears to be determined by both their starch and protein properties, while the quality of alkaline noodle and Chinese white salted noodle, made from wheats of higher protein content, are dependent upon protein content and its composition.[1] In the literature, the functional properties of wheat protein fractions are well defined for breadmaking and pasta cooking quality but not so for noodle quality. Assessing the cooking quality of both pasta and noodle is difficult because it is a perceived quality. Cooked pasta or noodle *al dente* quality has to meet certain consumer expectations and criteria, for example bright color, clean and smooth surface, firmness, springiness, lack of stickiness (to each other or to the lips and/or teeth), tolerance to moderate overcooking and to have a pleasant flavor.

In this study, thermostability measured by viscoelastic properties of water saturated gluten after heating, along with gluten composition, were assessed for their usefulness in evaluating gluten and flour quality, especially for Chinese white salted noodle application in the wheat breeding programs.

2 MATERIALS AND METHODS

Wheat samples were field-grown in four replicates at two locations (Swift Current and Indian Head) in Saskatchewan, Canada in 1996. Milling, extraction of proteins, evaluation of flour quality properties, gluten preparation and analysis, noodle preparation and its quality evaluation and statistical analysis were performed as described in a recent publication by Kovacs et al.[1]

Table 1 *Correlations between CGVW and CGVS and other tests used to predict noodle cooking quality (N = 20)*

Tests	CGVW Pearson	CGVW Spearman	CGVS Pearson	CGVS Spearman
FPC	0.23	0.38	0.44	0.57**
MTE	0.11	0.04	0.75**	0.67**
MBE	0.21	0.37	0.74**	0.65**
FABS	-0.25	-0.28	0.07	0.11
FDT	0.23	0.50*	0.75**	0.69**
FST	0.08	0.15	0.61**	0.46*
ALVW	0.21	0.42	0.81**	0.72**
SV	0.24	0.22	0.78**	0.60**
DNV	0.44	0.39	0.69**	0.51*
FNV	0.38	0.40	0.72**	0.55*
DNCF	0.40	0.49*	0.77**	0.61**
FNCF	0.29	0.48*	0.73**	0.67**
DSL	-0.11	-0.44*	-0.69**	-0.68**
MPRO	-0.48*	-0.15	-0.82**	-0.63**
ISG	-0.08	0.18	0.58**	0.53*

*,** Correlation significantly different from zero at 0.05 and 0.01 level of probability, respectively.

3 RESULTS AND DISCUSSION

The correlations between heated gluten viscoelasticity and other tests used to predict noodle cooking quality are shown in Table 1. Viscoelasticity of heated gluten extracted with salt solution (CGVS), showed significant correlations with most of the tests used to evaluate quality, whereas gluten extracted with water (CGVW) showed only few significant relationships. There is a weak relationship between flour protein content (FPC) and CGVS. Since the amount (volume) of gluten used to determine the viscoelasticity is constant, gluten quality is the determining factor of the viscoelastic properties. Among the mixograph and farinograph parameters, mixograph total energy (MTE), mixograph bandwidth energy (MBE) and farinograph dough development time (FDT) showed highly significant correlations with CGVS ($P \# 0.01$). Farinograph stability time (FST) - a very important quality parameter for noodle processing - showed a somewhat lower significant relationship and no relationship was evident with farinograph absorption (FABS). Alveograph work (ALVW) (energy input) value was highly correlated with CGVS. The parameters of the extensigraph also showed significant associations with CGVS. SDS sedimentation volume (SV) showed significant correlation with CGVS but not with CGVW (results not shown). Flour swelling volume (FSV) and the Rapid Visco-Analyser (RSV) parameters did not correlate with CGVW or CGVS. This was expected because flour swelling volume (FSV) and RVA tests measure flour quality for the lower protein type udon (Japanese white salted) noodle as determined also by its starch characteristics. The texture of dry and fresh noodle viscoelasticity (DNV, FNV) and dry and fresh noodle cutting force (DNCF, FNCF)), showed highly significant positive correlations with CGVS. Dough sheet length (DSL) showed a highly significant negative relationship with the viscoelasticity of cooked gluten, extracted with 2% salt solution. CGVS showed a highly significant negative correlation with the relative amount of monomeric proteins (MPRO), indicating that the monomeric proteins (low molecular weight) restrict or inhibit the formation of a strong gluten network (they have higher thermal stability), consequently it takes longer time to form the denatured gluten network (Figure 1), therefore allowing the starch granules to disintegrate during cooking, resulting in a soft noodle texture. On the other hand, the relative amount of insoluble glutenins (ISG) showed a positive relationship with CGVS. It can be postulated that the balance between the protein fractions, especially monomeric polypeptides and HMWGS, is one of the most important components determining the texture of cooked noodles.

4 GENERAL DISCUSSION

The change in the viscoelasticity of wet gluten upon heating can be used to predict noodle cooking quality. When the gluten is heat-denatured, it forms a network which resists water penetration and protects the starch granules from water uptake at temperatures where starch would begin to gelatinize. To ensure that the noodles remain firm and chewy, quick denaturation of protein around the starch granule is important to prevent the starch granules from excess water uptake and consequent gelatinization and disintegration of starch granules during cooking. If the starch granules disintegrate prior to the denaturation of the protein, then the noodle texture will be soft, especially if the flour has low protein content. Besides protein content, the relative amounts of gluten protein fractions with their different molecular sizes or weights are also important. The protein thermostability can be explained by its solvent accessible surface area as reported by Stellwagen and Wilgus[2] and the negative temperature coefficient of the carbon atom.[3]

Viscoelasticity of heat-treated gluten, isolated with 2% NaCl solution, significantly correlated with most of the tests used to measure dough and/or gluten strength and Chinese white salted noodle texture. The rate of thermal denaturation of proteins depends on M_r and packing density. High ratios of monomeric proteins such as gliadins and low M_r glutenin subunits to high M_r glutenin subunits increase the thermostability of the gluten. The measurement of viscoelasticity of heat denatured gluten can be a useful test to determine gluten quality. Our study showed that gluten viscoelasticity and most of the tests related to dough and/or gluten strength are independent of allelic variations of the high molecular weight glutenin subunits. This test has been developed for predicting white salted noodle quality.

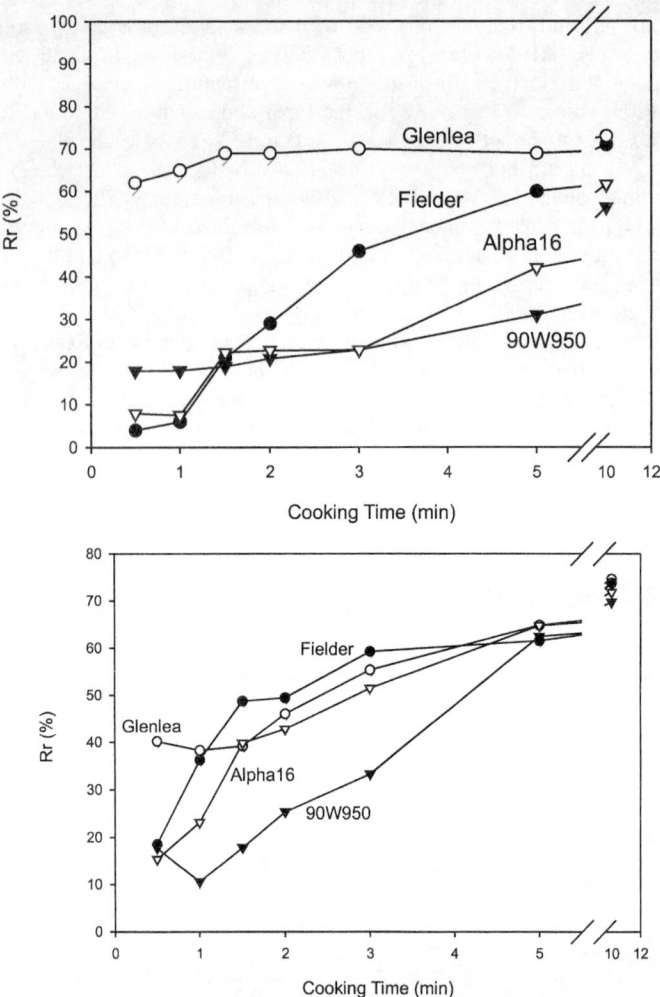

Figure 1 Effect of cooking time on the gluten viscoelasticity. (A) Gluten was extracted with water without salt or (B) with water containing 2% NaCl.

References

1   M.I.P. Kovacs, B.X. Fu, S.M. Woods and K. Khan, *J. Cereal Sci.* (in press).
2   E. Stellwagen and H. Wilgus, *Nature*, 1978, **275**, 342-343.
3   M.I.P. Kovacs, *Nature*, (submitted for publication).

WHY DOUGH INFLATION SHOULD HAVE TO MODIFY THE RATE OF GLUTEN PROTEIN THERMOSETTING

M. Pommet[1], A. Redl[1,2], S. Domenek[1], S. Guilbert[1] and M.-H. Morel[1]

[1] Unité de Technologie des Céréales et des Agropolymères, ENSA.M- INRA, 2 place Viala, 34060 MONTPELLIER, France
[2] Present address : Tate&Lyle, Amylum Europe N.V., Burchstraat 10, B-9300 AALST, Belgium

1 INTRODUCTION

Temperature plays an important role in the processing of wheat gluten based products as most of the technological processes involve heat-treatments. On a molecular level, conformation of proteins, their polymeric state and their interaction behaviour are affected by temperature. One of the most marked features of heat-induced gluten changes is the decrease in protein solubility in SDS-phosphate buffer. It has been shown that a rise of temperature causes protein unfolding which results in the exposure of hydrophobic protein zones.[1] In consequence, protein aggregation may occur primarily due to hydrophobic interactions. The unfolded state facilitates thiol / disulphide interchanges between exposed groups, which locks the protein into the denatured state due to the disulphide bond rearrangement.[2]

Processes such as extrusion, which is well studied in the field of food production, represents a real challenge to the production of biopolymers based on gluten.[3] To control the structural organisation and the properties of the final products, it is essential to get a better knowledge of the molecular changes in the proteins induced by shear on the one hand and by temperature on the other. This work focuses on the effect of temperature on gluten protein and attempts to model the kinetics of gluten protein aggregation.

2 METHOD AND RESULTS

2.1 Effect of simple heat treatment on gluten protein solubility[4]

Gluten (32.68 g) and glycerol (17.32 g) were hand-mixed in a mortar and after 20 min resting, the blend was hand-moulded to give disks of 2 mm thickness. The disks were sealed in a plastic bag under vacuum and incubated from 0 to 76h in a water bath at 70, 82, or 94°C. The samples were ground and protein was extracted with 1% SDS-phosphate buffer before being analysed by SE-HPLC.[5]

Upon heating, solubility loss of gliadin and glutenin macro-polymers is observed, but the former are less sensitive to heat (figure 1). In consequence the kinetics of solubility loss was modelled by considering two classes of protein (sol_1 and sol_2) disappearing according to first order kinetics and showing different loss rates (k_1 and k_2) (eq. 1).

$$\text{sol-P} = \text{NR-species} + sol_1 \exp(-k_1.t) + sol_2 \exp(-k_2.t) \quad \text{(eq. 1)}$$

Where sol-p is the content of soluble protein and NR-species is the content of non-reacting proteins (ω-gliadin ...).

All data were fitted together, in spite of the different temperatures used, by assuming that the rate constants followed the Arrhenius law (eq.2).

$k = A \exp(-Ea/RT)$ (eq.2)

The mathematical expressions(eq.1 and 2) allow a good fit of the experimental data (figure 2). An energy of activation of 177.7 kJ/mol was calculated.

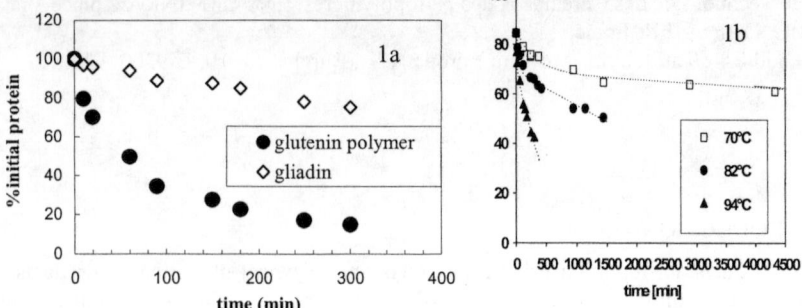

Figure 1a and 1b: (1a) Solubility loss of protein from gluten/glycerol blends heat treated at 82°C.(1b) Fitting of SDS-soluble protein drops using equations (1) and (2).

3.1 Effect of mixing at high temperature on gluten protein solubility[5]

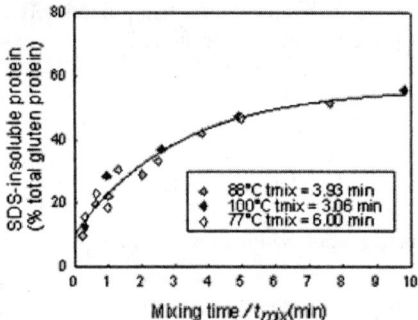

Figure 3 : Alignment of all experimental data due to reduced time.

Gluten/glycerol blends were mixed from 0 to 30 minutes at 100 rpm in a Brabender plasticorder regulated at 40, 60 and 80°C. Within 6 minutes mixing, the sample temperature levelledoff at 77, 88 and 100°C. Analysis in SDS-soluble protein content showed that mixing induced protein solubility loss, and that the phenomenon was enhanced by the mixing temperature. The timechanges in SDS-insoluble protein from the three data series, collapsed into a unique curve when considering the mixing time reduced by tmix, the time needed to reach a stable sample temperature during mixing (figure 3). This result proved that tmix was related to the temperature dependency of protein solubility loss during mixing. In consequence, an Arrhenius plot gave an estimation of the activation energy for protein aggregation during mixing (eq. 3).

$$1/tmix = A \exp(-Ea / RTmix) \quad (eq.3)$$

Where R is the gas constant and Tmix the mixing temperatures :77°, 88° and 100°C. An activation energy of 33.7 kJ/mol was calculated. This result, when compared with the 177 kJ/mol previously calculated, revealed that either the shear rate or the strain involved during mixing enhanced gluten protein aggregation upon heating.

4.1 Effect of oscillatory strain and temperature on gluten protein solubility

Gluten/glycerol blend prepared as previously (see 2.1) was laminated to 3 mm and cut into a disk (40mm diameter) which was placed in the sealed chamber of a Rubber Process Analyser® (RPA 2000 LV, Alpha Technologies, Akron, USA). The sample was submitted to an oscillatory strain (7 or 70%) at a frequency of 10.47 rads/s while submitted to various temperatures (90 to 150°C).

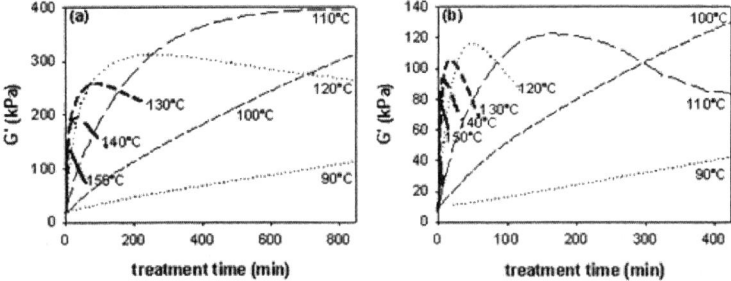

Figure 3 a and 3 b : Effects of heat on the G' modulus of gluten/glycerol blends submitted to 7 and 70% strain (3a and 3b). Modulus recorded at 10.47 rad/s.

Upon heating, G' modulus increased faster as the temperature increased. For the higher temperatures the increasing phase was followed by a decrease. For any given temperature, the time at which the conversion occurs was longer at 7% than at 70% strain. Also the decreasing phase was more abrupt at 70% strain. This indicates that higher strain accelerated the rate of gluten changes upon heating. Increase in G' modulus coincided with an increase in SDS-insoluble protein, whereas G' decrease was accompanied by the formation of small peptides, as revealed by SE-HPLC analysis from samples collected all along the 120°C treatment (not shown). Increase in G' modulus would reveal the aggregation of gluten protein, whereas the following decrease would correspond to the degradation of the previously aggregated gluten protein. In order to model the experimental data we assumed that change in G' modulus was similar to that of the formation of an intermediary product (B in the following sequence: A → B→ C) via first-order kinetics:

$$G' = G'_{max} \cdot k_{agg} \cdot [\exp(-k_{agg} \cdot t) - \exp(-k_{deg} \cdot t)] / (k_{agg} - k_{deg}) + G'_{ini} \cdot \exp(-k_{deg} \cdot t) \quad (eq.4)$$

Where G'max is the ultimate G' rise, G'ini is the starting G' value, k_{agg} and k_{deg} are the rate constants for gluten aggregation and gluten degradation mechanisms.

As before the rate constants were expressed in their Arrhenius form (eq.2) in order to fit the energies of activation. The following table(I) presents the results of the data fitting.

Table I : Changes in activation energy with strain constant

	Strain	
	7%	70%
Eagg (kJ/mol)	192.6	163.1
Edeg (kJ/mol)	208.3	106.8

Oscillatory strain has a major effect on gluten degradation, with a difference of about 100 kJ/mol. Gluten aggregation was also influenced by strain and it decreases by 30kJ/mol at 70% strain.

1 CONCLUSION

Heating provokes gluten protein aggregation, which was revealed by its loss of solubility in 1% SDS buffer or by the increase in G' modulus. Activation energy of the aggregation reaction was shown influenced by mixing shear and by strain. This means that during technological processes that involved shear treatment, like mixing and extrusion, gluten aggregation will progress more rapidly than we could have expected from its behaviour under simple heating.

But strain and shear are also encountered during the baking of leavened flour doughs. We can imagine that the bubble inflation which occurs during baking could also increase the rate of protein aggregation. As a test we modelled available data from Chevalier *et al.* on the changes in protein solubility during the baking of biscuit short-dough.[6] Activation energies ranging from 7 to 145 kJ/mol were fitted depending on the baking conditions. The highest value corresponded to the baking conditions leading to the slowest increase of biscuit thickness after 30s of baking (145 kJ/mol for 75% rise, 28 kJ/mol for 175% rise and 7kJ/mol for 250% rise). Even if this result must be confirmed on a more rigorous basis it could indicate that shear rate, like strain, affects the rate of gluten protein aggregation during heating.

References

[1] J. Schofield, R. Bottomley, M. Timms, M. Booth, *J. Cereal Sci.*, **1**, 241.
[2] P. Weegels, R. Hamer, 1998, *Interactions: the keys to cereal quality*, Eds R. Hamer & C. Hoseney, AACC, St Paul, Minnsota, USA.
[3] B. Cuq, N. Gontard N., S. Guilbert, *Cereal Chem.*, 1998, **75**, 1.
[4] S. Domenek, M.-H. Morel, A. Redl, S. Guilbert, *J. Agric. Food Chem.*, **50**, 5947.
[5] A. Redl, S. Guilbert, M.-H. Morel, *J. Cereal Sci.*, 2003, **38**, 105.
[6] S. Chevallier, G. Della Valle, P. Colonna, B. Broyart, G. Trystram, *J. Cereal Sci.*, 2002, **35**,1.

ON THE SWELLING PROPERTIES OF GLUTEN IN THE PRESENCE OF SALTS AND REDUCED pH

H. Larsson and U. Hedlund

Department of Food Technology, Lund University, PO Box 124, S-221 00 Lund, Sweden

1 INTRODUCTION

Common salt (NaCl) is used in the baking process to improve dough handling properties and bread flavour. Over the last decades, many experimental studies on wheat flour dough have investigated the influence of neutral salts on dough development and baking properties. Salt affects dough mixing or extensibility[1, 2, 3, 4, 5, 6] and baking properties [2]. Gluten proteins aggregate when salt is added to the wheat flour dough. The effect is dependent on the type of salt and its concentration[2]. . Low salt concentration (0.05-0.1M) increases dough strength properties. At higher concentrations (0.5-1.0M), salt effect on dough properties depends on salt type [5]. As the concentration increases, the ions act as counter ions shielding the charged areas of the protein, which usually keeps the molecules separated. Shielding induces protein aggregation that is due to increased hydrophobic interactions caused by reduction of protein net charge in the presence of counter ions [7]. In *The Hofmeister* series (or the lyotropic series), ions are ordered by their effect on protein solubility. Results from earlier work suggest that at low salt concentrations, the solubility and the aggregation properties of gluten proteins are more or less independent of anion type. At higher salt concentrations, hydrophobic interactions predominate and the type of anion is of vital importance for the result. At high concentrations, dough stability is decreased by chaotropic anions (I^-, SCN^-, ClO_4^-) and increased by kosmotropic anions (SO_4^{2-}, Cl^-, Br^-) [5].

This study is a part of a larger investigation on the swelling and viscoelastic properties of gluten at increasing concentration of Mg and Na salt, reduced pH and in the presence of the amino acids lysine and valine. Certain amino acids such as arginine, glutamic acid and glycine are kosmotropic and have similar effects on protein structure/function as the ions of the *Hofmeister* series [9]. Corresponding fresh gluten were isolated by ultracentrifugation from fresh dough mixed at different times [10]. Depending on wheat cultivar, mixing time and additives, the swelling of gluten in water is well defined and is independent of the dough water content [1, 10, 11, 12]. Viscoelastic properties of isolated gluten at small oscillating deformations gave a well-defined measure of gluten cross-linking density.

2 MATERIALS AND METHODS

2.1 Materials
A spring wheat cv. Vals (10.2% protein), supplied by Svalöf Weibull AB, Landskrona, Sweden was used for the salt and pH experiments. A strong and a weak commercial bread flour mixture (Vetemjöl special, 12.0% protein and Vetemjöl, 10.0% protein, respectively), from Kungsörnen AB, Sweden were used for the amino acid experiments. Sodium sulphate (Na_2SO_4) was from Merck, sodium chloride (NaCl), L-lysine monohydrochloride and DL-valine were from Sigma-Aldrich, and magnesium chloride ($MgCl_2$), magnesium sulphate ($MgSO_4$) and lactic acid (98%) were from VWR International.

2.2. Methods
Dough mixing was performed in the 10.0 g Mixograph (Reomix, Bohlin Rheology, Öved, Sweden). Water, corresponding to dough water contents (total basis) of 48.0% (cv.Vals), 45.4% (weak flour) and 45.7% (strong flour) was added to the flours. Salt was added in concentrations ranging from 0-0.4M. Dough pH was adjusted to 4.2 with 1M lactic acid by punching a pH-electrode directly into the dough.. Doughs were mixed for 8.0, 4.5 and 5.0 min (Vals, weak and strong flour, respectively).
Gluten was separated from the dough by ultracentrifugation (LE80, Beckman Instruments, Stockholm, Sweden) for 1 hour at 100,000 xg at 20°C. The separated gluten fraction was divided in two; one was used for water content determination[11] and the other for rheological measurements. The measurement of water content of the gluten fraction was replicated four times. The viscoelastic properties of the three different gluten samples were determined in triplicates. Frequency sweeps (0.01-10 Hz, 25°C) in the linear viscoelastic region were performed with a Bohlin VOR Rheometer using the plate-plate geometry (diameter = 15 mm, gap = 2 mm) (Metric Analys, Stockholm, Sweden). The maximum time between start of mixing, gluten isolation and end of rheological measurement was four hours.

3 RESULTS AND DISCUSSION

3.1 Effect of cation
The negative impact of sodium on health has led to studies on the effect of other cations in food systems. The effect of cations on wheat flour dough or protein functional properties are less studied compared to the anions [2, 13]. Here, the effect of Mg^{2+} was compared with Na^+ by adding the Cl^- salts to dough.

For NaCl, the water content decreased with increasing concentration of NaCl. The water content of gluten increased when NaCl was replaced with $MgCl_2$ (Figure 1a-b). However, the water content was similar to the gluten without salt added and not influenced by $MgCl_2$ concentration. For gluten viscoelastic properties, G' and G" increased for the Na^+-salt and decreased slightly for the more chaotropic Mg^{2+}salt.

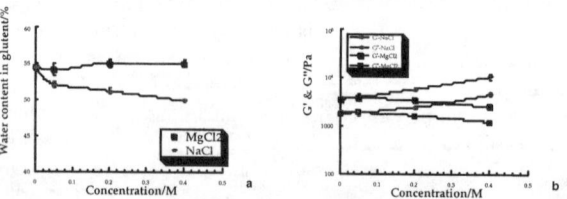

Figure 1. *Effect of cation (Mg^{2+}/Na^+) on the water content (a) and viscoelastic properties (b) of gluten.*

3.2 Effect of pH

Dough pH was reduced from 6.0 to 4.2 with lactic acid on *cv.Vals*. pH 4.2 resulted in sticky dough and an increased amount of unseparated phase after UC. It was interesting to see that gluten swelled and absorbed a larger amount of water at pH 4.2 (Figure 2). This was consistent with a reduction in G' and G" of gluten. In an earlier study, we determined the rheological properties of dough and were not able to detect any pH effect [1]. Only a moderate effect on farinograph parameters has been reported before [14].

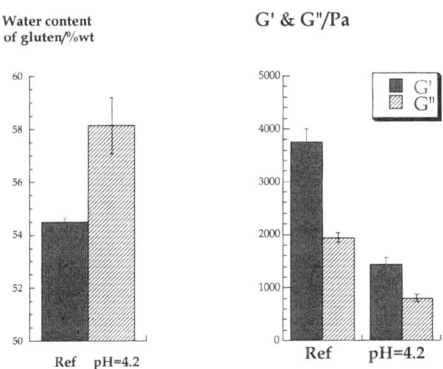

Figure 2. *Effect of reduced pH on the water content (a) and viscoelastic properties (b) of gluten.*

3.3 Effect of lysine

Lysine, an essential amino acid, is present in low amounts in foods. In the 50's, the addition of essential amino acids to cereals was investigated in order to improve the diet consumed by populations living more or less only on foods low in lysine [15, 16, 17]. The effect of adding equimolar amount of NaCl, lysine, and valine on the viscoelastic properties of gluten from weak and strong commercial flour mixture is shown in Fig 3. Lysine had a great effect on G' of gluten. The increase in G' was considerably greater than the effect of NaCl. Glutamic acid and glycine showed similar kosmotropic (stabilising) properties [9]. Dough pH and addition of valine have no effect. The effect on the viscoelastic properties was consistent with a decrease in the water content of gluten (Table 1). Similar effects were observed or the weak and the strong flour.

Table 1. *The effect of NaCl, lysine and valine on the water content of gluten from weak and strong flour.*

	Water in gluten (% total wt)	
	Weak	Strong
Reference	55.5±0.5	57.3±0.8
NaCl	50.4±0.6	48.8±0.5
Lysine	47.0±0.5	49.3±1.0
Valine	54.0±0.7	56.0±0.7

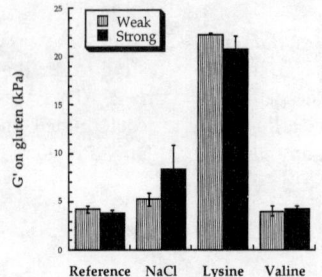

Figure 3. *The effect of adding NaCl and the amino acids lysine and valine on the viscoelastic properties of gluten.*

References

1. Larsson, H. *Cereal Chem* **79**, 2002, 544
2. He, H., R. R. Roach, R. C. Hoseney. *Cereal Chem* **69**, 1992, 366
3. Holmes, J. T., R. C. Hoseney. *Cereal Chem* **64**, 1987, 343
4. Kinsella, J. E., M. Hale. *J Agric Food Chem* **32**, 1984, 1054
5. Preston, K. R. *Cereal Chem* **66**, 1989, 144
6. Danno, G., R. C. Hoseney. *Cereal Chem* **59**, 1982, 196
7. Salovaara, H. *Cereal Chem* **59**, 1982, 422
9. Cacace, M. G., E. M. Landau, J. J. Ramsden. *Quart Rev Biophys* **30**, 1997, 241
10. Georgopoulos, T., H. Larsson, A.-C. Eliasson. *Food Hydrocolloids* **in press**, 2003.
11. Larsson, H., A.-C. Eliasson. *Cereal Chem.* **73**, 1996, 18
12. Larsson, H., A.-C. Eliasson. *Cereal Chem.* **73**, 1996, 25
13. Butow, B. J., P. W. Gras, R. Haraszi, F. Bekes. *Cereal Chemistry* **79**, 2002, 826
14. Doguchi, M., I. Hlynka. *Cereal Chem* **44**, 1967, 561
15. Deshpande, P. D., A. E. Harper, C. A. Elvehjelm. *J Nutrition* **62**, 1957, 503
16. Li-Chan, E., N. Helbig, E. Holbek, S. Chau, S. Nakai *J Agric Food Chem* **27**, 1979, 877
17. Belitz, H. D., W. Grosch,. In, Springer-Verlag Berlin, Heidelberg, Berlin, Germany, 1999

Acknowledgements

The Cerealia Foundation R&D supported this work.

STUDIES ON THE DIFFERING PROPERTIES OF GLUTEN AND DOUGH ON THE FORMATION OF THE GLUTEN NETWORK

R. Kieffer

German Research Centre of Food Chemistry, Lichtenbergstraße 4, D-85748 Garching, Germany

1 INTRODUCTION

Dough elasticity contributes to the baking quality of wheat flour (e.g. bread volume). Chemical modifications of gluten protein with reducing or oxidizing reagents affect not only elasticity but also bread volume. Elasticity, like other rheological characteristics, can be determined by rheological measurements. Since dough is a non-homogenous material, its protein and starch structures may change during analysis[1]. Correlation has not been found between elasticity and bread volume[2,3]. Until now, only poor correlations could be established between dough properties and bake volume.

In the present work, the discrepancy between gluten properties and dough behavior had been investigated using several rheological methods. Light microscopy of the developing dough structure and the behavior of gluten filaments were also studied.

2 MATERIAL AND METHODS

Seven wheat cultivars differing in protein content (N = 1.44 – 2.04%) and protein quality (gliadin / glutenin ratio = 1.72 – 2.61) belonging to the quality classes E (extra high) to C (no bread quality) were used[2]. Gluten and dough were prepared according to the AACC standard methods[4]. Demixed dough was rounded for 20 tours a Brabender dough-ball rounder (Brabender)[5].

Samples were measured using oscillatory tests (OSC), creep and creep recovery tests (C+CR) and by micro extension tests (EX) and the results correlated to the bread volume (Vol) generated in the micro baking tests[2,3]. G^*, G', G'' and $\tan \delta$ were determined by OSC. Total deformation γ [%], the irreversible deformation V, and the reversible deformation E were measured by C+CR. Maximum resistance (Re) was measured by micro extension tests. For light microscopy, 40 μm dough sections were
stained with Coomassie Blue to visualize protein. Transparency of starch was enhanced by adding glycerol.

3 RESULTS AND DISCUSSION

Correlation between gluten rheological properties and bread volume: Gluten resistance, Re, was highly correlated to all the C+CR and V/E properties of gluten and to the ratio of gliadin to glutenin. Extensibility showed poor correlations to other parameters. At low deformations, OSC parameters showed no correlation to any other rheological characteristic of gluten. The only parameter correlating gluten properties to bake volume was G* (r= -0.826).

Correlations between gluten and dough: When the dough cultivars were compared, tan δ showed very small difference and did not change when excess water was added. All C+CR parameters of gluten were correlated to the C+CR parameters of rounded dough (r= 0.731 to 0.922) but not to unrounded dough. OSC values were poorly correlated while Re of gluten and dough were not correlated at all.

Correlations between rounded and unrounded doughs to bread volume: Neither C+CR nor OSC parameters of rounded and unrounded dough were correlated. G* and G' (-0.85 <r> -0.82) of rounded doughs and Re of both types of dough were highly correlated to bread volume (r= 0.927 and 0.913). In addition, Re of the two doughs correlated with each other (r= 0.990).

There is no rheological parameter that showed the same ranking among the cultivars tested (rounded and not rounded dough). However, rounded dough exhibited gluten like properties due to the increased development of a cohesive protein network [1]. During rounding, there is aggregation of gluten proteins that is similar to the deformation occurring in the micro extension test. Such aggregations could be followed by microscopy of dough with stained proteins by gently moving the section between glass plates (Figure 1).

Elasticity at small (G') or large deformations (E) seems to be not involved in the expansion of wheat dough. This is evident by looking at the G* and Re values obtained.

Among the different rheological tests, only G* and Re had been found useful for the prediction of baking performance and evaluation of wheat quality. Since G* was negatively correlated to the volume while Re was positively correlated, these parameters represent different properties affecting baking performance. Re reflects the firmness of dough at the moment of rupture and mirrors the rupture of dough structures (Figure 2 A.b) during fermentation and during oven rise. G* expresses the resistance of gluten or dough against deformation and, therefore, against the increase in volume during fermentation. Low Re at normal dough Ex is an indication of a very weak network that can not support dough weight or a long-term stretching of during fermentation.

Although material firmness, Re, is the resulting force due to the following physical properties, viscosity, elasticity, and cohesivity. Cohesivity seems to affect dough inflation to maximum volume. These could be demonstrated by mechanically treating dough to get weak and strong gluten structures. Rounding an optimally developed dough produces strong structures[1]. Figure 2 shows the baking results while Figure 3 shows the extensigrams of these two doughs. Maximum possible contribution of gluten to the total Re is also shown if all the gluten forms one single pure and optimal cohesive strand.

Without rounding, bread shape was flat (Figure 2 A.a) while crumb cell walls were thin and weak and had the tendency to rupture during oven rise. Staining the proteins before kneading or after freeze cutting revealed that optimum mixing in the farinograph had dispersed gluten evenly (Figure 2 A.B.). After relaxation, gluten contracted to somewhat globular bodies linked together by thin filaments. The thin filaments were subjected to early rupture during fermentation and oven rise. When dough was rounded and shaped, bigger aggregates were formed (Figure 2 B.). Simultaneous demixing of starch and gluten occurred because they lacked affinity to each other and gluten tends to minimize its surface for energy reasons. Stronger strands together with high starch accumulations contributed to larger crumb pores, thicker cell walls and the round shape of the leavened product.

The formation of viscoelastic gluten network did not explain the low correlation among gluten, dough elasticity, and volume-yield. Rapid elongation of gluten filaments when stretched showed reversible deformation (Figure 4). When the extension was stopped, the stretched filaments rapidly relaxed and glided over each other until a small reversible elongation of few μm to 100 μm remained. The magnitude of the remaining reversible deformation seemed to depend on the viscous fractions of dough. The gliadin/glutenin ratio of the cultivars and addition of gliadin and soluble pentosans to the flour had a large effect on the relaxation of the fibrils. This relaxation was accompanied by shear thinning of the filaments. Further elongation ruptured the filament.

Stretching of the gluten network occurred mainly due to slipping of gluten bundles. This is the reason why elasticity of dough and gluten did not correlate to the bake volume. Rheological methods monitoring shear thinning (C+CR test or the resistance at the breaking point (Re) using the extension test), are better suited for predicting the baking result. Lack of correlation between Re of gluten and dough may be explained by the loss of other dough components interacting with gluten during isolation. Gluten also had a low contribution to the total Re of dough (Figure 3).

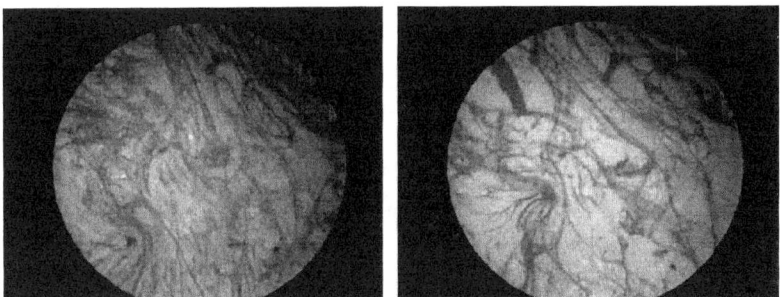

Figure 1 *Aggregation of gluten during rounding of dough. Protein stained by Coomassie blue*

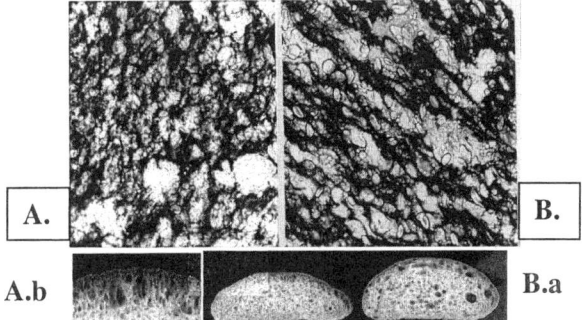

Figure 2 *A: not rounded dough; B: rounded dough; a: micro-baking test; b: weak crumb structure of not rounded dough*

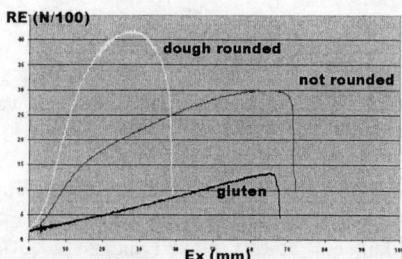

Figure 3 *Micro extensigrams of Dough and gluten (see text)*

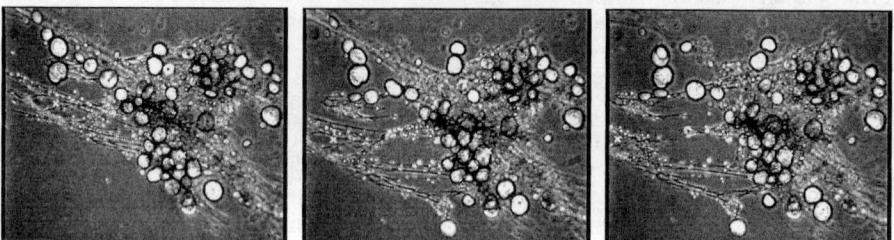

Figure 4 *Relaxation of gluten accompanied by rupture of the filaments*

4 CONCLUSION

Dough cohesivity can explain the high correlation between Re and Vol. A dense gluten network will allow long-distance stretching through slipping between gluten fibrils. The elastic properties of gluten and dough did not affect baking performance. However, dough elasticity may have a role in preserving the round-shaped surface of wheat bread.

References

1. R. Kieffer, N. Stein, *Cereal Chem.*, 1999, **76,** 688
2. Biczo G: *Diplomarbeit Fachhochschule NTA Prof.Dr.Grübler, Isny, Germany.* 2001
3. R. Kieffer, *Proceedings of the 3rd international symposium on food rheology and structure Zürich* ISBN:3-905609-19-3, 2003, 549
4. Approved Methods of the AACC, St. Paul, Minnesota (USA)
5. R. Kieffer, H. Wieser, MH. Henderson, A. Graveland, *J. Cereal Sci.*, 1998, **27,** 53

THE DYNAMIC DEVELOPMENT AND DISTRIBUTION OF GAS CELLS IN BREAD-MAKING DOUGH DURING PROVING AND BAKING

W. Li and B.J.Dobraszczyk .

The University of Reading, Department of Food Science and Technology, Whiteknights, PO Box 226, Reading RG6 6AP, UK.

1 INTRODUCTION

The unique property for wheat flour is able to form the visco-elastic dough and hold the gas during processing. The distribution and development of gas cells during proving and baking are of importance to bread volume and texture and closely related to flour components[1,2,3]. Although gas content and dynamic density of breadmaking dough during proving have been measured[4], the measurement of development gas cells in dough during proving and baking has not been achieved satisfactorily. The aim of the present work was investigate the dynamic development of gas cells in breadmaking dough during proving and their distribution in breadcrumb from flours with different baking performances using Image-Pro Plus analysis software with an Evolution LC colour digital Camera. This finding would be able to provide a clear figure of changes of gas cells during proving and baking and a better understanding of baking processing.

2 MATERALS AND METHODOLOGY

2.1 Materials

Bread-making flour and biscuit-making flour were kindly provided by a UK company, Heygate.

2.2 Methodology

The bread of both flours were made on the Chorleywood breadmaking process.
The dough samples (25 g ×5) of both flours obtained during processing were spread on the bottom of a sterilised Petri-dish (100×20 mm) and the pictures were shot as dough was proved for 10 min, 20 min, 30 and 40 min.
100 × 3 g of dough were proved and baked in proving and baking ovens respectively, the proving height was measured at 10 min, 20 min, 30 and 40 min of proving time. The bread height was measured after bread was cooled for 40 min .
Image analysis on 1.5 ×1.5 cm^2 of dough and bread slice was performed using Image-Pro Plus analysis software. The new mask was produced by the image with a certain range of

histogram, which was adjusted manually to satisfy the original picture. Once the black and white mask was obtained, the number of gas cells was counted and grouped into four groups according to the size of gas cells (0-0.1, 0.1-0.5, 0.5-2 and 2-50 mm). The area distribution of gas cells was also processed.

3 RESULTS

3.1 Image of proving dough and bread slice

The pictures of dough with different proving time from the breadmaking flour clearly show the development of gas cells during proving (Figure 1 a.10 min, b.20 min c. 30 min, d. 40 min). The similar results were obtained for the dough from biscuit-making flour (not shown). The pictures of bread crumb for both flours are shown in Figure 2. The texture of bread crumb from breadmaking flour (Figure 2 (a) was much finer than that from the biscuitmakiing flour (Figure 2 (c). Some coalescences of gas cells can be observed in bread crumb from breadmaking flour. Those pictures were imaged at different histograms and the new mask was obtained for analysis (Figure 2 (b) and (c) respectively for both flours.

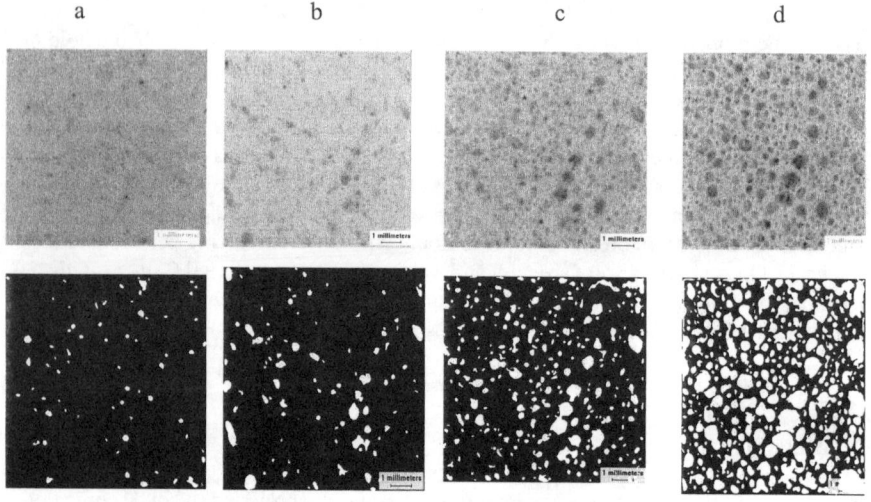

Figure 1 *The picture and image of dough with different proving times. (a) 10 min, (b) 20 min, (c) 30 min and (d) 40 min.*

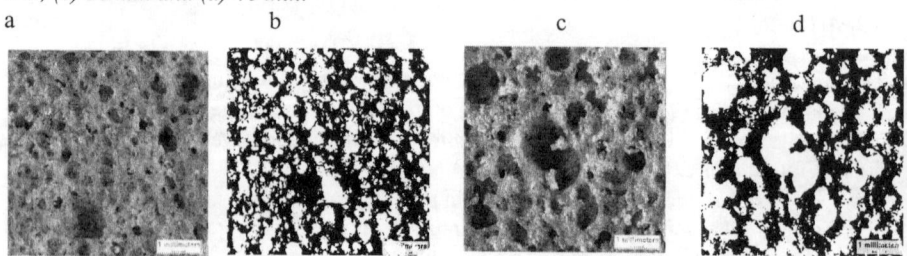

Figure 2 *The texture and image of bread crumb for bread-making (a and b) and biscuit-baking flours (c and d).*

3.2 The number of gas cells in proving dough and bread slice

The number of gas cells changed significantly during proving (Table I). The gas cells with the size ranges of 0-0.1 and 0.1-0.5 mm² mainly appeared in the first 30 min of proving time. Then, in last 10 min of proving time, the number of middle (0.5-2mm²) and large (2-50 mm²) gas cells increased at the expense of the reduction of the number of small size gas cells. Correspondingly, the proving height of dough for both flours increased as the gas cells changed. After baking, more large gas cells were observed in the bread slice compared with that in the corresponding dough at the end of proving for both flours. However, more small and middle gas cells were observed in the dough and bread slice for the bread-making flour than that for the biscuit-making flour. The height of bread loaf for the breadmaking flour was also much higher than the proving height of dough at the end of proving. This suggests that the coalescences of gas cells in dough happened in last 10 min proving and during baking was able to lead to the difference in bread volume and demonstrates that the more significant coalescence of gas cells happed in biscuit-making dough, resulting in a smaller proving height and baking height at the end of proving and baking.

Gas cell mm²	Bread-making flour					Biscuit-making flour				
	10 min proving	20 min proving	30 min proving	P40 min proving	Bread	10 min proving	20 min proving	P30 min proving	P40 min proving	Bread
0-0.1	85±1	305±26	603±97	347±35	138±3	80±7	295±5	395±38	218±5	60±12
0.1-0.5		8.5±2	125±3	104±22	33±13		48±4	95±3	101±2	16±4
0.5-2				35±10	13±4			3±1	1±0	29±2	10±6
2-50					3±2				6±1	19±7
Height of P & B (mm)	36±1	40±0	50±0	58±4	68±1	32±4	38±4	45±7	54±1	55±1

Table 1. *The number of gas cells in the dough with different proving time and in the bread slice (2.25 cm² area). The results are the means and ranges of duplicates.*

3.3 The area distribution of gas cells

The area distribution of gas cells was also processed using this software. A shift of gas cells from small size to middle and large gas cells during proving and baking can clearly seen in Figure.3 a and b, showing that the development and expansion of gas cells in the dough and bread during proving and baking. The gas cells with a size of 0.1-2 mm² contributed a higher proportion of the total area in the breadcrumb for bread-making flour (40%) than for biscuit-baking flour (25%). This indicates that those gas cells lead to a finer texture and a bigger volume of bread for bread-making flour. The larger size gas cells, therefore, are detrimental to bread-making performance.

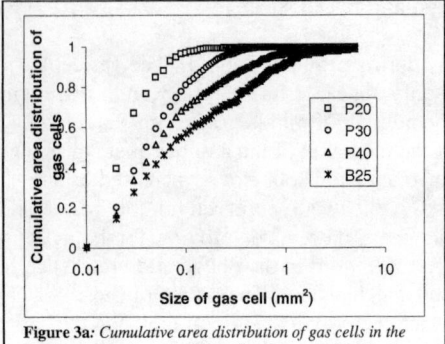

Figure 3a: *Cumulative area distribution of gas cells in the dough and bread crumb of Canadian flour.   P20, 20 min proving; o P30: 30 min proving Δ P40: 40 min proving; ⋇ B20: 20 min Baking.*

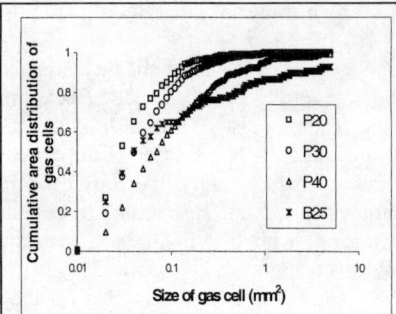

Figure 3b: *Cumulative area distribution of gas cells in the dough bread crumb of Riband flour.   P20, 20 min proving;; o P30: 30 min proving; Δ P40: 40 min proving; ⋇ B20: 20 min Baking.*

4 CONCLUSIONS

Coalescences of gas cells observed in the present study mainly occurred in the last 10 min of proving and during baking. More significant coalescence happed in biscuit-making bread than in bread-making flour, resulting in the difference in the proving height and the height if baking loaf and texture of bread crumb between the flours.

The gas cells with a size range of 0.1-2 mm^2 are of importance to the texture of bread and volume, but gas cells larger than 2 mm^2 is detrimental to bread quality.

References

1. W. Li and B.J.Dobraszczyk. *Young Cereal Scientist Workshop*, 2003. Valencia, Spain.
2. H.D. Sapirstein, R. Riller and W. Bushuk,. *Cereal Chem.*, 1994, **71**, 383
3. G.M. Campbell at al, Pages 207-231, 199, In *Bubbles in food*. Eds: G.M. Campbell, C. Webb and S.S.Pandiella. AM.Assoc. Cereal Chemistry . St Paul, MN.
4. G.M. Campbell, R. Herreo-Sanchez, R. Payo-Rodriguez and M.L. Mercham, *Cereal Chem.*, 2001, **78**, 272

Acknowledgments

The authors gratefully acknowledge funding by the Biotechnology and Biological Sciences Research Council (BBSRC). Grant Ref. 45/D15320.

VITAL WHEAT GLUTEN: CHEMICAL AND FUNCTIONAL ASPECTS

E. Marconi[1], M.C. Messia[1], M.F. Caboni[2], M.C. Trivisonno[3], G. Iafelice[1] and R. Cubadda[1]

[1]DISTAAM, University of Molise, Via De Sanctis, 86100-Campobasso, Italy
[2]Dipartimento di Scienze degli Alimenti, University of Bologna, Via Fanin, 40, 40127 Bologna, Italy
[3]PST Moliseinnovazione, Via De Sanctis, 86100-Campobasso, Italy

1 INTRODUCTION

Vital wheat gluten is widely used to improve the rheological and technological performances of weak flour and semolina and to counteract any functional changes caused by the incorporation of non-traditional raw materials (replacement of some or all the wheat flour/semolina with non gluten cereals or non cereal materials)[1].
Vital wheat gluten is the dried insoluble gluten protein of wheat flour/semolina from which starch and soluble components were removed by washing and which was reduced into a cream-coloured powder by drying and grinding. The drying process is the most critical phase since the gluten is readily denatured by heat with a drastic loss of functionality/vitality. Vital wheat gluten, when rehydrated, should have physical characteristics which closely resemble those of washed native gluten. The quality of dry gluten can vary on the basis of the type of wheat (species and variety) and the type of drying process (freeze drying, vacuum drying, spray drying, etc) used for their production[2,3,4].

In this work, commercial and laboratory-prepared (vacuum dried) vital wheat glutens were characterized for chemical, physical and rheological properties in order to assess their quality and the intensity of the drying treatment.

2 METHODS AND RESULTS

2.1 Material
Two commercial vital wheat gluten samples (from soft and durum wheat) and two vital wheat gluten samples prepared in laboratory from two Italian durum wheat varieties (Simeto=strong gluten and Ofanto=weak gluten) were used. The gluten of the latter samples was dried using a vacuum dryer pilot plant (Namad -Rome).

2.2 Analytical Methods
Protein was evaluated by Dumas combustion method (AOAC 990.03); amino acids were determined by HPAEC-PAD Dionex system (Dionex Corporation Sunnyvale, CA) according to the procedure reported by Messia et al.[5]; total lipids were determined by AACC method 30.10; free lipids were determined by Soxhlet procedure; bound lipids =

total lipids-free lipids; Gluten Index was determined by AACC method 38-20; L*, a*, b* values were evaluated by means of reflectance colorimeter (CR300 Chroma-meter, Minolta, Japan); furosine was determined by HPLC procedure reported by Resmini et al.[6]; sterols were determined by GC procedure reported by Pelillo et al.[7]

2.3 Results

Table 1 shows the chemical composition of commercial glutens, laboratory-prepared glutens and their respective semolina samples. The moisture content, an important parameter involved in the product costs and stability during storage[8], was very low in laboratory-prepared samples (<2.0%).

The protein content of both laboratory-prepared samples (82.4 and 83.3% d.b.) was higher than those of commercial soft and durum wheat glutens (79.7 and 72.2% d.b, respectively).

Lipids of gluten samples varied widely from 4.3 to 9.7% d.b., and were mainly in bound form; in fact the bound lipids were >80% in gluten versus 50% of the respective semolina samples (Figure 1). The ratio of bound to free lipids was highly significantly correlated with gluten vitality[9].

In Table 2 the sterol and amino acid content of wheat gluten is reported. Gluten samples were characterised by a higher sterol content than semolina samples. The sterol/stanol ratio, was significantly higher in soft wheat gluten than in durum wheat gluten. This ratio, was proposed as good marker for discriminating between hexaploid and tetraploid wheat species[7].

Lysine content was lower in gluten than in semolina, vice versa proline and glutamic acid were higher in wheat gluten than in semolina, since gluten proteins (gliadins and glutenins) are rich in proline and glutamic acid and poor in lysine.

The drying process, the most critical phase for vitality/functionality of the vital gluten, was characterized by the determination of gluten index, colour (L*, a*, b* values) and furosine (Table 3).

Table 1 *Chemical composition of commercial glutens, laboratory-prepared glutens and respective semolina samples (% d.b.)*

	Moisture	Protein	Ash	Lipid
Laboratory-prepared samples				
Durum wheat gluten (Ofanto)	2.0	82.4	1.86	8.8
Durum wheat semolina (Ofanto)	12.2	13.7	0.93	1.7
Durum wheat gluten (Simeto)	1.6	83.3	1.51	6.4
Durum wheat semolina (Simeto)	11.5	11.5	0.90	1.9
Commercial samples				
Soft wheat gluten	8.5	79.7	0.79	4.3
Durum wheat gluten	5.9	72.2	1.84	9.7

All gluten samples showed a good functionality as attested by gluten index values >40, except for commercial durum wheat gluten that showed a very poor quality (GI=2). The gluten index of both laboratory-prepared samples was >60 although the gluten quality of the respective semolina was very different (GI=3 and 60 for Ofanto and Simeto, respectively).

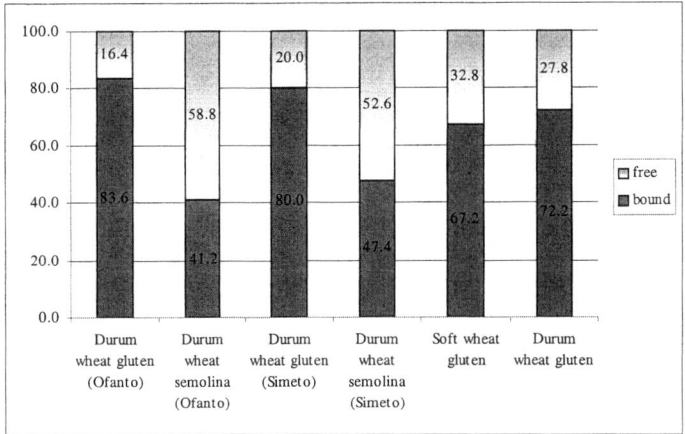

Figure 1 *Distribution (%) of bound and free lipids in gluten and semolina samples*

b* values (yellow index), as expected, were significantly higher in durum wheat glutens (from 11.2 to 21.9) than in soft wheat gluten (1.7), due to the higher carotenoid content of durum wheat. a* values (red index) were higher in commercial glutens than in laboratory-prepared glutens probably due to the different drying conditions used. In fact, heat treatment promotes the development of the non-enzymatic browning (Maillard reaction).

Table 2 *Total sterol and amino acid content of commercial glutens, laboratory-prepared glutens and respective semolina samples*

	Sterols (mg/g lipid)								Amino acids (g/16g N)		
	Campesterol	Campestanol	Stigmasterol	β-Sitosterol	β-Sitostanol	Other	Total	Sterols/stanols	Lysine	Proline	Glutamic acid
Laboratory-prepared samples											
Durum wheat gluten (Ofanto)	6.7	3.2	0.3	21.7	4.6	0.9	37.4	3.7	2.31	11.68	34.54
Durum wheat semolina (Ofanto)	5.0	3.1	0.2	14.2	3.6	0.6	26.7	2.9	2.63	9.98	35.76
Durum wheat gluten (Simeto)	8.0	4.3	0.2	24.7	5.3	1.1	43.8	3.5	2.41	11.75	35.38
Durum wheat semolina (Simeto)	3.7	2.0	0.1	11.5	2.6	0.5	20.4	3.7	2.65	10.18	35.64
Commercial samples											
Soft wheat gluten	7.0	2.0	0.4	24.7	3.1	1.4	38.6	6.3	--	--	--
Durum wheat gluten	5.3	2.8	0.4	17.8	4.3	1.6	32.1	3.3	2.07	11.63	37.97

The development of Maillard reaction was better monitored by using furosine assay, which is specific for ε-N-deoxy-ketosyl-lysine (Amadori compound) and is the most sensitive and most accepted method to determine the extent of "early stage" of the reaction[10]. As regards furosine content, commercial gluten samples showed values significantly higher (58.7 and

87.5 mg/100g protein) than laboratory-prepared glutens (12.6 and 16.3 mg/100g protein). The mild conditions adopted for preparing laboratory-prepared gluten (vacuum drying) slightly increased the furosine level found in the semolina samples 8.5 and 13.3 mg/100g protein, respectively.

3 CONCLUSION

In conclusion, the results show a wide variability in composition and rheological properties of different types of gluten, which attests the need to standardize and to define the characteristics of commercial vital wheat glutens.

Moreover, the intensity of drying process (heat damage) for vital wheat gluten production could be successfully assessed by furosine marker.

Table 3 *Qualitative characteristics of gluten samples and respective semolina samples*

	Gluten Index	Colour			Furosine (mg/100g protein)
		L*	a*	b*	
Laboratory-prepared samples					
Durum wheat gluten (Ofanto)	40	93.69	-2.79	21.86	12.6
Durum wheat semolina (Ofanto)	3	96.23	-3.18	19.7	8.5
Durum wheat gluten (Simeto)	68	96.22	-2.84	20.79	16.3
Durum wheat semolina (Simeto)	60	97.15	-3.21	18.45	13.3
Commercial samples					
Soft wheat gluten	43	97.22	0.30	1.7	58.7
Durum wheat gluten	2	94.83	-0.65	11.12	87.5

References

1 E. Marconi, M. Carcea, *Cereal Foods World*, 2001, **46**, 522.
2 F. Meuser, A. Kutschbach, R. Kieffer, H. Wieser and P. Schieberle, *Cereal Chem.*, 2002, **79**, 617.
3 P.L. Weegels and R.J. Hamer, *Cereal Food World*, 1992, **37**, 379.
4 R.A. Miller and R.C. Hoseney, *Cereal Food World*, 1996, **41**, 412.
5 M.C. Messia, G. Panfili, P. Manzi and V. Vivanti, in: Ricerche e innovazioni nell'industria alimentare, Proceedings of 5° Congresso Italiano di Scienza e Tecnologia degli Alimenti (CISETA), 2002, 1110.
6 P. Resmini, L. Pellegrino and G. Battelli, *Ital. J. Food Sci.*, 1990, **2**, 173.
7 M. Pelillo, G. Iafelice, E. Marconi and M.F. Caboni, *Rapid Commun. Mass Spectrom*, 2003, **17**, 2245.
8 M. Wooton, R.R. Reaoch, C.W. Wrigley and P.W. Gras, *Food Technol. Aust.*, 1982, **34**, 154.
9 C.K. Wadhawan and W. Bushuk, *Cereal Chem.*, 1989, **66**(6), 456.
10 E. Marconi, G. Panfili and R. Acquistucci, Ital, *J. Food Sci.*, 1997, **9**, 47.

CONFOCAL VISUALISATION OF MDD DOUGH DEVELOPMENT

M.P. Newberry, L.D. Simmons and M.P. Morgenstern

New Zealand Institute for Crop and Food Research, Private Bag 4704, Christchurch, New Zealand

1 INTRODUCTION

Dough mixing is a crucial stage in the production of bread, particularly in modern large scale processes that utilise high speed mixers to develop the dough. Mixing involves combining the flour, yeast, salt and other ingredients with water to form a single cohesive dough mass. As the mass is mixed, complex chemical and rheological changes occur that result in the formation of a thin gluten film that surrounds the starch granules. This film is derived from protein and lipid components in the flour and involves the interaction of proteins through disulfide bonds, association by secondary hydrogen and hydrophobic bonds, and the formation of lipid-protein complexes.[1] The formation of this film enables the dough to retain gas and contributes texture and volume to the finished product. However, the precise physical and chemical mechanisms of these protein-lipid-starch interactions remain unclear. The ability to observe these changes at a microscopic level may provide more information on the exact nature and effect of these changes. A variety of techniques have been applied to the study of dough microstructure including light microscopy, electron microscopy and more recently, confocal scanning laser microscopy. Traditional microscopic techniques may not present precise information because of a need to 'freeze' or 'fix' materials with an associated risk of subsequently viewing artefacts. Laser scanning confocal microscopy (LSCM) is a non-invasive technique, which allows materials to be visualised in three dimensions in a natural dynamic state.[2] Recently LSCM has been applied to the study of bread dough structure,[3,4,5] but at present the application of this technique to bread doughs is still evolving. The approach taken in this study was to investigate the relationship between dough components and dough development during mixing through LSCM using a dual staining that allowed the comparison of structural changes in the dough as it was mixed on a high speed mixer.

2 MATERIALS AND METHODS

Bread doughs were prepared using a commercially milled bakers flour, and were mixed on a 50 g Mechanical Dough Development mixer.[6] Doughs were mixed to the optimum water absorption of 62%.[7] The doughs were mixed to either optimal work input [8] (optimal development) of 10.7 Wh·kg^{-1}, or were under-mixed by receiving only 3.5 Wh·kg^{-1} of mixing. Two fluorescent dyes, Rhodamine (EC201-383-9, Sigma Chemical Co.) and Fluorescein (Sodium salt, EC 208-253-0, Sigma Chemical Co.) were added to the flour before mixing, both at 0.05% of flour weight. The Fluorescein stained the starch components visibly yellow, while the Rhodamine stained the protein structures visibly red. Images of each component could then be observed simultaneously. Imaging was carried out on a Leica TCS 4D confocal microscope (Leica Lasertechnik GmbH, Heidelberg, Germany) with a 16x oil immersion objective lens. The microscope was operated in fluorescent mode with excitation conducted at 488 nm using an air-cooled Ar/Kr laser. Dough samples were placed on glass cavity slides with a cover slip and observed under the microscope.

3 RESULTS AND DISCUSSION

Fluorescein clearly revealed the starch granules and vacuoles or cavities within the dough (Fig. 1), while the rhodamine highlighted the protein present in the dough — both that which is associated with the starch and that present as distinct strands or filaments (Fig. 2). The protein strands highlighted by the rhodamine in Figure 2, appear as "trenches" within the fluorescein images (Fig. 1) and clarified which dough components were being stained. The fluorescein also aided in the identification and monitoring of the behaviour of gas bubbles within the developing dough.[3]

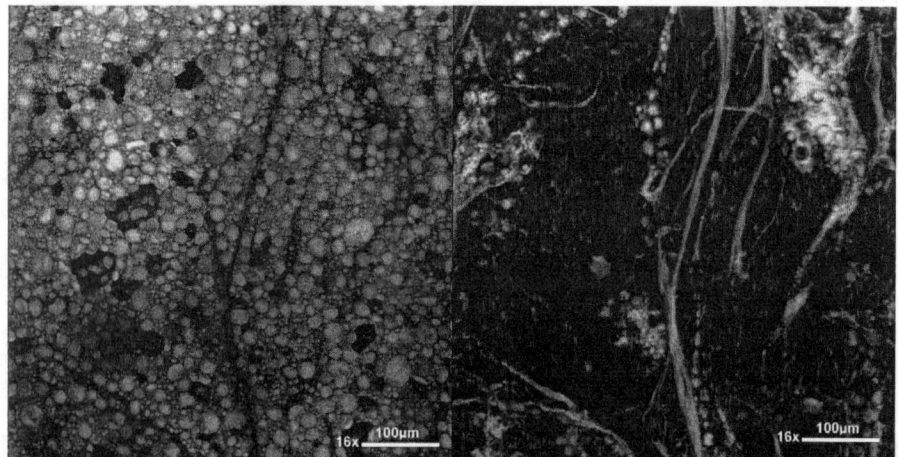

Figure 1. *Mixed dough, fluorescein image only. Note darker line or 'trench' in the middle of the image, corresponding to the protein strand in Fig. 2.*

Figure 2. *Mixed dough, rhodamine image only.*

Bread doughs are somewhat opaque to LSCM with the degree of penetration being limited by the scattering of the emission wavelength by the starch granules. This opacity restricts the scanned depth of images to ~20 μm. Nevertheless the images show a great deal about the dough structure, particularly the protein structure and how this changes with dough development.

The protein distribution in under-mixed dough was found to be more globular, with the protein closely associated with clusters of starch granules (Fig. 3). As mixing proceeded and the dough became more developed, the protein separated from the starch granules forming distinct protein strands (Fig. 4). The protein now appears as individual and apparently aligned strands, making the protein distribution more even throughout the dough, and providing the network that gives dough its characteristic ability to retain gas during fermentation and baking.

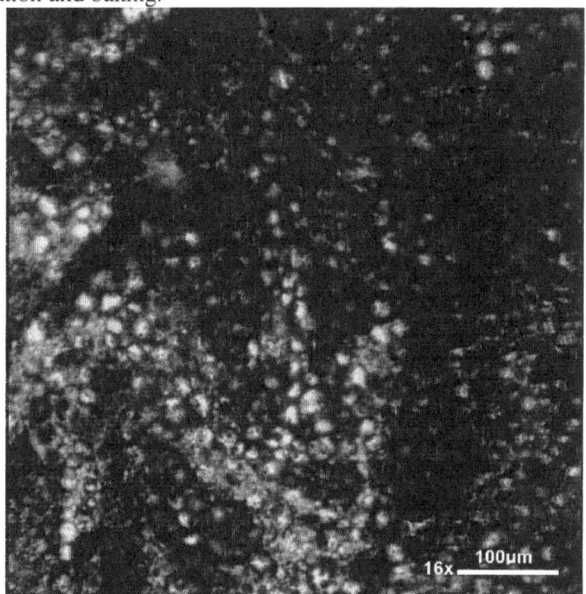

Figure 3. *Under-mixed dough. Protein is strongly associated with the starch granules and has not differentiated into a separate structure.*

Previous LSCM studies of dough have utilised fluorescein isothiocyanate, [4] which stains the protein by reaction with protein amide NH groups to give thiocarbamates, or autofluorescence, [5] which relies on the presence of aromatic amino acid residues but is relatively weak. Neither approach has shown protein strands as clearly as rhodamine staining.

4 CONCLUSION

Laser scanning confocal microscopy of under-mixed and optimally mixed doughs clearly shows the differences in the physical structure of the dough that occur during dough development. An increased amount of development showed a transition from an uneven distribution of protein associated with the starch granules to a more uniform distribution of separate protein filaments. Stains that mark different dough components enable the changes in relationship between those components that occur during dough development to be observed. Observations can be made continually, so that the technique has merit in tracing dynamic relationships such as bubble growth in doughs.[3]

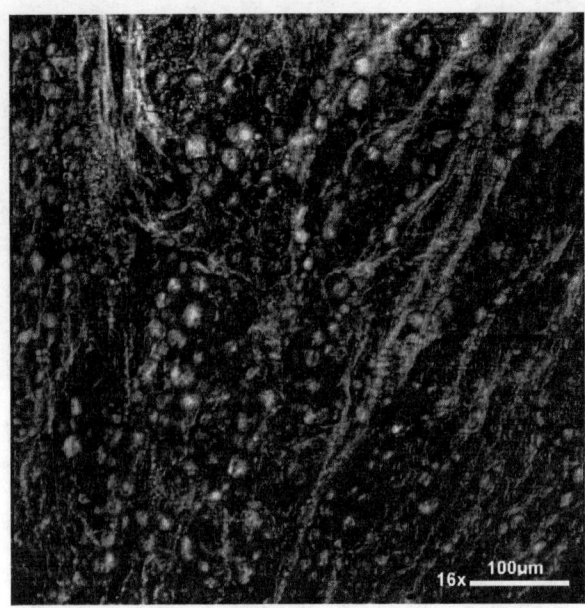

Figure 4. *Optimally mixed dough. Protein strands now form parallel lines of a sheet-like network.*

5 REFERENCES

1. F. MacRitchie, in *Chemistry and Physics of Baking,* eds. J.M.V. Blanshard, P.J. Frazier and T. Galliard, Royal Society of Chemistry, London, 1986, p. 132.
2. Y.Vodovotz, E.Vittadini, J.Coupland, D.J.McClements, and P.Chinachoti, *Food Technology*, 1996, **50**, 74.
3. K.H.Sutton, L.D.Simmons, M.P.Morgenstern, A.Chen and T.L.Crocker, *Proceedings of the 51st Australian Cereal Chemistry Conference,* eds M. Wootton, I.L.Batey and C.W.Wrigley, RACI, Melbourne, 2002, p. 272.
4. L.Lee, P.K.W.Ng, J.H.Whallon, and J.F.Steffe, *Cereal Chemistry*, 2001, **78**, 447.
5. B.A.Bugusu, B.Rajwa, and B.R.Hamaker, *Scanning,* 2002, **24**, 1.
6. T.A.Mitchell, in *Proceedings of the international symposium on advances in baking science and technology,* ed. C.C.Tsen, Kansas State University, Manhattan, 1984, p. 313.
7. N.G.Larsen and D.R.Greenwood, *J. Cereal Sci.,*1991, **13**, 195.
8. R.C.Hoseney, *Cereal Foods World,* 1985, **30**, 453.

DETERMINATION OF BREADMAKING QUALITY OF WHEAT FLOUR DOUGH WITH DIFFERENT MACRO AND MICRO MIXERS

S. Tömösközi[1], Á. Kindler[1], J. Varga[1], M. Rakszegi[2], L. Láng[2], Z. Bedő[2], O. Baticz[1], R. Haraszi[3] and F. Békés[3]

[1] Budapest University of Technology and Economics, Department of Biochemistry and Food Technology, H-1111 Budapest, Műegyetem rkp. 3. tomoskozi@mail.bme.hu
[2] Agricultural Research Institute of Hungarian Academy of Sciences, Martonvásár, H-2462 Martonvásár, Brunszvik 2, Hungary
[3] CSIRO Plant Industry, GPO Box 1600, Canberra ACT 2601, Australia

1 OBJECTIVES

Four different dough mixers – Farinograph, Valorigraph (a standardized Hungarian-type dough mixer), Mixograph and a micro Z-arm mixer – were used for the determination of rheological properties of wheat dough made from twelve different Hungarian wheat varieties (Triticum aestivum). The main goals were the following: i) to compare the mixing parameters obtained with a recently developed micro Z-arm mixer[1,2] with the widely used and standardized test methods; ii) to evaluate and compare the different mixing curves and parameters obtained with the above mentioned procedures to understand and explain the mechanical and/or molecular background of the different rheological properties of wheat dough; complex rheological characterisation of the Hungarian wheat cultivars.

2 MATERIALS AND RESULTS

The investigated wheat flours were milled on a laboratory-scale mill, type FQC 109 (Metefém Ltd, Hungary) according to the relevant standard procedure (Hungarian Standard MSZ 6367/-1989). The protein and moisture content of wheat kernel and their laboratory flour is summarized in Table 1. The applied experimental procedures and the evaluation modes of the different curves are summarized on Figure 1.
The shape and details of mixing curves are more informative for the researchers than the calculated mixing parameters, alone. All registered curves and evaluated parameters are shown on Figure 2, while the Table 2. consists the whole correlation matrix. All significant correlation cases are marked with red colour ($p < 0,05$, $N=12$).

3 CONCLUSION

The only detectable difference among the results from the Z-arm type mixer observed was the more rapid dough formation -shorter mixing time- in case of micro-scale studies. It appears mostly in case investigating stronger, very stable wheat flour samples, not effecting however the experimental results. The possible reason of this symptom is the faster water addition by the built-in automatic water pump. As it was mentioned in previous reports[3,4], pin mixers cause different mixing effects during formation and destabilisation period of mixing. Due to this fact, the registered curves at pin mixers, such as Mixograph, are markedly altered from the curves of Z-arm equipment. However, the significant correlation coefficients between different parameters, originated from different methods, suggest that more or less the same rheological behaviour are measured with the two types of methods. Further investigation is needed to clarify that what kind of rheological properties are covered in the different curves. There is a strong correlation between the water absorption values determined experimentally with Z-arm mixers and WA calculated from protein content in cases of pin mixers. Further studies are necessary to clarify this relationship and to evaluate the possibility of the future development and application of a simpler calculation procedure.

These facts underlay previous results[5]: the micro Z-arm mixer, a recently developed product of a Hungarian-Australian research cooperation, could be a useful tool not only for research studies, but also for any routine analytical application in breeding or in the grain industry. The types of Z-arm mixers, such as Farinograph, Valorigraph and micro Z-arm dough mixer gave similar curves and mixing parameters: significant correlations were found among all relevant parameters originated from these methods.

FARINOGRAPH	VALORIGRAPH	Micro Z-arm mixer	MIXOGRAPH
Method: ISO 5530 - 1	Method: ISO 5530 - 3	Method: according to ISO 5530 - 3	Method: according to AACC 54-40
Type of equipment: Farinograph® E Size of mixing bowl: 50g	Type of equipment: Valorigraph® FQA 205 Size of mixing bowl: 50g	Type of equipment: Micro Z-arm mixer (prototype) Size of mixing bowl: 4g	Type of equipment: A Swenson-Working Mixograph Size of mixing bowl: 10g
Evaluation of registered curves			
Statistical evaluation: In every cases three paralel measrement were evaluated. The correlation matrex was calculated wit Statistica 6. software (StatSoft Inc, USA)			

Figure 1 *Procedures for determination of mixing properties of wheat dough*

Name of wheat varieties	wheat kernels		laboratory flours	
	protein (% m/m,Nx5,8)	moisture (% m/m)	protein (% m/m, Nx5,8)	moisture (%,m/m)
MV Verbunkos	16,10	11,70	16,03	9,31
MV Palotás	15,28	10,79	14,68	10,95
MV Magvas	15,29	12,51	14,44	10,19
Bankuti 1201	16,25	12,24	16,27	10,03
MV Emma	15,82	12,41	15,24	10,66
MV Tamara	14,45	11,67	14,04	8,81
MV Emese	15,03	12,69	14,49	11,18
MV Suba	16,38	11,80	16,14	10,11
MV Palma	14,01	12,54	13,42	11,03
MV Magdalena	16,75	12,34	15,80	8,92
MV 16-2001	14,91	11,02	11,77	8,59
MV Martina	17,32	11,15	13,26	8,88

Table 1. *Chemical composition of Hungarian wheat varieties and their laboratory flours*

Gluten Rheology and Functionality

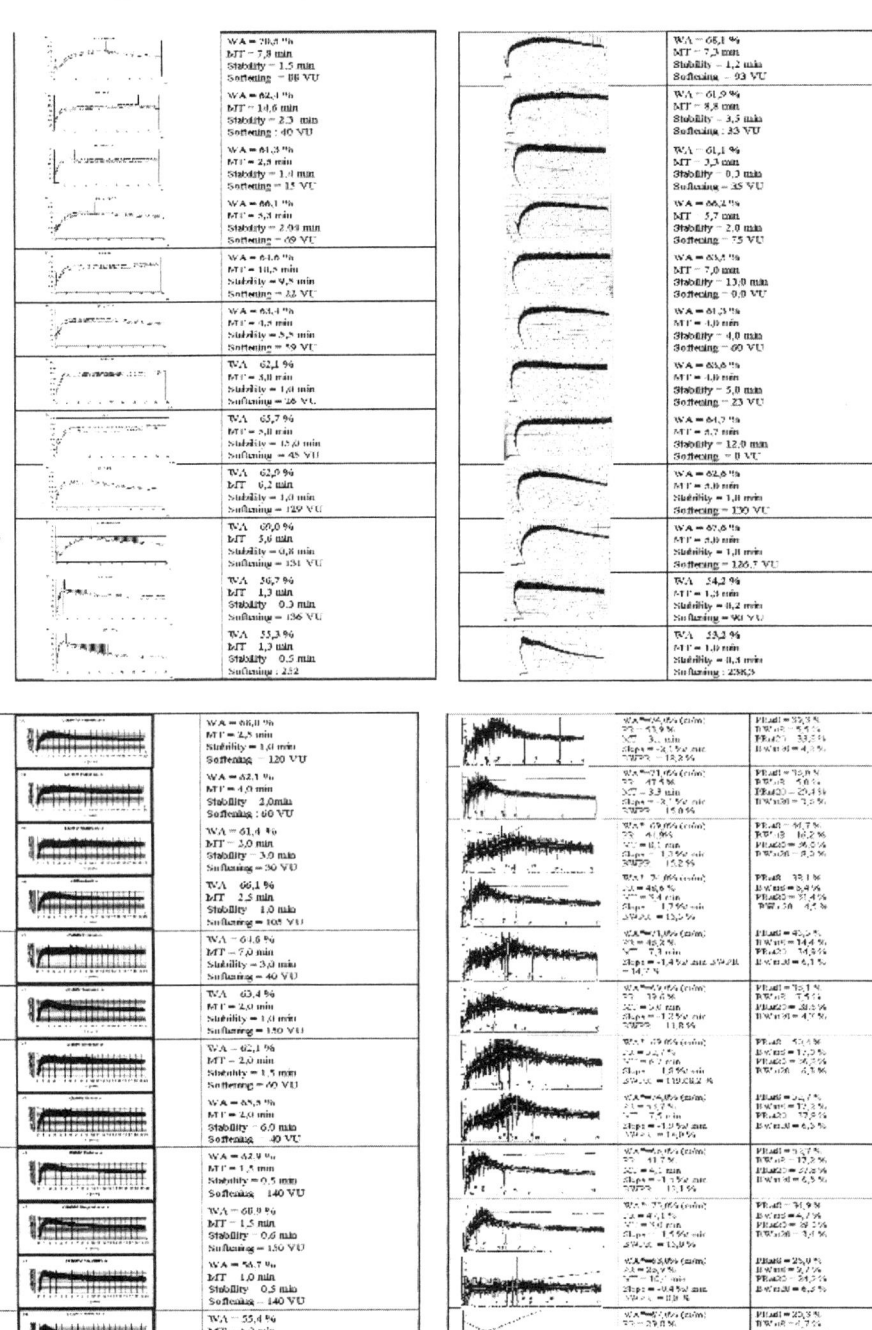

Figure 2 *Summary of the mixing curves and calculated rheological parameters*

Farinograph	Valorigraph			
	MT[min]	Stab [min]	BD [VU]	WA [ml]
FMT [min]	0,92	0,36	-0,39	0,45
FStab [min]	0,33	0,90	-0,58	0,24
FBD [FU]	-0,56	-0,52	0,96	-0,53
FWA [%]	0,71	0,23	-0,34	0,97
Valorigraph	Micro Z-arm mixer			
	MT[min]	Stab [min]	BD [VU]	WA [ml]
MT [min]	0,66	0,32	-0,54	--
Stab [min]	0,67	0,76	-0,57	--
BD [FU]	-0,56	-0,69	0,94	--
WA [%]	0,29	0,24	-0,45	--
Farinograph	Micro Z-arm mixer			
	MT[min]	Stab [min]	BD [VU]	WA [ml]
MT [min]	0,73	0,19	-0,40	--
Stab [min]	0,43	0,86	-0,47	--
BD [FU]	-0,54	-0,58	0,93	--
WA [%]	0,27	0,20	-0,33	--

Farinograph	Mixograph								
	PR [%]	MT [min]	Slope [%/min]	BWPR [%]	PRat8 [%]	BWat8 [%]	PRat20 [%]	BWat20 [%]	WA [%] (calculated)
MT [min]	0,46	-0,42	-0,47	0,35	0,22	-0,22	0,30	**-0,57**	0,45
Stab [min]	0,37	0,27	-0,19	0,33	**0,59**	0,52	0,47	0,00	0,43
BD [FU]	**-0,67**	-0,27	0,13	**-0,72**	**-0,85**	**-0,62**	**-0,91**	0,42	-0,40
WA [%]	**0,81**	-0,47	-0,47	**0,74**	0,51	-0,08	**0,63**	**-0,68**	**0,83**
Valorigraphh	Mixograph								
	PR [%]	MT [min]	Slope [%/min]	BWPR [%]	PRat8 [%]	BWat8 [%]	PRat20 [%]	BWat20 [%]	WA [%] (calculated)
MT [min]	**0,74**	-0,45	-0,61	**0,65**	0,48	-0,09	**0,56**	-0,68	**0,69**
Stab [min]	0,43	0,29	-0,22	0,41	**0,66**	**0,60**	0,52	-0,02	0,38
BD [FU]	**-0,59**	-0,43	0,04	**-0,66**	**-0,84**	**-0,70**	**-0,88**	0,36	-0,34
WA [%]	**0,90**	-0,45	-0,53	**0,84**	0,64	0,03	0,72	**-0,68**	**0,83**
Micro Z-arm mixer	Mixograph								
	PR [%]	MT [min]	Slope [%/min]	BWPR [%]	PRat8 [%]	BWat8 [%]	PRat20 [%]	BWat20 [%]	WA [%] (calculated)
MT [min]	0,34	0,02	-0,19	0,29	0,39	0,22	0,40	-0,20	0,34
Stab [min]	0,50	0,40	-0,25	0,51	**0,75**	**0,73**	0,64	0,10	0,42
BD [FU]	**-0,66**	-0,38	0,14	**-0,72**	**-0,87**	**-0,72**	**-0,90**	0,29	-0,37

Table 2 *Comparison of the mixing properties of wheat dough determined with different methods – correlation matrices (Marked correlations are significant at $p < 0,05$ $N=12$*

References

1 S.Tömösközi, J. Varga, C.W. Gras, C. Rath, A. Salgó, J. Nánási, D. Fodor and F. Békés, In: Wheat Gluten, Eds. Shewry, P.R. and Tatham, A.S., Royal Soc. Chem., Cambridge, UK., 2000, 321,
2 P.W.Gras, J, Varga, C. Rath, S. Tömösközi, D. Fodor, A. Salgó and F. Békés, *in Proceeding* of 11th International Cereal and Bread Congress, Broadbeach, Qld, Australia, 2000
3 J.A. Johnson, J.A. Shellenberger and C.O. Swanson, Cereal Chem, 1946, **23**, 388.
4 J.L. Hazelton and C.E. Walker, Cereal Chem., 1994, **71** (6), 632.
5 F. Békés and P.W. Gras, Cereal Foods World, 1999: **44** (8), 580-586.

Acknowledgments

This work has been supported by the Hungarian Scientific Research Found (OTKA, Project No.: T-034486) and by National Research and Development Programs (NKFP, Project No: 04/35/2001)

RHEOLOGICAL PROPERTIES AND MICROSTRUCTURE OF MIXED AND UNMIXED FLOUR-WATER SYSTEMS

L. Unbehend[1], M.G. Lindhauer[2] and F. Meuser[3]

[1]Faculty of Food Technology, Kuhaceva 18, 31 000 Osijek, HR
[2]Federal Center for Cereal, Potato and Lipid Research, Schützenberg 12, 32756 Detmold, D
[3]TU Berlin, Seestraße 13, 13353 Berlin, D

1 INTRODUCTION

To produce baked products of desired quality it is necessary to adjust recipe and processing technique to the quality of the flour being used. Since the preparation of dough has a considerable influence on the resulting product, it is very important to understand the physical and chemical background in dough during its development. Many research projects have been carried out and papers published to the subject of dough development[1-5]. It is known that dough development is influenced by the flour constituents, their and the forces acting between them, but mostly by insoluble wheat proteins, which form gluten during dough development. Gluten development is the most important process during dough development in determining dough properties. However its extreme heterogeneity, combined with its atypical physical properties, necessitated the use and development of innovative analytical techniques[6]. It must be recognised that such a protein network, even when formed from an optimal combination of polypeptides, is not sufficient to ensure good quality. Therefore determination of the structure of the gluten is one of the most formidable problems ever faced by the protein chemist. The aim of this research work was to obtain and characterise dough forming properties of gluten in its native state (preparing HUF-system = Hydrated Unmixed Flour) compared with the dough forming properties of gluten prepared under mixing and kneading conditions (preparing dough), using physical and microscopical techniques.

2 METHOD AND RESULTS

2.1 Preparation of HUF-system

Preparation of HUF-system was done at the room temperature. Filter paper was laid on a sieve and dampened with water. The flour was spread on the filter paper in a thin, uniform layer (0.5-1 cm). To prepare the HUF-system, distilled water was added in a ratio 1:5 (flour:water). After the water has wetted the flour thoroughly, the excess water was allowed to drain through the sieve. The wet system was taken from the sieve, carefully covered with a second filter paper and than wrapped in a dry cotton towel. The excess water in the system was absorbed by the filter paper and the towel. The developed HUF-system was weighed, to verify that the water content is in agreement with the water content according to ICC standard method No. 115 (water absorption).

2.2 Uniaxial extension measurements

The mixed systems, doughs, were used for measurements with the Extensograph following the ICC standard method No. 114. Because of salt addition during the preparation of doughs according to the ICC standard method, the HUF-systems preparation, described above, was modified for this test. The exact amount of salt was calculated and added to the water to such an amount, that the salt content in the final HUF-system was 2% of the weight of the flour corresponding to the dough. The developed HUF-system with the correct water- and salt-content was removed from the filter paper, carefully folded, placed and closed in the dough bracket of Extensograph. In this paper only the Extensograph results after a time of 45 min will be shown.

2.3 Dynamic testing

Dynamic oscillation measurements in the frequency sweep mode were done with a plate-plate-system (Gabo Eplexor). The flour-water-systems were prepared from 10 g flour and distilled water. The measurements were made with a constant strain of 0.5 % by increasing frequency from 0.1 to 100 Hz. The storage modulus (E´), the loss modulus (E´´), the complex modulus (E* = E+iE´´) and tan delta (E"/E') were registered.

2.4 Microscopy investigations

Both water-flour-systems were prepared from flour and distilled water. From the dough, gluten was isolated by washing. Isolation of gluten from the HUF-system was done without mechanical influence. The HUF-system was carefully put in cotton linen and tied. The tied cotton linen was set on the sieve and distilled water was discontinuously added through it to the HUF-system. The flow of water through the HUF-system removed starch. This was repeated until the water flow was free of starch granules. Pieces of dough, of HUF-system and of glutens were cut off, frozen and freeze-dried. For the microscopic investigation, specimens were attached to the sample stubs with silver conducting paint and coated with a layer of gold approximately 20-25 nm thick in a sputter coater. The coated samples were viewed in a scanning electron microscope Leitz AMR 1 6000 T at an accelerating potential of 20 kV and photographed.

2.5 Results of uniaxial measurement- extensogram results

As can be seen the dough of variety Bussard and of variety Contra had a lower resistance than the respective HUF-systems (Figure 1). The extensibility of the dough of both varieties was greater than the extensibility of the unmixed systems. The energies and extensogram maximums were higher with the unmixed systems. The ratios of resistance to extensibility were lower in investigated doughs than in HUF-systems.

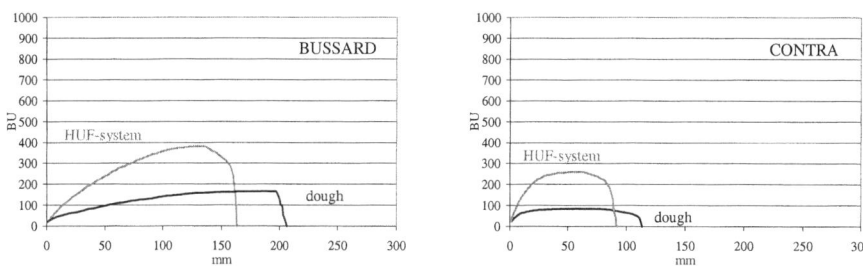

Figure 1 *Extensogram curves of unmixed (HUF-system) and mixed (dough) systems of different wheat varieties*

2.5 Results of dynamic oscillation measurement- frequency sweep results

At the beginning of the analysis the elasticity of the unmixed system of Bussard was higher than the elasticity of the dough and therefore Tan Delta values of the HUF-system were lower (Figure 2). The Tan Delta values of the HUF-system exceeded that of the dough at the frequency of 1.19 Hz. The values of Tan Delta of dough decreased and the respective values of the HUF-system were similar. The variety Contra showed the same differences as Bussard. The Tan Delta curve of the unmixed system lay above that of the mixed system. The decrease of the Tan Delta for both systems was similar.

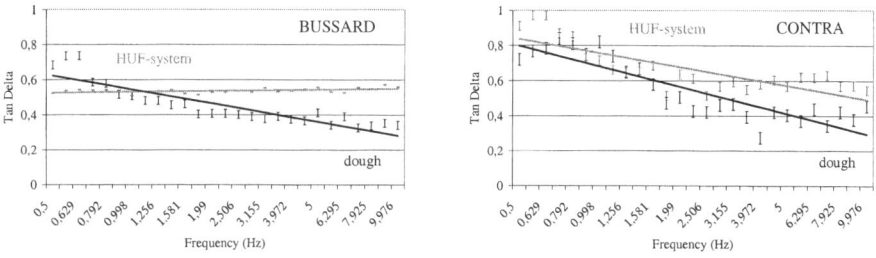

Figure 2 *Tan Delta of unmixed (HUF-system) and mixed (dough) systems of different wheat varieties*

2.6 Results of microscopical investigations

A change of the gluten matrix was evident when the results for the optimally kneaded dough and HUF-system were compared (Figure 3). The optimally kneaded dough showed a continuously interconnected gluten matrix surrounding most of the starch granules building gluten films. Gluten in HUF-system also showed a continuous matrix but not strong and irregularly connected. The starch granules appeared to be only partially covered by gluten aggregates.

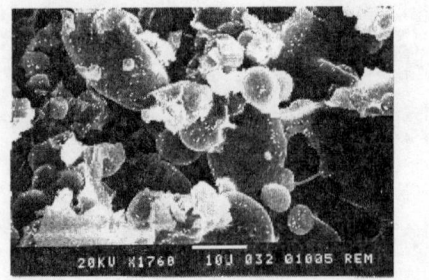 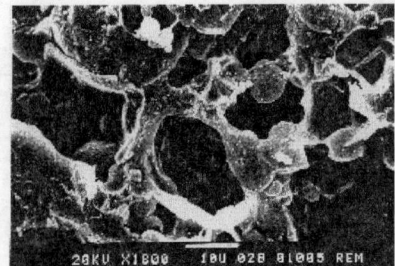

HUF-system   dough

Figure 3 *Scanning electron micrographs of dough and HUF-systems after freeze-drying*

3 CONCLUSION

The results have shown that aggregation potential of hydrated wheat proteins is strong enough to build a viscoelastic system, HUF-system, even without mixing and kneading. The results obtained in rheological investigations showed that dough had variety-dependent soft and extensible structure. However, the HUF-systems were more firm and less extensible than doughs specifically according to variety. Direct viewing of the dehydrated microstructure showed that the proteins in HUF-system appeared as a three-dimensional framework of aggregates extending between the starch granules and partly surrounding them in a sponge-like structure. On the other hand it has been demonstrated that protein films are predominant in an optimally mixed dough.

With this procedure the changes of flour-water-system properties during mixing and kneading can be shown and a new possibility for the determination of wheat quality has been opened.

References

1. O. Paredes-Lopez and W. Bushuk, *Cereal Chem*, 1982, **60**, 19.
2. O. Paredes-Lopez and W. Bushuk, *Cereal Chem*, 1982, **60**, 24.
3. T. Amend und H. D. Belitz, *Z Lebensm Unters Forsch*, 1990, **190**, 401.
4. T. Amend und H. D. Belitz, *Z Lebensm Unters Forsch*, 1990, **191**, 184.
5. T. Amend, *Getreide Mehl und Brot*, 1995, **49**, 359
6. C. W. Wrigley and J. A. Bietz, in *Wheat: Chemistry and technology,* 3rd Edn., ed.Y. Pomeranz, AACC , St. Paul, 1988, ch. 5, p. 159.

EVALUATION OF DURUM WHEAT QUALITY USING MICRO-SCALE AND BASIC RHEOLOGICAL TESTS

S. Uthayakumaran[1], D. Lafiandra[2] and M. C. Gianibelli[3]

[1]Food Science Australia, 11 Julius Avenue, North Ryde, NSW 2113, Australia.
[2]Università degli Studi della Tuscia, Dipartimento di Agrobiologia e Agrochimica, Via S. Camillo de Lellis, 01100 Viterbo, Italy.
[3]CSIRO Plant Industry, PO Box 1600, Canberra, ACT, 2601, Australia.
 e-mail: cristina.gianibelli@csiro.au

1 INTRODUCTION

Durum wheat is mainly used for the manufacture of pasta which is made by extruding semolina, a coarse flour. Agglomerating semolina produces another end product: couscous. Large strain descriptive rheological methods such as the Alveograph and Extensograph have been extensively used in the characterisation of the durum wheat semolina. The Alveograph of Chopin has been widely adopted for the evaluation of gluten strength especially in European countries such as Italy, France and Spain. Alveograph submits the dough to a biaxial extension and provides information about the strength of the dough and its extensibility. The breaking force required to rupture the dough, which is performed at 50 % of water absorption, is correlated to its strength. The Alveograph is one of the instruments more widely accepted for the evaluation of semolina dough strength and extensibility. However, the large amount of semolina required (250 g) makes its use inadequate when dealing with small samples, for instance breeding program lines or samples from reconstitution studies. Therefore alternative micro-scale tests such as the 2 g Mixograph (mixing properties), micro-scale extension tester need to be used to analyse physical dough properties. The use of micro-Mixograph and Extension Tester, which work with very small amount of sample (2 g each) has represented an improvement in the evaluation of dough flour properties when the amount of sample is limited. Basic rheological tests unlike the empirical tests have physical units which enable comparison of results from different tests. Very little work has been done on the characterisation of rheological properties in durum wheat and correlating these with results from empirical tests. The aim of this study was to evaluate a set of durum wheat semolinas with different dough properties (estimated by the Alveograph of Chopin) by small-scale devices and to investigate the relationships of empirical tests to basic dough rheology.

2 MATERIALS AND METHODS

A set of Italian durum wheats (*Triticum turgidum* var. *durum*) was evaluated in this study (Table 1). Protein content of semolina was determined by Kjeldahl's method. *Functional properties:* Alveograph measurements: ICC Standard N 121 was used to determine

Alveograph curves. The values considered in this study are W: total area under the curve and P/L ratio (P=tenacity and L=length of the curve).

Mixing and extension studies: All formulations were mixed in triplicate on a 2 g Mixograph™ (TMCO, Lincoln, NE, USA) using 3.5 g of dough. This small-scale machine has demonstrated a close correlation with conventional 35 g equipment[1]. Mixing time, peak resistance, resistance breakdown, bandwidth at breakdown were determined automatically using software developed by Gras and Bekes[2]. Extension tests were performed on 3.5 g dough mixed to peak dough development in a 2 g Mixograph. Dough samples (1.7 g / test) were moulded into cylinders approximately 6 mm in diameter. The moulded pieces, approximately 45 mm long, were mounted on a sample carrier and rested at 30°C and >90 % RH for 45 minutes before extension testing[2].

Basic rheological tests: The elongational properties of dough were studied using a constant-strain rate extension technique. All formulations were mixed in a 10-g Mixograph. The dough was mixed to peak dough development and suspended between a fixed and a moving grip both having a diameter of 30 mm. The dough sample was rested for 45 min before testing and moisture loss was prevented by applying a layer of petroleum jelly around the edge. The dough was pulled apart exponentially in a Universal Testing Machine (United Calibration Corp, Huntington Beach, CA), at a constant-strain rate (0.01 s^{-1}). Force and distance data collected by computer were used to calculate the rheological parameters of strain and eleongational viscosity (Pa.s). Viscometric properties were studied using shear viscometry[3]. The mixed dough was mounted on a controlled stress rheometer (Reologica Stresstech, Reologica Instruments AB, Lund, Sweden) in the parallel plate configuration (25 mm) diameter. A constant shear rate of 0.9664 s^{-1} was applied to the sample and viscosity was plotted against time, the maximum viscosity (Pa.s) during shear was determined. Frequency sweeps between 0.1 Hz – 20 Hz were conducted on a controlled stress rheometer (Reologica Stresstech, Reologica Instruments AB, Lund, Sweden) at a strain of 0.0005.

3 RESULTS AND DISCUSSION

Variation in protein content among cultivars was observed (from 12.2 to 14.2 %). High protein content has been associated with good cooking quality and has been generally recognised as one of the main factor associated with pasta cooking quality[4].

In the present study, the Alveograph clearly differentiated cultivars with diverse gluten strength (strong, moderate, weak and very weak types). It was also observed that strong gluten cultivars produced less extensible dough with a higher P/L ratio (tenacity/length of the curve) than cultivars with moderate to weaker gluten. Cultivar with high P/L ratio have been associated with high W values. However, in this study the cultivar Demetra has a unusual behaviour. Although Demetra has a high P/L ratio its W value was very low indicating a dough very inextensible with poor pasta quality.

Most of the analysed semolina showed an high dough strength as estimated by a long mixing time, peak resistance, bandwidth at peak resistance and also high tolerance to overmixing (decrease in RBD) (Table 1). Wide variation in the R_{max} and extensibility were observed among the cultivars analysed (Table 1). For instance, Preco had the most extensible dough with the lowest R_{max}, while Ares presented the opposite behaviour: highest R_{max} and lowest extensibility. The cultivar Demetra had the lowest area under the curve indicating a weak type of dough.

Table 1 *Protein content, Chopin Alveograph parameters, mixing and extension parameters of Italian cultivars*

Cultivars	Protein (%)	W	P/L	MT	PR	RBD	Ext	R_{max}
Ares	14.2	394.00	1.28	266.0	379.0	12.0	0.0688	0.8667
Baio	13.6	323.00	1.5	231.7	329.0	13.7	0.0723	0.7597
Demetra	12.9	106.00	1.59	121.0	293.7	27.0	0.0860	0.2647
Flaminio	14.2	410.00	1.11	204.0	394.0	12.3	0.0848	0.8295
Nefer	13.0	297.00	1.16	262.3	317.3	12.3	0.0750	0.6122
Preco	13.9	246.00	0.97	158.3	397.0	12.7	0.1268	0.3744
Appio	12.2	131.00	0.65	141.0	354.0	19.3	0.0815	0.4945
Grazia	12.9	311.00	0.98	113.7	194.3	9.2	0.0938	0.7291
Svevo	14.1	324.00	1.73	189.7	383.0	13.7	0.0845	0.5487

W = total area under the curve (joules x 10-4); P/L ratio (P=tenacity and L=length of the curve (height x 1.1/length)); MT = Mixograph mixing time (seconds); PR = peak resistance (AU), RBD = resistance breakdown (%); Ext = extensibility (mm); Rmax = maximum resistance (N).

Under uniaxial elongation, the elongational viscosity of the dough increased with extension, more rapidly at the highest strain levels. This sharp increase in elongational viscosity with increasing strain is known as strain hardening. A maximum viscosity value was reached during this strain-hardening stage, at which point the dough sample ruptured between the plates. Samples Appio and Preco had the highest elongational viscosity and greatest rupture strain and samples Nefer and Demetra had the lowest elongational viscosity and rupture strain (Figure 1). Sample Grazia had elongational properties in the middle of the range. During shear viscometry, unlike newtonian liquids, dough never reached a steady shear state. Instead, the viscosity increased with shearing, reaching a maximum at which point the sample fractured. Hence the maximum peak viscosity was used to compare the different samples. Based on the results of the maximum peak viscosity the Italian samples could be divided into two groups. Samples Ares, Baio, Demetra and Svevo had the highest maximum peak viscosity and the others had lower maximum peak viscosities (Figure 2). Strain sweeps conducted at 1 Hz revealed the linear visco-elastic region for all the flours studied were below a strain of 10^{-3}. Frequency sweep experiments were therefore performed at strain of 5×10^{-4}, within the linear region of the dough. Samples Appio, Grazia and Svevo showed high values for storage modulus.

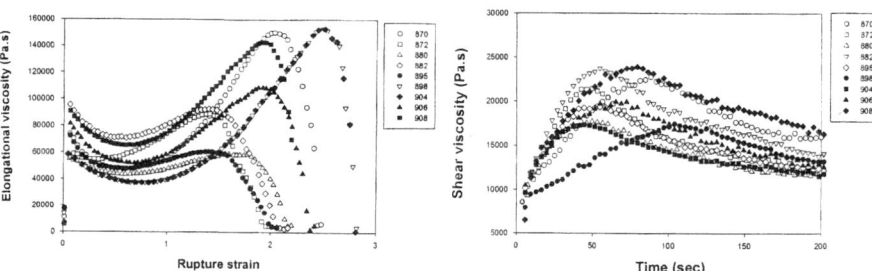

Figure 1 *Elongational properties of the nine semolina*

Figure 2 *Steady shear viscosity of the nine semolina*

Protein content was significantly correlated with maximum shear viscosity (Table 2). The W value, the most important parameter used for the evaluation of dough properties in durum wheat, was significantly associated with the R_{max}, area under the curve and maximum shear viscosity. These results indicate that small-scale extension testing as well as shear viscometry is a good approach to be used in the evaluation of durum wheat dough. They show high correlation with the International standard test (Chopin Alveograph) and can be performed with only 2 g of semolina. Further studies are being carried out to improve this method for the analysis for dough semolina. This is a worthwhile approach to be used in breeding programs of durum wheat, specially when improvement of durum wheat cooking quality is needed.

Table 2 *Correlation of parameters from empirical and basic rheological tests*

	W	P/L	Protein	MT	PR	BWPR	RBD	BWBRD	TMBW	MBW	Ext
W	1.000										
P/L	0.165	1.000									
Protein	*0.760*	0.452	1.000								
MT	0.647	0.244	0.503	1.000							
PR	0.221	0.069	0.589	0.445	1.000						
BWPR	0.333	0.001	0.603	0.554	**0.982**	1.000					
RBD	**-0.819**	0.219	-0.454	-0.423	-0.011	-0.167	1.000				
BWBRD	**-0.919**	0.045	-0.468	*-0.672*	0.010	-0.145	0.859	1.000			
TMBW	0.658	0.159	0.446	**0.991**	0.361	0.488	-0.494	-0.727	1.000		
MBW	0.354	0.018	0.630	0.514	**0.985**	**0.994**	-0.160	-0.144	0.442	1.000	
Ext	-0.238	-0.305	0.079	-0.548	0.081	0.036	-0.074	0.361	-0.532	0.033	1.000
Rmax	**0.865**	-0.053	0.421	0.593	0.010	0.143	-0.711	**-0.931**	0.631	0.163	-0.522
Time to Rmax	0.218	-0.357	0.336	-0.164	0.205	0.246	-0.546	-0.125	-0.125	0.237	**0.860**
Area under the cur	**0.817**	-0.287	0.462	0.201	-0.006	0.103	**-0.834**	**-0.808**	0.252	0.144	0.021
EV	0.065	-0.371	0.205	-0.086	0.439	0.467	-0.268	0.048	-0.083	0.468	0.321
RS	-0.376	-0.612	-0.226	-0.470	0.143	0.119	0.036	0.411	-0.447	0.110	0.572
SV	**0.833**	0.508	*0.712*	0.501	0.186	0.232	-0.477	*-0.666*	0.463	0.282	-0.485
G'	-0.036	-0.135	-0.258	-0.167	-0.005	-0.001	-0.022	0.036	-0.183	0.042	-0.331

significant at 0.01
significant at 0.05

4 CONCLUSION

The small-scale tests used in this study have been useful to discriminate among durum wheat cultivars with different strength (scored with traditional large-scale tests). The Extension testing and the shear viscometric test appear as the most suitable small-scale tests to select durum wheat cultivars with adequate dough strength for the production of high pasta quality.

References

1. C.R. Rath, P.W. Gras, C.W. Wrigley and C. Walker *Cereal Foods World* 1990, **35**,572
2. Gras, P.W. and Bekes, F. In Proceedings of the 6th Gluten Workshop, RACI, Australian Cereal Chemistry Conference, Melbourne, Australia, 1996, 506.
3. S. Uthayakumaran, M. Newberry, F.L. Stoddard and F. Bekes, In *Proceedings of the Wheat Gluten Workshop,* ed. P.R. Shewry ans A.S. Tatham, Royal Society of Chemistry, Cambridge, UK. 2000, 396
4. B.A. Marchylo, J.E. Dexter, J.M. Clarke, and N. Ames. *Proceedings Symposium On Wheat Protein and Marketing,* University Extension Press, University of Saskatchewan, Saskatoon, SK. Eds. D.B. Fowler, W.E. Geddes, A.M. Johnston and K.R. Preston. 1998, 53

GLUTEN EXTENSIBILITY: A KEY FACTOR IN URUGUAYAN WHEAT QUALITY

D. Vázquez[1,2] and B. Watts[2]

[1]INIA, La Estanzuela, CC 39173. Colonia. CP 70000. URUGUAY. dvazquez@inia.org.uy
[2]Department of Human Nutritional Sciences, University of Manitoba. Winnipeg, MB. R3T 2N2. CANADA.

1 INTRODUCTION

Highly elastic doughs are essential for production of pan bread, but the elasticity must be balanced by extensibility for best volume and quality. This balance is even more critical for the production of high quality hearth breads. Gluten proteins, responsible for both the elasticity and extensibility of bread doughs, have been widely studied. However, factors responsible for strength have received much more attention than those that responsible for plasticity. Lack of extensibility, resulting in "short doughs", is a major problem with Uruguayan wheats. The objective of this study was to evaluate the factors that contribute to this problem.

2 MATERIAL AND METHODS

2.1. Wheat samples

Wheat samples were obtained from the INIA breeding program. Uruguayan wheat is divided in two classes according to growing cycle: short and long cycle. Eight short cycle advanced lines and fifteen long cycle ones were evaluated.

2.2. Flour analyses

Flour was extracted using a Brabender Quadrumat Jr. laboratory mill (AACC method 26-50[1]) and tested for protein (AACC method 46-11[1]), SDS sedimentation volume[2], and for mixogram (35g-bowl, AACC method 54-40[1]) and alveogram (250-bowl, AACC method 54-30[1]) parameters. All analyses were carried out in duplicate.

2.3. Breadmaking

French-type breads were baked following a typical Uruguayan procedure. Flour (100g) was mixed with yeast (2.2g), salt (2.5g) and water (adjusted amount). After 90min fermentation at 28°C, molding and proofing (90min), loaves were baked 20 min at 212°C. Volume, height and width of cooled bread were determined. All samples were baked on two different days. Single measures on the replicates were averaged.

3 RESULTS

Flour quality tests results are summarised in Table 1. The data indicated that that all wheat flour samples were suitable for hearth loaves, since they all had high Falling Number, as well as appropriate protein content and gluten strength..
Bread results are summarised in Table 2. The wide range in loaf volumes indicated that there was considerable variability within both sets of samples.
Correlation coefficients (Table 3) showed that loaf height was significantly correlated with SDS sedimentation volume ($r=0.73$ or 0.87; $P<0.01$). Loaf volume, the most important bread quality parameter, was significantly correlated with the alveogram P/L (tenacity:extensibility) ratio ($r=-0.52$ and -0.92; $P<0.05$ and $P<0.01$ respectively). Gluten strength, as measured by SDS sedimentation volume, alveogram W and mixogram parameters, was not correlated with loaf volume.

Table 1. *Flour Quality Characteristics of Long and Short Cycle Wheat Genotype Groups*[1]

Wheat type	FN (sec)	Prot (%)	MixH (cm)	MixT (min)	SDSS (ml)	W (j/10000)	P/L
Long cycle genotypes[2]							
Maximum	448	14.1	6.8	7.6	21.5	407	3.5
Minimum	275	11.4	4.6	3.8	11.0	207	0.8
Mean	387	12.4	5.3	5.7	15.3	278	1.9
Standard deviation	44	0.7	0.6	1.2	2.7	55	0.8
Short cycle genotypes[3]							
Maximum	419	12.9	5.5	5.0	17.0	301	3.7
Minimum	313	11.4	4.5	4.0	10.5	120	0.7
Mean	368	12.0	5.1	4.4	14.8	222	1.9
Standard deviation	36	0.6	0.4	0.4	1.9	59	1.1

[1] Falling Number (FN), protein content (Prot), mixogram height (MixH), mixogram mixing time (MixT), SDS sedimentation volume (SDSS) and alveogram parameters (W and P/L).
[2] Average of 15 samples.
[3] Average of 8 samples.

Table 2. *Bread Loaf Properties for Long and Short Cycle Wheat Genotype Groups*[1]

	LH (cm)	LW (cm)	H/W	LV (ml)
Long cycle genotypes[2]				
Maximum	7.7	6.3	0.9	750.0
Minimum	6.2	5.2	0.8	412.5
Mean	6.8	5.6	0.8	573.8
Standard deviation	0.5	0.4	0.1	90.6
Short cycle genotypes[3]				
Maximum	7.8	5.9	0.9	630.0
Minimum	6.2	4.1	0.5	434.4
Mean	7.0	5.0	0.7	559.3
Standard deviation	0.6	0.6	0.1	69.2

[1] Loaf height (LH), loaf width (LW), height/width ratio (H/W) and loaf volume (LV).
[2] Average of 15 samples.
[3] Average of 8 samples.

Table 3. *Correlation Coefficients between Quality Parameters and Bread Loaf Properties of Long and Short Cycle Wheat Genotype Groups*[1].

	Prot (%)	MixH (cm)	MixT (min)	SDSS (ml)	W (j/10000)	P/L
Long cycle genotypes[2]						
LH (cm)	-0.11	-0.35	0.06	0.73**	-0.30	-0.24
LW (cm)	0.24	0.16	-0.24	0.40	-0.16	-0.25
H/W	-0.36	-0.52*	0.32	0.32	-0.14	0.03
LV (ml)	0.32	0.26	-0.42	0.41	-0.25	-0.52*
Short cycle genotypes[3]						
LH (cm)	0.60	-0.17	0.36	0.87**	0.40	-0.08
LW (cm)	0.26	0.26	-0.29	-0.46	-0.50	-0.88**
H/W	0.21	-0.21	0.39	0.81**	0.58	0.46
LV (ml)	0.68*	0.02	0.13	0.12	-0.26	-0.91**

[1] Abbreviated names are as given in Tables 1 and 2.
[2] Average of 15 samples.
[3] Average of 8 samples.
*: significant at $P<0.05$
**: significant at $P<0.01$

4 CONCLUSIONS

According to these results extensibility appears to be a key factor in Uruguayan wheat breadmaking quality. Based on this information, further studies on the relationship between gluten composition, strength and extensibility, and their effects on hearth bread quality are being conducted.

References

1. AACC. Approved methods of the American Association of Cereal Chemists. Methods, The Association: St. Paul, MN, USA, 1993, methods 26-50, 46-11, 54-30 and 54-40.
2. R.J. Peña, A.Amaya, S.Rajaram, and A. Mujeeb-Kazi, J.Cereal Sci., 1990, **12**, 105.

Gluten Polymers and Tools to Investigate their Structure

LINKING GLUTENIN PARTICLE SIZE TO HMW GLUTENIN SUBUNIT COMPOSITION.

C. Don[1], G. Mann[2], F. Bekes[2], and R.J. Hamer[1,3,4]

[1]TNO Nutrition and Food Research, Zeist, Netherlands; [2]CSIRO Plant Industry, Canberra, ACT, Australia; [3]Wageningen Centre for Food Sciences, [4] Wageningen University, Wageningen, Netherlands.

1 INTRODUCTION

The very high molecular weight fraction of glutenins is considered important in determining wheat flour quality. The quantity and size of this fraction is likely to be determined by both genetic factors and growing conditions. Although the genetic basis of HMWGS is clear, there is as yet little information how specific GS combine to form large aggregated networks. The use of dedicated sets of wheat lines[1] and methods to characterize the very high molecular weight glutenin macropolymer fraction (GMP) could help unravel the relationship between glutenin composition, size and function. Since we recently discovered that GMP consists of glutenin particles, and have suggested that these derive from protein particles in the endosperm, we focus on the possible link between glutenin composition and the ability of glutenin proteins to form such particles.

2 METHOD AND RESULTS

2.1. Characteristics of eight near isogenic wheat lines

A set of eight near isogenic lines derived from the cross Olympic x Gabo was used. The set has been described previously [1, 2] and has been generated to have variations only in the Glu-A1, -B1 and -D1 loci (see figure 1). The samples appeared quite homogenous, the triple null sample (figure 1a, no 3) showed some contamination and sample 4 that was provided as a '+-+' was scored '-++' on the basis of SDS-PAGE and HPLC analysis.

2.2. Ability to form very large glutenin aggregates and particles

For each sample the SDS insoluble glutenin macropolymer fraction (GMP) was isolated as described earlier and compared with the amount of Unextractable Polymeric Protein (%UPP) as determined earlier[2]. Table 1 demonstrates excellent correspondence between these two parameters, reflecting the fact that both measure similar fractions of glutenin. More importantly however, table 1 demonstrates that HMWGS[1] are a prerequisite to the

---

[1] HMWGS: High Molecular Weight Glutenin Subunits

formation of GMP. Interestingly, the presence or absence of the 1-A1 subunit hardly affects the quantity of GMP present. In contrast, the presence of 1B-17/18 or 1D-5/10 subunits allows for GMP to be formed albeit in smaller quantities. CSLM demonstrates the presence of glutenin particles in these samples (data not shown). In contrast, the triple null (---) and the 1A1 (+--) lines did only contain minute quantities of particles, presumably resulting from contamination with other lines (see also figure 1, lanes 3 and 8).

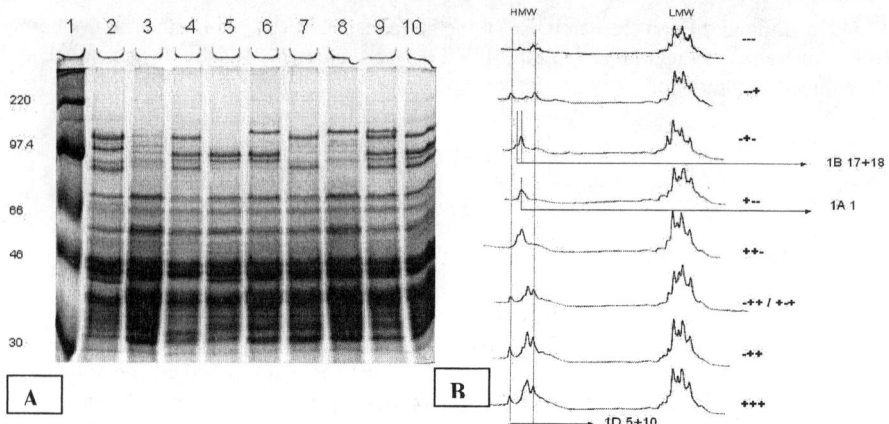

Figure 1 *Characterisation of Olympic x Gabo near isogenic lines. A. SDS PAGE: Lane 1: marker; lane 2 Soissons (ref); lane 3-10: near isogenic lines differing only in HMWGS: 3: ---; 4: -++; 5: -+-; 6: ++-; 7: --+; 8: +---; 9:+++; 10: -++. The first symbol represents the presence (+) or absence (-) of HMWGS-1A1, the second for HMWGS-1B 17+18, and the third for HMWGS 1D 5+10. B. RP-HPLC traces of samples 3-10.*

2.3. Composition of GMP particles

GMP particles were analysed by RP-HPLC to characterise their composition in terms of HWMGS and LMWGS. Figure 1b clearly demonstrates the presence of both HMWGS and LMWGS in GMP. Interestingly, in spite of the large variation in HMWGS only minor variations in LMWGS composition were observed. This could indicate the validity of a model in which LMWGS form larger polymers or clusters that become attached to a HMWGS 'backbone'. The HPLC trace for the triple null (---) line shows some contaminating HMWGS and a composition of LMWGS like in flour.

The HMW/LMW ratio of GMP correlates to that of flour ($r^2=0,9$), although the ratio in GMP is about 1,6 times higher, reflecting a higher content of HMWGS.

2.4. Size of GMP particles

The average voluminosity of GMP particles was measured by viscometry of dispersions of GMP in SDS-solution and by Coulter laser analysis[3]. Both parameters represent a measure for the average size of the particles and are highly correlated ($r^2=0,91$).

Table 1 *Properties of GMP particles*

Type	η^1 (L/g)	$D3,2^2$	% UPP^3	wet weight GMP gel (g)
---	0,10	4,3	15,3	0,6
-++[4]	0,24	5,5	38,7	2,8
-+-	0,11	4,5	27,7	2,8
++-	0,15	5,0	30,2	2,2
--+	0,16	5,1	25,0	1,7
+--	0,13	4,9	21,1	0,8
+++	0,28	7,2	53,8	3,5
-++	0,27	6,6	45,3	3,0

[1]: intrinsic viscosity; [2]:surface to volume ratio; [3]:unextractable polymeric protein (data taken from [2]). [4]:relabeled, originally provided as +-+ sample

It is clear that the presence of certain subunits is required to form larger particles (Table 1). The samples containing 1D-5+10 have η values between 0,24 and 0,28 L/g, whereas samples lacking this pair of HMWGS have η values of ca 0,14 L/g. The presence of 1D-5+10 only does not lead to large particles; this seems to require the presence of also 1B-17+18. It stands to reason that the ability to form large particles leads to more GMP. The data show however, that GMP quantity and GMP particle size are only related in part (r^2=0,58). Apparently, a large quantity of GMP does not always indicate a large GMP particle size. With this set of samples a relation was found between HMW/LMW ratio in flour and the average size of the particles (r^2=0,75). This is in agreement with results obtained with commercial wheat varieties (data not shown), where a higher ratio between HMW/LMW subunits corresponded to a larger average size.

3 CONCLUSIONS

The formation of glutenin particles, stable in SDS-solution, requires the presence of specific HMWGS. Within the set used, HMWGS from either Glu-1B or Glu-1D loci proved essential. The size of the particles seems related to the presence of 1D-5+10. The largest particles were observed with a combination of GS from 1A, 1D/1B, 1D and 1A, 1B, 1D. Our results indicate a particle backbone structure consisting of HMWGS to which polymers of LMWGS become attached. Finally, the combination of GMP particle analysis and the use of near-isogenic wheat lines open new possibilities to understand factors determining glutenin structure and aggregate size.

References

1 S.Uthayakumaran, H.L.Beasley, F.L.Stoddard, M.Keentok, N.Phan-Thien, R.I.Tanner, and F.Bekes, *Cereal Chem.*, 2002, **79**, 294.
2 H.L.Beasley, S.Uthayakumaran, F.L.Stoddard, S.J.Partridge, L.Daqiq, P.Chong, and F.Bekes, *Cereal Chem.*, 2002, **79**, 301.
3 C.Don, W.Lichtendonk, J.J.Plijter, and R.J.Hamer, 2004, *J. Cereal Sci.*,**38**, 157.

GLUTEN MACROPOLYMER IN WHEAT FLOUR DOUGHS: STRUCTURE AND FUNCTION FOR WHEAT QUALITY

R. Kuktaite[1], H. Larsson[2], S. Marttila[1], K. Brismar[1], M. Prieto-Linde[1] and E. Johansson[1]

[1] Department of Crop Science, The Swedish University of Agricultural Sciences, Box 44, SE-230 53 Alnarp, Sweden
[2] Department of Food Technology, Lund University, Box 124, SE-221 00 Lund, Sweden

1 INTRODUCTION

The largest protein molecules in nature occur in wheat flour. The main protein components are the monomeric gliadins and the polymeric glutenins. The gliadin fraction is a complex mixture of largely single-chain, hydrophobic polypeptides that have intramolecular disulphide bonds. Glutenin consists of subunits ranging in size up to 100,000, linked by intermolecular disulphide bridges. Gliadins and glutenin subunits combine to form gluten polymers, which gives wheat dough its unique viscoelastic properties. Glutenin polymers play a key role in bread-making procedure[1, 2].

The amount of polymeric protein in flour, the allelic variation at the high (HMW-GS) and low molecular weight (LMW-GS) glutenin subunit loci, and the ratio between HMW-GS and LMW-GS are the main sources of the variation in bread-making quality[3]. The variation of HMW-GS and LMW-GS has been extensively studied, and its correlation with bread-making quality is quite well established[4]. However, relatively little is known of subunit interactions during gluten macropolymer formation[3].

The aim of this study was to better elucidate the structure, formation, and functional properties of wheat gluten polymers in dough using electrophoresis, size-exclusion high-performance liquid chromatography (SE-HPLC), and microscopy.

Abbreviations used: HMW = high molecular weight glutenin subunits; LMW = low molecular weight glutenin subunits; LUPP = large unextractable polymeric protein in total large polymeric protein;

2 MATERIALS AND METHODS

2.1 Materials

Commercially available Swedish wheat flour mixtures, ranging from weak to strong, and Swedish wheat cultivars were investigated. Flours were mixed with water under chosen times producing undermixed and overmixed doughs.

2.2 Ultracentrifugation

Freshly mixed dough samples were centrifuged for 1h at 100,000 x g in an ultracentrifuge. Dough was separated into five phases: liquid, gel, gluten, starch and bottom phase. The gluten phase was studied for protein polymer analysis[5].

2.3 Microcopy

Freshly collected gluten phases were examined directly in a scanning electron microscope (SEM) at low acceleration voltage (2 kV) with secondary electron detector. Gluten phases were used to visualise structural differences in the protein network due to applied mixing methods. All microscopy pictures were modified in Adobe Photoshop 6.0.

2.4 SE-HPLC

A two-step extraction procedure, followed by SE-HPLC, was used to investigate the amount and size distribution of protein polymers and monomers [6].

3 RESULTS AND DISCUSSION

3.1 Microscopy

To study the structure of gluten macropolymers in dough, microscopy (SEM) was performed on gluten phases from flours of different strength, mixed for various times. SEM showed that protein film in undermixed dough generally had a rougher structure than that in overmixed dough (Figure 1a-b). Starch granules were unevenly embedded in the gluten network of gluten phases in undermixed dough. With prolonged mixing, starch granules were more evenly spread and covered by the gluten film. Dough elasticity increased with increasing mixing time, resulting in a smoother gluten network. According to earlier results, prolonged mixing untangles interchain disulphide linkages, modifies chain orientation, ruptures bonds between low molecular weight (LMW) and high molecular weight (HMW) glutenin subunits and promotes new polymer-polymer interactions. A relation is also found with loss of resistance to extension forces of polymeric protein during dough overmixing[7]. A relationship between the gluten macropolymer rheological properties, mixing properties and specific glutenin composition has been detected[8].

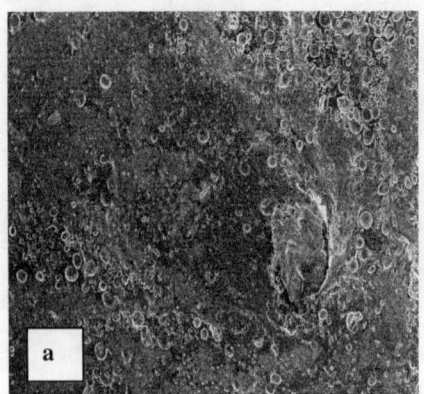 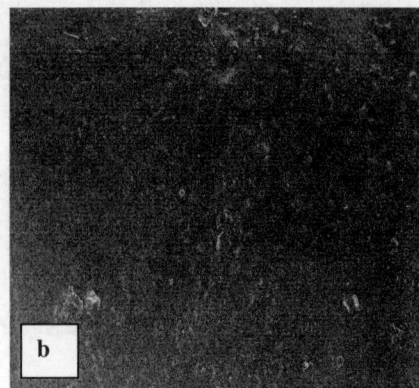

Figure 1a-b. *Protein structure in gluten phase of a strong flour in (a) undermixed, and (b) overmixed doughs; magnification for both images is 40X.*

3.3 SE- HPLC

SE-HPLC showed that the percentage of large unextracted polymeric protein in total large polymeric protein (LUPP) increased with flour strength. The lowest LUPP value in dough was for the weak flour, and the highest was for the strong flour (Figure 2).

Comparing LUPP values of gluten phases with different mixing times, we found that the highest percentage of LUPP occurred near the optimal dough mixing time for a flour. Strong flours had highest LUPP values during long mixing, while weak flours had highest values during short mixing (Figure 3).

4 CONCLUSIONS

Increased knowledge of the structure and function of gluten proteins and of the formation of gluten polymers during dough development is important for production of bread products of high and even quality.

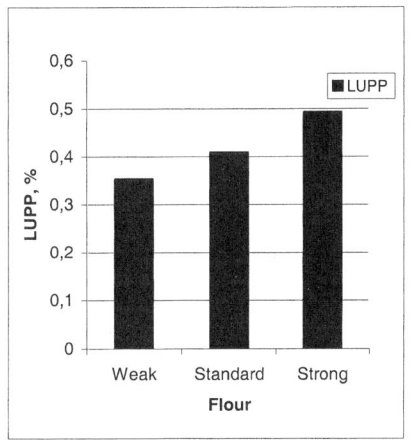

Figure 2. *Percentage of LUPP in fours having different strengths.*

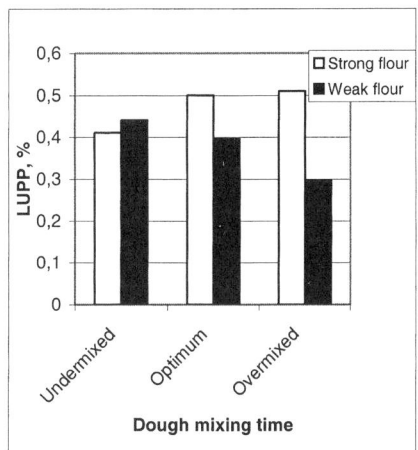

Figure 3. *Percentage of LUPP in gluten phases of different flours, after different mixing times; LUPP = large unextractable polymeric protein in total large polymeric protein.*

References

1. C.W. Wrigley, *Nature*, 1996, **381**, 738.
2. P. L. Weegels, A. M. Van de Pijpekamp, A. Graveland, R. J. Hamer and J.D. Schofield, *J. Cereal Sci.*, 1996, **23**, 103.
3. G. Tranquilli, M. Cuniberti, M.C. Gianibelli, L. Bullrich, O.R. Larroque, F. MacRitchie and J. Dubcovsky, *J. Cereal Sci.*, 2002, **36**, 9.
4. D. Lafiandra, S. Masci, R. D'Ovidio and B. Margiota, *Proceedings of the 7th International Workshop Gluten*, 2000, 3.
5. R. Kuktaite, H. Larsson and E. Jahansson, *Acta Agronomica Hungarica*, 2003, **51**, 163.
6. E. Johansson, M.-L. Prieto-Linde & J. Jönsson, *Cereal Chem.*, 2001, **78**, 19.
7. K.M. Tronsmo, E.M. Færgestad, Å. Longva, J.D. Schofield and E.M. Magnus, *J. Cereal Sci.*, 2002, **35**, 201.
8. H. Singh and F. MacRitchie, *J. Cereal Sci.*, 2001, **33**, 231.

Acknowledgements

This work was supported by the Swedish University of Agricultural Sciences, the Swedish Research Council for Environment, Agricultural Sciences and Spatial Planning and Royal Physiografic Society.

CRITICAL FACTORS GOVERNING GLUTEN PROTEIN AGGLOMERATION ON A MICRO-SCALE

B. Schurgers, W.S. Veraverbeke, E. Dornez and J.A. Delcour

K.U. Leuven, Laboratory of Food Chemistry, Kasteelpark Arenberg 20, B-3001 Heverlee, Belgium

1 INTRODUCTION

Different systems are used for the industrial separation of gluten and starch[1-3]. They all include a step in which gluten proteins agglomerate to form a network from which starch is subsequently washed out using excess water. However, the systems differ in the way gluten proteins are allowed to agglomerate (dough or batter) and in the way washed-out starch is separated from the agglomerated gluten.

Many different factors, both at the level of the flour (protein quantity and quality, non-protein components) and the process (process conditions, addition of processing aids) affect the process of gluten agglomeration and as a consequence gluten-starch separation. However, at present very little is known concerning the gluten protein agglomeration process.

In this study, a micro-scale gluten-starch separation system was used to investigate the effect of different process parameters on gluten protein agglomeration.

2 MATERIALS AND METHODS

2.1 Micro-scale gluten-starch separation system

A micro-scale gluten-starch separation system was developed for fast evaluation of gluten protein agglomeration in flour batters. With this system, water and wheat flour are mixed in a Rapid Visco Analyzer (RVA) (Newport Scientific, Australia) to produce flour batters. The RVA allows controlling batter formation temperature (T), mixing speed (MS) and mixing time (MT). Batters with different dry matter content (DMC) were produced by changing the water-to-flour ratio, while always keeping a constant total batter weight of 25 grams. The resulting flour batters were poured over two consecutive sieves (400 and 125 μm). The gluten retained on the sieves was washed with water (300 mL) and dried (2h at 130°C). Protein contents of the isolated gluten were determined with the Dumas method (protein = N x 5.7).

2.2. Evaluation of the effect of process parameters on gluten agglomeration

The micro-scale gluten-starch separation system was used to study the effects of the above mentioned process parameters on the gluten protein agglomeration index (GPAI), the gluten protein recovery (GPR) and the gluten yield (GY). GPAI was defined as the ratio (in %) of the amount of protein on the 400 μm sieve to the amount of protein on the 400 and 125 μm sieves; GPR was defined as the percentage of flour protein recovered in the gluten; and GY was defined as percentage of flour dry matter recovered in gluten.

Response Surface Methodology (RSM)[4] was used to develop models for predicting GPAI, GPR and GY from process parameters DMC, T, MS and MT. Experiments were designed according to a "Central Composite Design" augmented with "Star Points" and "Centers of Edges". DMC, T, MS and MT were varied in the ranges of 25-40 %, 20-40 °C, 80-640 rpm and 5-30 min, respectively. A good breadmaking quality wheat flour (Legat) was used for the experiments. RSM models were validated by comparison of actual and predicted values for eight new (random) process parameter combinations.

3 RESULTS AND DISCUSSION

Validation of the RSM models for GPAI, GPR and GY showed linear relationships between actual and predicted values with slopes of 1.03, 0.99 and 0.94, respectively and R^2-values of 0.88, 0.93 and 0.94, respectively. Model predictions of the effects of process parameters on GPAI, GPR and GY are discussed below.

3.1 Effects of process parameters on GPAI

Large variations in GPAI were found in the ranges tested for the different process parameters. Analysis of the model for GPAI revealed that increasing DMC or T invariably had a significant positive effect on GPAI. This is illustrated in Table 1 for the maximal GPAI (GPAI$_{max}$) that can be reached at a certain DMC and T combination by choosing an optimal combination of MS and MT. As shown in Table 1, increasing T at a constant DMC always increases GPAI$_{max}$. The same is true for increases of DMC at constant T. A positive effect on GPAI of increases of DMC and T was not only observed for the MS and MT combinations that give rise to GPAI$_{max}$ but also for all other MS and MT combinations. Furthermore, it was shown (for example from the data in Table 1) that DMC and T affected GPAI synergistically.

It was further found that both the type and the strength of the effects of variations in MS and MT are strongly dependent on DMC and T. The type of the effect of MS is generally an effect characterised by a maximum value of GPAI at a certain MS within the range of tested MSs (this effect is indicated by "maximum" in Table 1). This implies that working above or below this "optimal" MS value lowers GPAI values. As shown in Table 1, the "optimal" MS value decreased with DMC and increased with T. With respect to the type of effect of MT, different types were observed for different DMC-T combinations. These different types are indicated in Table 1 with "minimum", "plateau" and "linear +". "Minimum" indicates an effect that is characterised by a minimum value for GPAI at a certain MT within the range of tested MTs. "Plateau" indicates an increase of GPAI as a function of MT that is characterised by lower increases of GPAI as a function of MT at higher MT values until a plateau value is reached for GPAI at a MT value within the range of tested MTs.

Table 1 *Type and strength of the effects of mixing speed (MS) and mixing time (MT) on GPAI and maximal gluten protein agglomeration index (GPAI$_{max}$) as a function of dry matter content (DMC) and temperature. Weak, medium and strong effects are indicated in italic, regular and bold, respectively. Values for MS (in rpm) and MT (in min) that produce minimal or maximal values for GPAI are indicated between brackets.*

DMC	Temperature		
	20 °C	30 °C	40 °C
25 %	MS: maximum (400) *MT: minimum (17)* GPAI$_{max}$ = 21	MS: maximum (440) **MT: linear +** GPAI$_{max}$ = 34	**MS: plateau (440)** MT: linear + GPAI$_{max}$ = 44
32.5 %	MS: maximum (250) *MT: linear +* GPAI$_{max}$ = 34	*MS: maximum (300)* MT: linear + GPAI$_{max}$ = 59	*MS: maximum (370)* MT: plateau (30) GPAI$_{max}$ = 68
40 %	MS: maximum (230) **MT: plateau (30)** GPAI$_{max}$ = 50	*MS: linear +* MT: plateau (30) GPAI$_{max}$ = 86	*MS: linear +* **MT: plateau (30)** GPAI$_{max}$ = 98

"Linear +" indicates a linear increase of GPAI as a function of MT for the whole range of tested MTs. Variations in the strength of the effects of MS and MT as a function of DMC and T are also indicated in Table 1. At lower DMC and lower T, MS is more important in determining GPAI; at higher DMC and higher T, however, MT becomes most important. A visual illustration of the strong dependency of both the type and strength of the effects of MS and MT on DMC and T is given in Figure 1.

3.2 Effects of process parameters on GPR and GY

In analogy with GPAI, large variations were found in GPR and GY in the ranges tested for the different process parameters. The effects of the different process parameters on GPR and GY were also all very similar to the effects on GPAI. This indicates that, for the gluten-starch separation system used in this study, GPAI is a good indicator for the efficiency of the gluten-starch separation.

4 CONCLUSIONS

With the micro-scale gluten-starch separation system used in this study, variations in process parameters greatly impacted gluten protein agglomeration behaviour. Furthermore, all parameters used to describe gluten protein agglomeration, i.e. GPAI, GPR and GY, were affected in a similar way. It was found that (1) increasing DMC or T increases gluten protein agglomeration at all MS and MT combinations, (2) DMC and T act synergistically, and (3) the type and strength of the effects of MS and MT are largely dependent on DMC and T.

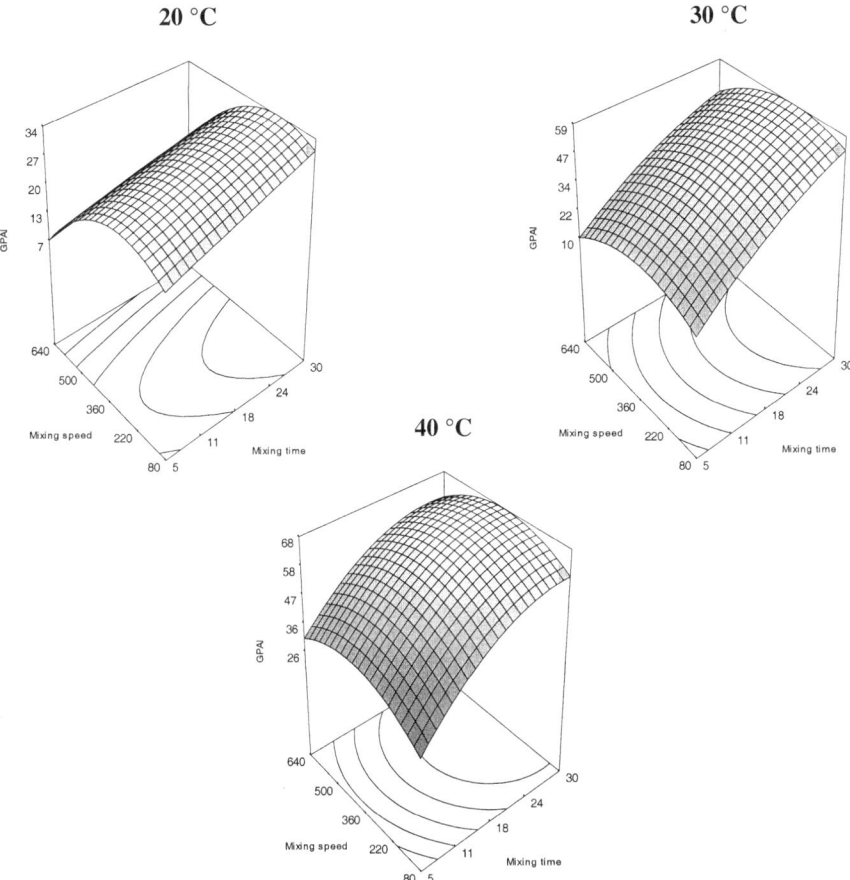

Figure 1 *3D-plots of gluten protein agglomeration index (GPAI) as a function of mixing speed (rpm) and mixing time (min) for different temperatures (20, 30 and 40 °C) at a dry matter content of 32.5 %.*

References

1. R.A. Anderson, V.F. Pfeifer and E.B. Lancaster, *Cereal Chem.*, 1958, **35**, 449.
2. B. Godon, M.-P. Leblanc and Y. Popineau, *Qual. Plant. Plant Foods Hum. Nutr.*, 1983, **33**, 161.
3. P.L. Weegels, J.P. Marseille and R.J. Hamer, *Starch/Stärke*, 1988, **40**, 342.
4. G.E.P. Box, W.G. Hunter and J.S. Hunter, in *Statistics for Experimenters*. John Wiley and sons, New York, USA, 1978, pp 510-539.

Acknowledgements

Financial support by Amylum (Tate & Lyle, Aalst, Belgium) and helpful discussions with Dr Stefaan Roels (Amylum) are gratefully acknowledged

APPLICATION OF FLOW FIELD-FLOW FRACTIONATION AND MULTIANGLE LASER LIGHT SCATTERING FOR SIZE DETERMINATION OF POLYMERIC WHEAT PROTEINS

K.R. Preston, S.G. Stevenson, S. You and M.S. Izydorczyk

Grain Research Laboratory, Canadian Grain Commission, 1404-303 Main Street, Winnipeg, Manitoba, Canada, R3C 3G8 email: kpreston@grainscanada.gc.ca

1 INTRODUCTION

Both the amount and size distribution of the polymeric proteins appear to play a major role in the processing characteristics of wheat flour[1]. Recent studies involving the use of SE-HPLC equipped with refractive index and multiangle laser light scattering (MALLS) detectors indicate that the larger polymeric proteins eluting at or near the void volume have weight average molecular weight (M_w) values of 10 million or more[2,3]. Studies have shown that flow field-flow fractionation can improve the resolution of these proteins[4-6] due to an effectively limitless void volume. In addition, recent studies indicate that wheat polymeric proteins eluting at the void volume in SE-HPLC may be complexed with lower molecular weight material which influences size measurements[7]. With flow FFF these complexes appear to dissociate at the very low concentration used to obtain the fractograms.

In this report, we review recent work in our laboratory on the use of flow FFF combined with laser light scattering to assess the size distribution of wheat polymeric proteins extracted in dilute acetic acid with sonication from flour and dough following removal of monomeric protein with the same solvent.

2 METHODS

Straight grade flour from Katepwa, a Canadian hard red spring bread wheat, was obtained from advanced plant breeder testing trials. For dough preparation[8], flour and water were mixed in a farinograph at optimum absorption to 0.5x peak, peak and 4x peak. Pieces were cut from the dough, rested 15 min, frozen in liquid nitrogen and freeze dried. After drying, samples were ground to a powder prior to extraction.

Extraction of flour and dough in duplicate was carried out using dilute acetic acid to remove the monomeric enriched protein then dilute acetic acid with sonication to remove the polymeric enriched protein[7]. Flow FFF of protein extracts was carried out in a conventional frit inlet/ frit outlet (FI/FO) channel using hydrodynamic relaxation with $V_s=0.2$, $V_f=1.4$ and $V_c=5$ ml/min and 0.05 M acetic acid containing 0.002% FL-70 as eluent[5]. The protein eluted from the channel outlet was concentrated approximately 8x due to removal of most eluent through the FO. This eluent was monitored by absorbance at 210 nm, refractive index and MALLS using a DAWN DSP Laser Photometer (Wyatt Technology). The relationship between elution time and Stokes diameter (d_s) was determined using standard proteins as previously described[5]. The distribution of size fractions (monomeric and small, large and very large polymeric) was determined by integration of peak areas at OD_{210} from FFF fractograms using both extracts as outlined previously[7].

Weight average molecular weight was determined with ASTRA software V4.72 from Wyatt using the equation $R_\theta/K^*c = MP(\theta)-2A_2cM^2P^2(\theta)$ from Zimm[9] where the second virial coefficient (A_2) was assumed to be zero. Z-average root mean square values for radius of gyration (R_g) were determined from the equation $P(\theta) = 1+16\pi^2<R_g^2>\sin^2(\theta/2)/3\lambda^2_0$ using ASTRA software. The Debye fit method was used for the monomeric protein extract and the Berry fit method was used for the polymeric protein extract. Values of dn/dc for determining the optical constant (K^*) were measured by dilution after determining protein concentration by combustion nitrogen analysis. Values of dn/dc for the monomeric and polymeric protein extracts in elution buffer were 0.208 and 0.169 mL/g, respectively. Further details are given in previous papers[2,3,10].

3 RESULTS AND DISCUSSION

The Katepwa flour monomeric protein extract obtained with dilute acetic acid accounted for 82.5% of the total extractable protein while the polymeric protein extract obtained with the same solvent using sonication for 30 sec represented 16.3% of the total extractable protein. Flow FFF profiles were obtained at several concentrations to determine the effect of loading (data not shown). For the polymeric protein extracts, overloading was evident at injection levels above approximately 2 µg protein as shown by a shift of the absorbance or RI peak to earlier elution times and a more rapid increase in M_w values with increasing elution time relative to lower injection volumes. At low concentration (< approximately 0.5 µg protein), the light scattering intensity was too weak to obtain a reliable M_w profile.

Figure 1 shows the flow FFF RI, M_w and R_g profiles for the polymeric extract. The peak at about 17 min showed a M_w value close to 300,000, suggesting the predominance of polymers composed of a relatively small number of disulfide linked low and/or high molecular weight glutenin subunits. At later elution times M_w increased to values approaching 10,000,000. These values are probably an underestimate of the upper limit since the sonication step used to extract these polymeric proteins causes some cleavage of the largest polymers[7]. These values are generally consistent with M_w values obtained for polymeric wheat protein by MALLS at or near the exclusion limit of SE-HPLC columns[2,3]. There was also a significant amount of protein eluting at M_w values less than 100,000, indicating the presence of monomeric proteins. Previous studies indicate that these proteins may form complexes with the large glutenin polymeric proteins eluting at the void volume in SE-HPLC columns[7]. The low protein concentrations used with flow FFF may enhance the dissociation of these complexes. Once dissociation occurs, separation in the column due to large differences in diffusion coefficients would further reduce the reverse reaction.

The initial decrease in R_g values during the first 20 min elution is likely due to the sensitivity of this parameter to the low light scattering intensity of the monomeric and small polymeric proteins. After the peak, R_g showed a small but consistent increase from about 28nm at 20 min to about 36nm at 40 min. The slope of log R_g vs log M_w, which provides an estimate of the conformational properties of polymers[11], decreased from about 0.39 at 20 min to 0.17 at later elution times. These results suggest that polymeric wheat proteins become more compact as they increase in size, consistent with recent SE-HPLC/MALLS results obtained by Carceller and Aussenac[3].

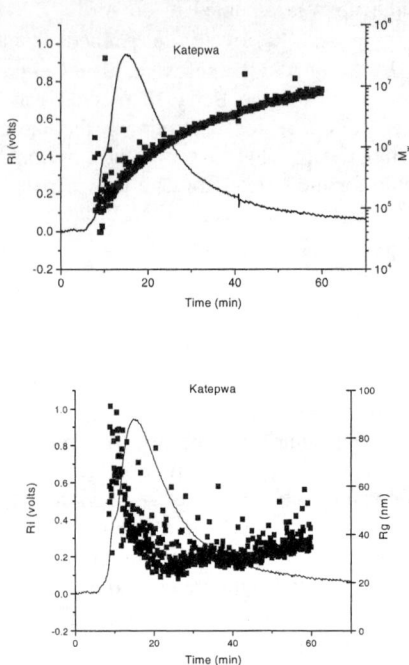

Figure 1 *Flow FFF RI, M_w and R_g profiles for the Katepwa polymeric protein extract*

Flow FFF M_w profiles of flour and dough mixed to 0.5x peak, peak and 4x peak at approximately equal protein loading did not show any apparent difference (data not shown). However, mixing resulted in a large significant (p<0.05) increase in the extractability of proteins in acetic acid without sonication and a concomitant large decrease in the amount of protein extracted in acetic acid with sonication. For flour, 83.2% of the total extractable protein was present in the dilute acetic acid extract. During mixing the dilute acetic extractable protein increased to 94.4, 95.4 and 95.4% for 0.5 peak, peak and 4x peak, respectively. Based on Stokes diameter, these changes were accompanied by a significant decrease in the proportion of small (8-19 nm), large (19-37.5 nm) and very large polymeric protein (>37.5 nm) fractions and an increase in the amount of the monomeric protein (<8 nm) fraction as shown in Table 1. Although the smaller polymeric protein fraction showed the largest decrease (35.7 to 32.4%) in the amount of polymeric protein with mixing, the larger polymeric protein size fractions showed much larger proportional decreases. These results suggest that the larger

Table 1 *Effects of mixing on distribution of FFF size fractions in % of total extractable protein[a]*

	Monomeric	Small Polymeric	Large polymeric	Very large polymeric
Flour	56.7	35.7	6.1	1.5
0.5X peak	61.6	33.6	4.1	0.7
peak	62.4	33.1	3.9	0.6
4X peak	63.0	32.4	3.9	0.7

[a] values based on size fractions determined in both monomeric and polymeric protein extracts

polymeric proteins are most sensitive to shear during mixing as expected from polymer theory[1]. Since most of the changes in size fractions occurred during the early mixing stage, it appears that shear induced de-polymerization occurs rapidly during the mixing process. The much smaller impact of shear with continued mixing suggests that there may be considerable variability in the structural stability of polymeric wheat proteins or that changes during mixing such as increased inter-protein interactions may stabilize structural integrity. These changes would be expected to have important effects on processing quality.

References

1. F. MacRitchie, Adv. Food Nutr. Res. 1992, **36**, 1.
2. S. R. Bean and G. L Lookhart, Cereal Chem. 2001, **78**, 608.
3. J. –L. Carceller and T. Aussenac, J. Cereal Sci. 2001, **33**, 131.
4. S. G. Stevenson and K. R. Preston, J. Cereal Sci. 1996, **23**,121.
5. S. G. Stevenson, T. Ueno and K. R. Preston, Anal. Chem. 1999, **71**, 8.
6. K.–G. Wahlund, M. Gustavsson, F. MacRitchie, T. Nylander and L. Wannerberger, J. Cereal Sci. 1996, **23**, 113.
7. T. Ueno, S. G. Stevenson, K. R. Preston, M. J. Nightingale and B. M. Marchylo, Cereal Chem. 2002, **79**, 155.
8. K. R. Preston, S. G. Stevenson and K. Takatsu, in *Cereals 2000 Proceedings of the 11th ICC Cereal and Bread Congress and 50th Australian Cereal Chemistry Conference, Surfers Paradise, Australia, September 8-15, 2000*, eds M. Wootton, I. L. Batey and C. W. Wrigley, Royal. Aust. Chem. Instit., Melbourne, 2001, p. 386.
9. B. H. Zimm, J. Chem. Phys., 1948, **16**, 1093.
10. S. G. Stevenson, S. You, M. S. Izydorczyk and K. R. Preston, J. Liquid Chrom. & Rel. Topics, In press.
11. P. J. Wyatt, Anal. Chim. Acta, 1993, **272**, 1.

CONFOCAL SCANNING LASER MICROSCOPY OF GLUTEN PROTEINS AND LIPIDS IN BREAD DOUGH

W. Li[1], B.J. Dobraszczyk[1] and P.J. Wilde[2]

[1]The University of Reading, School of Food Biosciences, UK
[2]Institute of Food Research, Norwich, UK.

1 INTRODUCTION

It has been well accepted that gluten proteins, which mainly contain gliadin and glutenin, are closely related to flour quality and baking performance[1]. In breadmaking, the network structure of gluten proteins is vital to avoid rupture of gas cell walls of dough[2,3]. It was also suggested that the surface properties of gliadin gas contribute to the stability of gas cells[4]. However, little is known about the molecules of those components as aspects of dough structure and composition of gas cell walls in bread dough.

The aim of this study was to investigate the surface and rheological properties of the purified gliadin and glutenin proteins, their locations in bread dough as well, and then to compare locations of gluten proteins with those of polar lipid (phosphoglyceride) and nonpolar lipid (dedocanoic acid) in bread dough. The results can provide a better understanding of the functionality of those components in the expansion capacity of gas cells and the rheological properties of bread dough.

2 MATERIAL AND METHODS

2.1 Materials

The UK breadmaking flour, *cv.* Hereward harvested in 1999

2.2 Methods

2.2.1 Preparation of purified gliadin and glutenin proteins
Gluten sample was washed out from Hereward flour and crude gliadin and glutenin proteins were fractionated from chloroform defatted gluten sample and then purified by gel filtration chromatography as described by Li [5].

2.2.2 Measurement of surface and rheological properties of proteins

Small deformation dilation test: A small deformation dilation test (strain 5%) was carried out using Kokelaar's method [6].
Large deformation compression and expansion test: the method was used as described by Gunning et al [7].

2.2.3 Fluorescence labelled proteins
The purified gliadin and glutenin proteins were labelled with rhodamine B (Sigma, UK). 100 mg of freeze-dried protein sample was dyed with 50 ml 0.01% (w/v) rhodamine B for 6 hrs, α-amino groups react with the rhodamine B to form a covalent bond. This solution was dialysed against distilled water for 48 hrs at 4^0 C and then freeze dried.

2.2.4 Preparation of dough sample
2g of flour was mixed with or without 5 mg of fluorescence labelled proteins, 2mg of fluorescence labelled polar lipid (phosphoglyceride, NBD-HPC, Molecular Probes, Cambridge) or fluorescence labelled nonpolar lipid (NBD-dodecanoic acid, Molecular Probes, Cambridge) and 1.2 mL distilled water on a 2g-direct drive Mixograph (National Manufacturing TMCO, Lincoln, NE, USA) to the peak time. Dough was loaded on the confocal microscopy system and scanning was carried out during proofing at room temperature.

3. RESULTS

3.1 The surface and rheological properties of purified gliadin and glutenin

The dynamic surface tension and surface rheological properties of purified gliadin and glutenin spread films were measured both in small deformation dilation and large deformation compression and expansion tests. The results in Figure 1 show that the purified gliadin decreased the surface tension rapidly, when it was spread on the water-air interface. This spread film also had a high elastic modulus. In contrast, the purified glutenin was not able to decrease the surface tension as far as gliadin under the same conditions, and the elastic modulus of its spread film was not as high as that of gliadin either. The results are agreement with those from previous work[4].

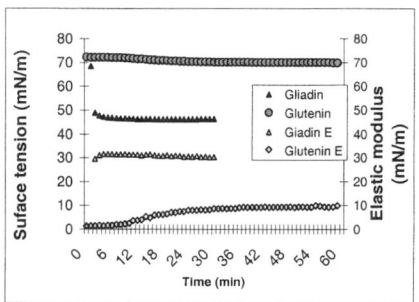

Figure 1. *The dynamic surface tension and elastic modulus of purified gliadin and glutenin spread films in small deformation test. (E: elastic modulus)*

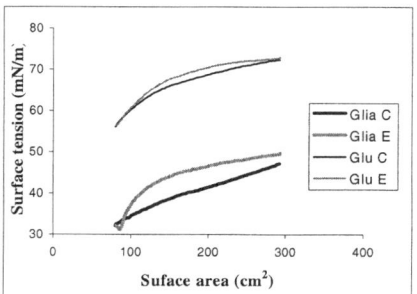

Figure 2. *The surface tension of gliadin and glutenin spread films in large deformation compression and expansion test (GluC: glutenin in compression; GluE: glutenin in expression. GliaC, Gliadin in compression; Glia E, Gliadin in extesion.*

In the large deformation compression and expansion test (Figure 2), the surface tension for gliadin and glutenin spread films all decreased during the compression process, but the surface tension of glutenin film increased immediately as the expansion proceeded. At the end of expansion, it reached to its initial value. In contrast, the surface tension of gliadin film recovered relatively slowly. There was a difference in the values for the end of expansion and the beginning of compression. This suggested that the glutenin film was more elastic than that of gliadin in this large deformation compression and extension test.

3.2 Locations of gluten proteins in breadmaking dough

The locations of labelled gliadin in bread dough (Figure 3 a.) were found not only in the dough bulk (not shown), but also at the gas cell walls. However, the labelled glutenin was mainly observed in the bulk of dough (Figure. 3 b). Based on the z-axis scanning on bread dough enriched with labelled gluten proteins, a real 3-D matrix structure was revealed (Figure 3 c). It was observed that gluten proteins form matrix and film separating and surrounding gas cells in bread dough.

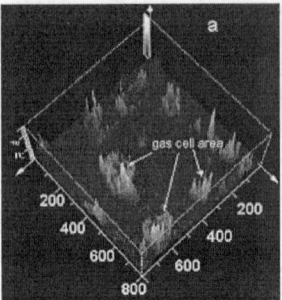 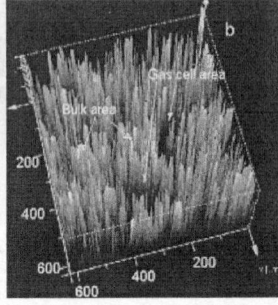 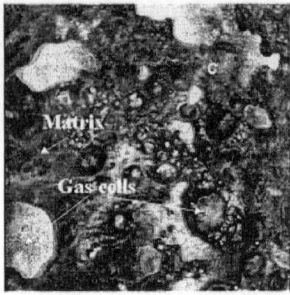

Figure 3. *The locations of purified gliadin, glutenin and gluten proteins in bread dough. a. The fluorescence intensity of labelled gliadin in bread dough; b. The fluorescence intensity of labelled glutenin in bread dough; c. The matrix structure of gluten proteins in bread dough.*

3.3 The locations of added polar lipid and non polar lipids

The locations of polar lipid (phosphoglyceride) were found to be very similar to that of the labelled gliadin. The average fluorescence intensity of polar lipid from the dough surface to 0.12 mm of depth (Figure 4) showed that polar lipid occurred around gas cell and in the area of bulk of dough. The fluorescence labelled nonpolar lipid was found mainly on the surface of some particles in the bulk of dough, probably, some as small vesicles (Figure 5 a).

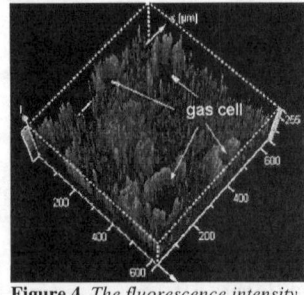 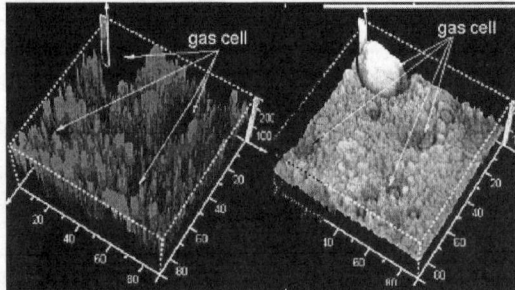

Figure 4. *The fluorescence intensity of labelled polar lipid in bread dough.*

Figure 5. *The nonpolar lipid in bread dough. a The fluorescence intensity of labelled nonpolar lipid, b. 3-D picture of dough under the transparent light.*

The locations of gliadin and polar lipid were also identified, when both were added in the same dough. The locations of polar lipid and gliadin were shown in Figure 6 a and b respectively. The analysis of fluorescence intensity by overlay of those two pictures (Figure 6 c) showed that both polar lipid (light grey) and gliadin protein (dark grey) were found in the matrix structure and on gas cell walls. Stronger light gray colour on some particles was observed. Therefore, it can be suggested that this polar lipid well incorporated with gliadin in the bread dough, forming matrix strands and occurring on gas cell walls. This finding is good evidence to prove that gliadin is a component of gas cell wall.

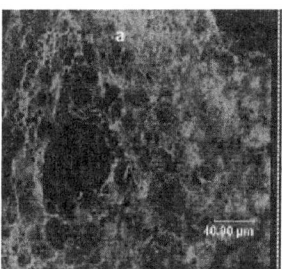
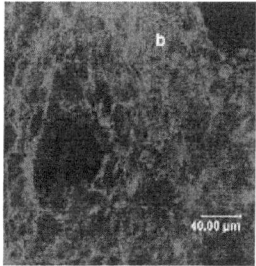
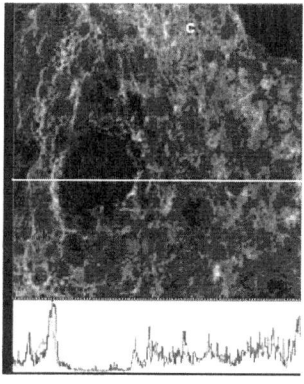

Figure 5 *The locations of labelled polar lipid and labelled gliadin in bread dough during proving. a. labelled polar lipid; b labelled gliadin. c. the overlay of a and b, the florescence intensity of polar lipid and gliadin were indicated by light grey and dark grey.*

4. CONCLUSION

Gluten proteins including gliadin and glutenin with polar lipid form a three-dimensional matrix, surrounding and separating the gas cells and entrapping the starch granules in the breadmaking dough.

Glutenin was only found in the bulk matrix of dough, gliadin was both in the bulk matrix and at the gas cell surface. The polar lipid had very similar distribution to gliadin, being in the matrix and on the gas cell walls. Moreover, the polar lipid was also observed on the surface of some particles. Most nonpolar lipid was found to be on the surface of particles or as lipid vesicles in the dough.

References

1. J.D.Schofield, In *Chemistry and Physics of Baking*, Eds Blanshard JMV, Frazier PJ, and Galliard T. Royal Society of Chemistry, London pp14,1986.
2. B. J. Dobraszczyk and C.A. Roberts, *J. Cereal Sci.* **20**: 256,1994.
3. W. Li, B.J. Dobraszczyk and J.D. Schofield, *Cereal Chem.* **80**:333-338,2003.
4. J. Örnebro, T. Nylander and A.C. Eliasson, *J Cereal Sci.*, **31**:195, 2000.
5. W. Li, *Ph.D thesis*, The University of Reading, Library, pp 67-68, 2001.
6. J. J. Kokelaar, *Ph D thesis*, The University of Wageningen, Library, pp 37, 1994.
7. P. A. Gunning, A. R. Mackie, P. J. Wilde and V. J. Morris, *Langmuir* **15**: 4636, 1999.

Acknowledgement

The authors gratefully acknowledge funding by the Biotechnology and Biological Sciences Research Council (BBSRC). Grant Ref. 45/D15320.

DYE LIGAND CHROMATOGRAPHY IN THE PURIFICATION OF WHEAT FLOUR PROTEINS

G. Alberghina[1], M.E. Amato[1], S. Fisichella[1], D. Lafiandra[2], D. Mantarro[1], A. Palermo[1], A. Savarino[1] and G. Scarlata[1]

[1] Department of Chemistry, University of Catania, V.le A. Doria 6, 95125 Catania (Italy)
[2] Department of Agrobiology and Agrochemistry, University of Tuscia, 01100 Viterbo (Italy)

1 INTRODUCTION

Gluten proteins have a major influence on the dough physical properties and the breadmaking quality of wheat flour. These proteins are gliadins and glutenins. Gliadins are monomeric proteins; while glutenins are polymeric proteins composed of high and low molecular weight subunits (HMW and LMW), linked together by disulphide bonds.

The variation in the amount and in the composition of the HMW subunits is related to the breadmaking quality of different wheat varieties[1]. The studies of the properties of HMW glutenin subunits have been limited by the difficulty of isolating single subunits in large amounts from complex mixtures of gluten proteins present in wheat. Therefore, we have undertaken a study by a simple chromatographic technique, namely Dye-Ligand Chromatography (DLC), This technique offers high specificity, purity and recovery in a single chromatographic step and provides clear advantages in terms of economy, safety and adsorbent capacity[2]. In this paper we report the DLC application for HMW glutenin subunits separation by using three different resins (Reactive Red 120 and Reactive Yellow 86 immobilized on Agarose, Cibacron Blue F3GA immobilized on Sepharose CL-6B) because the nature of the binding mechanism of the dyes varies significantly from protein to protein and from dye to dye[3]. These resins were utilized to purify some HMW glutenin subunit mixtures, like 1Dx5+1Dy10, 1Bx7 and 1Dx2+1Dy12, 1Bx7 and 1Dy12 from WRU 6979, Alisei 1 and Alisei 2 flours, respectively.

2 METHOD AND RESULTS

2.1 Preparation of HMW glutenin subunit mixtures

HMW glutenin subunit mixtures were prepared using a modified extraction procedure of Marchylo et al.[4]; the total extract of HMW glutenin subunits was then dissolved in 0.05 M Tris-HCl buffer, pH 7.6 containing 50% (v/v) propan-1-ol and 4M urea; 0.04 M dithiothreitol was added and the mixture was stirred for 1h at 60°C and then for 2 h at room temperature. To avoid the re-oxidation of SH groups, HMW glutenin subunits were blocked by adding 0.08 M 4-vinylpyridine to the mixture by stirring for 15 min at 60°C.

The solution containing the blocked subunits was dialyzed against 1% (v/v) acetic acid for 72h.

2.2 Dye-Ligand Chromatography

A sample of the reduced and blocked protein extract (about 3 mg) from each flour, Alisei 1, Alisei 2 and WRU 6979, was loaded onto columns packed with Reactive Yellow 86-Agarose (8 cm x 1 cm I.D.) and with Cibacron Blue F3GA-Sepharose (8 cm x 1 cm I.D.), whereas a sample of the reduced and blocked protein extract (about 3 mg) from Alisei 1 and WRU 6979 flour was loaded onto a column packed with Reactive Red 120-Agarose (8 cm x 1 cm I.D.).

The columns, equilibrated with 0.01 M sodium acetate pH 4.5 (for Reactive Yellow 86-Agarose and Reactive Red 120-Agarose resins) and with 0.01 M sodium acetate pH 3.5 (for Cibacron Blue F3GA-Sepharose resin), were eluted at room temperature by using different SDS concentrations in 0.01M sodium acetate at a flow rate of 0.5 ml/min. Eluates were monitored at 224 nm with a Jasco 1575 UV detector.
The peaks collected from each Dye-Ligand Chromatography were dialyzed against 1% (v/v) acetic acid for 72 h, freeze-dried and identified by reversed-phase HPLC and SDS-PAGE analysis.

The present study has shown that, by employing the Reactive Red 120, Cibacron Blue F3GA and Reactive Yellow 86 dyes, the separation of HMW subunit mixtures by Dye-Ligand Chromatography was successfully achieved by choosing the appropriate dye. In case of the Alisei 1 sample, Cibacron Blue F3GA Sepharose CL-6B allowed a complete separation of the 1Dx2 and 1Dy12 subunits, while the 1Bx7 subunit was always obtained together with 1Dx2 subunit traces.

Figure 1 shows the chromatographic profile obtained from this column characterized by three peaks. The first peak, eluted with 0.01 M acetate pH 3.5 within the column void volume, contained proteins that did not show any interaction with the immobilized dye. The second peak, eluted with 0.05% (w/v) SDS, was collected in two parts; initial fractions were identified as the 1Dx2 subunit, while the last fractions were identified as a mixture of 1Bx7 and 1Dx2 subunits. The third peak, eluted with 0.15% (w/v) SDS, was identified as 1Dy12 subunit.

Reactive Red 120 was utilized to separate a mixture of 1Dx5+1Dy10 subunits coming from WRU 6979 flour; it proved useful, both to discard the impurities of LMW glutenin subunits, and to obtain the 1Dx5 and 1Dy10 subunits in a pure form.

Figure 2 shows the chromatographic profile obtained from this column characterized by four peaks. The first peak, eluted with 0.01 M acetate pH 4.5, contained impurities; the second peak, eluted with 0.07% SDS, contained LMW glutenin subunits that were not discarded during the Marchylo extraction. The 1Dx5 subunit was eluted with 0.08% (w/v) SDS, the 1Dy10 subunit was eluted with 0.12% (w/v) SDS.

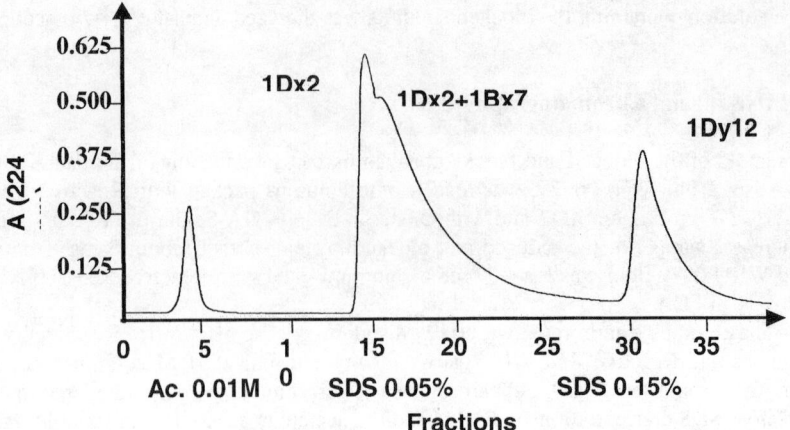

Figure 1 *DLC on Cibacron Blue F3GA-Sepharose CL-6B of Alisei 1 flour extract*

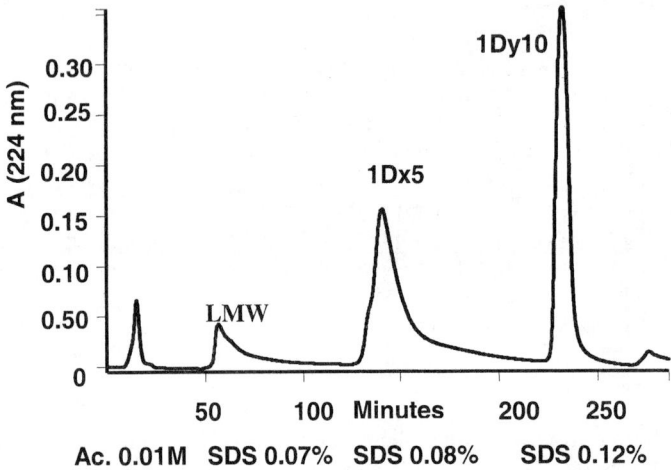

Figure 2 *DLC on Reactive Red 120-Agarose of WRU 6979 flour extract*

Finally, a sample of Alisei 2 extract was analysed on the Reactive Yellow 86-Agarose column; in this case, the 1Bx7 subunit was separated and the 1Dy12 was obtained with only small traces of the 1Bx7 subunit. Figure 3 shows the chromatographic profile obtained from this column characterized by three peaks. The first peak, eluted with 0.01 M acetate pH 4.5, contained impurities; the second peak obtained by eluting the column with 0.05% SDS, contained the 1Bx7 subunit. Eluting the column with 0.12% SDS, the 1Dy12 subunit was obtained with traces of 1Bx7 subunit (third peak).

3 CONCLUSION

Dye Ligand Chromatography provides an efficient method to purify HMW glutenin subunits mixtures because it enables high specificity, purity and recovery in a single chromatographic step; moreover, the purification is improved by choosing the appropriate dye.

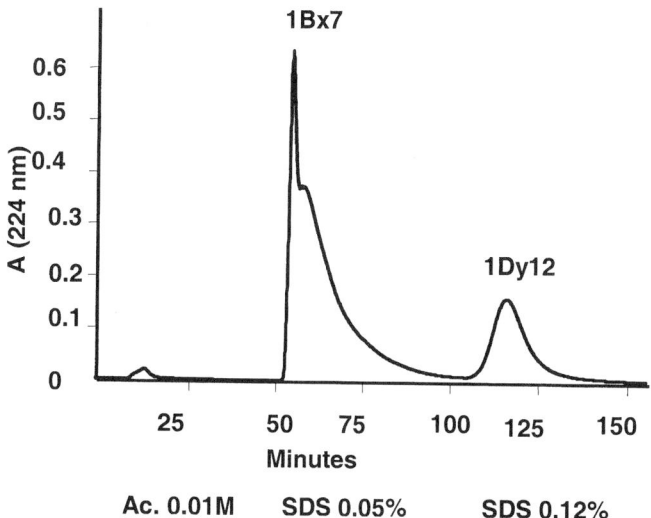

Figure 3 *DLC on Reactive Yellow 86-Agarose of Alisei 2 flour extract*

References

1. P.I. Payne, L.M. Holt, A.F. Krattiger, and J.M. Carrillo, *J. Cereal Sci.,* 1988, **7**, 229.
2. A. Denizli and E. Pişkin, *J. Biochem. Biophys. Methods* ,2001, **49**, 391.
3. P. Santambien, S. Sdiqui, E. Hubert, P. Girot, A.C. Roche, M. Monsigny, and E. Boschetti, *J. Chromatogr.,* 1995, **664**, 241.
4. B.A. Marchylo, J.E. Kruger and D.W. Hatcher, *J. Cereal Sci.,* 1989, **9**, 113.

STRUCTURAL STUDIES OF WHEAT FLOUR GLUTENIN POLYMERS AND HIGH MOLECULAR WEIGHT SUBUNITS BY CIRCULAR DICHROISM SPECTROSCOPY

M.E. Amato[1], S. Fisichella[1], D. Lafiandra[2], D. Mantarro[1], A. Palermo[1], A. Savarino[1] and G. Scarlata[1]

[1] Department of Chemistry, University of Catania, V.le A. Doria 6, 95125 Catania (Italy)
[2] Department of Agrobiology and Agrochemistry, University of Tuscia, 01100 Viterbo (Italy)

1 INTRODUCTION

Wheat flour contains a complex mixture of proteins such as albumins, globulins, gliadins and glutenins[1]. The glutenin proteins are composed of high and low molecular weight subunits (HMW and LMW) that form large polymers stabilised by inter-chain disulphide bridges[2].
HMW glutenin subunits that have been reduced and blocked, have been extensively studied because relationships were identified between these subunits and wheat quality characteristics [3]. The variation in HMW subunits sequences has been associated with variation in wheat quality, and therefore, the knowledge of HMW structure is important in order to understand the role of these proteins in determining the functional properties of wheat flour. In this work, we report the conformational studies of some HMW glutenin subunits by Circular Dichroism (CD)spectroscopy in the presence of some chemical denaturanting agents such as urea and sodium dodecyl sulphate (SDS).
Recently, polymeric glutenins have been shown to have a cause-and-effect relationship with flour properties, including baking quality[4].
The characterization of glutenin polymers is difficult because the complete extraction is only possible by using reducing agents in combination with denaturing agents. However, such reduced proteins are useful only for the study of the individual subunits. Here, we report a complete extraction procedure for unreduced glutenin polymers and their characterization in the presence of some chemical denaturants, such as urea and SDS, in order to obtain information about their intrinsic stability.

2 METHOD AND RESULTS

2.1 Preparation of HMW glutenin subunits

HMW glutenin subunit mixtures were prepared using a modified extraction procedure of Marchylo et al.[5]; the total extract of HMW glutenin subunits was then dissolved in 0.05 M Tris-HCl buffer, pH 7.6, containing 50% (v/v) propan-1-ol and 4M urea; 0.04 M dithiothreitol was then added and the mixture was stirred for 1h at 60°C and then for 2 h a room temperature. To avoid the re-oxidation of SH groups, HMW glutenin subunits were blocked by adding 0.08M 4-vinylpyridine to the mixture by stirring for 15 min at 60°C. The solution containing the blocked subunits was dialyzed against 1% (v/v) acetic acid for 72h. The HMW glutenin subunits were separated by preparative RP-HPLC using a Vydac C_{18} semi-preparative column (10 mm x 25 cm); peaks were detected at 224 nm and analytes were eluted with a 40 min linear gradient of 27-45% (v/v) acetonitrile (containing 0.05% TFA) at a flow rate of 3 ml/min.

2.2 CD measurement of HMW glutenin subunits

The conformational changes of some HMW glutenin subunits were studied by Circular Dichroism in 0.01 M sodium-acetate pH 3.5, at room temperature. CD spectra of the subunits were obtained at different urea concentrations (0-8 M), and at different urea concentrations (0-8 M) in the presence of 1% SDS. All the data were converted to give molar ellipticity values ($\Delta\varepsilon$) based upon the concentrations of the samples and expressed in cm^2*mmol^{-1} units. In each sample, the protein was dissolved at a concentration of about 0.4mg/ml, in 10 mM sodium acetate, pH 3.5. In 0.01 M sodium-acetate pH 3.5, 1Dx2 subunit, obtained from Alisei 1 flour, showed a mainly disordered structure characterized by a negative maximum below 200 nm and by a weak shoulder in the 215-230 nm range (see Fig.1). The CD spectrum (Fig.1) changed markedly in the presence of 1% SDS; the negative maximum below 200 nm decreased in intensity and shifted to higher wavelength, while the shoulder at 215-230 nm became more evident. Addition of 1M urea to the 1% SDS-acetate solution of 1Dx2 subunit eliminated the effect of SDS, as shown in Fig. 2; higher urea concentration (2-6 M) caused the decrease of the intensity of the shoulder in the 215-230 nm range until it disappeared, while at 7-8 M urea the spectrum was unchanged. Addition of 1M urea to the acetate solution of 1Dx2 subunit caused the disappearance of the shoulder in the 215-230 nm range and the spectrum was unchanged up to 2-4 M urea; higher urea concentration (5-6 M) induced the appearance of new weak positive band at 230 nm and the intensity did not change at urea concentration higher than 6M. The CD spectra results of HMW glutenin subunits show that the SDS promotes ordered structures; the addition of urea to the SDS-acetate solution of subunits destroys the ordered structure promoted by SDS, while its addition to the acetate solution of proteins does not cause an unfolding process, but induces conformational transitions to form a poly-L-proline II like structure[6].

2.3 Preparation of glutenin polymers

Alisei 1 flour was washed with 0.5 M sodium chloride by magnetic stirring for 1h at room temperature. After centrifuging at x5,000g for 15 min, the supernatant, containing albumins and globulins, was discarded and the residue was washed four times with distilled water to remove the salt. After centrifuging at x5,000g for 15 min, the residue was washed again by magnetic stirring for 1h at 60°C with 50% (v/v) propan-1-ol (30 ml) and centrifuged at x5,000g for 15 min to eliminate the gliadins.
The glutenin polymer was extracted with 2% SDS in 0.05 M sodium phosphate buffer, pH 6.9 by sonication for 1 h at room temperature. After centrifuging at x5,000g for 15 min, the supernatant was dialyzed against water for 10 days and freeze dried.

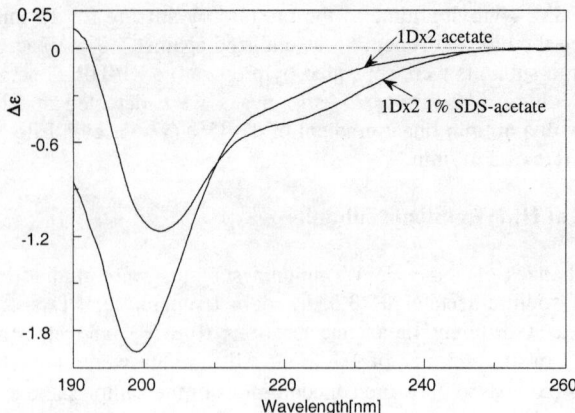

Figure 1 *CD spectra of 1Dx2 glutenin subunit at room temperature in 10 mM sodium acetate pH 3.5 and in 1% SDS-acetate*

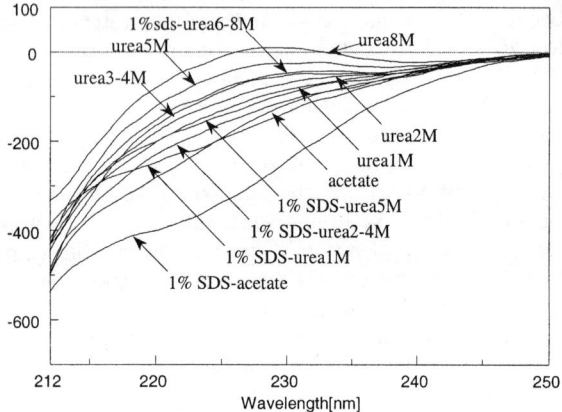

Figure 2 *CD spectra of 1Dx2 glutenin subunit at room temperature in the presence of urea (0-8M) and urea (0-8M) with 1% SDS*

2.4 CD measurement of glutenin polymers

The CD spectra of the Alisei 1 glutenin polymer (0.133% w/v) were recorded in 0.01 M sodium-acetate, pH 4.5 at room temperature at different urea concentrations (1-8 M) and at different urea concentrations (1-8 M) in the presence of 1% SDS. All the data were converted to give specific ellipticity values [ψ] based upon the sample concentrations and expressed in deg cm^2 decagram^{-1} units.

Addition of urea to the 0.01 M sodium-acetate, pH 4.5 solution of Alisei 1 sample, caused two subsequent conformational transitions, as can be observed by plotting the Specific Ellipticity against urea concentration (Fig. 3).

In the presence of 1% SDS-acetate, the CD spectrum of Alisei 1 changed markedly compared with that in sodium acetate; the negative maximum below 200 nm decreased in intensity and shifted to higher wavelength, while the shoulder at 215-230 nm became more

evident (Fig.4). Finally, the effect of urea addition to the 1% SDS-acetate solution induced an unfolding process through a two-step transition, as can be observed in Fig. 3.
The CD spectra results of the glutenin polymers show that all the changes induced by urea and by SDS follow a multi-step transition process.

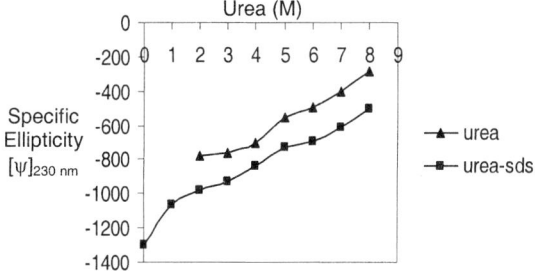

Figure 3 *Specific Ellipticity at 230 nm of Alisei 1 glutenin polymer against urea concentration in the presence and in the absence of 1% SDS*

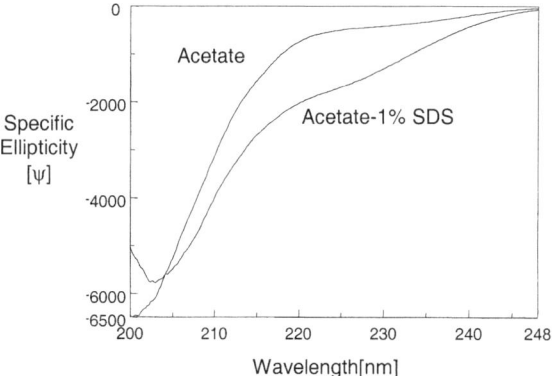

Figure 4 *CD spectra of Alisei 1 glutenin polymer at room temperature in 0.01M sodium acetate pH 4.5 and 1% SDS-acetate*

References

1 P.I. Payne, M.A. Nightingale, A.F Krattinger and L.M. Holt, *J. Sci. Food Agric.*, 1987, **40**, 51.
2 J.M. Field, P.R. Shewry, S.R. Burgess, J. Forde, S. Parmarand and B.J. Miflin, *J. Cereal Sci.*, 1983, **1**, 33.
3 G.J. Lawrence, F. MacRitchie and C.W. Wrigley, *J. Cereal Sci.*, 1988, **7**, 109.
4 P.L Weegels, R.J. Hamer and J.D. Schofield, *J. Cereal Sci.*, 1997, **25**, 165.
5 B.A. Marchylo, J.E. Kruger and D.W. Hatcher, *J. Cereal Sci.,* 1989, **9**, 113.
6 S. Fisichella, G. Alberghina, M.E. Amato, M. Fichera, D. Mantarro, A. Palermo, A. Savarino and G. Scarlata, *Biopolymers*, 2002, **65**, 142.

PREDICTION OF GRAIN PROTEIN CONTENT THROUGH PORTABLE FT-NIR MEASUREMENT OF DEVELOPING WHEAT

D.G. Bhandari[1], S.J. Millar[1] and J.C. Richmond[2]

[1] Campden & Chorleywood Food Research Association, Chipping Campden, Gloucestershire GL55 6LD, UK.
[2] Bruker Optics Ltd, Banner Lane, Coventry CV4 9GH.

1 INTRODUCTION

Wheat growers have great difficulty in accurately targeting nitrogen (N) fertiliser applications to meet market needs and minimise N pollution. The use of extra N fertiliser above that required for yield can increase the risk of nitrate leaching and pollution of surface and ground waters. Although late foliar urea application can boost wheat protein levels[1], this may not always be necessary or cost-effective[2] and currently there is no reliable, practical method of predicting the end protein content during the crop's lifetime. The recent advances in portable FT-NIR (Fourier Transform) technology now provide the means of addressing this problem. FT-NIR spectrometers feature very high frequency accuracy and a high resolution[3], and have a major advantage of allowing calibration transfer between instruments to be performed easily.

This 4-year project aims to produce a fast and accurate on-farm method of predicting grain protein content and quality using readily transportable high precision FT-NIR technology with an integrating sphere module. This will allow accurate decision making for the application of late N fertiliser. More specifically, this study will develop robust NIR calibrations necessary for accurate prediction purposes, using a wide range of both hard and soft milling varieties, grown over three seasons with a range of N fertiliser rates.

2 MATERIALS AND METHODS

2.1 Agronomic trials

In the first year of the study, sets of 52 growing crop at growth stage (GS) 71 and harvest samples, covering 17 different varieties and a range of protein contents, were obtained from sites in the UK at Boxworth, Terrington, High Mowthorpe and Rosemaund. None of the crop received foliar urea fertiliser after the sampling at GS 71.

2.2 Methods

Fresh, immature whole plant (above ground) and ears fractions were chopped into approximately 4 cm lengths and FT-NIR spectra over the range 900-2800nm were acquired using the Matrix-I spectrometer (Bruker Optics Ltd) featuring a quartz window with a rotating sampling cup and an integrating sphere detection system for reflectance analysis. Both sets of immature material were divided into two portions for drying using 2 regimes: 1) a two-stage heat treatment, at 40°C and 60°C in a conventional grain-drying oven; 2) 6 minutes at full power in a domestic type microwave oven (750W Matsui 170TC). FT-NIR spectra were acquired from this dried material and from the corresponding harvest grain ground using a Perten KT3100 Falling Number mill. Protein and moisture contents of the wheat samples were determined using Dumas (Nx5.7) and oven drying procedures respectively.

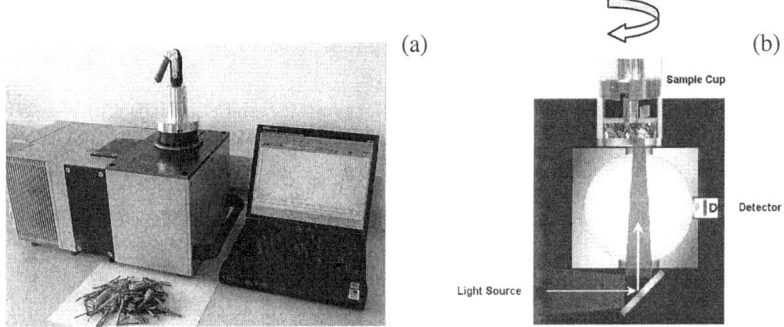

Figure 1 *The Matrix-I (a) and the integrating sphere assembly (b)*

3 RESULTS AND DISCUSSION

Oven dried immature whole plant and ear samples had moisture contents ranging from 9-11% and 9.7-12.3%, respectively. The microwave dried immature whole plant and ear samples had moisture contents ranging from 13-28% and 22-38%, respectively. The FT-NIR calibrations were produced against protein content in immature and mature samples using multiple linear regression analysis. The calibration performance for fresh whole plant in predicting mature grain protein was poor (squared correlation coefficient, $R^2 = 0.12$), as

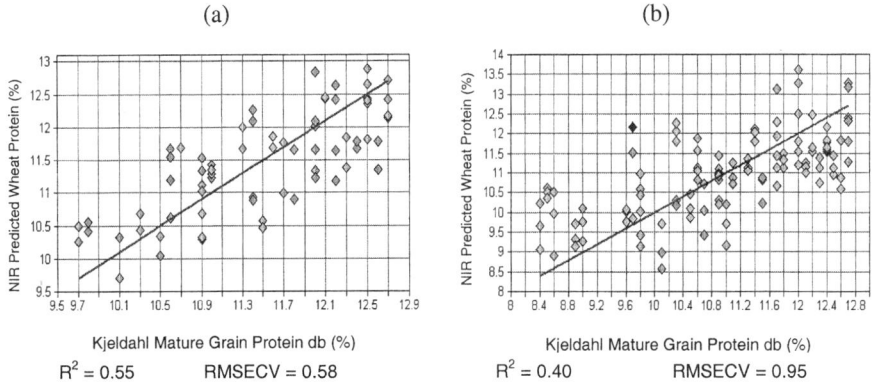

Figure 2 *NIR calibration for mature grain protein content using immature wheat spectra from microwave dried ears (a) and fresh ears (b)*

were those for microwave dried whole plant and oven dried ears ($R^2 = 0.14$ and $R^2 = 0.20$, respectively). The calibration performances for microwave dried ears and fresh ears in predicting mature grain protein content ($R^2 = 0.55$ and $R^2 = 0.40$, respectively) demonstrated some potential (see Figure 2).

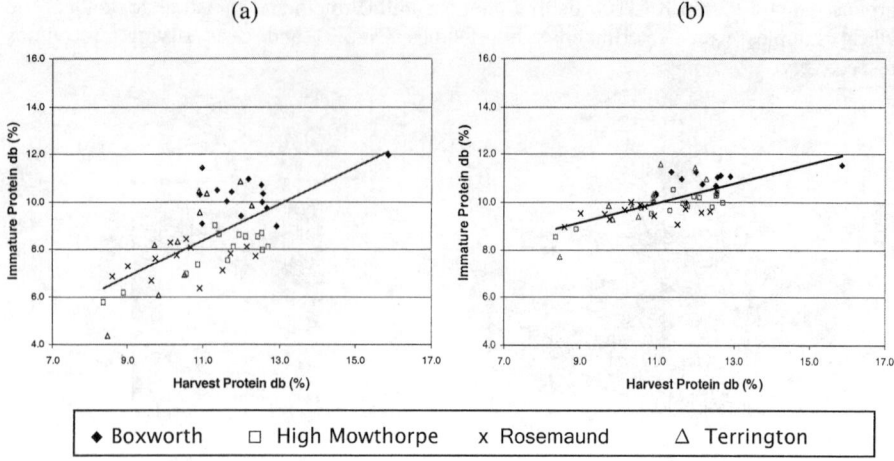

Figure 3 *Relationship between mature (harvest) grain protein content across 4 growing sites and immature whole plant protein content (a) and ear protein content (b)*

The relationship between mature grain protein and the developing wheat protein taken from all trial sites is shown in Figure 3. The protein contents ranged from 4.3 –12.3% in the immature whole plants and ranged from 7.8-14.2% in the ears. The overall R^2 was 0.43 for whole plant and 0.50 for ears when considering all sites. However, a consistent site effect was evident when data for both sets of samples were analysed based on their origins. Table 1 shows that the Terrington and High Mowthorpe sites had the highest R^2 values, while that for Rosemaund was the lowest. The NIR calibrations for mature grain protein content using immature sample spectral data will be performed on an individual growing site basis in order to establish the influence of environmental conditions.

Table 1 *Relationship (R^2) between mature grain protein content and immature whole plant and ear protein content determined for individual growing sites*

	Boxworth	Terrington	High Mowthorpe	Rosemaund
Whole plant	0.14	0.81	0.74	0.10
Ears	0.38	0.79	0.77	0.08

4 CONCLUSIONS

The preliminary results of this on-going study demonstrate that a portable FT-NIR spectrometer with an integrating sphere sampling assembly can provide good spectral data from crudely processed, heterogeneous developing wheat material. The dried immature ears yield spectra exhibiting features associated with analysis of mature wheat. In particular, the calibration performance for microwave dried ears show potential for the prediction of harvest protein. There was evidence for an effect of site on the relationship between protein content measured in immature wheat material and that measured in the harvest grain. The impact of this factor will be assessed through the analysis of the following second and third year trial samples.

References

1. M.J. Gooding and W.P. Davies, *Fertilizer Research*, 1992, **32**, 209.
2. P.M.R. Dampney, *HGCA Workshop: Wheat for the Milling Industry*, 1999, 9.
3. R.P Wayne, *Chemistry in Britain*, 1987, **23**, 440.

Acknowledgements

This work was sponsored by Defra through the Sustainable Arable LINK Programme and by HGCA. The authors wish to acknowledge the Project Research Partners: ADAS Consulting Ltd, Rothamsted Research, University of East Anglia and Heygates Ltd.

STRUCTURAL ANALYSES OF TWO HETEROLOGOUSLY EXPRESSED NATIVE AND MUTATED LOW MOLECULAR WEIGHT GLUTENIN SUBUNITS

V. Consalvi[1], R. Chiaraluce[1], C. Patacchini[2], D. Lafiandra[2], R. D'Ovidio[2] and S. Masci[2]

[1]Dipartimento di Scienze Biochimiche "A. Rossi Fanelli", Università degli Studi di Roma "La Sapienza", Piazzale Aldo Moro 5, 00185 Roma, Italy
[2]Dipartimento di Agrobiologia e Agrochimica, Università della Tuscia, Via S. Camillo de Lellis, 01100 Viterbo, Italy

1 INTRODUCTION

Low molecular weight glutenin subunits (LMW-GS) are the some of the most important components in determining visco-elastic properties of durum wheat, and exert also a significant role on bread-making characteristics of hexaploid wheats. Although there is strong evidence that the amount of LMW-GS is important, structural differences may not be excluded. The number and position of cysteine residues are likely to be the structural characteristics that mostly influence glutenin polymer organisation, and whose size is directly correlated with dough visco-elastic properties[1]. Here, we have performed a structural characterisation of two LMW-GS. These differ in the presence/absence of the first cysteine residue, which are involved in intermolecular disulphide bond, as well as six other amino acid substitutions, present mostly in the repetitive domain. The two polypeptides have been expressed heterologously in *E. coli*, purified and submitted to analyses using UV-visible, fluorescence and CD spectroscopy.

2 METHODS AND RESULTS

2.1 Expression and Purification of LMW1B and LMW1B⁻

The two polypeptides, whose amino acid sequences are reported in Figure 1, have been expressed heterologously in *E. coli* and purified[2] (Figure 1). The two LMW-GS have been termed LMW1B, corresponding to the wild type LMW-GS with eight cysteine residues, while LMW1B⁻ corresponds to the mutated polypeptide, with seven cysteine residues. Mutations were randomly caused during PCR amplification. The hydrophobicity patterns, relative to the two deduced sequences, showed that two amino acid substitutions caused significant changes in the repetitive region (Figure 2).

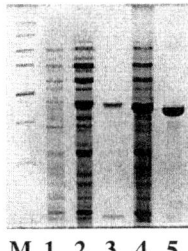

M 1 2 3 4 5

Figure 1 *SDS-PAGE showing* E.coli *heterologous expression and purification of LMW-GS 1B and 1B⁻. Lane M: mol. weight marker; lane 1, total cell proteins before induction of heterologous expression; lanes 2, 4, total bacterial proteins extracted after induction of expression of LMW1B and LMW1B⁻. On the right side of each of these lanes, the purification of the heterologous protein is reported (lanes 3 and 5, respectively).*

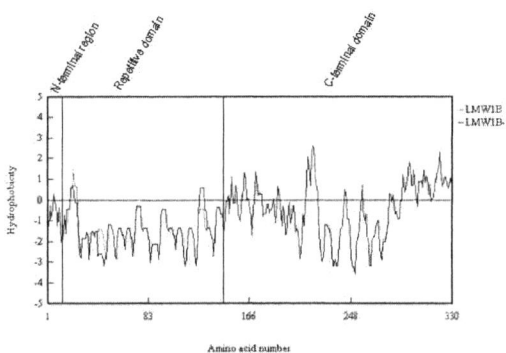

Figure 2 *Comparison of hydrophobicity levels of LMW1B (dotted line) and LMW1B⁻ (solid line) sequences plotted on the basis of their deduced amino acid sequences*

2.2 Spectroscopic Analysis

The spectral properties of LMW1B and LMW1B⁻ were analyzed in 70% ethanol at 20°C. The second derivative of the UV-visible spectra were similar, except for a peak at 291 nm which, for LMW1B⁻, is opposite in sign compared to that of LMW1B, suggesting a perturbation in the region of the Trp probably due to the L25W mutation. Notably, the aromatic residues of LMW1B and LMW1B⁻ are not locked in rigid tertiary contacts, as suggested by the lack of signal in the near-UV CD spectra of both proteins (data not shown). However, the Trp residues are shielded from the solvent, as indicated by the intrinsic fluorescence spectra characterized for both proteins by the same maximum emission wavelength at 347 nm, upon excitation at 295 nm (Figure 3).

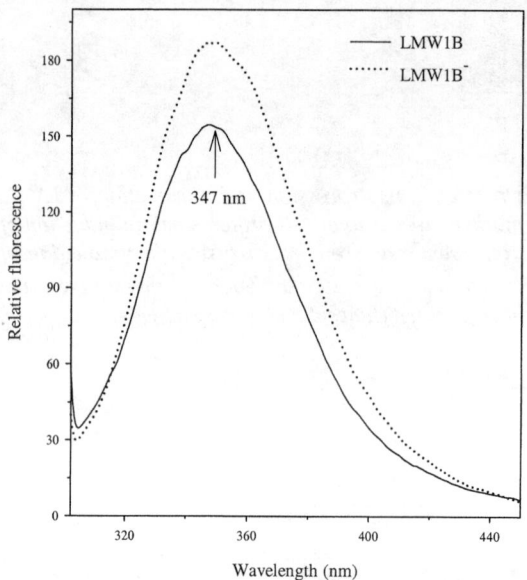

Figure 3 *Intrinsic fluorescence emission spectra (295 nm excitation) of LMW1B (continuous line) and LMW1B⁻ (dotted line). Spectra were recorded at 20°C in 70% ethanol at 0.20 mg/ml protein concentration.*

The far-UV CD spectra of LMW1B and LMW1B⁻ showed some differences indicative of different arrangements of the secondary structure elements (Figure 4). These spectra were simultaneously analyzed by several deconvolution methods[3] to estimate their secondary structure content, but most of these programmes failed to converge to a common solution. On the other hand, secondary structure consensus predictions performed by alternative programmes[4-5] predict around 10% helices, 2% strands and 88% of other structures for LMW1B and 11% helices, 2% strands and 87% loops for LMW1B⁻.This supports the evidence for a different secondary structure content as revealed by the inspection of the far-UV CD spectra of the two proteins, but not confirmed by spectral deconvolution. Non-regular secondary structure regions have been identified between residues 1 and 165 for LMW1B, and 8 and 147 for LMW1B⁻, according to Liu et al.[5].

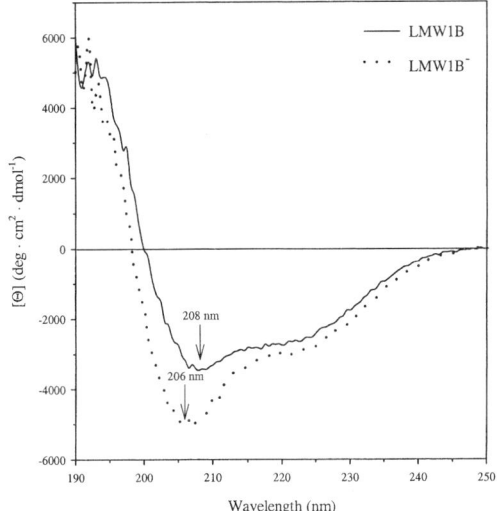

Figure 4 *Far-UV CD spectra of LMW1B (continuous line) and LMW1B⁻ (dotted line). Spectra were recorded in a 0.1 cm quartz cuvette at 20°C in 70% ethanol, 0.35 mg/ml protein concentration*

3 CONCLUSION

The random mutations in LMW1B⁻ did not significantly affect the amount of secondary structure. In general, the discrepancy between the secondary structure elements predicted from primary sequence and those measured from spectral deconvolution suggests that an alternative mutagenesis approach is required. This discrepancy might be due to different oxidation states of the two proteins, occurring during the purification process. Site-directed mutagenesis of all cysteine residues will be attempted to obtain more consistent information about the native secondary structure elements of LMW-GS.

Acknowledgements

Research was supported by the Italian Ministery for University and Research (MIUR), project "Aspetti biochimici, genetici e molecolari delle proteine della cariosside dei frumenti in relazione alle caratteristiche nutrizionali e tecnologiche dei prodotti derivati" (PRIN 2002).

References
1. T. Dachkevitch and J.C. Autran, *Cereal Chem.*, 1989, **66**(6), 448
2. C. Patacchini, S. Masci, R. D'Ovidio, D. Lafiandra, *J. Chromat. B*, 2003, **786**, 215
3. G. Deléage and C. Geourjon, *Comp. Appl. Biosc.*, 1993, **9**, 197.
4. B. Rost, *Methods Enzymol.*, 1996, **266**, 525-39
5. J. Liu, H. Tan and B. Rost, *J. Mol. Biol.*, 2002, **322**, 53-64.

GLUTEN SURFACE HYDROPHOBICITY OF BREAD AND DURUM WHEAT

N. Guerrieri

Dipartimento Scienze Molecolari Agroalimentari, University of Milan, via Celoria 2, 20133 Milano, Italy.

1 INTRODUCTION

The main protein constituent of wheat flour is gluten, a protein complex formed by gliadins and glutenins polypeptides. The network of these polypeptides determines the technological properties of the wheat flour. Some polypeptides, in particular the glutenins, are linked by disulphide bridges and interact with each other, and with the gliadins via hydrophobic interactions and hydrogen bonds. In this work, we investigated the hydrophobic surface of the gluten complex purified by *Triticum aestivum* and *durum* using a fluorescent hydrophobic probe ANS (8-anilino1-naphthalene sulphonate). A stepwise reduction was also performed; these experiments allowed the information on accessibility of the disulphide to reducing agent TCEP (tris (2-carboxyethyl) phosphine hydrochloride) and its effect on surface hydrophobicity.

2 METHODS AND RESULTS

2.1 Methods

2.1.1 Materials
Gluten, prepared by hand-washing, freeze dried and lyophilised, was isolated from different varieties and commercial flour. The samples studied were commercial bread wheat: Arabo, Northern Spring, Astron, Bussard, Zentos, Australian, Manitoba, Utility, Mac-Centauro, Hereward, Riband, the durum wheat: Capeiti, Creso, Grazia, Latino, Ofanto, Duilio, Colosseo, Simeto, Vitron 42 and Vitron 45. All chemical were analytical grade.

2.1.2 Protein preparation
The glutens (0.2-0.3 mg/ml of protein) were solubilised in 0.05M acetic acid, pH 3.1 stirring at 25°C for 30 minutes., The protein concentration was determined spectrophotometrically, as described by Eynard[1] et al. The yield of extracted proteins in the condition used approached 100% as determined by a Carlo Erba NA 1500 automatic nitrogen analyser.

2.1.3 Fluorescence measurement

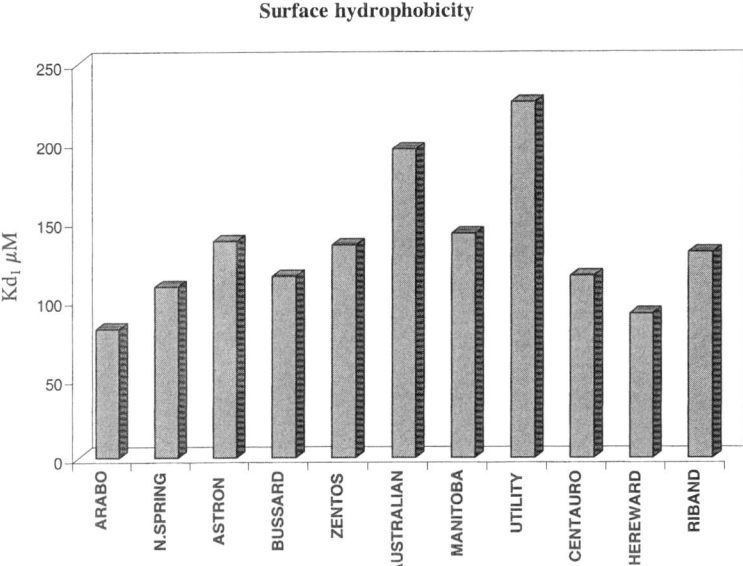

Figure 1 *Dissociation constants (Kd, μM) for ANS bound to gluten from* Triticum aestivum *varieties*

The fluorescence measurements were done with a Perkin Elmer Lambda 2B linked to a peltier cell-holder. Fluorescence due to the binding of ANS to the gluten was measured at λem 480nm, using λex 380nm. The ANS binding to the proteins was evaluated by titration. Increasing amounts of ANS of between 1 to 10mM were added to the solubilised gluten at 25°C with magnetic stirring, until saturation was reached.

2.1.4 Data analysis

The observed fluorescence, normalised for the protein concentration, reflects a co-operative binding by the hydrophobic probe. The resulting curve was de-convoluted using the Peakfit software (Jandel, Germany) into two components corresponding to high and low reactivity for the probe[2]. The determined dissociation constant (Kd, μM) is related to the surface hydrophobicity of the gluten analysed.

2.1.5 Stepwise reduction

Gluten samples solubilised in 0.05 M acetic acid, pH 3.1 were treated with TCEP, a reducing agent which can act in acid medium[3]. the, Different amount of TCEP were added to a final concentration of 1, 25, 50, 100 mM. The reduction was performed at 25°C for 30 min and followed by a titration with ANS as describe in 2.1.2.

2.2 RESULTS AND CONCLUSION

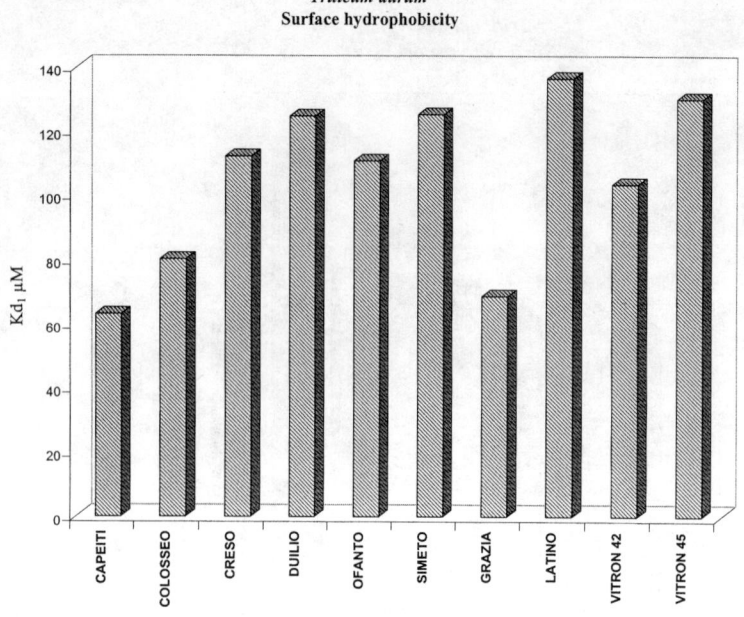

Figure 2 *Dissociation constants (Kd, µM) of ANS bound to gluten from* Triticum durum *varieties*

The dissociation constants (Kd) of the ANS bound to the gluten, solubilised from T*riticum aestivum* and T*riticum durum* varieties, are reported in Figure 1 and in Figure 2, respectively. In each figure, a distribution is present, and the overall values of the two types of wheat are significantly different. The ANS bound to the gluten of durum wheat is bound in the hydrophobic areas more tightly. In the bread wheat, the glutens that bound most strongly to the hydrophobic probe are Arabo and Hereward, in contrast to Australian and Utility, which bound the weakliest. In the durum wheat, the gluten of Capeiti and Grazia exhibited the lowest Kd values, while those of Latino and Vitron 45 exhibited the highest. There were no significant correlations between the values of surface hydrophobicity and some technological parameters (namely, alveograph and farinograph). However, a good correlation was found with the bread volume.
Further studies to achieve a better understanding of gluten properties and their relationship with molecular behaviour and technological properties are in progress.

Reference

1 L. Eynard, N. Guerrieri and P. Cerletti, *Cereal Chem.*, 1994, **71**, 434.
2 E. Sironi, N. Guerrieri and P. Cerletti, *Cereal Chem.*, 2001, **78**, 476.
3 J.C. Han and G.Y. Han, *Anal. Biochem.*, 1994, **220**, 5.

RELATIONSHIP BETWEEN FUNCTIONAL PROPERTIES OF WHEAT DOUGH AND THE RELATIVE PROPORTION OF THE POLYMERIC FRACTION

I. Király[1], O. Baticz[2], O. Larroque[3], A. Juhász[3,4], S. Tömösközi[2], F. Békés[3], A. Guóth[2], T. Abonyi[2] and Z. [4]Bedő

[1] Eötvös Loránd University, Department of Plant Physiology, H-1117 Budapest, Pázmány P. stny. 1/C, Hungary e-mail: ikiraly@ludens.elte.hu
[2] Budapest University of Technology and Economics, Department of Biochemistry and Food Technology, H-1111 Budapest, Műegyetem rkp. 3., Hungary
[3] CSIRO Plant Industry, GPO Box 1600, Canberra ACT 2601, Australia
[4] Agric. Res. Inst. Hungarian Acad. Sci., H-2462 Martonvásár, Brunszvik 2, Hungary

1 INTRODUCTION

The main functional component of wheat the gluten is organized at many different levels. Without neglecting the importance of the hard to estimate secondary interactions[1], it can be stated that the first order the intermolecular bounds between high molecular weight glutenin subunits (HMWGS) – and the low molecular weight glutenin subunits (LMWGS) – determine the functional properties of dough.

Presently UPP% seems to be one of the biochemical parameters useful in predicting baking quality. UPP% is the quotient of the phosphate/SDS soluble and total glutenin fractions of floor calculated on the base of the corresponding SE-HPLC peak areas. Thus the UPP% reflects the relative amount of macropolymer (GMP) in the flour and therefore a simple measure of the size distribution of polymeric glutenin. Because of their relatively superior contribution to form large polymers, UPP% is related to the rate of the HMWGS incorporated in the GMP.

Some steps during the sample preparation to determine UPP% such as the extraction of gliadins, and the use of sonication may influence the results and make disputable any comparison. Therefore we omitted the sonication, and extracted the soluble fraction on a strictly reproducible way during the preparation of the samples in this work. Our aim was to get information on the HMWGS and LMWGS composition of the GMP and to compare it to that of the flour. Here we present data on the glutenin subunit composition of the above phosphate/SDS extractable- and insoluble fractions, measured by RP-HPLC.

2 METHODS AND RESULTS

2.1 Wheat species investigated

Flours from 12 modern cultivated Hungarian wheat varieties (Table 1) were investigated. The grains were harvested in 2001 and 2002 on the experimental farm of the Agricultural Research Institute of the Hungarian Academy of Sciences in Martonvásár.

2.2 Sample preparation

Soluble fraction: 50mg flour was extracted with 1ml of phosphate-SDS (PS) buffer during 10 min (150rpm). Samples were centrifuged (5min, 10000g), – the pellet was kept for the preparation of the insoluble fraction – and 400μl of the supernatant was precipitated by 1000µl n-propanol. The samples were kept on 10°C for 10 min, and then centrifuged (12000g for 10 min). The precipitate was re-dissolved in propanol-urea-DTT buffer (PUD: propan-1 ol 50% v/v, urea 2M, tris-HCl 0,2 M pH 6.6, DTT 0,1 %) and incubated in water bath (60°C) for an hour. After adding 10µl of 4-vynil-pyridin the samples were incubated for an additional 15 minutes, than centrifuged at 12000g for 10 minutes. Before the chromatographic separation, the samples were filtered through a 45µm PVDF filter.

Insoluble fraction was prepared from the first pellet by dissolving it in PUD buffer and reducing the proteins as above. After centrifugation the samples were filtered into the sample vials.

Total fraction was prepared for RP-HPLC separation by the modified method of Marchylo et al.[2]. Total, soluble and insoluble fractions were analyzed by RP-HPLC.

2.3 RP-HPLC chromatography

The analytical separation was carried out on a Vydac C18 column using Perkin-Elmer LC 200 chromatograph equipped with autosampler, column thermostat (65°C) and DAD (diode array detector). The gradient elution was performed by 0.1(v/v)% TFA in acetonitril–water.

2.4 Measurement of physical characteristics of dough

Farinograph data (DDT=dough development time, S=stability, B=breakdown and WA=water absorption) was measured according to the International Standard ISO 5530-1:1988.

2.5 Statistics

Total-Chrom software package (Perkin-Elmer Instruments, USA) was used for processing the chromatograms and Statistica for Windows 6.0 (StatSoft, USA) for statistical evaluation.

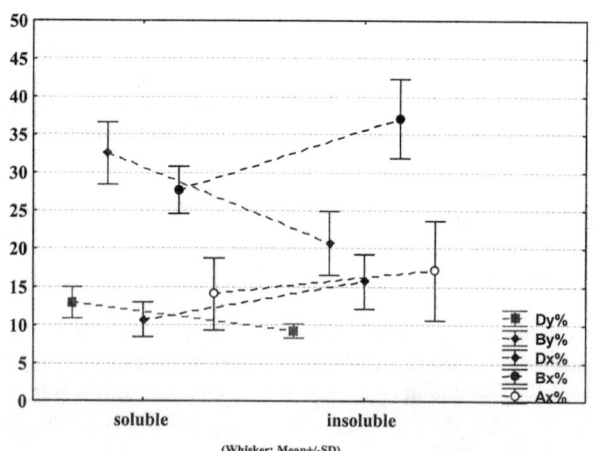

Figure 1 *The rate of the HMWGSs (Dy, By, Dx, Bx and Ax) in the soluble and insoluble fractions.*

Table 1 *Result summary of the standard farinograph measurements.*

Sample	DDT (min)	S (min)	B (FU)	WA (ml)
Bánkúti 1201	5.3	2.4	69.0	33.2
MV Magvas	2.5	1.4	15.0	30.8
MV Pálma	6.2	1.0	129.0	31.3
MV Palotás	14.6	2.3	40.0	31.2
MV Tamara	4.5	5.5	59.0	31.6
MV Verbunkos	7.8	1.5	88.0	35.1
MV Martina	1.4	0.5	258.0	27.6
MV Magdaléna	5.6	0.8	131.0	34.5
MV Emma	10.5	9.5	22.0	31.9
MV Emese	3.0	1.0	26.0	31.0
MV 16	1.2	0.3	136.0	28.3
MV Suba	5.0	15.0	45.0	32.1

Table 2 *Correlation between RP-HPLC and farinograph data for the insoluble fraction (significant correlation is highlighted).*

	DDT (min)	S (min)	B (FU)	WA (ml)
Dy%	-0.1106 p=0.732	-0.1661 p=0.606	-0.1110 p=0.731	0.3416 p=0.277
By%	0.5053 p=0.094	-0.0590 p=0.856	0.0232 p=0.943	0.2524 p=0.429
Dx%	-0.1326 p=0.681	-0.3821 p=0.220	0.2141 p=0.504	**-0.6440 p=0.024**
Bx%	-0.3460 p=0.271	0.0486 p=0.881	0.0508 p=0.875	-0.0920 p=0.776
Ax%	0.0398 p=0.902	0.2316 p=0.469	-0.1565 p=0.627	0.2128 p=0.507

3 DISCUSSION

The proportions of the HMWGSs were different in the insoluble and soluble fractions investigated (Fig. 1). This later fraction represents the gluten of lower polymerization degree, while the insoluble fraction consists of the GMP. The rate of y-type HMWGSs was higher in the soluble fraction while the x-type HMWGSs enriched in the insoluble fraction. The rate of By was considerably lowered in the insoluble fraction compared to the soluble fraction. Comparing the same way the rate of the Bx subunit is markedly higher in the insoluble fraction, which means that it has been preferentially incorporated during the formation of the macropolymer. Some alleles of HMWGS encoded on chromosome D play an important role in determining dough properties[3]. In our experiments the enrichment of Dx-type subunit in the insoluble fraction was moderate. The changes of the other HMWGSs (Ax, and Dy) were also modest.

The functional properties of the samples investigated are summarized on Table 1. Table 2 shows the results of correlation analysis of the relative proportions of HMWGSs and the data measured by standard farinograph method. It was found only one weak significant effect: a negative correlation between Dx% and WA (water absorption) in the insoluble fraction (p=0.024). This and the lack of other correlations frequently published may be due to the problems of sample preparation.

4 CONCLUSION

Most of the results obtained are in agreement with the present knowledge on the gluten macropolymer (GMP). The constancy of the LMW rate in the 3 fractions investigated (results not shown) seems to imply a primary role of LMWGSs in the formation of the gluten network. However further experiments are needed to figure the changes of the rates of HMWGSs and the HMW/LMW ratio in the 3 fractions. We consider that the method of sample preparation and sample handling is to be evolved.

References

1 G. Danno, R.G. Hoseney Cereal Chem. 1982, **59**, 196.
2 B.A. Marchylo, J.E. Kruger, D.W. Hatcher 1989, J. Cereal Sci., **9**, 113.
3 P.R. Shewry, N.G. Halford, A.S. Tatham 1992, Cereal Sci. **15**, 105.

ANALYSIS OF SULPHUR IN GLUTEN PROTEINS BY X-RAY ABSORPTION NEAR EDGE STRUCTURE (XANES) SPECTROSCOPY

A. Prange[1], B. Birzele[1], H. Modrow[2], J. Hormes[3] and P. Köhler[4]

[1]Department of Agricultural and Food Microbiology, Institute for Plant Diseases, University of Bonn, Meckenheimer Allee 168, D-53115 Bonn, Germany. E-mail: A.Prange@gmx.de
[2]Institute of Physics, University of Bonn, Nussallee 12, D-53115 Bonn, Germany
[3]Center for Advanced Microstructures and Devices (CAMD), Louisiana State University, 6980 Jefferson Highway, Baton Rouge, LA 70806, USA
[4]German Res. Centre of Food Chemistry, Lichtenbergstr. 4, D-85748 Garching, Germany

1 INTRODUCTION

X-ray absorption near edge structure (XANES) spectroscopy using synchrotron radiation is an excellent method for examining crystalline and non-crystalline biological samples *in situ*. It combines the penetration strength of X-rays and the advantage of being a local probe technique which does not require long range order. XANES allows insight into the electronic and geometric structure of the environment of the absorbing atom, e.g. the valence of the excited atom as well as the electronegativity of neighbouring atoms even if it is only used as a 'fingerprint method'.[1] Using this type of approach, the XANES spectrum of the sample to be analysed is compared with those of suitable model compounds yielding 'qualitative information'. To describe quantitatively the contributions of different constituents in a mixture to the XANES spectral features, a 'quantitative analysis' can be performed.[2,3] A review on X-ray absorption spectroscopy (XAS) and its application in biological, agricultural and environmental research has been published recently.[3] In this paper, we will give a concise introduction to the method and will summarise briefly our results on different aspects of the 'system gluten proteins' obtained by XANES spectroscopy at the sulphur K-edge.

2 METHOD AND RESULTS

2.1 Basics of X-Ray Absorption Spectroscopy

The physical entity which is measured when performing a XAS experiment is the dependence of the cross section of the photoabsorption process (i.e. the probability that it occurs) on the energy of the incoming photon. If the energy of the photon is varied in rather large steps over a large energy range, the behaviour shown in the insert of the top region of Figure 1 is observed: Jumps in the cross section are observed whenever the photon energy gets sufficient to excite electrons from a deeper core level; at the highest energy from the 1s level, proceeding with decreasing photon energy to 2s, $2p_{1/2}$ and $2p_{3/2}$ and so on. However, when scanning an absorption edge in very small energy steps, a behaviour as shown in Figure 1 can be observed.

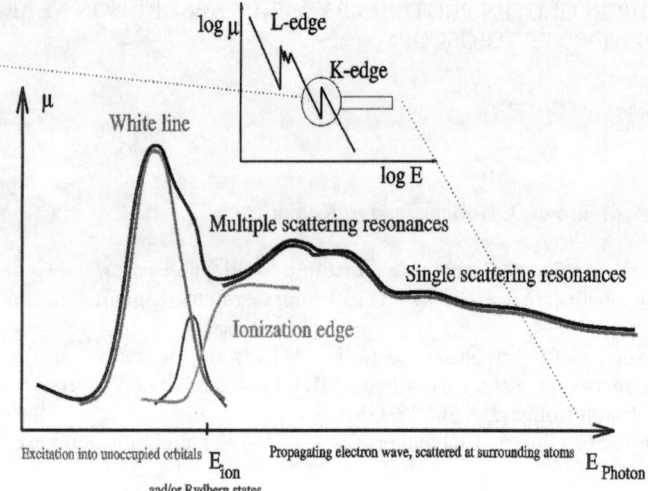

Figure 1 X-ray absorption fine structures and their (dominant) cause

At lower energies than the ionisation edge, transitions into bound unoccupied states exist, often leading to pronounced features called 'white line'. The oscillatory behaviour of the absorption spectrum above the ionisation threshold can be understood in a simple picture as interference structure of the outgoing photoelectron wave and its backscattering from atoms located around the absorbing atom. Closely behind the absorption edge, there are strong oscillations due to multiple scattering effects, and some 50 eV beyond it, single scattering processes dominate. The transitions into bound states and the multiple scattering region are known as X-ray Absorption Near Edge Structure (XANES), the latter known as the Extended X-ray Absorption Fine Structure (EXAFS), which will not be discussed further here.

What is so fine about all this fine structure? Chemistry modifies the valence state and thus the electronic structure of the elements, so one can learn a lot about the local chemical environment of the absorber atom in the sample. On the other hand, the interference pattern should be characteristic for a given arrangement of the atoms which surround the absorbing atom and allow for the extraction of information on its coordination geometry. Due to the intrinsic penetration strength of X-rays, there is no need for an Ultra High Vacuum (UHV) apparatus, so that *in situ*, phase-independent measurements are possible. As the technique is local, a tedious creation of crystalline samples is not necessary. By simple addition of two different atomic environments, the spectrum of a mixture of such compounds is reproduced. By choosing the photon energy, the element of interest is selected out of all different types of atoms, even if it is sitting somewhere in the bulk, buried under a pile of material, and one does not get additional, potentially confusing contributions from other elements.

2.2 XANES Spectroscopy of different Sulphur-containing Substances

To illustrate this principle, the S-K-XANES spectra of some substances (see figure caption) with sulphur in different oxidation states are shown in Figure 2 (left). The spectra differ in form and energy positions of the white lines and thus can be easily distinguished.

Evidently, there is a clear trend for the absorption edge in these reference spectra to shift to higher energy with an increasing formal oxidation state (Figure 2 (left)).

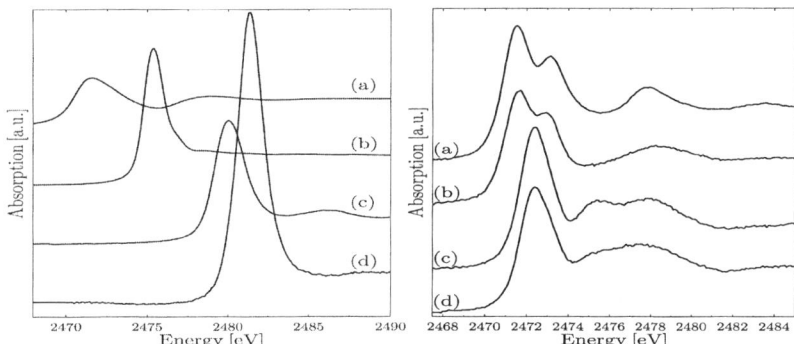

Figure 2 Sulphur K-edge XANES spectra of: <u>left:</u> (a) sulphur (S_8-rings), (b) dimethylsulphoxide, (c) cysteic acid and (d) zinc sulfate; <u>right:</u> (a) cystine, (b) glutathione (oxidised), (c) glutathione (reduced) and (d) cysteine.

This behaviour is known as "chemical shift". However, the resolving power of this method with respect to the chemical environment of the absorbing atom is much higher: in Figure 2 (right), the spectra of glutathione (reduced)/glutathione (oxidised) vs. cysteine/cystine are displayed to illustrate the influence of different organic rest groups. Evidently, it is easily possible to distinguish between different types of S-H and S-S containing compounds.

2.3 Sulphur in Wheat Gluten: *In situ* Analysis by XANES Spectroscopy

The sulphur containing gluten proteins largely determine the baking quality of wheat. XANES spectroscopy was applied to probe the speciation of sulphur in gluten proteins *in situ* (2.3.1-2.3.3). The spectra were recorded at beamline BN3 using synchrotron radiation from the ELectron Stretcher Accelerator ELSA at the Institute of Physics, Bonn. For a detailed description of the experimental setup, sample preparation, and calibration of the spectra the reader is referred to Bucher *et al.*[4] and Prange *et al.*.[5]

2.3.1 Speciation of sulphur in gluten proteins. Gluten proteins (gluten, gliadin, high molecular weight (HMW) and low molecular weight (LMW) subunits of glutenin), stored glutenin subunits as well as flour were investigated using the 'XANES fingerprint approach'.[5] The spectra confirmed the existence of disulphide bonds in oxidised (oxygen stream) glutenin subunits *in situ* for the first time, supporting their significance for the formation of gluten networks. Additionally, glutenin subunits, which were stored under ambient air and temperature conditions for 2 years, predominantly contained sulphur of higher oxidation states (sulphoxide, sulphonic acid state) and less sulphur in the disulphane state.[5,6]

2.3.2 Sulphur speciation of LMW subunits of glutenin after reoxidation with KIO_3 and $KBrO_3$ at different pH-values. The sulphur speciation in reduced LMW subunits of glutenin after reoxidation with KIO_3 and $KBrO_3$ at pH 2, 4, 6 and 8 has been investigated. The S-K-XANES spectra were analysed quantitatively to provide a relative percentage contribution of different sulfur species occurring in the samples. For a quantitative analysis of spectra, an interactive fitting and plotting package using the function minimisation tool

'MINUIT' (part of the CERNlib; wwwinfo.cern.ch/asdoc/minuit/node2.html) was applied.[6] Details on the quantitative analysis can be deduced e.g. from Prange et al..[2] Using KIO_3 and $KBrO_3$ for reoxidation of reduced LMW subunits of glutenin led to disulphane states but also to higher oxidation states (sulphoxide state, sulphonic acid state). At low pH-values, the strongest oxidation occurred. The choice of the oxidising reagent seems to be of minor importance.[6]

2.3.3 Influence of moulds on the sulphur speciation of gluten proteins. To characterise the influence of different moulds on a molecular level of the gluten network, LMW subunits of glutenin isolated from different wheat samples infected with *Fusarium* spp. and storage fungi (*Penicillium/Aspergillus* spp.) were investigated.[7] S-K-XANES measurements and a quantitative analysis of the spectra (cf. 2.3.2) revealed that the sulphur speciation is not changed by *Fusarium* spp. in comparison with control samples, which was reflected in baking tests. Storage fungi, however, appear to have a strong influence on the sulphur speciation.[7] In a first experiment, a sample highly infected with storage fungi was investigated by S-K-XANES. A significantly increased occurrence of higher oxidation states was observed as well as a drastic reduction of the loaf volume in corresponding baking tests. This interesting finding will be elucidated in further investigations.

3 CONCLUSIONS

XANES spectroscopy at the sulphur K-edge was successfully applied to different aspects of the system 'gluten proteins'. Our studies prove its suitability to investigate -qualitatively and quantitatively- the sulphur speciation in gluten proteins *in situ*. They exemplify a new application of XANES spectroscopy for investigating biological systems in a non-destructive way and offer perspectives for future *in situ* analyses.

References

1 A. Bianconi, in *X-ray absorption: principles, applications, techniques of EXAFS, SEXAFS and XANES*, ed. D.C. Koningsberger, R.C. Prince, John Wiley, New York, 1988, ch. 11, p. 573.
2 A. Prange, R. Chauvistré, H. Modrow, J. Hormes, H.G. Trüper and C. Dahl, *Microbiol.-UK*, 2002, **148**, 267.
3 A. Prange and H. Modrow, *Re/Views Environ. Sci. Bio/Technol.*, 2002, **1**, 259.
4 S. Bucher, J. Hormes, H. Modrow, R. Brinkmann, N. Waldöfner, H. Bönnemann, L. Beuermann, S. Krischok, W. Maus-Friedrichs and V. Kempter, *Surface Sci.*, 2002, **497**, 321.
5 A. Prange, N. Kühlsen, B. Birzele, I. Arzberger, J. Hormes, S. Antes and P. Köhler, *Eur. Food Res. Technol.*, 2001, **212**, 570.
6 A. Prange, B. Birzele, J. Krämer, H. Modrow, R. Chauvistré, J. Hormes and P. Köhler, *J. Agric. Food Chem.*, 2003, in press.
7 A. Prange, B. Birzele, J. Krämer, A. Meier, H. Modrow and P. Köhler, *Food Control*, 2003, in press.

MOLECULAR CHARACTERISATION OF WHEAT GLUTEN AND BARLEY PROTEIN

A.A. Tsiami, W. Li, D.L. Pyle and J.D. Schofield

School of Food Bioscience, The University of Reading, Reading RG6 6AP, UK

1 INTRODUCTION

The rheological properties of dough are of paramount importance for its performance during the bread making process. Gluten protein is the significant component, which gives to the dough the viscoelastic properties. The rheological properties of glutenin have been attributed to the molecular size (up to $10(^{\wedge}8)$ Da)[1-3] of the polymer. It was observed some time ago that barley hordein showed a similar polymer size distribution by size exclusion chromatography, but hordein lacks the functional properties of gluten[4]. Lately, the availability of hull-less barley in commercial quantities [5-6] as well as the health-promoting effects of barley β-glucan[7] have stimulated interest in barley.

Our research is focused on the protein characterisation of wheat and barley

2 METHODS AND RESULTS

The wheat flour used for the comparison with the barley flour was Bargain's Basta, grown in Sweden (Cerealia, Sweden) and the hull-less barley was the cultivar SW 1290 from Svalöf Weibulls AB, Sweden. SW 1290 was grown in 2000. The characterisation of the flour, milling fractions are presented by Anderson et al[8]

The M_w distributions and cumulative molar masses of each gluten protein fraction from the wheat flour are presented in Figure 1. Glutenin fraction R2, has two peak, one contained extremely large polymers of 5×10^8 at cumulative weight fraction (CWF) of 0.6 and a second peak contained smaller polymers the average size of which was 2×10^6 at CWF of 0.3. The other large molecular weight fraction R3, also a two peaks. The M_w distribution of R3 was lower than fraction R2 as expected. The first peak contained ultra high M_w polymers of 5×10^8 at a CWF of 0.4. The second peak a M_w of 5×10^6 centred at a CWF of 0.5. Fraction R4 had two distinct peaks, one of low M_w (about $1*10^5$) and one of higher M_w (1×10^8) at a CWFs of about 0.75 and 0.25 respectively of the total injected mass. Fraction R5, which was relatively monodisperse, had a M_w of 6×10^4 at a CWF of 0.90, and may correspond to ω–gliadins. This fraction had a small proportion of larger glutenin polymers (M_w of 1×10^7), which accounted for almost 10% of the injected mass. Fraction R6 was also a relatively monodisperse fraction with 95 % of the total injected mass having a M_w of 5×10^4 and probably represented gliadins, less than 5 % of total injected mass comprised glutenin polymers of very high M_w. Fraction R7 had a M_w of 5×10^4. Fraction R7 is gliadin according to Graveland et al.[9].

The M_w distributions and cumulative molar masses of each hordein protein fraction from the barley flour are presented in Figure 2.14. Hordein fraction R2, had three peaks. One contained extremely large polymers of 1×10^9 at a cumulative weight fraction (CWF) of 0.2 and the second peak contained large polymers the average size of which was 2×10^8 at a CWF of 0.3. The small M_w peak had an average M_w of 2×10^6 at a CWF of 0.5. The next large molecular weight fraction R3 also had two peaks. The M_w distribution of R3 was lower than fraction R2 as expected. The peak that contained the low molecular weight polymers had a M_w distribution of 1×10^6 at a CWF of 0.8. The high M_w peak had an average M_w of 3×10^7 at a CWF of 0.2. The medium and low molecular weight fractions (R4 to R7) were very similar in molecular weight size. Fraction R4 was a relative monodisperse fraction, 95 % of the total injected mass having M_w of about 2×10^5. The rest of fraction R4 has a M_w of about 1×10^7. Fraction R5, which was relatively monodisperse, had a M_w of 7×10^4 at a CWF of 0.90. This fraction had a small proportion of polymers (M_w of 1×10^6), which accounted for almost 10% of the injected mass. Hordein fraction R6 had a similar pattern, but had a lower M_w distribution of the R5 as it was expected, 90% of the injected mass has a M_w of 4.5×10^4 of the total injected mass; and the rest of the fraction had higher M_w polymers. Fraction R7 was also a relatively monodisperse fraction with 95 % of the total injected mass having a M_w of 4×10^4; less than 5 % of total injected mass comprised polymers of very high M_w.

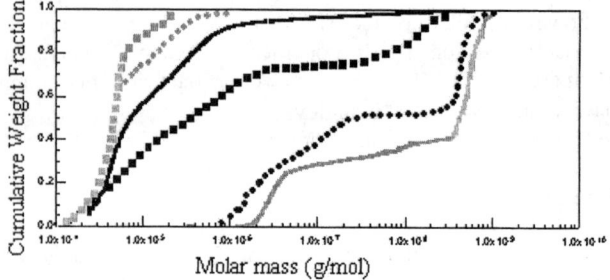

Figure 1. *The molecular weight distributions and cumulative weight fractions of wheat flour for each protein fraction applying the Debye theory, where: continues grey line R2 fraction; (●)R3 fraction, (■) R4 fraction, (continues black line) R5 fraction, (●) R6 fraction and (■) R7 fraction.*

The PSSG levels were also determined for the protein fractions obtained using the fractionation technique of Tsiami et al.[1].It was observed (Figure 3) that the medium and lower Mw fractions (R4-R6) of barley contained higher amount of protein-bound glutathione, than the analogous wheat fractions. Exceptionally high levels of PSSG were observed for fraction R2, and the wheat R2 fraction had a higher PSSG than the R2 fraction of barley. The GSH level (data not presented) of the barley flour is higher than that of wheat, which may lead to cleavage of SS bonds during dough mixing with barley flour. Further more the higher PSSG levels in the main barley protein fractions may prevent the protein form associating to form higher Mw polymers. This is consisten with the higher levels of PSSG found with low Mw barley fractions. Similarly, the levels of PSSC was higher in all the protein fractions of barley than the analogous fractions of wheat (data not presented). The PSSC levels in the medium and low Mw fractions of barley were double those of wheat, and this would also be likely to prevent the protein associating into higher Mw proteins.

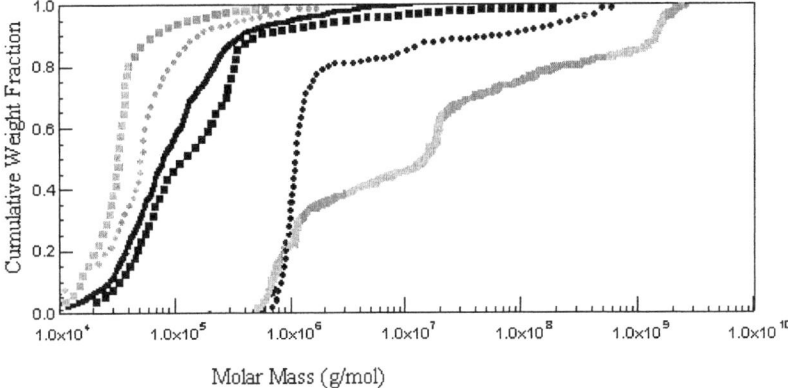

Figure 2. *The molecular weight distributions and cumulative weight fractions of barley flour for each protein fraction applying the Debye theory where: continues grey line R2 fraction; (●)R3 fraction, (■) R4 fraction, (continues black line) R5 fraction, (●) R6 fraction and (■) R7 fraction.*

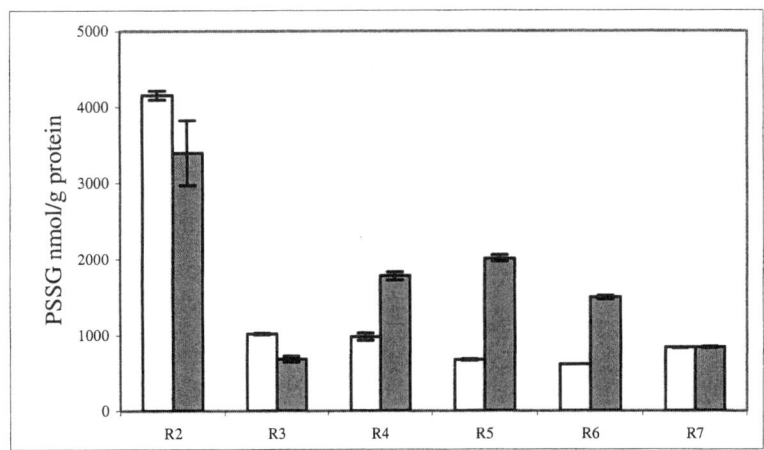

Figure 3. *Protein-bound glutathione contents of wheat (□)and barley(■) protein fractions.*

It is known that the size of the polymer determines its rheological properties according to the theory of synthetic polymers and computer simulations and it is known the relation of the glutenin size on its rheological properties[1-2]. Small deformation rheological tests were therefore performed to characterise the protein fractions of wheat and barley. The rheological parameters determined are presented in Table 1. With the decrease of the

Mw of the polymer (Fraction R2 towards R7 and fraction B2 towards B7), there is a decrease of the G' value and an increase of the tan δ value. The wheat glutenin fractions (R2-R5) showed that had predominantly elastic behaviour whereas the analogous barley protein fractions had predominantly viscous properties which we relate to the lack of higher Mw proteins in the barley.

Table 1. *Dynamic storage and loss modules values and tan d for gluten and hordein polymer fractions at frequency of 1Hz and temperature 20°C.*

Fractions (% dry)	Wheat gluten fractions G' (Pa)	G" (Pa)	tan δ		Barley hordein fractions G' (Pa)	G" (Pa)	tan δ
R2 (20%)	3720	541	0.14	B2	2980	1290	0.43
R3 (20%)	3552	530	0.15	B3	2670	4470	1.67
R4 (20%)	3236	1128	0.34	B4	380	756	1.99
R5 (30%)	504	400	0.79	B5	58	100	1.72
R6 (30%)	50	99	1.98	B6	134	150	1.11
R7 (30%)	18	48	2.7	B7	32	79	2.49

3 CONCLUSION

Analogous protein fractions obtained from barley had lower Mw distributions than those obtained from wheat, and the differences in Mw distribution were reflected in the rheological properties of the fractions. The levels of GSH and PSSG in wheat are much lower than the barley straight-run white flour. Also the low Mw polymer of barley contain higher amount of GSH and PSSG than the analogous fractions of wheat. This probably prevents the barley proteins from associating to give higher Mw polymers.

References

1. A.A. Tsiami, A. Bot, W.G.M. Agterof and R.D. Groot, *J. Cereal Sci.* 1997a, **26**, 15.
2. A.A. Tsiami, A. Bot and W.G.M. Agterof, *J. Cereal Sci.* 1997b, **26**, 279.
3. A.A. Tsiami. D. Every and J.D. Schofield ,In: Wheat Gluten, Eds. P.R. Shewry and A.S. Tatham, Royal Society of Chemistry, Cambridge, 2000, 244.
4. J.D. Schofield, In *Chemistry and Physics of Baking*, Eds J.M.V. Blanshard, P.J. Frazier, and T. Galliard, Royal Society of Chemistry, London, 1989. 14.
5. R.S. Bhatty, *Cereal Chem.* 1997, **74**, 693.
6. R.S. Bhatty, *Cereal Chem.* 1999, **76**, 588.
7. R.K. Newman, C.W. Newman, and H. Graham, *Cereal Foods World* 1989, **34**, 883.
8. A.A.M. Anderson, C.M. Courtin, J.A. Delcour, H. Fredriksson, J.D. Schofield, I. Trogh, A.A. Tsiami, and P. Åman, *Cereal Chemistry* (2003), accepted.
9. A. Graveland, M.H. Henderson, M. Pâques and P.A. Zandbelt, In: Wheat SAtructureBiochemistry and Functionality, Ed, J.D. Schofield, Royal Society of Chemistry, Cambridge, 1995, 90.

Acknowledgment

The research was financially supported by European Community (SOLFIBREAD EC contract QLRT-1999-30324).

Gluten Interactions with Exogenous and Endogenous Components

ASSOCIATION OF NON-PROTEIN COMPONENTS IN WHEAT GLUTEN WITH ITS QUALITY

L.Day[1], M.Augustin[1], I.L.Batey[2,3] and C.W.Wrigley[2,3]

[1] Food Science Australia, 671 Sneydes Road, Werribee, VIC 3030, Australia
[2] Value Added Wheat CRC, 1 Rivett Road, North Ryde, NSW 1670, Australia
[3] Food Science Australia, 11 Julius Ave., North Ryde, NSW 1670, Australia

1 INTRODUCTION

Wheat gluten has been produced commercially, primarily as a fortifier of low-protein wheat flour for baked and processed foods. Its use as a water-binding agent or protein enhancer/replacer in other food products, such as breakfast cereals, meats, cheese, snack foods and texturised meat analogues, is also increasing. Modifying gluten, either by acid or enzymatic hydrolysis, can improve its emulsifying, foaming and solubility properties, making it more suitable for using in functional food products such as sports drinks and milk-protein replacers.

Considerable research on the structure/function relationship of gluten proteins has been carried out, but much of this has been concentrated on amino-acid sequences and the polymeric nature of glutenin and their role in dough-based products. Our research is focused on investigating the interactions between the gluten proteins and non-protein components, particularly the lipids, to investigate their contributions to wheat-gluten quality. The lipid fraction in gluten has also been considered as a significant source of colour, odour and taste problems, which currently limits the use of these proteins as a food ingredient. Polyphenol oxidise (PPO), an enzyme known to cause wheat-flour darkening, was also investigated for its effects on undesirable gluten colour.

2 THE EFFECT OF LIPIDS ON GLUTEN COLOUR

The lipid content of each flour, gluten control (hand-washed gluten from each flour) and low-fat gluten (hand-washed gluten from chloroform-defatted flour[1]), is shown in Table 1. Flour 008 (typically used for commercial gluten production) has slightly higher lipid content than a breadmaking flour (cv. Sunco) or a cookie flour (cv. Bowie). A mill fraction (H/J) from the late stage of milling contained twice the amount of lipid in a mill fraction (X1-X2) from the first break. (The mill fractions were produced as part of commercial milling for gluten production.) These results reflect the contamination of the H/J stream with lipids from bran and germ, in contrast to the relatively pure endosperm flour X1-X2. The differences of lipid content in the gluten control were greater than in flour samples. The removal of lipids from flour using chloroform was shown to be effective since all the low-fat gluten produced from the defatted flour contained only traces of lipids.

Table 1: *The amount of lipids in flour as extracted by chloroform, gluten control (hand-washed gluten from each flour) and low fat gluten (hand-washed from chloroform-defatted flour). The amount of lipid in gluten was determined by acid hydrolysis.*

Sample	Total lipids (% w/w)		
	Flour	Gluten control	Low-fat gluten
Sunco	1.2	4.1	0.5
Bowie	1.4	5.1	0.6
Flour 008 (Fl 008)	1.5	6.2	0.3
Mill fraction X1-X2	1.4	7.7	0.5
Mill fraction H/J	2.8	11.3	0.5

The colours of the flour and the gluten samples (L, a and b values) were measured using a Minolta Colour Analyser. There were no differences in L (100=white, 0=black) values between the gluten control and the low-fat gluten from each flour, although flour H/J and flour 008 and their glutens (with high lipid content) had relatively lower L values than the others. However, removal of lipid reduced the b (yellowness (+), blueness (-)) values of the flour as well as the gluten, i.e. the low fat gluten had lower b values (Fig. 1) than the gluten controls. But the level of reduction did not correlate with the amount of lipid removal from each flour, suggesting that although flour lipids had a strong influence on the appearance of gluten, it may not be the only factor.

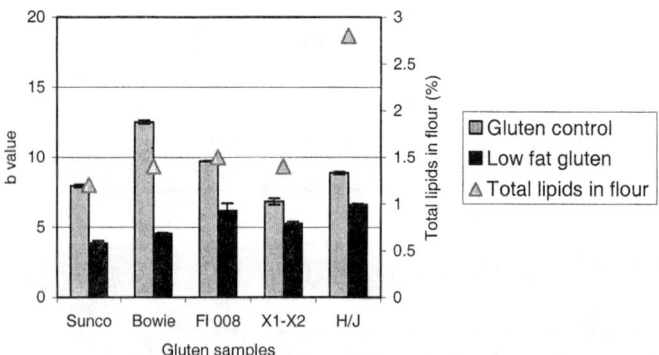

Figure 1: *The b values (yellowness (+), blueness (-)) of gluten control and low-fat gluten from wheat flours containing different amounts of lipid.*

3 THE EFFECT OF LIPID ON RHEOLOGICAL PROPERTIES OF GLUTEN

The rheological properties of gluten were measured using a Paar Physica Modular Compact Rheometer 300. The creep and recovery test was carried out by applying constant shear stress of 200 Pa until the sample reached a deflection (shear-angle) of 5° (equivalent to strain of 0.54), and the shear time (seconds to reach a certain deflection) was recorded. This is a measure of the flow of the gluten sample under the applied stress[2]. The shear time becomes shorter with more stretchable samples. After the sample reached the deflection angle of 5°, the shear stress was taken off to allow the sample to recover. The recovery was monitored for 100 s for all samples. The low-fat gluten produced from chloroform-defatted flour 008 had shorter shear time and better recovery than the gluten control, although there was no significant difference between the gluten control and the low-fat gluten produced from flours Sunco and Bowie (Fig. 2). It was evident that after the removal of lipid, gluten was more elastic, hence having better recovery than the gluten control, particularly for the poor-quality flours such Fl 008 and the mill fraction H/J. It is possible that lipids may interfere with non-covalent bonds between or within gluten-protein subunits, thereby reducing interactions between/within protein subunits. Removal of lipids may also increase the proportion of active sorption sites of protein for water, i.e., locations from where the lipids have been removed become available for water molecules to adsorb, thus facilitating non-covalent hydrogen bonds responsible for the elasticity of the gluten[3].

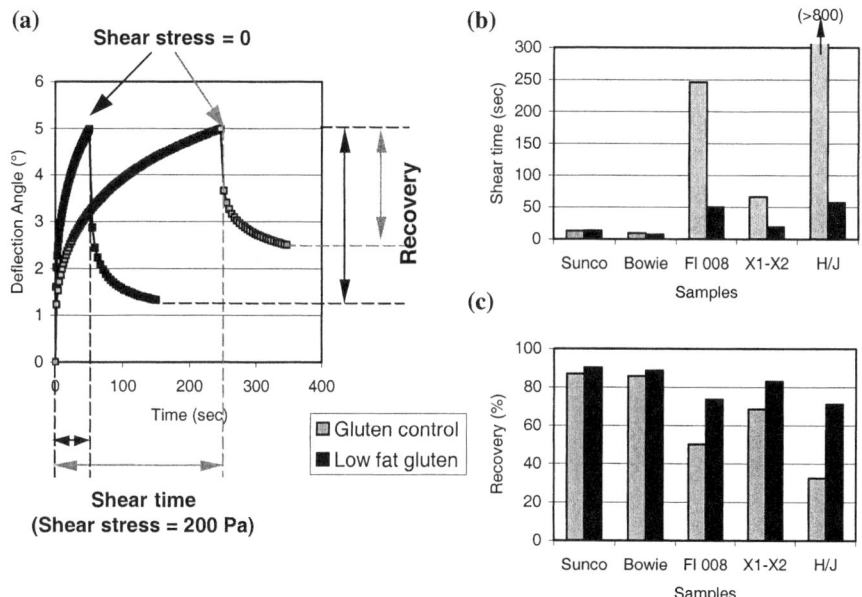

Figure 2: *The effect of lipids on gluten rheological properties as measured by the creep-and-recovery test (a) and the comparison of the shear times (b) and the recovery (c) of gluten before and after removal of lipids from flour.*

4 THE EFFECT OF POLYPHENOL OXIDASE IN FLOUR ON GLUTEN COLOUR

Polyphehol oxidase is known to induce darkening in wheat-based food products[4]. The enzyme activity of PPO during dough mixing and resting, part of gluten processing, may also contribute to gluten final colour. The levels of PPO in the wheat flours were assayed using a modified method of Bernier and Howes[5]. The preliminary results showed that the levels of PPO in flour had inverse correlations with the L values (Fig. 3). High levels of PPO were detected in flour H/J, which had the lowest L value. These results demonstrate that PPO activities in flour may also have a significant influence on the colour of gluten.

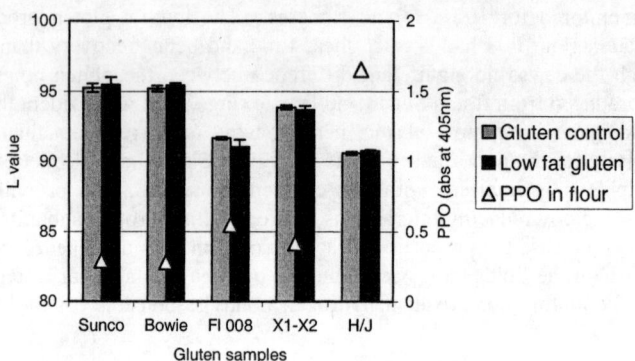

Figure 3: *The L values (100=white, 0=black) of gluten control and low fat gluten from wheat flours containing different amounts of lipid and the levels of PPO activities in the corresponding flours.*

References

1. F. MacRichie, *J. Cereal Sci.*, 1987, **6**, 259.
2. J.F. Steffe, in *Rheological Methods in Food Process Engineering*, Freeman Press, Michigan, USA, 1992, p.175.
3. P. Belton, *J. Cereal Sci.*, 1999, **29**, 103.
4. D.J. Mares and A.W. Campbell, *Aust. J. Agric. Res.*, 2001, **52**, 1297.
5. A.-M. Bernier and N.K. Howes, *J. Cereal Sci.*, 1994, **19**, 157.

ASCORBATE IMPROVER EFFECTS, DOUGH MIXING PROPERTIES AND BREAD QUALITY INTERDEPENDENCE ON FLOUR PROTEIN COMPOSITION

D. Every, W.B. Griffin, L. D. Simmons and K. H. Sutton

NZ Institute for Crop & Food Research Ltd, Private Bag 4704, Christchurch, New Zealand.

1 INTRODUCTION

It has been repeatedly demonstrated that dough mixing, rheological properties and bread quality are related to protein composition.[1,2] The best relationships deal with the amounts and molecular weight distribution of the polymeric glutenin proteins. Except for one early study by Pomeranz[3] in 1965, the effects of exogenous oxidizing agents on these relationships have not been considered. In this paper, a set of 27 wheat cultivars and breeding lines with a wide variation in the baking improvement response to ascorbate, dough mixing work input, and bread quality were analysed for protein composition and protein thiol distribution. An analysis was also made of the effect of AA on the protein composition and protein thiol distribution in dough samples at different stages of mixing and proofing.

2 METHODS AND RESULTS

2.1 Methods

SDS-extractable proteins and SDS-unextractable proteins were determined by size-exclusion HPLC.[4] From these determinations, the flour (F) contents of the following protein groups were made: total protein (P), unextractable polymeric (glutenin) protein (UPP), extractable polymeric (glutenin) protein (EPP), total polymeric protein (PP), and gliadin (Gli). Exposed cysteine content of protein was measured as ppm using bromobimane labelling of cysteine and SE-HPLC.[4] Water absorption, work input and baking properties were determined by a mechanical dough development system using 125 g of flour.[5] Each flour sample was baked with or without 100 ppm AA. Bake score values (BS) were derived from a formula that combined loaf volume and crumb texture determinations.[5] The bake score improver response to AA (BIRA) and the volume improver response to AA (VIRA) were measured by the difference in quality parameter between loaves baked with and without AA. Protein composition and protein thiol distribution were also measured in doughs mixed in a 1 kg MDD Morton mixer (double Z-blade) with 1 min slow mixing and 2.5 min fast mixing.

Table 1. Correlation coefficients for chemical composition and baking properties.

	Volume no AA	Volume AA	VIRA	BS no AA	BS AA	BIRA	Wh/kg	WA
%P in F	0.33	**0.60**	0.27	0.33	**0.53**	0.01	0.19	**0.64**
%PP in F	0.45	**0.50**	0.01	**0.52**	0.46	-0.23	0.45	**0.53**
%UPP in F	**0.53**	0.34	-0.26	**0.70**	0.32	**-0.53**	**0.75**	0.31
%EPP in F	0.14	**0.51**	0.40	0.06	0.48	0.28	-0.15	**0.62**
%Gli in F	0.23	**0.60**	0.40	0.17	**0.53**	0.19	-0.04	**0.66**
%UPP in PP	**0.50**	0.09	**-0.52**	**0.72**	0.08	**-0.74**	**0.88**	-0.09
%PP in P	0.30	-0.23	**-0.63**	0.46	-0.16	**-0.62**	**0.63**	-0.32

[a]Significant correlation coefficients equal or greater than 0.5 are in bold.

2.2 Relationships between protein composition and baking parameters

Table 1 and Fig. 1 show that BS of bread made without AA had moderately positive linear relationships to %UPP in F and %UPP in PP. The texture component of BS contributed to these relationships more than the volume component. Correlations to other protein composition parameters were weak to poor. In contrast, the BIRA had a moderately negative linear relationship to %UPP in PP and weak to poor correlations to other protein composition parameters. The interaction of these negative and positive relationships resulted in weak to poor linear relationships between the BS of bread made with AA and all the protein composition parameters; only %P in F and %Gli in F had weak positive linear relationships to BS (Table 1). However, Fig.1 indicates that the BS of bread made with AA may have a non-linear relationship to %UPP in PP, with optimum BS at intermediate values of %UPP in PP.

One of the aims of this work was to develop small scale methods that could identify wheat that would produce maximum bread quality using minimum WI in the MDD-short fermentation system. The relationships between WI and BS properties (Fig. 2) almost mirror those between %UPP in PP (Fig. 1). This was probably as a consequence of the moderately strong positive relationship between WI and %UPP in PP (Table 1, Fig. 3). Since commercial bread is generally made with AA as an oxidative improver, the relationships that are most useful are those between BS with AA, WI and %UPP in PP (Figs. 1-3). It seems clear from this work that obtaining high BS and low WI wheat can only be achieved with wheat of high BIRA and low %UPP in PP. In fact, it seems that the degree of BIRA is largely dependent on the %UPP in PP.

It might also be expected that BIRA has some relationship to the redox status of flour – the free cysteine and glutathione, the cysteine content of protein, and the redox enzymes such as dehydroascorbate reductase. Only a negative correlation of BIRA with cysteine content in SDS-extractable polymeric protein (EPP) was found (data not shown). The only other correlations with respect to redox status of flour were a positive correlation of BS (no AA) with cysteine in EPP, a positive correlation of WI with cysteine in EPP, and a negative correlation of WI with glutathione.

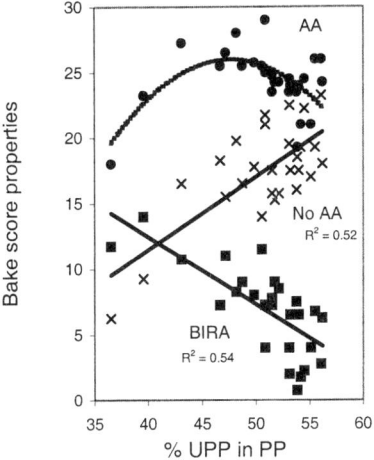
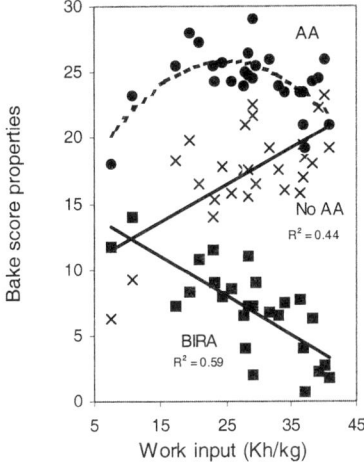

Figure 1. *Relationships between % UPP in PP and bake score properties*

Figure 2. *Relationships between work input and bake score*

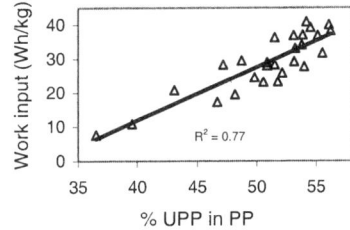

Figure 3. *Relationships between % unextractable protein polymer in total protein polymer (% UPP in PP) and work input.*

2.3 Effect of AA on the polymeric protein composition and thiol content in dough.

Figure 4 shows that AA has insignificant effect on %UPP in PP during mixing, but considerably increases the % UPP in PP during proofing. Figure 5 shows that the exposed cysteine content of EPP declines during mixing and proofing, and AA has an insignificant effect on this process. In contrast, the exposed cysteine content of UPP remains unchanged during mixing, either with or without AA, but was slightly reduced by AA during proofing. The lack of change of cysteine in UPP suggests that with this type of mixer (1 kg MDD Morton mixer) disulphide bonds (SS) in UPP are not mechanically ruptured as much as they appear to be with a 10 g MDD Mitchell type mixer[4]; the intensity of the mechanical forces of the Morton mixer appears to be somewhere between the Mitchell type mixer and the sheeting process.[4] The decline of cysteine in EPP probably resulted from SH/SS interchange reactions with oxidised glutathione, oxidation of thiols by oxygen and dehydroascorbate, and free radical reactions. The decline in %UPP in PP during mixing probably resulted from some mechanical ruption of SS bonds in UPP, cleavage of SS bonds in UPP by SH/SS interchange reaction with oxidised glutathione, and reduction of SS bonds in UPP by glutathione and other reducing substances in flour.

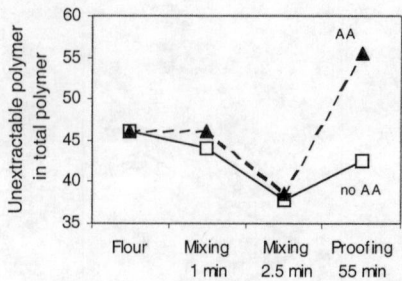 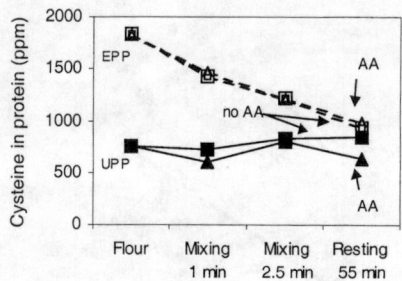

Figure 4. *Effect of AA on the molecular weight distribution of polymeric proteins in dough during mixing and proofing.*

Figure 5. *Effect of AA on the cysteine content of SDS-extractable polymeric protein (EPP) and SDS-unextractable polymeric protein (UPP) during mixing and proofing.*

3 CONCLUSION

The main aim if this work was to find methods that could identify wheat with maximum baking quality and minimum processing (WI, additives) requirements. Our results show that wheat with these properties invariably has a high BIRA, so reducing oxidative improver additives (AA) is not feasible. To achieve high BIRA and low WI, it seems that the wheat should have as low %UPP in PP as possible, without compromising baking quality. Analysing these parameters could help breeding programmes for best quality bread wheat.

The results show that AA increases the %UPP in PP and slightly reduces the exposed cysteine content of UPP during proofing of dough. This may partly explain the mechanism of the BIRA. That is, oxidised AA (dehydroascorbate) oxidises and crosslinks the exposed thiols of EPP to form more UPP, which is the protein most highly associated with bread quality. Protein disulphide isomerase may also have a role in BIRA by catalysing the oxidation of thiols by dehydroascorbate and isomerising the SS bonds into structural configurations of UPP that are optimal for functionality.

References

1. J. D. Schofield, in *Wheat Production, Properties and Quality*, eds. W. Bushuk and V.F. Rasper. Blackie Academic and Professional, Glasgow, 1994, ch. 7, p. 73.
2. M. Southern and F. MacRitchie, *Cereal Chem.*, 1999, **76**, 827.
3. Y. Pomeranz, *J. Sci. Fd. Agric.*, 1965, **16**, 586.
4. K.H. Sutton, N.G. Larsen, M.P. Morgenstern, M. Ross, L.D. Simmons and A.J. Wilson, *J. Cereal Sci.*, 2003, (in press).
5. W.H. Swallow and D.W. Baruch, *Wheat Research Institute Report No. WR 86/103*, New Zealand DSIR, Christchurch, 1986, p. 1.

UTILIZATION OF RAPID VISCO ANALYZER FOR ASSESSING THE EFFECT OF DIFFERENT LEVELS OF TRANSGLUTAMINASE ON GLUTEN QUALITY

A. Basman[1], H. Köksel[1] and P.K.W. Ng[2]

[1] Department of Food Engineering, Hacettepe University, 06532 Beytepe, Ankara, Turkey
[2] Department of Food Science & Human Nutrition, Michigan State University, East Lansing, MI, 48824-1224

1 INTRODUCTION

Transglutaminase (TG: protein-glutamine γ-glutamyl transferase, EC 2.3.2.13) is an enzyme capable of catalyzing the formation of nondisulfide covalent crosslinks between peptide-bound glutaminyl residues and ε-amino groups of lysine residues in proteins. Application of this crosslinking to wheat gluten proteins would be of particular interest because of their high glutamine content. The formation of protein polymers as the result of TG action has the potential to modify the rheological properties of gluten[1].

The Rapid Visco Analyzer (RVA) is an instrument most commonly used for assessing starch properties[2]. Various cereal researchers have also used it for different purposes such as assessing the quality potential of barley prior to malting[3], detecting sprout damage in wheat[4], assessing noodle quality[5], predicting the quality of different types of flat breads from tests on flour samples[6,7], and monitoring and optimizing process conditions during extrusion cooking[8]. More recently, preliminary methods to assess proteins such as wheat gluten have been developed using the RVA[9,10]. This method depends on the increase in suspension viscosity of gluten proteins when dispersed in dilute lactic acid. The viscosity values at 3 and 10 minutes and the breakdown viscosity measured using the RVA are indicators of flour quality.

The aim of the work presented here is to use the RVA for assessing the effect of different levels of transglutaminase on the gluten quality of two wheat cultivars, as the RVA results are expected to relate with some rheological properties such as extensibility test results.

2 MATERIALS AND METHODS

In this study, flours of wheat cultivars Roane (soft red winter wheat) and Sharpshooter (hard red spring wheat), and a bacterial TG (Ajinomoto, Paramus, NJ) were used. Samples were assessed in an RVA 4 (Newport Scientific, Warriewood, NSW, Australia) equipped with data analysis software (Thermocline, Warriewood, NSW, Australia). Control and TG-supplemented flour samples (15.0 g flour, 12% moisture basis) were dispersed in 22.5 mL distilled water and incubated in a water bath (45°C) for 30 min. 2.5 mL 1.0 M lactic acid was added just prior to testing by RVA, which brought the final lactic acid concentration of the slurry to 0.1 M. The RVA testing profile for wheat gluten acid method is given in Table 1. The viscosity values at 3.0 min (Visc3) and 10.0 min (Visc10) were determined from the RVA curves. The breakdown viscosity was calculated as the difference between viscosities at 3 min and 10 min. The tests were performed in duplicate and the average results are reported. Extensibility test results (Rmax and Extensibility) of the same samples at 45 and 90 minutes were determined with a Texture Analyzer[11] (Texture Technologies Corp., Scarsdale, NY/ Stable Microsystems, SMS, Godalming, Surrey, UK) equipped with a Kieffer extensibility rig and the results were reported in our previous study[12]. RVA data (Visc3, Visc10 and breakdown viscosity values) and extensibility test results were correlated. All statistical analyses were performed using MS Excel.

Table 1 *Rapid Visco Analyzer Wheat Gluten Acid Profile*

Time (min:sec)	Value (temperature or mixing speed)
00:00	25.0°C
00:00	1000 rpm
01:00	160 rpm
05:00	25.0°C
07:00	50.0°C
10:00	50.0°C (end of test)

3 RESULTS AND DISCUSSION

The effects of different levels of TG on viscosity values of Roane and Sharpshooter flours are presented in Figure 1. Increasing levels of TG resulted in an increase in Visc3, Visc10, and breakdown viscosity values of both cultivars except for the Visc3 and breakdown viscosity values at the 1% TG addition level for both cultivars.

The correlation coefficients between RVA data (Visc3, Visc10 and breakdown viscosity values) and rheological properties (extensibility test results at 45 and 90 minutes) for cv. Roane are generally higher than the correlation coefficients for cv. Sharpshooter (Table 2). Extensibility at 45 min, extensibility at 90 min and Rmax at 90 min values of Roane samples were highly correlated with RVA parameters with correlation coefficients higher than 0.85. Breakdown viscosity has been found to give higher correlation coefficients with extensibility test results of soft wheat flour as compared to those of hard wheat flour. These results indicate good potential of the "RVA wheat gluten acid method" to give an idea about some of the rheological properties of a particular wheat flour.

In the present study, TG addition to the flour samples increased the viscosity of the flour suspensions in lactic acid. This is probably due to the increase in the average molecular weight of gluten proteins as a result of TG-catalyzed crosslinking. Viscosity values obtained using RVA were quite high, especially at higher TG addition levels. This was more evident for cultivar Sharpshooter. In one of our previous studies, it was observed that TG utilization at a wide range of addition levels (0.1%-1.5%) resulted in improving effect at lower doses and deleterious effect at higher doses on rheological and bread properties[12]. At lower TG addition levels, the baking quality of a weak flour can be improved to a level of quality that might be achieved with a stronger flour. However, addition of TG to the weak and strong wheat flours studied influenced their rheological properties in different manners and to varying extents. The deleterious effect of TG addition was observed for the strong flour at a lower addition level than that for the weak flour due to excessive crosslinking. As the results of the present study were evaluated together with previous results[12], the increases in viscosity at higher TG addition levels may not directly indicate an improvement in the wheat quality. Therefore, an individualized optimum enzyme level should be estimated for different flour samples, depending on the flour quality, to obtain better bread properties.

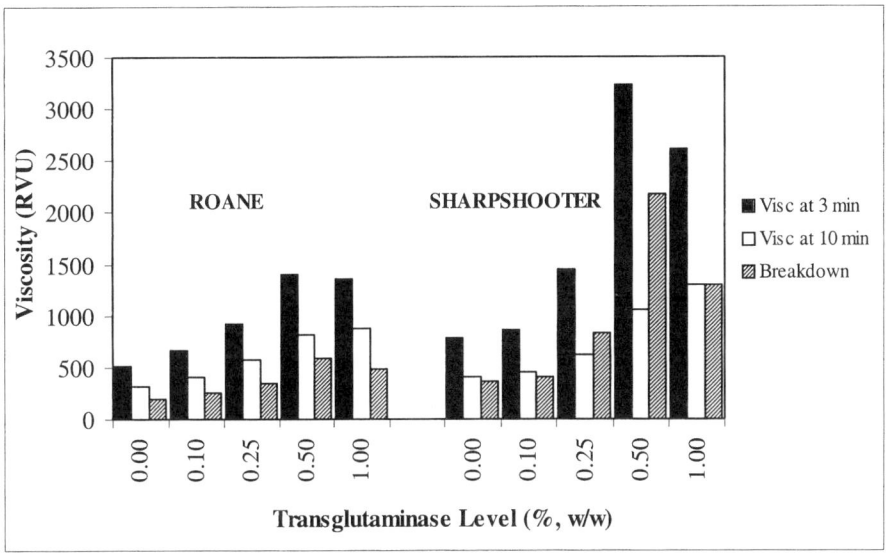

Figure 1 *Effects of Different Levels of Transglutaminase on Viscosity Values (RV units) of Roane and Sharpshooter Flours.*

Table 2 *Correlation coefficients (r value) between various RVA parameters and extensibility test results of wheat flours supplemented with various levels of Transglutaminase*

ROANE	Viscosity at 3 min	Viscosity at 10 min	Breakdown viscosity
Rmax at 45 min	0.75	0.82	0.63
Extensibility at 45 min	-0.94	-0.97	-0.87
Rmax at 90 min	0.95	0.98	0.89
Extensibility at 90 min	-0.97	-0.98	-0.92
SHARPSHOOTER	Viscosity at 3 min	Viscosity at 10 min	Breakdown viscosity
Rmax at 45 min	0.89	0.87	0.83
Extensibility at 45 min	-0.71	-0.78	-0.62
Rmax at 90 min	0.85	0.97	0.73
Extensibility at 90 min	-0.75	-0.84	-0.65

A comparison of the parameters obtained using RVA showed that Visc10 had a continuous increasing trend while Visc3 and breakdown viscosity values decreased at 1% TG addition level. Considering the deterioration in rheological and baking qualities at high TG addition levels, the Visc3 and the breakdown viscosity values seem to be useful indicators of quality. The overall results indicated that RVA shows great promise as a quick method for evaluation of effect of transglutaminase enzyme on gluten quality.

References

1 H. Köksel, D. Sivri, P.K.W. Ng and J. F. Steffe, *Cereal Chem*, 2001, **78**, 26.
2 C.W. Wrigley, R.I. Booth, M.L. Bason, C.E. Walker, *Cereal Foods World,* 1996, **41**, 6.
3 M. Glennie Holmes, *J. Inst. Brew.,* 1995, **101**, 29.
4 A. S. Ross, C. E. Walker, R. I. Booth, R. A. Orth and C. W. Wrigley, *Cereal Foods World,* 1987, **32**, 827.
5 J. F. Panozzo and K. M. McCormick, *J. Cereal Sci.,* 1993, **17**, 25.
6 J. Qarooni, E.S. Posner and J.G. Ponte Jr, *Lebensm.-Wiss. u.-Technol.,* 1993, **26**, 93.
7 J. Qarooni, E.S. Posner and J.G. Ponte Jr, *Lebensm.-Wiss. u.-Technol.,* 1993, **26**,100.
8 G. H. Ryu, P. E. Neumann and C. E. Walker, *J. Food Sci.*, 1993, **58**, 567.
9 N. E. Turner, R. W. Sleigh and M. L. Bason, in *Gluten '96, Proc. 6th International Gluten Workshop*, ed. C. W. Wrigley, Royal Aust. Chem. Inst., Melbourne, 1997, p. 489.
10 N. Turner and M. Bason, *RVA World,* 1997, **11**, 2.
11 J. Suchy, O.M. Lukow and M.E. Ingelin, *Cereal Chem.,* 2000, **77**, 39.
12 A. Basman, H. Köksel and P.K.W. Ng, *Eur. Food Res. Tech.,* 2002, **215**, 419.

GLIADINS AND POLYSACCHARIDES INTERACTION

N. Guerrieri[1], P. Cerletti[1], F. Secundo[2]

[1]Dipartimento Scienze Molecolari Agroalimentari and Centro per lo Studio della Celiachia, University of Milan, via Celoria 2, 20133 Milano, Italy.
[2]Istituto di Chimica del Riconoscimento Molecolare, CNR, via Mario Bianco 9, 20131 Milano, Italy

1 INTRODUCTION

We utilised the enzyme amyloglucosidase as a probe of polysaccharides accessibility, and we reported a reduction in accessibility of the enzyme in gliadins-starch and gliadins- beta dextrin models[1]. In the present study, we examined the gliadins-polysaccharides interaction in a model system with polysaccharides of different botanical origin: wheat, rice, potato, maize and commercial dextrin. We found a good model in the gliadins-dextrin sample, and we analysed its molecular properties with different techniques: HPLC-SE, surface hydrophobicity with the hydrophobic probe ANS (8-anilino1-naphthalene sulphonate), Circular Dichroism (CD), ATR-FT/IR, Elisa test for gliadins.

2 METHODS AND RESULTS

2.1 Model systems and amyloglucosidase accessibility

Gliadins from commercial sources or purified from Hereward wheat flour were a total protein extract (31.000 Da). In a closed system, we heated to 100°C for 10 min a mixture of 10% gliadins, 90% starch of different botanical origin (rice, potato, wheat and maize) and 60% water. The samples were freeze dried and lyophilised. We compared starch accessibility to the enzyme versus time of digestion with the amyloglucosidase assay[1], A sample with only starch was a reference (Figure 1). No difference occurred with rice, potato and maize starch. A decrease in the polysaccharide accessibility was noted for wheat starch and dextrin which indicates a possible interaction between gliadins and wheat starch or dextrin.

We chose the gliadins-dextrin system. Commercial dextrin has a MW of 15.000-3.000 Da and is smaller than starch. The model system was modified to increase the product. The ratio gliadins : dextrin was 1:1 with 60% water at 100°C temperature. The

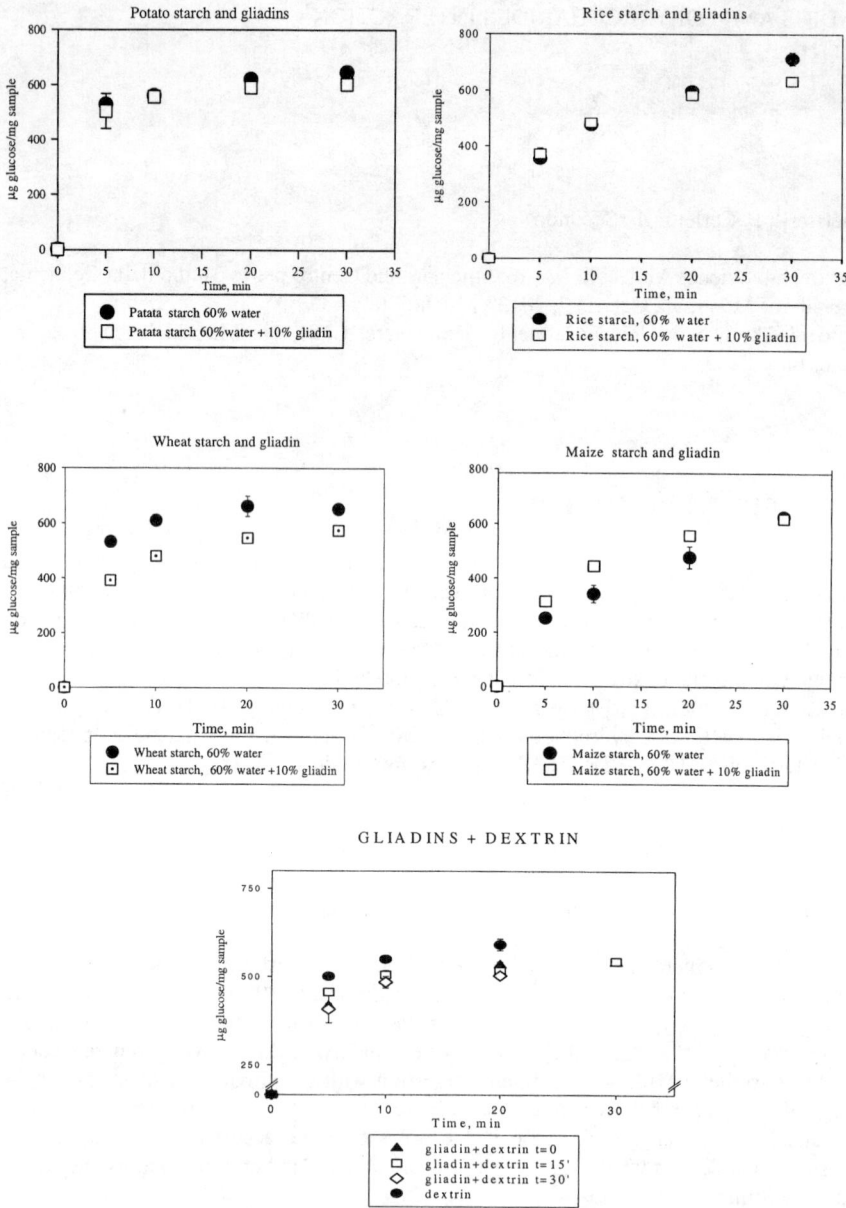

Figure 1 *Amyloglucosidase accessibility to different model systems*

reaction was followed in a close system versus time: 0, 15, 30, 60, 90, 120 minutes (Figure 1 gliadins + dextrin). A decrease in enzyme accessibility occurred without heat treatment (t=0), and after 15, 30 and 60 min the decrease was similar to wheat starch. After 90 min at 100°C, the sugar was instable and other components appeared (data not shown) evidenced by HPLC-SE.

2.2 Molecular properties of the gliadins-dextrin samples

For the first 30 min, a component with different molecular mass from the gliadins and dextrin was identified by HPLC-SE (Ultrogel 250), an acetone purification allowed to remove the free dextrin (data not shown).

To understand the nature of the interaction, we analyzed the surface hydrophobicity of the gliadins with the fluorescent probe ANS (8-anilino-1-naphthalene sulfonate). The surface hydrophobicity of the gliadins-dextrin samples were measured by a titration with the ANS[2]. The results displayed a decrease versus reaction time between gliadins and dextrin. The samples loose hydrophobicity and probably become more hydrophilic (Figure 2). A control of heat denaturation of the gliadins sample showed a different behaviour in regard to gliadins-dextrin system.

The CD spectra in the far UV showed a different behaviour of the gliadins-dextrin sample at t=0 respect to gliadins. An increase in alpha helix and a decrease in the beta region was evidenced. With the heat treatment, we have a partial lost of structure but the gliadins are not denatured (data not shown).

FT/IR spectra were recorded by FT/IR-600 spectrophotometer (Jasco). The samples were prepared as films in the attenuated total reflectance (ATR) cell and equilibrated at aw 0.06. The comparison of ATR-FT/IR spectra of dry (aw: 0.06) films of gliadins and gliadins + dextrin showed differences in the amide I and amide II region, which might be indicative of hydrogen bonds formation between gliadins and dextrin (data not shown).

The ELISA test (Redascreen gliadins) utilised to identify and quantify gliadins for celiac patients showed a reduction of the immunochemical response due to the presence of the dextrin. The dextrin alone gave a negative response (Figure 3). Part of the gliadins could interact (covalently or not) with the dextrin and modify the toxicity of the protein.

3 CONCLUSION

The amyloglucosidase assay provided information on accessibility of the polysaccharide in the model system employed. With gliadins-rice, -potato and -maize starch, no change was observed with polysaccharide. Instead in the model system gliadins-wheat starch and gliadins-dextrin, polysaccharide accessibility decreased. An interaction between the two macromolecules can occur. This interaction is quite stable in very dilute solutions in water, in buffers and changing the pH of the medium.

Studies on the gliadins-dextrin system showed formation of a new product identified by HPLC-SE, which decreased surface hydrophobicity and decreased the response by ELISA gliadins quantification. The CD spectra of the gliadins-dextrin sample showed a change in the alpha helix structure and a minor effect of heat treatment which did not unfolded the gliadins probably due to the presence of the polysaccharide. Further studies on the spectroscopic behaviour are in progress.

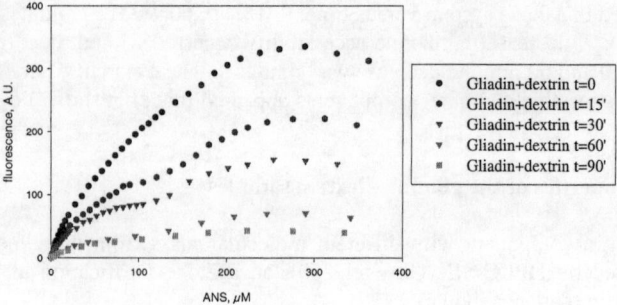

Figure 2 *Surface hydrophobicity of the gliadins-dextrin samples*

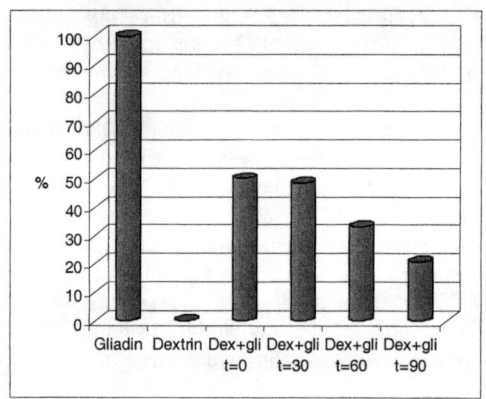

Figure 3 *ELISA (Ridascreen) response of the gliadins-dextrin samples, gliadins and dextrin.*

References

1 N. Guerrieri, L. Eynard, V. Lavelli, P. Cerletti, *Cereal Chem.*, 1997, **74**, 846.
2 E. Sironi, N. Guerrieri, P. Cerletti, *Cereal Chem.*, 2001, **78**, 476.

STRUCTURE-FUNCTION RELATIONSHIPS OF PHOSPHOLIPIDS IN BREADMAKING

G. Helmerich and P. Koehler

German Research Centre of Food Chemistry, Lichtenbergstraße 4, D-85748 Garching, Germany

1 INTRODUCTION

Phospholipids can act as emulsifiers which influence the baking performance of wheat doughs[1,2]. These polar lipids, e.g. lecithin, can be isolated on an industrial scale from plant sources (soybean, rapeseed, sunflower) and used as improvers for breadmaking, normally in concentrations up to 0.6 % (w/w). Lecithin improves the fermentation behaviour of yeasted doughs[3], the loaf volume of bread[4] and the structure of bread crumb[5]. Commercial lecithin contains several fractions that can be further subdivided into numerous individual components. The effects of the commercial product as well as fractions thereof are well known[6,7]. Since lecithin is a product from natural sources, the isolation of homogeneous components is extremely difficult. Isolation of individual phospholipids, the preparation of mixtures and the comparison of their improver effects with commercial lecithin was thought to be a promising way to get insight into the effect of lecithin.

2 METHOD AND RESULTS

2.1 Materials

Crude and deoiled lecithins from sunflower, rapeseed and soybean were provided by Degussa Texturant Systems (Hamburg, Germany). Phosphatidic acid (PA), phosphatidylcholine (PC), phosphatidylethanolamine (PE), phosphatidylinositol (PI), phosphatidylserine (PS), and the corresponding lyso-compounds (LPA, LPC, LPE), were purchased from Sigma-Aldrich Chemie GmbH (Deisenhofen, Germany). The German winter wheat variety Flair from the 2000 harvest was used for micro-scale baking tests.

2.2 Analytical methods

Phospholipid classes in lecithins were determined by ^{31}P NMR spectroscopy. 25 to 75 mg of lecithin was dissolved in deuterochloroform/methanol 2+1 (v+v; 2 mL), 1 mL of an aqueous caesium ethylenediamine tetraacetic acid solution (0.2 mol/L; pH 10.5) was added and the sample was shaken vigorously. After phase separation, an aliquot of the lower deuterochloroform phase containing the phospholipid classes free from disturbing polyvalent inorganic cations was mixed with 2 µL of triethyl phosphate as a calibration standard. NMR spectra were recorded on a Bruker AC 250 spectrometer. Quantification was performed using commercially available phospholipid classes as external standards. All phospholipids and lysophospholipids that had been applied were resolved (Figure 1). The phospholipid content of the lecithin samples are summarised in Table 1.

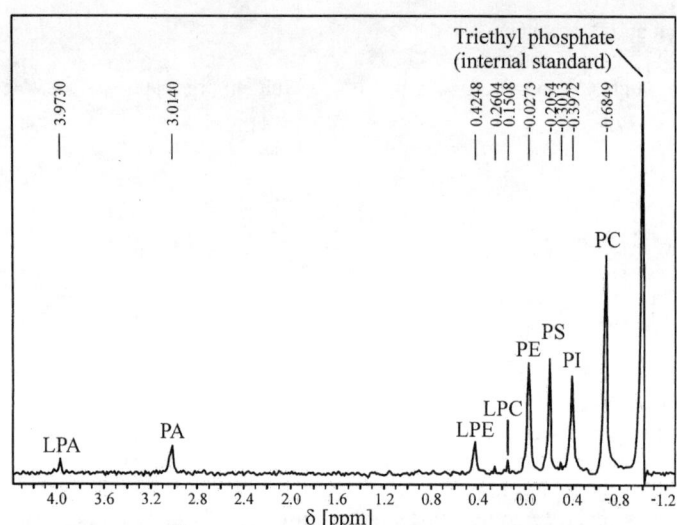

Figure 1 ^{31}P NMR spectrum of crude rapeseed lecithin.

Table 1 *Phospholipid content [%, w/w] of lecithin samples as determined by ^{31}P NMR.*

Lecithin sample	PC	PE	PI	PA	PS	LPE	Total
Soybean, crude	23.0	12.3	11.4	6.6	1.8	4.6	59.7
Soybean, deoiled	23.2	14.1	13.4	6.7	1.9	4.9	64.2
Rapeseed, crude	17.0	8.4	10.6	5.6	n.d.	5.7	47.3
Rapeseed, deoiled	24.5	12.2	12.5	7.2	2.9	11.8	71.1
Sunflower, crude	18.1	7.6	12.6	6.6	n.d.	n.d.	44.9
Sunflower, deoiled	30.2	11.3	20.9	7.8	n.d.	n.d.	70.2

2.3 Isolation of phospholipid classes

PC, PE, PI, and PA were isolated in high purity by preparative thin layer chromatography (TLC). 250 µl (50 mg) of deoiled soybean and sunflower lecithin solution in chloroform were applied onto a silica 60 TLC plate with concentrating zone. After a chloroform run, the separation of phospholipid classes was performed by using chloroform/methanol/water 65+45+4 (v+v+v) as mobile phase[8]. The separated phospholipid bands were then sharpened by another chloroform run. After scraping out the zones containing the phospholipid classes, they were extracted from the silica gel by using tetrahydrofuran and methanol. The purity of each phospholipid class was verified by analytical TLC[9]. Phospholipid classes were recovered in yields between 41 and 78 % (Table 2). To obtain reasonable quantities (approx. 750 mg) of each phospholipid class for micro-scale baking test, 60 to 250 TLC plates had to be run (Table 2). Phosphatidylcholine was also isolated in a 1 - 2 g scale from egg lecithin by column chromatography on aluminium oxide[10] by using methanol/chloroform 1/1 (v/v) as eluent. The yield was 90 % with a purity of 98% determined by ^{31}P NMR spectroscopy.

Table 2 *Isolation of phospholipid classes by preparative thin layer chromatography.*

Phospholipid class	Rf value	Amount [mg/50 mg lecithin]	Yield [%]
PC	0.43 to 0.50	11	78
PI	0.54 to 0.59	7	41
PA	0.64 to 0.69	3	75
PE	0.75 to 0.81	5	42

2.4 Micro-scale baking test

The micro-scale baking test with 10 g of flour was carried out according to Koehler and Grosch[11]. For the commercial lecithins, the optimum concentration was 0.8 to 1 % based on flour. Within this range, the loaf volume increased by 22 to 37 % depending on the type of lecithin (Table 3). Baking tests with isolated phospholipid classes showed for the first time that PI is the only phospholipid class which is active in very low concentrations (0.02 to 0.1 %). None of the other phospholipid classes were active in this concentration range. Other phospholipids were only active in concentrations above 0.1 %. In higher concentrations, most of them showed a negative effect on loaf volume (Table 4). The effects of individual phospholipid classes on baking quality were compared to those of phospholipid mixtures with a defined composition. A mixture of PI/PE/PA in a 2/1/1 (w/w/w) ratio was most effective in breadmaking with an increase in loaf volume of up to 55 % (Table 5).

Table 3 *Micro-scale baking test with commercial lecithins. Increase of the loaf volume compared to bread without additive.*

Commercial lecithin	Concentration (% w/w, based on flour weight)				
	0.2	0.4	0.6	0.8	1.0
Soybean, crude	+ 7	+ 11	+ 25	+ 33	+ 33
Soybean, deoiled	+ 9	+ 26	+ 34	+ 34	+ 37
Sunflower, crude	+ 2	+ 7	+ 12	+ 22	+ 22
Sunflower, deoiled	+ 5	+ 12	+ 21	+ 25	+ 24
Rapeseed, crude	+ 7	+ 16	+ 26	+ 24	+ 23
Rapeseed, deoiled	+ 7	+ 20	+ 34	+ 33	+ 35

3 CONCLUSION

^{31}P NMR was found to be effective for the quantitative analysis of phospholipids in lecithins. The activity of the lecithin samples was determined by micro-scale baking tests. Among the lecithin samples, it was shown that soybean and deoiled rapeseed lecithin have the highest baking activity. Pure phospholipid classes were isolated by TLC in a semi-preparative scale (1 – 2 g) from crude lecithins and have been used as reference compounds in baking and rheology. Baking tests with these phospholipid classes showed for the first time that PI is the only phospholipid class which is active in very low concentrations (0.02 to 0.1 %). The effect of individual phospholipid classes were compared to those of mixtures to identify those phospholipid mixtures with a high baking activity. A mixture of PI/PE/PA in a 2/1/1 (w/w/w) ratio was found to be most effective in breadmaking with increase in loaf volume of up to 55 %.

Table 4 *Micro-scale baking test with isolated phospholipid classes from deoiled sunflower lecithin. Increase of the loaf volume compared to bread without additive.*

Phospholipid classes	Concentration (% w/w, based on flour weight)						
	0.02	0.04	0.06	0.08	0.1	0.2	0.4
Phosphatidylcholine	± 0	- 2	+ 1	±0	- 2	± 0	+ 4
Phosphatidylethanolamine	+ 1	± 0	- 1	- 1	+ 2	+ 5	- 3
Phosphatidylinositol	+ 12	+ 13	+ 8	+ 4	- 7.5	- 9	- 10
Phosphatidic acid	+ 1	+ 5	+ 7	- 2	- 7.5	- 7	- 2

Table 5 *Micro-scale baking test with mixtures of isolated phospholipid classes from deoiled sunflower lecithin. Increase of the loaf volume compared to bread without additive.*

Mixture	Ratio (w/w)	Concentration (% w/w, based on flour weight)				
		0.2	0.4	0.6	0.8	1.0
PC/PE/PI/PA	2/1/1/0.5	+ 7	+ 14	+ 22	+ 23	+ 26
PC/PE/PI/PA	2/1/2/0.5	± 0	+ 5	+ 6	+ 6	+ 9
PE/PI/PA	1/2/1	+ 22	+ 26	+ 33	+ 34	+ 55

References

1. Y. Pomeranz, M. Shogren and K.F. Finney, *Food Technol.*, 1968, **22**, 324.
2. Y. Pomeranz, M.J. Carvajal, M.D. Shogren, R.C. Hoseney and K.F. Finney, *Cereal Chem.*, 1970, **47**, 429.
3. L. Wassermann, *Fette Seifen Anstrichmittel*, 1983, **85**, 117.
4. A. Radoev and M. Kalichkov, *Khranitelna Prom.*, 1963, **12**, 13.
5. W. Adams, A. Funke, H. Gölitz and G. Schuster, *Getreide Mehl Brot*, 1991, **45**, 355.
6. E. Mettler, W. Seibel, J.-M. Brümmer and K. Pfeilsticker K, *Getreide Mehl Brot*, 1992, **46**, 43.
7. H.D. Jodlbauer, W. Freund and J. Senneka, *Getreide Mehl Brot*, **1992**, 46, 174.
8. R. Lange and H.-J. Fiebig H-J, *Fett/Lipid*, 1999, **101**,77.
9. G. Helmerich and P. Koehler, *Getreide Mehl Brot*, 2002, **56**, 195.
10. D.N. Rhodes and C.H. Lea, *Biochem. J.*, 1955, **60**, 353.
11. P. Koehler and W. Grosch, *J. Agric. Food Chem.*, 1999, **47**, 1863.

Acknowledgement

This research project was supported by the FEI (Forschungskreis der Ernährungsindustrie e.V., Bonn), the AiF and the Ministry of Economics and Technology. Project No. 12637 N.

EFFECT OF ASCORBIC ACID IN DOUGH: REACTION OF OXIDISED GLUTATHIONE WITH REACTIVE THIOL GROUPS OF WHEAT GLUTELIN

P. Koehler

[1]German Research Centre of Food Chemistry, Lichtenbergstraße 4, D-85748 Garching, Germany

1 INTRODUCTION

Ascorbic acid (AA) is widely used as an improving agent for the production of various baked goods. Under the conditions normally used for dough mixing, this ingredient can have considerable effects on both dough and bread properties. Except for the last step of the reaction, the mechanism of action of ascorbic acid has been proven. In the glutenin proteins, only the amount and the positions of the thiol groups blocked by oxidised glutathione (GSSG) generated after the addition of AA have not been established. In the present study, free SH-groups of the glutenins have been quantified, and their reaction with GSSG has been established.

2 METHOD AND RESULTS

2.1 Wheat samples

Flours of the wheat cultivars Glockner (1996 harvest), Flair (1996 harvest), Contra (1998 harvest), Rektor (1991 harvest), Kanzler (1995 harvest), Astron (1998 harvest), Soissons (1998 harvest), Apollo (1987 harvest), Monopol (1990 harvest) and the wheat class CWRS (1990 harvest) were used.

2.2 Determination of the concentration of free thiol groups in glutenin

The concentration of free SH groups is important, because they are able to react with GSSG generated by the addition of AA and can therefore inhibit depolymerisation of glutenin. Flour was reacted with Ellman's reagent (DTNB), excess reagent was removed by dialysis and the residue was freeze-dried and extracted by a micro Osborne fractionation yielding an albumin/globulin, a gliadin, and a glutenin fraction. The latter was reduced to release reduced Ellman's reagent (NTB), which was quantified by high-performance liquid chromatography (HPLC) and UV detection at 327 nm without disturbance by co-eluted peptide or protein material. The concentration of NTB was proportional to the concentration of free thiol groups in glutenin.

The concentration of free SH groups in the glutenin fraction was in the range of 0.22 to 0.33 µmol/g of labelled flour corresponding to 5.6 to 8.2 µmol/g of protein (Table 1). These values are lower than the total SH content of flour (1 to 1.5 µmol/g of flour)[2]. They show that approximately 30 % of the free SH groups of flour can be assigned to the glutenins. However, the data are in the same range as the SH-concentrations of dough (2 to 4 µmol/g of protein) determined by Andrews et al[3].

Table 1 *Concentration of SH groups in the glutenins isolated from flours of ten different wheat cultivars*[a]

	APO	AST	CON	CWR	FLA	GLO	KAN	MON	REK	SOI
µmol/g of protein	7.3± 0.7	6.0± 0.5	8.2± 0.6	6.3± 0.5	6.0± 0.3	6.2± 0.5	6.5± 0.6	5.6± 0.5	5.7± 0.4	7.5± 0.8
µmol/g of flour	0.26± 0.026	0.24± 0.021	0.22± 0.017	0.33± 0.026	0.23± 0.012	0.31± 0.027	0.33± 0.030	0.28± 0.024	0.26± 0.019	0.28± 0.029

[a] mean values of duplicate determinations ± standard deviation. Wheat cultivars: APO: Apollo; AST: Astron; CON: Contra; CWR: CWRS; FLA: Flair; GLO: Glockner; KAN: Kanzler; MON: Monopol; REK: Rektor; SOI: Soissons.

2.3 Concentration of free SH groups in glutenin as affected by ascorbic acid

Dough samples containing 0, 20, 50, 75, 100, 125, 150, and 200 mg AA/kg of flour were prepared, immediately frozen, and lyophilised. Free SH groups in the glutenin fraction were then determined as described above. The results are summarised in Table 2. Increasing AA concentrations up to 100 mg/kg of flour caused an increase in the concentration of free SH groups by approximately 35 %. Higher concentrations of AA (200 mg/kg) led to a slight decrease. This result is surprising and is not in accordance with both hypotheses about the mechanism of AA. The hypothesis of Grosch[4-6] postulates a reduction of the amount of free SH groups on reaction of proteins with GSSG and CSSC. The same is true for the hypothesis of Every[7,8] in which free SH groups of gluten proteins are directly oxidised to disulphides by dehydroascorbic acid. At the moment, no explanation can be given for this effect of AA on the glutenins.

Table 2 *Wheat class CWRS. Concentration of SH-groups in the glutenins isolated from dough as affected by different concentrations of ascorbic acid*[a]

	Concentration of ascorbic acid [mg/kg of flour]				
	0	20	50	100	200
µmol/g of protein	5.2±0.5	5.9±0.4	5.9±0.4	7.1±0.4	6.5±0.5
µmol/g of flour	0.27±0.026	0.31±0.020	0.31±0.022	0.37±0.019	0.34±0.028

[a] mean values of duplicate determinations ± standard deviation.

2.4 Location of free SH groups in glutenin reacting with GSSG

Flour of the wheat class CWRS was mixed on addition of AA (125 mg/kg of flour). A small amount of [^{35}S]-labelled glutathione (GSH) was added as a tracer (0.5 % of endogenous GSH). The dough was washed with distilled water and residual gluten was extracted with 70 % (v/v) aqueous ethanol (pH 5.5). The residue (glutenin) was partially hydrolysed with thermolysin, the peptide mixture was pre-separated by gel permeation chromatography on Sephadex G25 and separated by HPLC. Radioactively labelled peptides were identified by scintillation analysis, isolated by HPLC, and sequenced by automated Edman degradation. The resulting amino acid sequences were assigned to known sequences of gluten protein components.

Altogether, 5 groups of peptides linked with GS* were found (Table3). The length of the sequences ranged from 4 up to 21 amino acid residues. All isolated peptides were derived from LMW subunits of glutenin and contained the cysteine residues C^{b*} and C^x (nomenclature of Koehler et al[9]). The peptides P2-10-2a, P2-10-2b and P2-8-6 contained the cysteine residue C^{b*} as part of the characteristic tripeptide motif PCS which is present in the s-type of LMW subunits of glutenin. Cystine peptides containing this tripeptide as one part of the molecule have already been identified in thermolytic glutenin digests[9,10]. Peptides of the same type containing GS have been identified by Huettner and Wieser[11] after addition of [^{35}S]-GSH during mixing of wheat dough without AA. The same is true for the cysteine residue C^x, which was identified in the two peptides P4-5-3a and P4-2-6.

Table 3 *Wheat class CWRS. Amino acid sequences^a and origin of glutenin peptides linked to GS* isolated from dough mixed with 125 mg ascorbic acid/kg of flour.*

Peptide	Position of the amino acid in the peptide 1 . . . 5 . . . . 10 . . . . 15 . . . . 20 25	Cys^b
P2-10-2a	I Q Q Q P Q P F P Q Q P P C S Q Q Q Q P P	C^{b*}
P2-10-2b	V Q Q Q P Q P F P Q Q P P C S Q Q Q Q P P	C^{b*}
P2-8-6	F S Q Q Q P C S Q Q Q Q Q P	C^{b*}
P4-5-3a	L G Q C V	C^x
P4-2-6	L G Q C	C^x

^{a} gaps in the amino acid sequence were assigned to cysteine residues (C).
^{b} designation of cysteine residues of LMW subunits according to Koehler et al[9].

The cysteine residues C^{b*} and C^x of LMW subunits of glutenin have been postulated to form intermolecular disulphide bonds and are thought to be more reactive than other cysteine residues of gluten proteins. Evidence for this hypothesis has been provided by the addition of GSH during mixing[11], which reacted almost exclusively with these two cysteine residues. The results of the present study lead to the assumption that these cysteine residues are at least partly present in flour in the thiol form and are quickly converted to disulphides by thiol/disulphide interchange reactions and oxidation during mixing, corresponding to an increase of the molecular mass of the glutenin polymer. In case of the addition of AA during mixing, GSSG formed of GSH is a potential reaction participant in the thiol/disulphide interchange reaction with the free cysteine residues C^{b*} and C^x of LMW subunits of glutenin. This reaction seems to compete with the exchange reactions caused by other disulphides present in the dough system, e.g. protein disulphide groups. Another possibility is the reaction of GSSG with high-molecular weight aggregates containing LMW subunits of glutenin that still have one free thiol group.

3 CONCLUSION

The results of this study give evidence that AA acts as an improver according to the hypothesis of Grosch[4-6] because mixed disulphides formed by reaction of GSSG and gluten proteins with free thiol groups were identified. This reaction was only possible after conversion of endogenous GSH to GSSG by DHA formed after oxidation of added AA. Direct oxidation of protein thiols by DHA, as proposed by Every et al[7,8], would not have yielded peptides containing radioactively labelled GS. The peptides contained exclusively the cysteine residues C^{b*} and C^x present in LMW subunits of glutenin, which are proposed to form intermolecular disulphide bonds. These cysteine residues seem to be at least partly present in the thiol form in flour. During dough mixing they are converted to protein-protein disulphides, however, a small portion remains in the reduced state and is able to react with GSSG to glutathione-protein mixed disulphides.

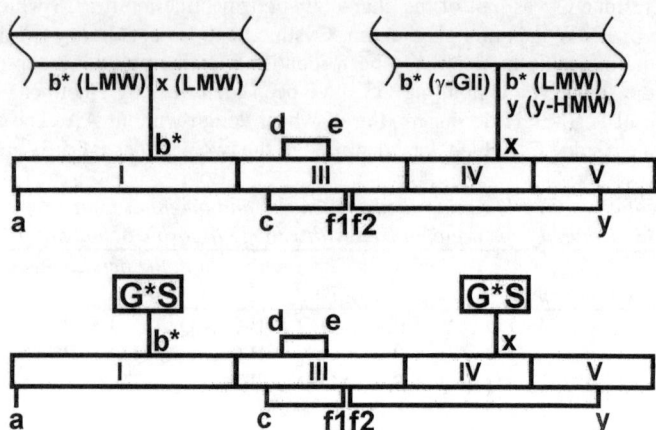

Figure 1 Wheat class CWRS. Top: LMW-subunit of glutenin as part of the glutenin polymer. Bottom: Mixed disulphide of $[^{35}S]$ glutathione (G*S) and LMW subunit of glutenin formed after reaction of GSSG with free thiol groups.

References

1. H. Wieser, S. Antes and W. Seilmeier W, *Cereal Chem.*, 1998, **75**, 644.
2. S. Antes and H. Wieser 'Quantitative determination and localisation of thiol groups in wheat flour' in *Wheat Gluten*, eds. P.R. Shewry and A.S. Tatham, The Royal Society of Chemistry, Cambridge, U. K., 2000; pp. 211-214.
3. D.C. Andrews, R.A. Caldwell and K.J. Quail, *Cereal Chem.*, 1995, **72**, 326.
4. W. Grosch and H. Wieser, *J. Cereal Sci.*, 1999, 29, 1.
5. R.G. Sarwin, G. Laskawy and W. Grosch, *Cereal Chem.*, 1993, **70**, 553.
6. B. Hahn and W. Grosch, *J. Cereal Sci.*, 1998, **27**, 117.
7. D. Every, L. Simmons, M. Ross, P.E. Wilson, J.D. Schofield, S.S.J. Bollecker and B. Dobraszczyk, 'Mechanism of the ascorbic acid improver effect on baking' in *Wheat Gluten*, eds. P.R. Shewry and A.S. Tatham, The Royal Society of Chemistry, Cambridge, U. K., 2000; pp. 277-282.
8. D. Every, L. Simmons, K.H. Sutton and M. Ross, *J. Cereal Sci.*, 1999, **30**, 147.
9. P. Koehler, H.-D. Belitz and H. Wieser, *Z. Lebensm. Unters. Forsch.*, 1993, **196**, 339.
10. B. Keck, P. Koehler and H. Wieser, *Z. Lebensm. Unters. Forsch.* 1995, **200**, 432.
11. S. Huettner and H. Wieser, *Eur. Food Res. Technol.*, 2001, **213**, 460.

EFFECT OF PENTOSANS ON GLUTEN FORMATION AND PROPERTIES

M. Wang [1,2], T. van Vliet [1,3] and R.J. Hamer [1,3]

[1] Department of Agrotechnology and Food Sciences, Centre for Protein Technology, Wageningen, the Netherlands; [2] Wuhan Polytechnic University, Wuhan, P.R. China; [3] Wageningen Centre for Food Sciences, Wageningen, the Netherlands

1 INTRODUCTION

Wheat flour contains an important non-starch carbohydrate fraction, called pentosans. This fraction originates from different botanical parts of the kernel and can be distinguished into a soluble fraction WEP (Water extractable pentosans) and WUS (Water unextractable solids) that is not water extractable. Pentosans have been studied for about 50 years since they were discovered. Nowadays, wheat pentosans are becoming more and more important in view of the increased use of 'dark' (high pentosans containing) flour for the production of wheat bakery and gluten products. Cereal scientists have markedly increased their knowledge of the structure, properties and functionality of pentosans. Several theories exist to explain the effects of pentosans (and pentosan modifying enzymes), but some of them are disputed. Also, the effects of pentosans on gluten formation and properties are rarely studied.

2 METHOD AND RESULTS

2.1. Effect of pentosans on gluten yield

Gluten was extracted from flour-water-salt dough using a modified Glutomatic 2200 system (Perten, modified by TNO) as described previously[1]. This allows us to do experiments on a 4 gram scale. We determined the gluten yield using Soissons flour, as affected by addition of different pentosans, ferulic acid (FA) and xylanase (Table 1). Addition of water unextractable solids (WUS) or water extractable pentosans (WEP) or WEP treated by xylanase (WEP_x) during gluten separation decreased gluten yield. In contrast, addition of free FA together with WUS or WEP or WEP_x could correct almost completely for the lower gluten yield in the presence of WEP or WEP_x, but not in the presence of WUS. Addition of xylanase significantly increased gluten yield in all cases.

2.2. Effect of pentosans on gluten properties

We also studied possible effects of pentosans on the rheological properties of the resulting gluten samples. For this purpose we used a Kieffer extensibility rig as described previously

[1] (Table 1). Addition of WUS or WEP typically produced a gluten with a higher R_{max} and a smaller E at R_{max}. In contrast, addition of xylanase resulted in a larger E at R_{max}. Adding FA together with WUS or WEP or WEP_x all gave a gluten with a comparable R_{max} and a larger E at R_{max} than if WUS or WEP or WEP_x were added alone.

2.3. Voluminosity and aggregation properties of glutenin macropolymer (GMP)

In order to find an explanation for the different gluten properties observed, we further studied characteristics of its aggregated glutenin polymer fraction (GMP). The average voluminosity and aggregation properties of GMP isolated from the resulting gluten samples were measured by viscometry of dispersions of GMP in SDS^2 (Table 1). Addition of WUS or WEP led to an increase in [η] and a decrease in K' compared to the control, indicating a larger specific volume of GMP particles and a lower tendency to aggregate. On the other hand, addition of xylanase or of FA together with WUS or WEP resulted in a decrease in [η] and an increase in K' compared to their control, respectively, pointing at a smaller specific volume of the GMP particles and a larger tendency to aggregate. No large differences in [η] were observed among GMP_{WEPx}, $GMP_{WEPx+FA}$ and the control. However, K' shows some interesting differences, particularly on addition of FA.

Table 1 *The yield and properties of gluten and GMP for one wheat cultivar*

Sample Name	Gluten Yield (%, dm)	R_{max} (N)	E at R_{max} (mm)	[η] (dL/g)	K'
Soissons Control	10.6 ± 0.20	0.29 ± 0.02	97 ± 5	3.22 ± 0.01	1.02 ± 0.06
2%WUS	8.8 ± 0.16 [a]	0.42 ± 0.03 [a]	75 ± 4 [a]	3.37 ± 0.06	0.7 ± 0.13
2%WUS+30ppmFA	9.9 ± 0.17 [a]	0.41 ± 0.03 [a]	87 ± 5	3.32 ± 0.01	0.78 ± 0.09
100ppm xylanase	11.3 ± 0.13 [a]	0.27 ± 0.02	116 ± 5 [a]	3.09 ± 0.06	1.39 ± 0.05
0.3%WEP	9.7 ± 0.10 [a]	0.39 ± 0.02 [a]	77 ± 4 [a]	3.33 ± 0.01	0.75 ± 0.09
0.3%WEP+30ppmFA	10.3 ± 0.20	0.38 ± 0.03 [a]	91 ± 4	3.29 ± 0.00	1.03 ± 0.02
0.3%WEP_x	9.8 ± 0.18 [a]	0.30 ± 0.03	88 ± 4	3.22 ± 0.04	0.81 ± 0.03
0.3%WEP_x+30ppmFA	10.5 ± 0.15	0.31 ± 0.03	98 ± 5	3.19 ± 0.04	1.22 ± 0.09

Data are mean ± S.D.
[a] means significant difference from control at $p < 0.05$.

2.4. Effects of WUS and xylanase on different wheat cultivars

The effects of WUS and xylanase on three wheat cultivars (Scipion, Soissons, and Amazon) were also studied in terms of yield and properties of gluten and GMP. The volume surface diameter ($D_{3,2}$) of GMP dispersed in SDS was measured by Coulter laser analysis [2] (Table 2). The same trend was found with three wheat cultivars of very different qualities ranging from weak (Scipion) to strong (Amazon). We observed however that Amazon was clearly less affected by addition of pentosans or xylanase than the other two cultivars. The very higher protein content, the lower pentosans content, and the very high $D_{3,2}$ and K' of Amazon flour (data not shown) may explain why pentosans and xylanase have a small effect on its gluten yield and properties. Clear correlations between gluten properties and GMP particle properties could help explain why also gluten rheological properties were changed upon addition of pentosans. Maximum resistance against extension (R_{max}) of gluten is best predicted by [η] ($R^2=0.85$) and $D_{3,2}$, ($R^2=0.71$) while extensibility at maximum resistance (E at R_{max}) of gluten is best predicted by K' ($R^2=0.82$).

Table 2 *The yield and properties of gluten and GMP for three wheat cultivars*

Sample name	Gluten Yield (%, dm)	R_{max} (N)	E at R_{max} (mm)	[η] dL/g	K'	$D_{3,2}$ (µm)
Scipion control	8.8 ± 0.15	0.36 ± 0.02	89 ± 5	3.18	1.04	10
2%WUS	7.5 ± 0.10 [a]	0.49 ± 0.03 [a]	68 ± 3 [a]	3.34	0.69	16
100ppm xylanase	9.5±0.16 [a]	0.30 ± 0.02 [a]	110 ± 6 [a]	3.00	1.23	6
Soissons control	10.7 ± 0.18	0.38 ± 0.02	93 ± 5	3.26	1.13	20
2%WUS	9.3 ± 0.17 [a]	0.50 ± 0.03 [a]	70 ± 4 [a]	3.41	0.76	26
100ppm xylanase	11.5 ± 0.14 [a]	0.32 ± 0.02 [a]	116 ± 6 [a]	3.05	1.51	15
Amazon control	20.8 ± 0.30	0.51 ± 0.03	110 ± 6	3.57	1.62	26
2%WUS	19.5 ± 0.23 [a]	0.54 ± 0.04	100 ± 5	3.78	0.98	38
100ppm xylanase	21.7 ± 0.31	0.49 ± 0.03	115 ± 6	3.39	1.83	24

Data=mean ± S.D.
[a] means significant difference from control at $p < 0.05$.

3 CONCLUSIONS

Based on our observations, we propose a possible explanation for the effect of pentosans on gluten formation and properties. Both a physical effect and a chemical effect are involved. The physical effect is related to viscosity and likely also depletion attraction between protein particles. Viscosity is a general effect limiting aggregation rate and hence gluten yield. Depletion attraction is related to the ratio in size of pentosans on the one hand and GMP particles on the other. The chemical effect is related to pentosan bound FA and 'controls' the tendency of the particles to aggregate (K') and hence also gluten yield. We assume K' reflects the physical interaction e.g Van der Waals attraction and steric repulsion between GMP particles. In our explanation pentosans do not so much affect the growth of these particles directly after mixing, but hinder the further agglomeration of especially smaller particles to end up in the gluten. The partial agglomeration of GMP particles caused by pentosans will result in GMP with different particle size distribution and properties and gluten with changed rheological properties. This theory allows us to better understand how pentosans affect gluten formation and properties and can be used to optimize gluten processing, and also to further improve the baking process.

References

1 Wang, M.W., Hamer, R.J., van Vliet, T., Oudgenoeg, G., *Journal of Cereal Science*, 2002, **36,** 25.
2 Don, C., Lichtendonk, W., Plijter, J.J., Hamer, R.J., *Journal of Cereal Science*, 2003, **38,** 157.

Nutritional Aspects, Intolerances and Allergies

THE STRUCTURAL AND BIOLOGICAL RELATIONSHIPS OF CEREAL PROTEINS INVOLVED IN TYPE I ALLERGY

E.N.C. Mills[1], J.A. Jenkins[1], S. Griffiths-Jones[2], P.R. Shewry[3]

[1]Institute of Food Research, Norwich Laboratory, Norwich Research Park, Colney, Norwich NR4 7UA, UK.
[2] Sanger Institute, Hinxton, Cambridge, UK
[3]IACR-Long Ashton Research Station, Bristol, UK.

1. INTRODUCTION

Allergic reactions can occur following a variety of environmental challenges, including inhalation of dusts and particles such as pollen, and ingestion of foods. They generally involve generation of an IgE antibody response towards an environmental agent, known as an allergen. IgE binds to the surface of histamine containing mast cells which bind allergen, becoming cross-linked in the process. This triggers the release of mediators, such as histamine, from the mast cells, which proceed to cause the acute inflammatory reactions which are manifested in symptoms. These can include respiratory (asthma, rhinitis), cutaneous (eczema, urticaria) or gastrointestinal (vomiting, diarrhoea) symptoms which may occur alone or in combination. A rare but very severe reaction is anaphylactic shock characterised by respiratory symptoms, fainting, itching, urticaria, swelling of the throat or other mucous membranes, and a dramatic loss of blood pressure. Symptoms appear rapidly in allergic reactions, in contrast to food intolerances which are reproducible, some times non-immune-mediated, reactions whose symptoms can sometimes take days to manifest themselves, an example of which is the gluten intolerance syndrome, Coeliac's disease. This review is restricted to IgE-mediated allergies, which essentially develop in two phases – (1) sensitisation and (2) subsequent reaction on re-exposure to an allergen. Both stages are triggered by allergens which are almost always proteins.

2. STRUCTURAL RELATIONSHIPS OF ALLERGENS

2.1 Common Properties of Allergens

For a food allergen to sensitise via the gastrointestinal (GI) tract it must possess certain structural and biological attributes which preserve its structure from the destructive effects of low pH, proteolysis, and surfactants such as bile salts. Whilst there are some exceptions the vast majority of allergens thought to sensitise via the GI tract belong either to the prolamin superfamily (comprising the prolamin storage proteins of cereals, ns LTPs, 2S albumins and the α-amylase inhibitors) or the cupin superfamily (comprising the 11S legumin-like and 7S vicilin-like seed storage globulins)[1]. All these allergens generally share two properties which enable them to survive digestion: abundance and structural stability. This combination of properties may help to ensure that sufficient protein survives in an immunologically active form to be taken up by the gut and sensitise the mucosal immune system. However, whilst abundance is an important factor, it is probably secondary to protein stability. Thus ribulose-1, 5-bisphosphate carboxylase/oxygenase (usually abbreviated to rubisco), accounts for about 30-40% of total leaf protein in most species, it has never been found to be an allergen, while ns LTPs have been designated as pan-allergens and yet are generally minor components in edible plant tissues.

The factors involved in sensitisation and elicitation of allergic reactions via the lungs are less complex than for exposure via the GI tract, as the allergens are not modified by food processing and digestion. Inhalant allergens are generally readily soluble in dilute salt solutions, allowing allergen solubilisation from inhaled particulates in the liquid layers lining the lungs. This may be accompanied by proteolytic activity which assists allergens to enter the body by permeabilising the lung lining. Some types of allergens, such as such as the α-amylase/trypsin inhibitors and 2S albumins members of the prolamin superfamily, also possess these characteristics and hence lend themselves to being both inhalant and "true" food allergens (i.e. able to sensitise via the gastrointestinal tract)[1]. Another group of allergens sensitise via inhalation and can go on to trigger food allergies. This is best illustrated by the Bet v 1 family of pollen-fruit cross-reactive allergens[2] which are abundant in pollen and can go on to trigger food allergies because conserved homologues are also present in fruits and vegetables. Similarly latex allergens may initially sensitise via contact with tissues (e.g. during surgery and catheterisation) and then lead to dietary allergies to related proteins in fruits and vegetables[3]. It has previously been noted that a remarkable large number of plant food allergens, including the α-amylase/trypsin inhibitors, ns LTPs and Bet v 1 homologues amongst many others, are also involved in plant protection, including the pathogenesis-related (PR) proteins[2]. It is possible that PR proteins are particularly allergenic as they must be stabile in order to survive digestion by proteases secreted by the fungi at the site of infection. As the majority of cereal allergens belong to the prolamin superfamily, and the remainder of this article will focus on this superfamily, with brief reference to other cereal allergens which have been characterised.

3. ALLERGIES TO CEREALS

Cereals have been found to trigger two types of allergic disease, the occupational allergy known as Baker's asthma, which results from inhalation of flour particles in dusty working environments such as bakeries, and as a consequence of ingestion of cereal containing foods. There also appear to be some individuals who react to wheat proteins as a result of prior sensitisation to grass pollen who are serologically distinct from those who are exposed to flour in a work environment. Only around 7 types of foods are responsible for causing the majority of food allergies, including wheat[4]. However, IgE-mediated allergy to wheat products does not appear to be as widespread as allergies to foods such as egg and peanut, despite a public perception that wheat allergy is prominent. Diagnosis of wheat allergy is further complicated by the low solubility of cereal seed storage prolamins in the dilute salt solutions routinely used in clinical diagnosis, which may mean that cereal allergy may remain undiagnosed.

3.1 Cereal allergens of the prolamin superfamily

Members of the prolamin superfamily in cereals include the seed storage proteins, α-amylase/trypsin inhibitors, puroindolines (PINs) and non-specific lipid transfer proteins (ns LTPs). Despite the diversity of their functions' these proteins all share a conserved pattern of cysteine residues containing a characteristic Cys-Cys and Cys-X-Cys motifs, where X represents any other which can be defined by the formula:

Cys-(X=7-13)-Cys-(X=8-26)-Cys-Cys- (X=8-30)-Cys-X-Cys-(X=20-48)-Cys

For most members of the prolamin superfamily the conserved cysteine skeleton accounts for almost the whole protein. The exceptions are the prolamin storage proteins, which contain a repetitive domain inserted into the skeleton at either the *N*- or *C*- terminal end. Whilst the overall degree of sequence identity between the conserved regions of various members of the prolamin superfamily is low, a comparison of known 3-dimensional structures of prolamin superfamily members demonstrates striking similarity at the structural level. Structures have been determined for the α-amylase/trypsin inhibitors and the nsLTPs and they share a related fold consisting of bundles of four α-helices stabilized by disulphide bonds. It is not possible to describe the full range of plant food allergens in this short article but the vast majority fall into the classes discussed above, the cupin/prolamin superfamilies which have storage and/or protective functions and predominantly sensitise via the GI tract[1].

3.1.1 Seed storage prolamin allergens

Whilst generally associated with triggering the food intolerance syndrome, Coeliac disease, the seed storage prolamins of cereals have also been found to trigger allergies to cereals, by ingestion and through inhalation (Baker's asthma)[5]. Wheat can cause atopic dermatitis and a condition known as exercise-induced anaphylaxis (EIA). The latter is a severe allergic reaction that certain patients experience only when taking exercise following consumption of a problem food; allergens have been described as triggering such reactions include polymeric HMW and LMW subunits of glutenin, a γ-, α- and a ω-gliadin.

3.1.2 α-Amylase/trypsin inhibitor allergens

Most subunits of the α-amylase/trypsin inhibitors present in seeds of barley, wheat and rye react with IgE from allergic patients with Bakers' asthma, when tested *in vitro*. However, the relative activities of the different subunits vary greatly, with the most active being glycosylated forms of subunits present either as monomeric and tetrameric α-amylase inhibitors in barley, or as tetrameric α-amylase inhibitors in wheat. Members of this same protein family have been identified as food allergens, including a single M_r ~15,000 subunit identified as an allergen, whilst another inhibitor, termed CM3, has been identified as an allergen triggering atopic dermatitis.

3.1.3 Non specific lipid transfer protein (ns LTPs) allergens

NsLTPs are sparsely abundant proteins and yet have been found to be highly potent allergens and appear to be responsible for more severe allergies, particularly anaphylaxis, in fruits such as peach and more recently in hazelnuts. Whilst the ns LTP of wheat has not to date been directly characterised as an allergen, members of the same protein family have been identified in maize, with a significant degree of cross-reactivity with LTPs from a variety of fruits. ns LTPs have also been shown to be allergens in beer, possibly originating from both barley and wheat grains used in brewing and was responsible for a cross-reactive occupational allergy to spelt reported in one individual who experienced symptoms on consuming cereal foods.

3.2 Other cereal allergens

There are a number of other allergens which have been identified in individuals suffering from Baker's asthma. These include a protein termed Tri a Bd 17K which has now been identified as a glycosylated peroxidase[6]. Other allergens that have been implicated in Baker's asthma include wheat germ agluttinin and a putative trypsin inhibitor. A recent proteomic analysis using 2D PAGE immunoblotting identified around 100 IgE-binding polypeptides of which around 9 predominated, although the clinical relevance of all these proteins has not been demonstrated. Allergens identified included a serpin, and metabolic enzymes such as triosephosphate isomerase and glyceraldehydes-3-phosphate dehydrogenase.

4. CONCLUSIONS

In the same way that only a limited number of foods are responsible for the majority of food allergies, the majority of allergens responsible also belong to only a limited number of protein families, and in cereals it appears that almost all allergens belong to the prolamin superfamily. It would seem therefore that certain structural features and properties regarding abundance of an allergen in a food, its overall thermal and proteolytic stability play a role in predisposing certain proteins to becoming allergens However, rather than any intrinsic ability to elicit an IgE-response, these properties probably determine the exposure of the immune system to immunologically active protein or derived fragments. Such properties may relate more to the need for allergenic proteins to function as effective antigens via the lungs or gastrointestinal tract than intrinsic ability to trigger IgE responses.

References

1. E.N.C. Mills, C. Madsen, P.R. Shewry and H.J. Wichers, *Trends Food Sci Technol*, 2003, **14,** 145.
2. K. Hoffmann-Sommergruber, *Biochem Soc Trans,* 2002, **30,**930.
3. S. Wagner and H.Breiteneder, *Biochem Soc Trans.*, 2002, **30,** 935.
4 R.K.Bush and S.L. Hefle, CRC *Critical Reviews in Food Science and Nutrition*, 1996, **36 (Suppl),** S119.
5, E.N.C. Mills, J.A. Jenkins, M.J.C. Alcocer, Shewry, PR, *CRC Crit Rev Food Sci Nutr* 2004 (in press)
6. H. Yamashita, Y. Nanba, M. Onishi, M. Kimoto, M. Hiemori and H. Tsuji, *Biosci Biotechnol Biochem.* 2002, **66,** 2487.

Acknowledgements

This work was partly supported by EU FAIR CT-98-4356 (Protall) and the competitive strategic grant from BBSRC to IFR (ENCM, JAJ) and Res (PRS).

COELIAC DISEASE – SPECIFIC TOXICOLOGICAL AND IMMUNOLOGICAL STUDIES OF PEPTIDES FROM α-GLIADINS

H. Wieser[1], W. Engel[1], J. Fraser[2], E. Pollock[2], H.J. Ellis[2] and P.J. Ciclitira[2]

[1]German Research Centre of Food Chemistry, Lichtenbergstrasse 4, D-85748 Garching, Germany
[2]Rayne Institute, St. Thomas' Hospital, Lambeth Palace Road, London SE1 7EH, United Kingdom

1 INTRODUCTION

Coeliac disease (CD) may be defined as a primary immune response of the upper small intestine to the storage proteins of wheat, rye and barley (= antigen) and to tissue transglutaminase (= autoantigen) in genetically susceptible individuals. The disease is characterised by a flat mucosa and, in consequence, by the generalised malabsorption of nutrients. The CD lesion is mainly induces by production of interferon γ (IFN) by gluten specific T cells. Peptides corresponding to amino acid residues 57-68, 62-75 and 57-73 of α-gliadins have been proposed as the immunodominant epitopes in the majority of CD patients[1,2]. Deamidation of glutamine residue at position 65 (Q65) to glutamic acid (E65) was necessary for optimal T cell activation. We wished to investigate whether such immunoactive epitopes exacerbate CD in vivo and which residues are essential for T cell stimulation.

2 MATERIALS AND METHODS

2.1 Production of antigens

The stepwise enzymatic hydrolysis of gliadin from the wheat cultivar 'Rektor' was performed with pepsin and trypsin both attached to agarose (→ PTG). Peptides were synthesised by Fmoc chemistry and a solid-phase peptide synthesiser. The crude peptides were purified by two steps of reversed-phase liquid chromatography[3]. All purified peptides were chromatographically pure and had masses corresponding to the theoretical values.

2.2 Instillation test

Four adults with CD, all of whom were volunteers and on a gluten-free diet, underwent three challenges. 1 g PTG served as a positive control and was given first to confirm a positive reaction. After a period of several weeks for recovery of the duodenal mucosa, 100, 50 or 20 mg of gliadin peptide G8 and finally, 100, 50 or 20 mg of the negative control, casein peptide C1, were instilled into the duodenum. Biopsies were taken before

Table 1 *Origin and amino acid sequences (singe letter code) of synthetic peptides*

Peptide	Origin[1]	Sequences
G8	α2 (56-75)	LQLQPFPQPQLPYPQPQLPY
G9	α2 (56-75, E65)	LQLQPFPQPELPYPQPQLPY
G5	α2 (56-68, E65)	LQLQPFPQPELPY
G4	α2 (62-75, E65)	PQPELPYPQPQLPY
C1	β-casein (53-72)	AQTQSLVYPFPGPIPNSLPQ

the infusion, 2, 4 and 6 h after commencing the infusions, using a Quinton hydraulic multiple biopsy capsule. The biopsy specimens were assessed blindly for villus height to crypt depth ratio (VH:CD), enterocyte cell height (ECH) and intraepithelial lymphocyte (IEL) count. The Mann-Whitney U test with 95 % confidence intervals was used for statistical analysis.

2.3 T cell stimulation assays

T cells were isolated from small intestinal biopsies obtained from coeliac patients and cloned. Antigens were incubated with antigen-presenting cells (APC) prior to addition of T cells. Following incubation for 18-48 h, ^{3}H-thymidine was added. After 18 h, the stimulatory effect was determined by the measurement of radioactivity incorporated into T cells and cytokine concentration using ELISA kits. The stimulation index (SI) was calculated by dividing the mean counts per minute (cpm) for T cells plus APC plus antigen by the mean cpm for T cells plus APC. An SI of 2 and more was considered positive.

3 RESULTS AND DISCUSSION

3.1 In vivo toxicity of gliadin peptide G8

Peptide G8 corresponding to residues 56-75 (Table 1) was chosen for in-vivo challenge as incorporates immunodominant peptides from α-gliadins (residues 57-68, 62-75 and 57-73) described in the literature[1,2]. Additionally residue 56 (L56) of α-gliadins was included, because N-terminal glutamine (Q57) would have been partially transformed to pyroglutamic acid, which would have complicated peptide purification. The effects of G8 were compared with those of the positive control PTG and the negative control peptide C1 corresponding to residues 53-72 of β-caseins. The latter comprised 20 amino acid residues incorporating all residues present in peptide G8. Four unrelated CD patients all having the tissue type HLA DQ$_2$ were studied. PTG (1 g) was given first to confirm a positive reaction to a known CD activating substance and to prime the small intestine T cells. All subjects developed significant changes in the VH:CD ratio, ECH and IEL counts within 4 h (Table 2). After several weeks for recovery, 100 mg of peptide G8 were given to the first patient, however, the patient developed diarrhoea and faecal incontinence on the following day, thus the dose was reduced to 50 or 20 mg for the subsequent patients. Similar to PTG this peptide caused a significant reduction in VH:CD ration and ECH, as well as an increase in IEL count 4 h after instillation when compared to the initial biopsy material. In

Table 2 Results of in vivo challenge (0h vs. 4h)[a]

Challenge	Patient 1		Patient 2		Patient 3		Patient 4	
	0 h	4 h	0 h	4 h	0 h	4 h	0 h	4 h
ECH[b]								
PTG	35,5	12,4***	33,7	11,4***	30,7	24,8***	24,4	17,6***
G8	29,8	23,2***	30,1	25,3***	27,5	23,7***	28,1	13,8***
C1	35,2	36,9^{ns}	34,6	35,0^{ns}	28,5	28,6^{ns}	23,8	24,8^{ns}
VH/CD[b]								
PTG	1,90	1,27**	1,37	0,74***	1,49	0,87***	1,57	0,99**
G8	1,82	1,28**	2,27	1,69*	0,90	0,60*	1,63	0,44***
C1	1,98	1,98^{ns}	2,11	2,02^{ns}	1,05	1,15^{ns}	1,27	1,21^{ns}
IEL[b]								
PTG	7	17	15	25	11	28	14	23
G8	16	25	13	24	15	30	14	23
C1	9	10	12	13	15	11	22	14

[a] Mean values of 30 (ECH), 10 (VH/CD) and 1 (IEL) determinations, *** $p < 0.001$, ** $p < 0.01$, * $p < 0.05$, ns non-significant.
[b] ECH = enterocyte height (μm), VH/CD = villus height/crypt depth, IEL = intraepithelial lymphocyte counts

contrast, the control peptide C1 did not affect any of those parameters significantly. Thus it could be demonstrated that peptide G8 corresponding to residues 56-75 of α-gliadins exacerbates coeliac disease in vivo.

3.2 Immunodominant epitopes of gliadin peptide G8

T cells were isolated from biopsies of CD patients and tested for reactivity to α-gliadin peptides including epitopes of residues 56-75. Because deamidation of glutamine at position 65 has been shown to be important for T cells activation[1], all peptides synthesised were modified with glutamic acid at that position (E65). Firstly, the stimulatory effects of peptide G9 (α56-75/E65) and the overlapping fragment peptides G5 (α56-68/E65) and G4 (α62-75/E65) were studied. The results demonstrated that SI and IFN production were highest for G9, followed by G4; the values for G5 were significantly lower. For further studies, the sequence of peptide G4 was modified by an alanine residue in each position except E65 (peptides G4-1A – G4-3A, G4-5A – G4-14A). Immunoassays demonstrated that while substitutions G4-11A through G4-14A had no effect on T cell stimulation, other substitutions had a profound influence (Table 3). Substitutions at positions G4-3A through G4-10A abolished stimulation with the exception of position 9 (G4-9A), where there was residual interferon production. Substitutions at positions 1 and 2 led to a partial down regulation of T cell stimulation. Thus the results indicated that sequences including residues 62-71/E65 (PQPELPYPQP) were important for an immunological effect.

Table 3 Reactivity of T cell clones with alanine substituted analogues of peptide G4[a]

Peptid	Clone no. 4		Clone no. 6/8		Clone no. 9	
	SI	IFN	SI	IFN	SI	IFN
G4	3,7	580	14,3	402	6,6	36
G4–1A	1,1	28	1,9	60	<1	*
G4–2A	<1	*	1,0	42	<1	*
G4–3A	<1	*	<1	*	<1	*
G4–5A	<1	*	<1	*	<1	30
G4–6A	<1	*	<1	*	<1	*
G4–7A	<1	*	<1	*	<1	*
G4–8A	<1	*	<1	*	<1	*
G4–9A	1,3	15	<1	31	2,0	48
G4–10A	1,0	*	<1	*	1,0	*
G4–11A	3,7	640	13,7	205	7,5	180
G4–12A	4,1	540	13,0	225	7,5	100
G4–13A	3,1	350	12,2	600	6,7	140
G4–14A	3,9	670	13,6	600	7,1	210

[a] SI = stimulation index, IFN = interferon γ concentration (pg/mL),
* = below detection limit of 15 pg/mL (clone no. 4) and 30 pg/mL (clone no. 6,8,9).

4 CONCLUSIONS

The instillation of gliadin peptide G8 (α56-75) into the small intestine of four CD patients has demonstrated a significant and specific damage of mucosa. Immunological studies on T cells of CD patients have shown a significant and specific stimulatory effect of gliadin peptide G9 (α56-75/E65) and overlapping fragment peptides G4 (α62-75/E65) and G5 (α56-68/E65). The residues 62-71 of the α-gliadin sequence are essential for stimulation.

References

1 H. Arentz-Hansen, R. Korner, O. Molberg et al., *J. Exp. Med.*, 2000, **191**, 603
2 R.P. Anderson, P. Degano, A.J. Godkin et al., *Nat. Med.*, 2000, **6**, 337
3 H. Wieser and W. Engel, in: *Proceedings of the 17th Meeting of the Working Group on Prolamin Analysis and Toxicity (M. Stern, ed)*, 2003, pp. 121-125

Acknowledgements

This work was supported by the German Federal Ministry of Education and Research (project number 03 12246D).

MASS SPECTROMETRY AS A TOOL FOR PROBING GLUTEN PEPTIDE MODIFICATIONS RELEVANT TO CELIAC DISEASE

G. Mamone, P. Ferranti, D. Melck, F. Tafuro, F. Addeo

Institute of Food Science - CNR .
via Roma 52, 83100 Avellino, Italy;
e-mail: mamone@isa.cnr.it

1 INTRODUCTION

Transglutaminase(TG)-catalyzed transamidation of gliadin peptic-tryptic digest has been evaluated by HPLC and nES-MS/MS analysis. By tagging with mono-dansylcadaverine (MDC) as acyl-acceptor, transamidated fluorescent peptides were identified as α/β-, γ-gliadin, and LMW glutenin fragments. The sequence consensus for transamidation and deamidation was Q-X-P and Q-X-Q. The benefit of the procedure lies in its vast applicability for defining Q susceptibility pathways of any dietary protein to TGase.

2 METHODS

2.1. Peptic and Tryptic digestion

1mg of purified gliadin was dissolved in 1ml of 5% formic acid (pH 2.0) and incubated in a 37 °C water bath with pepsin (1:100 enzyme:substrate ratio) for 2h. The reaction mixture was lyophilized, dissolved in 1ml of 100mM ammonia and digested with trypsin (1:50 enzyme:substrate ratio); after incubation at 37°C for 4h, the reaction was stopped by freezing at −20°C.

2.2. Sample incubation with tTGase

1mg of PT was incubated with tTGase (50:1 peptide/enzyme, w/w), at 37°C for 0, 2, 4h in 5mM Tris-HCl buffer, pH 7.6, containing 2mM MDC, 5mM $CaCl_2$, 10mM NaCl, and 10mM DTT (final volume 1ml). The tTG-catalysed reaction was stopped by adding 8μl 0.4M EDTA. For deamidation, PT or synthetic peptides were incubated with tTGase (1:1 peptide/enzyme, w/w) in transamidation buffer (pH 6.8). The reaction was stopped as above.

2.3. HPLC analysis and nES-MS/MS analysis

Liquid chromatography was performed using a 2.1 mm i.d.x250mm, C18, 5μm reverse-phase column with a flow rate of 0.2 ml/min on a Agilent 1100modular system. Solvent A was 0.1% TFA (v/v) in water; solvent B was 0.02% TFA in acetonitrile. Separation of the peptides was effected with a gradient 5-70% solvent B over 90min. The column effluent was monitored both by a UV detector (absorbance at 220nm) and by fluorimetric detector (λexcitation, 338 nm; λemission 500 nm).

HPLC fractions monitored by fluorescent detector were manually collected and analyzed by nESI-MS/MS. A Q-STAR mass spectrometer (PE Sciex), equipped with a nanospray interface was used. Samples were desalted using a Zip-Tip$^{(TM)}$C18 micro-column and sprayed from a gold-coated 'medium length' borosilicate capillary (Protana). The capillary voltage used was 800V. The determined amino sequence was used as the input to search the corresponding peptides in NCBI and Swiss-PROT database. The search was carried out using the protein prospector/ms-pattern software (http://prospector.ucsf.edu).

3 RESULTS AND DISCUSSION

Coeliac disease (CD) diagnosis was made by detecting tTGase as the most important, if not sole, serum endomisial autoantigen.[1] tTGase, a calcium-dependent ubiquitous intracellular enzyme, catalyses the covalent and irreversible cross-linking of proteins and peptides with endogenous proteins. tTGase can also convert a large number of peptide-bound Q-residues into negatively charged glutamic acid or Q-K isopeptide links[1,2]. The Q residues, selectively modified by tTGase, belong to the -Q-X-P- sequence consensus.[2] Our hypothesis is that some digestive enzyme-resistant oligopeptides escape to transamidation by traversing the lamina propria. They are blocked by tTGase, through Q-residue deamidation, suitable to bind the celiac-specific HLA-DQ2 and -DQ8 molecules through E-residue anchors.[1] The final objective of this work was to determine the Q residues modified by tTGase for predicting *in vivo* toxicity of a dietary protein. For this purpose, MDC-tagged fluorescent peptides were separated by HPLC, and characterized by nES-MS/MS. The Q sites occupying a -2 position with respect to a P residue were preferentially transamidated. Application of this strategy is reported herein, with reference to the PT digest of the whole gliadin fraction, known as a good substrate of tTGase. Gliadin, one of the two protein gluten families composed of more than fifty components,[3] generate an extraordinary number of closely-related PT peptides (Figure 1a), none of which have been retrieved in Data Bank archives. Incubation of PT with MDC produced a limited number of intense fluorescent peaks (Figures 1b and 1c) and some Q-MDC-tagged peptides were identified by MS/MS (Table 1). In Figure 2, as an example of the keen analytical capability of tandem MS/MS technique, the mono MDC derivative mass spectrum of 14-mer peptide (Figure 2), is shown. The 18-mer peptide was a N-terminally truncated form of the 25-mer peptide (peak 6, Q^{16} MDC-derivative) (Table 1). It was contained within the 33-mer peptide, previously isolated from the digest of recombinant α2-gliadin by gastric and pancreatic enzymes.[4] The 33-mer peptide has been indicated as a possible initiator of the inflammatory response to gluten in CD patients.[4] This peptide lacked the gliadin PT digest of the flour sample as the L^7-Q^8 bond was sensitive to pepsin, while the recombinant gliadin peptide counterpart, having P^7 residue, was not. Peptides 1, 3, 4, and 6 gave each two Q endoresidues MDC-tagged (lines 7, 8, 9, and 10 in Table 1). Eleven out of thirteen Q residues were transamidated, ten and two belonging to -Q-X-P- and -Q-X-Q- sequence consensus, respectively, Table 1. Previous data, showing that tTGase preferred mostly Q-X-P to Q-P or Q-X-X-P (X represents any amino acid residue)[2] are consistent with our findings. Hence, a strict specificity of tTGase towards the Q-X-P sequence consensus was found.

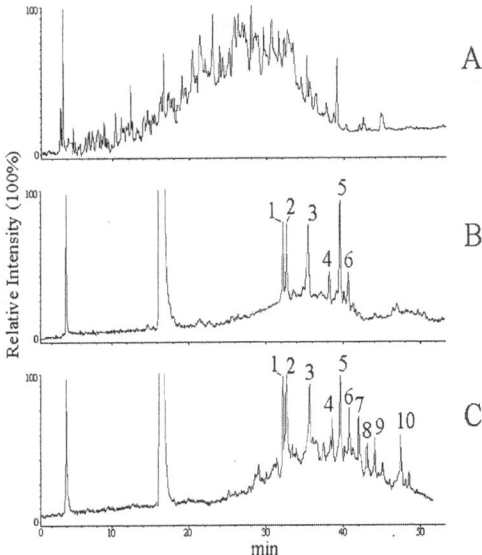

Figure 1: *HPLC- chromatogram of PT-gliadin incubated with tTGase and MDC, detected by UV (A) and by fluorescence (B) (2h-), and (C) (4h-incubation time).*

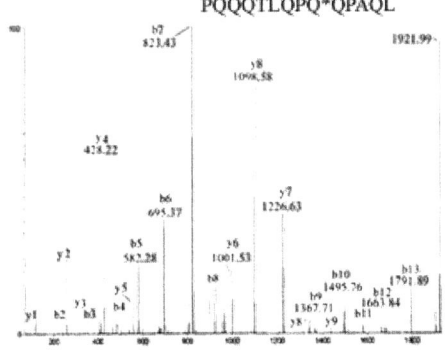

Figure 2: *Deconvoluted nES-MS/MS spectrum of peptide HPLC peak 1. The Q* indicates Q residue cross-linked with MDC.*

The peptides derived from α/β-, and γ-gliadin, one from a LMW-glutenin (Table 1) occurred in SWISS-PROT DataBank differing by at least one amino acid substitution or deletion. The toxic component was the 19-mer peptide, though all isolated peptides were affected by transamidation and deamidation. Propensity to deamidate peptides was also evaluated at lower pH values using a PT gliadin digest as the substrate and tTGase as the substrate modifier. A HPLC profile similar to that shown in Figure 1a was obtained (not shown), which made it difficult to resolve individual peptide species that differed by either one molecular mass unit or multiple values. Some synthetic peptides mimicking natural components of the gliadin PT digest (Table 1) were used as tTGase substrate and the deamidated Q residues belonged to the -Q-X-P- sequence consensus. By nES-MS/MS, peptide 1 (Table 1) gave a spectrum (not shown) consistent with the conversion of native Q^2 and Q^9 into E^2 and E^9 residues.

HPLC peak	Peptide identified by nES/MS-MS	identification SwissProt
1	P Q Q Q [T/F] L [Q/P] P Q* Q P [A/Q] [Q/Q] L	γ-gliadin P08079
2	S H I [L/P] G [P/L] E R P S Q Q* Q P L P P Q Q T L	LMW glutenin; Q8W3W9
3	P Q Q P F P [S/-] Q Q Q* Q P L I	γ-gliadin Q94G96
4	L G Q Q* Q P F P P Q Q P Y P Q P Q P F	α/β gliadin; P04721
5	Q P Q [-/L] P Y P Q P Q* L P Y P Q P Q P F	α/β gliadin P18573
6	P Q L Q P F [L/P] Q P Q [-/L] P Y P Q P Q* L P Y P Q P Q P F	α/β gliadin P18573
7	P Q* Q Q [T/F] L [Q/P] P Q* Q P [A/Q] [Q/Q] L	γ-gliadin P08079
8	P Q* Q P F P [S/-] Q Q Q* Q P L I	γ-gliadin Q94G96
9	L G Q Q* Q P F P P Q* Q P Y P Q P Q P F	α/β gliadin; P04721
10	[P/L] Q* L P Q F [L/P] Q P Q [-/L] P Y P Q P Q* L P Y P Q P Q P F	γ-gliadin P0879

Table 1: *Gliadin peptides from the PT digest, identified by n-ES-MS/MS, containing Q-residue MDC tagged. The Q* indicates Q residue cross-linked with MDC.*

The procedure outlined here could have a general application to identify any tTGase susceptible peptide triggering genetically predisposed subjects to autoimmune disease.

References

1. L.M.. Sollid, *Nature Reviews*, 2002, **2**, 647.
2. B. Fleckenstein, O. Molberg, S.W. Qiao, D.G. Schimdt, F. Von der Muller, K. Elgstoen, G. Jung and L.M. Sollid, *J. Biol. Chem.*, 2002, **277**, 34109.
3. G. Mamone, P. Ferranti, L. Chianese, L. Scafuri and F. Addeo, *Rapid Commun. Mass Spectrom.*, 2000, **14**, 897.
4. L Shan, O. Molberg, I. Parrot, F. Hausch, F. Filiz, GM. Gray, LM. Sollid and C. Khosla, *Science*, 2002, **297**, 2275.

FORMULATION OF GLUTEN FREE BREAD USING RESPONSE SURFACE METHODOLOGY

D.F. McCarthy[1], E. Gallagher[1], T.R. Gormley[1], T.J. Schober[2,3] and E.K. Arendt[3]

[1] Teagasc, The National Food Centre, Ashtown, Dublin 15, Ireland.
[2] National Food Biotechnology Centre, [3] Department of Food and Nutritional Sciences, University College Cork, National University of Ireland, Cork, Ireland.

1 INTRODUCTION

Coeliac disease (CD) is an inflammatory disease of the upper small intestine which results from gluten ingestion in genetically susceptible individuals.[1] The prevalence of CD in the European population is estimated to be as high as 1 in 130-300, making it one of the most prevalent genetic diseases.[2,3] Treatment consists of the permanent withdrawal of gluten from the diet.[1] Gluten is the main structure forming protein in flour, and is responsible for the viscoelastic characteristics of dough and contributes to the appearance and crumb structure of bread.[5] The formulation of high quality gluten free (GF) breads therefore presents a formidable challenge to both the cereal technologist and the baker. Many commercially available GF breads are based on wheat starch that has been rendered GF. However, studies have shown that the low level of toxic protein that remains can cause persistent symptoms in treated CD patients. To ensure complete recovery, a diet containing naturally GF ingredients should be followed.[6] Rice flour is a naturally GF ingredient. However, a gluten replacement, such as hydroxypropylmethylcellulose (HPMC) is necessary to provide structure and gas retaining properties in the dough.[5,7,8,9] The objective of this study was to apply Response Surface Methodology (RSM) to optimise a wheat starch-free gluten free bread formulation.

2 MATERIALS AND METHODS

The formulation (based on flour/starch weight) was 50% rice flour, 50% potato starch, 10% skim milk powder, 6% vegetable oil, 5% fresh yeast, 5% sugar and 2% salt. A central composite design was prepared (Table 1), consisting of two variables: HPMC and water. Following preliminary screening trials, five levels of each variable were chosen with analysis of 13 combinations of these variables being performed. Assessment of error was derived from 5 replications of one treatment combination. From the data obtained, optimal ingredient levels were determined. Eight baking trials were carried out for evaluation of the optimised GF formulation and for short-term shelf life analysis. Ingredients were weighed according to the formulation, and levels of HPMC and water required per treatment (Table 1). Dry ingredients were mixed for 1 min at a low speed. The yeast was dissolved in water at 35°C before being added with the oil during mixing for a further minute. All ingredients were then mixed at medium speed for 2 min., 450g of batter was scaled into 1lb tins, proofed at 40°C until the batter had reached the top of tins and baked at 230°C for 25 min. Loaves were cooled to room temperature and heat sealed in polyethylene bags until tested.

Analyses were performed 24h after baking and four loaves from each treatment were tested. In the shelf life study (on the optimised formulation), two loaves from each replicate were packaged in a 60% N_2/40% CO_2 atmosphere and were tested on days 1, 4 and 7. Loaf parameters measured were specific volume, loaf height, crust and crumb colour (Minolta Chromameter CR-100,). Crust and crumb characteristics were assessed using a texture analyser (TA-XT2i, Stable micro Systems). The crust penetration test was carried out using a 6mm cylindrical aluminum probe. A 20mm cylindrical perspex probe was used for crumb texture profile analysis (TPA). Crumb moisture was determined in a one stage drying process for 2h at 130°C. Crumb grain was assessed using a digital image analysis system.[10] Sensory analysis was conducted using an untrained panel of 20. Panelists were asked to assess the breads for acceptability and to mark a 5 cm line (0 = unacceptable, 5 = very acceptable) in accordance with their opinion.

Table 1: *Experimental design*

Treatment no.	Coded Levels	
	HPMC	Water
1	-1	-1
2	+1	-1
3	-1	+1
4	+1	+1
5	-1.414	0
6	+1.414	0
7	0	-1.414
8	0	+1.414
9	0	0
10	0	0
11	0	0
12	0	0
13	0	0

Variable levels (% flour weight basis):
HPMC: -1.414 = 0.50, -1 = 0.79, 0 = 1.50, +1 = 2.21, +1.414 = 2.50;
Water: -1.414 = 70, -1 = 73.66, 0 = 82.50, +1 = 91.34, +14.41 = 95.

3 RESULTS

Specific volume and loaf height increased as water levels increased ($P < 0.01$). However, breads with the highest specific volume had large holes in the crumb. While crumb firmness decreased as water levels increased ($P < 0.01$, Fig. 1), higher levels of HPMC increased crumb firmness. The softening effect of high water levels can be attributed to the higher specific volume and less dense crumb structure reported in these breads. Crumb grain analysis showed that the number of small cells (0.05 – 4.00 mm^2) increased as water and HPMC increased ($P < 0.01$). However, at the higher levels of addition, the number of small cells decreased (Fig. 2). This may be due to the presence of large holes in breads with high water addition. The number of large cells (> 4.00 mm^2) decreased as both water ($P < 0.05$) and HPMC increased. This may be due to the presence of a large hole instead of many large cells in breads with highest water addition. Optimisation was based on the generation of the best results for these responses. Other responses included crumb colour

(CIE L^*), which increased as HPMC increased ($P < 0.05$). As expected, crumb moisture increased as water addition increased ($P < 0.0001$). Higher levels of water deceased crust firmness ($P < 0.05$), which again may be attributed to the less dense crumb structure.

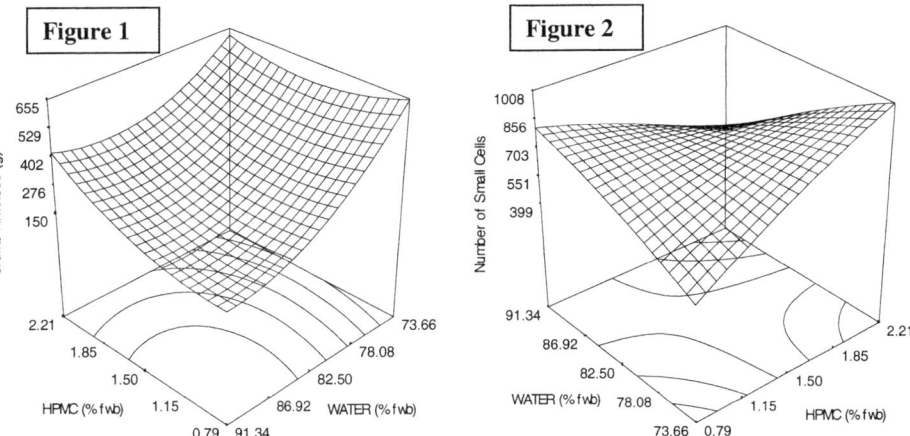

Figures 1 and 2: *3-D surface plots of crumb firmness (Fig. 1) and image analysis data (Fig. 2) of gluten free breads.*

The optimised levels obtained were 2.21% HPMC and 79.05% water. Specific volume and loaf height compared favourably to the predicted values (Table 2). Crumb firmness values were lower than those predicted. Crumb grain analysis revealed a lower number of small cells and a higher number of large cells than those predicted.

Textural changes during the shelf life study are outlined in Table 3. Crumb firmness increased over time, while crust firmness, crumb springiness and resilience decreased. The decrease in springiness and resilience indicate that the bread became more brittle during storage. These changes were perceived in the tendency of the bread to crumble when sliced on day 7. A decrease in crumb moisture occurred over the 7-day testing period (Table 3). This is attributed to the equilibration of moisture between crumb and crust.

The optimised GF bread was rated 1.8 on a scale of 0 – 5 in sensory tests. It must be noted that no panelist was a CD sufferer. Therefore, most would be unfamiliar with breads based on rice and potato starch, thus partly accounting for the low score.

4 CONCLUSION

This study showed that RSM can be successfully applied to optimise HPMC and water levels in a non wheat starch based GF bread formulation.

Table 2: *Comparison of predicted and measured values for the optimised gluten free formulation*

Parameter	Predicted value	Measured value
Specific volume (mlg^{-1})	3.08	3.03 ± 0.16
Loaf height (mm)	99	104 ± 10
Crumb firmness (g)	461	313 ± 49
Number of small cells	851	773 ± 37
Number of large cells	21	24 ± 2

Table 3: *Analysis of crust and crumb characteristics, and crumb moisture for the optimised formulation over a 7-day testing period*

Parameter	D1	D4	D7
Crumb firmness (g)	365 ± 65	514 ± 89	565 ± 105
Crust firmness (g)	539 ± 111	472 ± 118	425 ± 75
Crumb springiness	0.81 ± 0.03	0.77 ± 0.03	0.75 ± 0.04
Crumb resilience	0.36 ± 0.03	0.30 ± 0.04	0.27 ± 0.03
Crumb moisture (%)	47.50 ± 0.22	46.60 ± 0.33	45.99 ± 0.56

References

1 C. Feighery, *BMJ*, 1999, **319**, 236.
2 P. Seraphin and S. Mobarhan, *Nutrition Reviews*, 2002, **60**, 116.
3 A. Fasano and C. Catassi, *Gastroenterology*, 2001, **120**, 636.
4 D. Schuppan, *Gastroenterology*, 2000, **119**, 234.
5 E. Gallagher, T.R. Gormley and E.K. Arendt, *Trends Food Sci. Tech.* (in press).
6 T. Thompson, *J. Am. Diet. Assoc.*, 2001, **101**, 1456.
7 G. Ylimaki, Z.J. Hawrysh, R.T. Hardin and A.B.R. Thomson, *J. Food Sci.*, 1988, **53**, 1800.
8 A. Haque and E. Morris, *Food Res. Int.*, 1994, **27**, 379.
9 L. Cato, G. Rafael, J. Gan and D.M. Small, Proceedings of the 51st Australian Cereal Chemistry Conference, 2002, 304.
10 P. Crowley, H. Grau and E.K. Arendt, *Cereal Chem.*, 2000, **77**, 370.

INVOLVEMENT OF LIPID TRANSFER PROTEINS IN FOOD ALLERGY TO WHEAT

F. Battais[1], J.P Douliez[1], D. Marion[1], Y. Popineau[1], G. Kanny[2], D.A. Moneret-Vautrin[2] and S. Denery-Papini[1]

[1] INRA Research Unit on Plant Proteins and their Interactions (URPVI) - Rue de la Géraudière, BP 71627 44316 Nantes – France - denery@nantes.inra.fr
[2] Department of Internal Medicine, Clinical Immunology and Allergology, University Hospital - 29 avenue de Lattre de Tassigny - 54035 Nancy – France - s.barrat@chu-nancy.fr

1 INTRODUCTION

Lipid transfer proteins are soluble plant proteins that comprise an ubiquitous multigenic family. In wheat seeds, they are mainly found in the aleurone layer and represent 1 to 4 % of soluble proteins. These proteins are classified into two main families according to their molecular mass and amino acid sequences : LTP1 for proteins around 9 kDa and LTP2 for proteins around 7 kDa[1]. LTPs have a compact 3D structure composed of four helical bundles and a C-terminal arm formed by turns and display a hydrophobic cavity for lipid binding. The structure is stabilised by four disulfide bridges that are strictly conserved among LTPs. Plant LTPs are known to be resistant to heating and to digestive proteases as with several other food allergens[2]. Indeed, LTPs are described allergens in several foods including fruits, vegetables and nuts. They are of major importance in allergy to Rosaceae (apple, peach, apricot) not related to pollen in the Mediterranean population[3-6]. LTP1s have also been identified as major allergens for patients with a food allergy to maize [7] and for patients with an allergy to beer [6, 8].

Moneret-Vautrin *et al.* (in preparation) indicated that the prevalence of food allergies to wheat has increased for about ten years among children, as well as among adults, and reached a level comparable to that of coeliac disease. Prolamins, as well as some albumins/globulins are involved in this allergy [9]. In a previous study (Battais *et al.*, in preparation), we reported, for the first time, the presence of IgE antibodies to LTP1 in about 30% of patients suffering of food allergy to wheat. The objective of this study was to better characterise the IgE reactivity of these sera and to observe possible cross-reactions in the family of cereal LTPs by using several LTPs purified from wheat, barley and maize seeds.

2 MATERIALS AND METHODS

Twenty three sera from patients with a food allergy to wheat, a positive response to purified LTP or albumins/globulins, and 2 controls without food allergies were used in this study. Food allergy was established on the basis of IgE dependent sensitisation (prick tests and RAST cap system) and by standardised oral challenge (double or single blind, placebo controlled, food challenges). LTPs from wheat, barley and maize were purified by ion exchange and RP-HPLC and the purity of these fractions was controlled by mass spectrometry [10, 11]. The reactivity of IgE antibodies against purified LTPs was analysed by ELISA using a fluorescent substrate (F-ELISA).

3 RESULTS

3.1 Reactivity of wheat LTP1 positive patients with various wheat protein fractions

In a previous study on 60 patients suffering from food allergy to wheat, we reported that profiles of IgE binding to prolamin fractions and albumin/globulin extract differed according to age and symptoms of the patients (Battais et al., in preparation). Some differences were also observed in the frequency and intensity of IgE-binding to the purified wheat LTP1 (WLTP1). Indeed, among the 17 patients positive to the WLTP1, 11 were children and 6 adults. Twelve (10 children and 2 adults) of these patients were suffering from atopic dermatitis, sometimes with asthma (Figure 1). In this group, the intensity of IgE responses to WLTP1 were often high. These patients also reacted strongly with the albumin/globulin extract. Most of them also displayed IgE responses to the α-gliadin and glutenin fractions and several of them to the γ and ω-gliadin fractions. In this group, one child (25) reacted only to WLTP1 and albumin/globulin extract.

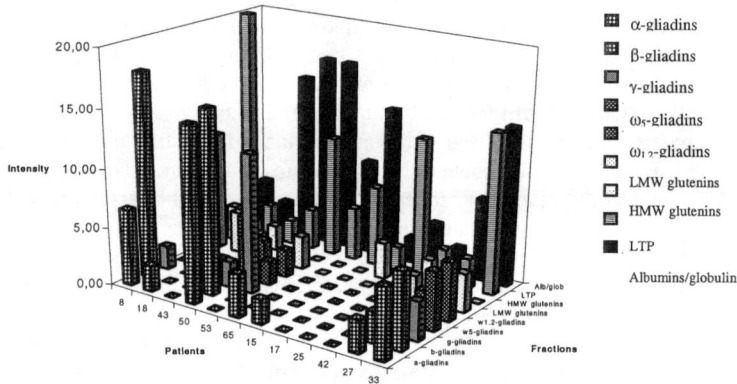

Figure 1 *Intensity of IgE responses to wheat proteins analysed by F-ELISA for 12 patients positive to WLTP1 and suffering from AD.*

Five other WLTP1 positive patients were suffering from anaphylaxis, urticaria or wheat dependant exercise induced anaphylaxis (WDEIA). Among them, one child (41) reacted strongly to the WLTP1 as well as with all the other fractions tested. The 4 other patients were adults which reacted only weakly to the WLTP1 and the albumin/globulin extract ; as shown previously, three of them had IgE-binding to ω5-gliadin, a major allergen for patients with WDEIA[12].

3.2 Reactivity of wheat LTP1 positive patients with other cereal LTPs

Fifteen of the WLTP1 positive sera, 8 WLTP1 negative sera but positive to albumin/globulin extract and 2 control sera from patients without food allergy were analysed in F-ELISA against the WLTP1, a purified wheat LTP2 (WLTP2) and LTP1 purified from barley (BLTP1) and maize (MLTP1). None of the WLTP1 were negative (except one with a very low response for the MLTP1) and control sera had IgE binding to any LTPs. One of the previously WLTP1 with a weakly positive serum was no longer positive in the test.

The other WLTP1 positive sera could be classified into three groups according to their reactivity to the four cereal LTPs (Table 1). Two patients reacted with the four LTPs tested ; one (25) of them detected the four proteins with a strong intensity ; however, both patients displayed a response of higher intensity for the MLTP1 than for the other proteins. Three other sera reacted with the two wheat LTPs and more weakly, or not at all, with the MLTP1 but not with the BLTP1. For these three sera, the intensity of IgE responses to WLTP2 was higher than that of IgE responses to WLTP1. Eight sera reacted to two or three proteins in the LTP1 family but never to the WLTP2. Two of them reacted with similar intensities with wheat and barley LTP1 and more weakly with the MLTP1, one reacted more strongly with WLTP1 than with BLTP1 and 5 reacted only weakly with wheat and maize LTPs. The two remaining sera reacted only weakly with WLTP1.

Table 1 *Reactivity in F-ELISA of wheat LTP1 positive sera with other cereal LTPs*

Patients (symptoms)	Wheat LTP1	Wheat LTP2	Barley LTP1	Maize LTP1
25 (AD +asthma)	28	33	30	65.5
63 (WDEIA)	3.3	2.5	7.5	21
33 (AD +asthma)	8.5	16	-	2
41 (AS)	11	40	-	4
65 (AD)	2.2	4	-	-
8 (AD)	36	-	46	2.7
18 (AD)	6.5	-	7	4.5
50 (AD)	3.5	-	3	2
53 (AD)	12.5	-	2.5	-
17 (AD)	2.1	-	-	3
45 (U)	2.2	-	-	3
36 (U)	2	-	-	3.1
27 (AD+A)	2.6	-	-	2.9
42 (AD+A)	2.9	-	-	-
43 (AD)	2.1	-	-	-

AD : atopic dermatitis ; As : anaphylactic shock ; U : urticaria ; WDEIA : wheat-dependent exercise-induced anaphylaxis.

4 DISCUSSION

These results highlight the diversity of IgE responses to cereal LTPs with possible cross-reactions between wheat, barley and maize LTPs. Sequence homologies exist between WLTP1 and BLTP1 (72 % of homologies) and to a lesser extent between these proteins and MLTP1 (around 55%). These proteins also show very similar 3D structures. Cross-reactions in the LTP1 family could thus be explained by the existence of both linear and conformational common epitopes. However, there is only 18 % sequence homology between WLTP1 and WLTP2, 10 % of which being linked to the very well conserved cysteine pattern. Possible cross-reactions between these two proteins could be explained by homologies in their folding.

Most of the WLTP1 positive sera displayed a broad spectrum of reactivity towards several prolamin fractions as well as albumins and globulins. The majority of these patients are very young children suffering from multiple food allergies. In these patients, the high intestinal permeability may allow the immune system to be stimulated by a larger number of proteins. However, in some cases, cross-reactions in IgE responses may occur in relation with the conserved sequences around cysteines in LTPs and in the non-repetitive domains of prolamins or in other cysteine-rich albumins belonging to the prolamin super-family [13].

References

1. J. Douliez, T. Michon, K. Elmorjani, D. Marion. *Journal of Cereal Science* 2000; **32**:1-20.
2. R. Asero, G. Mistrello, D. Roncarolo, S.C. de Vries, M.F. Gautier, C.L. Ciurana, et al. *Int. Arch. Allergy Immunol.* 2000; **122**:20-32.
3. E.A. Pastorello, L. Farioli, V. Pravettoni, C. Ortolani, M. Ispano, M. Monza, et al. *J. Allergy Clin. Immunol.* 1999; **103**:520-6.
4. Pastorello EA, D'Ambrosio FP, Pravettoni V, Farioli L, Giuffrida G, Monza M, et al. *J. Allergy Clin. Immunol.* 2000; **105**:371-7.
5. R. Asero, G. Mistrello, D. Roncarolo, M. Casarini, P. Falagiani. *Ann. Allergy Asthma Immunol.* 2001; **87**:68-71.
6. R. Asero, G. Mistrello, D. Roncarolo, S. Amato, R. van Ree. *Ann. Allergy Asthma Immunol.* 2001; **87**:65-7.
7. E.A. Pastorello, L. Farioli, V. Pravettoni, M. Ispano, E. Scibola, C. Trambaioli, et al. *J. Allergy Clin. Immunol.* 2000; **106**:744-51.
8. G. Garcia-Casado, J.F. Crespo, J. Rodriguez, G. Salcedo. *J. Allergy Clin. Immunol.* 2001; **108**:647-9.
9. F. Battais, F. Pineau, Y. Popineau, C. Aparicio, G. Kanny, L. Guerin, D.A.Moneret-Vautrin, S. Denery-Papini. *Clin. Exp. Allergy* 2003; **33**:962-70.
10. J.P. Douliez, S. Jegou, C. Pato, C. Larre, D. Molle, D. Marion. *J. Agric. Food Chem.* 2001; **49**:1805-8.
11. J.P. Douliez, C. Pato, H. Rabesona, D. Molle, D. Marion. *Eur. J. Biochem.* 2001; **268**:1400-3.
12. M. Lehto, K. Palosuo, E. Varjonen, M.L. Majuri, U. Andersson, T. Reunala, et al. *Clin. Exp. Allergy* 2003; **33**:90-5.
13. P.R. Shewry, F. Beaudoin, J. Jenkins, S. Griffiths-Jones, E.N.C. Mills EN. *Biochem. Soc. Trans.* 2002; **30**:906-10.

PROPERTIES OF PROTEINS IN IMMATURE WHEAT GRAINS RELEVANT TO THEIR TRASFORMATION

F. Bonomi [1], S. Iametti [1], M. Zardi[2], M. A. Pagani[2], M. G. D'Egidio [3]

[1]Dipartimento di Scienze Molecolari Agroalimentari, University of Milan, Via Celoria 2, I-20133 Milan, Italy
[2]Dipartimento di Scienze e Tecnologie Alimentari e Microbiologiche, University of Milan, Via Celoria 2, I-20133 Milan, Italy
[3] Istituto Nazionale per la Cerealicoltura, via Cassia 176, I-00191 Roma, Italy

1 INTRODUCTION

Some properties of immature durum wheat grains suggest that they may represent a valuable ingredient for the production of special foods, as also indicated by technological tests. Immature grains were found to have a high content of fructo-oligosaccarides, fructose-rich polymers with important biological functions[1]. These compounds are no more present in the ripe grains. The total content and composition of protein fraction also vary during grain ripening. Of particular interest is the development of the gliadin and glutenin fractions involved in the formation of the gluten network during bread and pasta-making processes, and responsible of gluten intolerance. Biochemical studies on the protein components during the different ripening phases have been limited to coarse chemical studies and to approaches based on protein fractionation.

Inter-protein interactions play a prominent role as for determining the suitability to technological transformation in most food systems. The standard chemical/technological indexes for wheat proteins are not always sufficient for defining the relevance of hydrophobic interactions and of interprotein disulfide exchange to the formation of an interprotein network[2, 3]. In this work we have determined the properties of proteins in immature grain by detecting the number of accessible thiols and the surface hydrophobicity properties on ground immature wheat grains at different times after flowering, without prior protein fractionation, and in conditions apt at assessing the accessibility of thiol groups in the protein interior. We also evaluated gluten proteins development, both by separation techniques and by immunochemical detection with anti-gliadin antibodies.

2 METHODS AND RESULTS

2.1 Protein evolution in developing grains

In the time frame considered in this study (from 9 to 28 days after flowering), the total protein content in immature grains on a dry matter basis of the durum wheat cultivar Ofanto showed only minor variations, remaining in the range of 13-14 %. Protein solubility was determined by suspending flours in denaturing and non-denaturing buffers (50 mM phosphate, 0.1M NaCl, pH 7.0 in the presence or in the absence of 8M urea and

10 mM DTT). Flours were obtained by milling grains at different days after flowering (daf). The amount of protein dissolved in each solvent system was determined by a colorimetric method[4], and was found to undergo major changes, as summarized in Figure 1. In particular, a decrease in albumins and globulins was accompanied by a noticeable increase in protein aggregates requiring urea and DTT for solubilization, that leveled off at 17 daf.

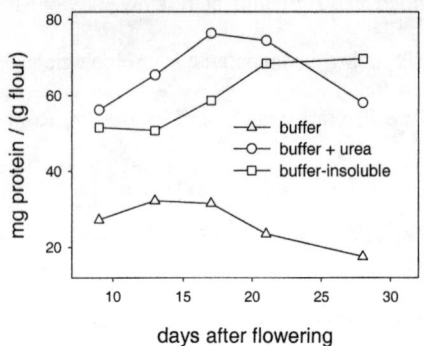

Figure 1 *Time course of the protein solubility changes in developing grains*

The proteins solubilized at different times in denaturing buffer were also characterized by SDS-PAGE and immunoblotting performed by using commercial anti-gliadin antibodies. Both approaches confirmed that synthesis of the major gluten polypeptides begins at 17 days after flowering, and that no immunoreactive gluten material was present at 9 or 13 daf.

2.2 Structural indexes for proteins in developing grains

To avoid possible perturbation of the interprotein network in the system being investigated, determinations aimed at defining structural features of proteins were performed on flour suspensions in non denaturing buffer.

Thiols were detected by a specific colorimetric reagent[5], added in excess during preparation of the suspension. To estimate thiol accessibility, this determination was carried out also in the presence of 1% sodium dodecyl sulfate, a detergent that loosens protein structure but does not disrupt disulphide bond. The results of these determinations as a function of ripening time are reported in Figure 2, and indicate a marked decrease of both total thiols and of thiol accessibility at the onset of gluten synthesis (17 daf).

These findings suggest that the proteins in this system acquire a compact aggregation state. To further investigate this hypothesis, we determined protein surface hydrophobicity by a ligand-binding approach, developed in previous studies[2], that allows to determine the amount of a specific hydrophobic probe (1, 8, anilino-naphthalene-sulfonate, ANS) remaining bound to the buffer-insoluble protein fraction.

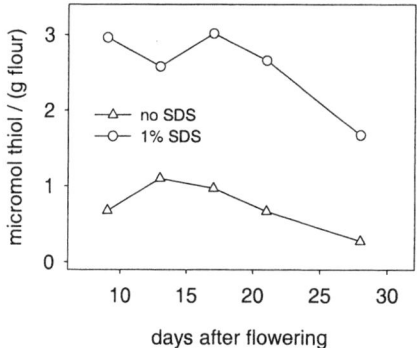

Figure 2 *Time course of thiol accessibility changes in developing grains*

As shown in Figure 3, accessibility of the probe to exposed hydrophobic regions on the protein surface drops dramatically after 17 daf, suggesting that major compaction of the protein structure begins, as expected when gluten proteins become the most abundant protein component. However, changes in the mutual protein relationships occur before the actual synthesis of gluten protein begins, as indicated by the data in the lower panel of Figure 3.

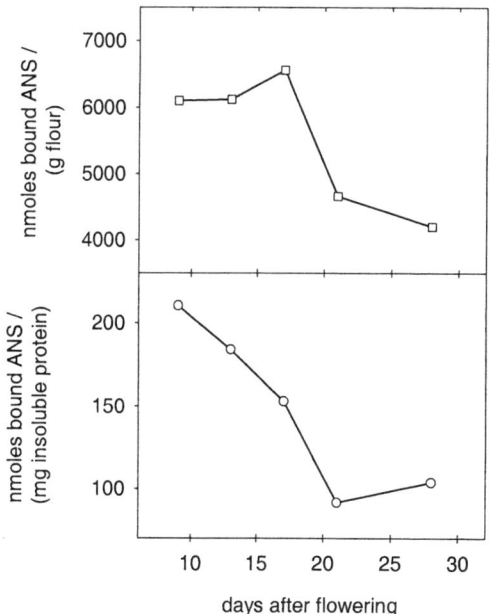

Figure 3 *Time course of surface hydrophobicity changes in developing grains*

3 CONCLUSION

These results suggest that immature grain has good potential for the production of functional food or gluten-free food. However, given the abundance of albumins and globulins before the onset of gluten synthesis, transforming immature grains into food commodities will require particular attention as for the choice of the most appropriate time for grain harvesting. This parameter is critical, also in view of the changes in overall organization of the protein network occurring prior to the onset of gluten synthesis, as indicated by our surface hydrophobicity data.

In this frame, it appears worth investigating what pattern of events is observed in other cultivars, that have a different use in the pasta-making industry, and that may show a different temporal pattern of protein deposition and of accumulation of other nutritionally relevant components. These studies are being currently carried out in our laboratories.

References

1. M.G. D'Egidio, T. Cervigni *Le Scienze*, 1998, n. 363, 108
2. M. A. Pagani, S. Iametti, M. G. D'Egidio and F. Bonomi In *"Plant proteins from European crops: food and non-food applications" INRA, Nantes*, 1998, 225.
3. A. Rondanini, F. Bonomi, S. Iametti, M. Lucisano, M.A. Pagani, P.P. Rovere *High Pressure Res.*, 2000, **19**, 167.
4. M.M. Bradford *Anal. Biochem.*, 1976, **72**, 248.
5. G.L. Ellman, *Arch. Biochem. Biophys.*, 1959, **82,** 70.

BINDING OF GLUTEN PEPTIDES TO THE COELIAC DISEASE-ASSOCIATED HLA-DQ2 MOLECULE BY COMPUTATIONAL METHODS

S. Costantini[1§], G. Colonna[1], M. Rossi[2] and A.M. Facchiano[1,2]

[1] CRISCEB – Research Center of Computational and Biotechnological Sciences, Second University of Naples, via Costantinopoli 16 – 80138 Naples, Italy.
[2] Institute of Food Science, CNR, via Roma 52 A/C – 83100 Avellino, Italy.
[§] PhD fellowship supported by E.U.

1 INTRODUCTION

Celiac disease (CD) often starts shortly after the first introduction of wheat into the diet. Symptoms include diarrhea, malabsorption, and failure to thrive, which is due to a inefficient uptake of nutrients by a flattening intestinal epithelium. Removal of gluten from the diet, at present the only treatment for the disease, is an effective way to stop the disease process, and reintroduction of gluten in the patient's diet invariably leads to the reappearance of the symptoms [1-2]. In recent years some different gluten-derived peptides have been identified that are recognized by T cell clones isolated from biopsies of CD patients [3-5]. CD is limited to genetically predisposed individuals expressing HLA-DQ2 (DQA1*0501/B1*0201) and/or -DQ8 (DQA1*0301/B1*0302) heterodimers. The molecular mechanism is considered to involve the DQ2 and DQ8 molecules for binding gluten peptides and presenting them to T cells in the small intestine. Peptide binding to HLA-DQ2 and HLA-DQ8 molecules is most efficient when negatively charged amino acids are present at anchor positions in the peptide. Gluten contains very few negatively charged amino acids, but many glutamines, which may be deamidated to the negatively charged glutamate. This reaction may be catalyzed by the tissue transglutaminase (tTG) enzyme [2-3].

In our work, we used bioinformatics and biocomputing approaches to predict the three-dimensional structure of HLA-DQ2 molecule, simulate its complex with gluten peptides, and evaluate the energies and the molecular details of the interaction.

2 METHODS

2.1 Protein modeling of DQ2 structure

DQ2 protein sequences used refer to the following SwissProt entries: DQA1*0501 chain, P01909 and DQB1*0201 chain, P01918. The very high degree of similarity of these two chains with the corresponding chains in DQ8 allowed us to use the comparative modelling approach to create a 3D model of DQ2. The DQ8 structure from the DQ8-insulin B9-23 complex [6] (PDB code: 1JK8) was used as template. We used an homology modelling procedure already used with success and described in our previous papers [7,8]. In brief, searches for sequence similarity within databases and the alignment of the selected sequences were performed with the BLAST program [9]. The programs MODELLER [10] and Quanta (Accelrys, San Diego, CA) were used to build 10 full-atom models for each protein chain under prediction. For selecting the best model among them, their stereo chemical quality were verified with the program PROCHECK[11]. The quaternary structure of the DQ2 dimer was assembled with the software InsightII (Accelrys, San Diego, CA) by superimposing the two chains to the DQ8 chains, then by minimizing the interaction between the two chains.

2.2 Simulation of DQ2-peptide complexes

The localization of peptide binding site in DQ2 was based on the DQ8-peptide interaction in the PDB reference model used, and also in according to the anchor positions known by literature [1,2]. We used four natural gluten peptides, i.e. alpha I (fragment 60-68 from alpha 9 gliadin), alpha II (fragment 62-70 from alpha 2 gliadin), alpha III (fragment 67-75 from alpha 2 gliadin), gamma I (fragment 115-123 from gamma 5 gliadin), being demonstrated the interaction of such peptides with DQ2 [3,5]. Moreover, we created additional complexes by modifying specific peptide residues, in order to obtain models for deamidated peptides. For each DQ2 - peptide complex, the global structure was optimized with the software InsightII (Accelrys, San Diego, CA) by using 500 steps of energy minimization under conjugate gradient algorithm. After minimization, the energy of interaction with the DQ2 molecule was evaluated for each peptide.

3 RESULTS

3.1 Modelling of DQ2

The sequences of both DQ2 subunits have been analyzed by computer programs in order to find similar sequences in databases and perform structural predictions. The sequences of DQ8 chains were found to be the most similar to DQ2 chains, showing 91% of sequence identity for both chains. In these conditions, homology modelling strategy can be used with very good results. Therefore, the models of DQ2 alpha and beta chains were created by using the homology modelling strategy on the DQ8 template (PDB code: 1JK8). The assembly of DQ2 dimer was obtained by superimposing the two modelled chains to the corresponding DQ8 chains, in order to obtain the same relative orientation of the two subunits. The resulting dimer has been optimized for conformation by energy minimization.

3.2 Simulation of DQ2-peptide interaction and effects of side chain modifications

We used the model of DQ8 complexed with an insulin peptide to identify the peptide binding site in our DQ2 model and create the complex DQ2 - peptide. The anchor positions known by previous studies and reported in literature have been considered in order to find the better interaction. Different gluten peptides are reported in literature to be related to celiac disease, and affinity studies describe their interaction with DQ2. For most of them, deamidation of specific glutamine was suggested as a reaction which increases the affinity for DQ2. We used four natural gluten peptides, i.e. alpha I, alpha II, alpha III, and gamma I from gliadins. We aligned these peptides to the insulin peptide (see Figure 1), on the basis of literature information concerning the location of specific residues in anchor sites.

Nutritional Aspects, Intolerances and Allergies 393

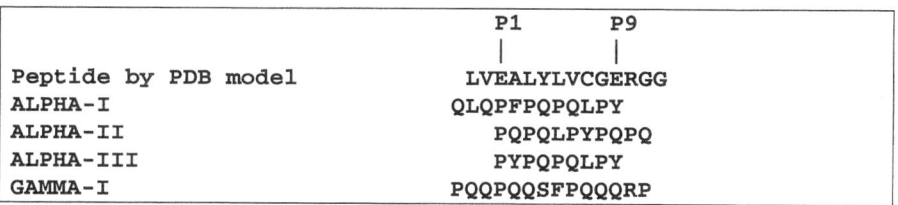

Figure 1. *The sequences of gluten peptides used in our simulations, aligned to the insulin peptide present in the DQ8 template model (PDB code: 1JK8). Anchor positions P1 and P9 are evidenced.*

Moreover, we simulated the glutamine deamidation and other substitutions in different peptides and positions, so that a total number of 11 peptides has been created. The simulation of their interaction with DQ2 allowed to verify the agreement to reported experimental results as well as to have a more detailed analysis of DQ2-peptide interaction. Energy values are shown in Figure 2.

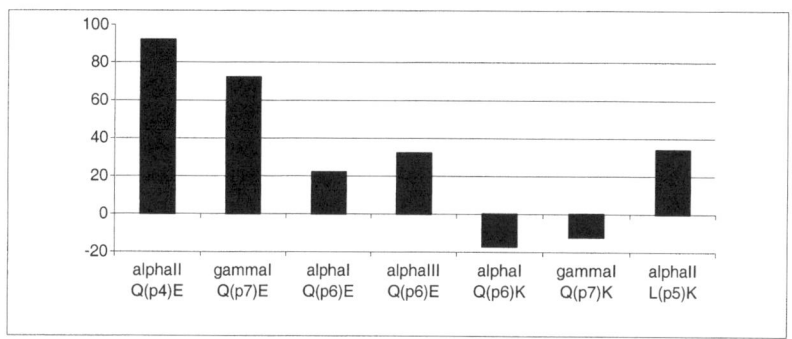

Figure 2. *Effects of peptide mutations on the energy of interaction between DQ2 and the peptide. It is reported the difference (expressed in Kcal/mole) between the mutated peptide and the natural gluten peptide. Positive values indicate better interactions with DQ2.*

Substitution of Q (glutamine) with E (glutamate) improves the interaction in p4, p6 and p7 positions. Positive charge of K (lysine) in p6 and p7 decreases the interaction, while a subtle increase is observed when K substitutes L (leucine) in p5. These results confirm with computational approaches the preference of p4 and p7 anchor positions for the negatively charged side chain of glutamate in comparison to the glutamine side chain. The lower improvement by substitution in p6 may be due to the consideration that a different residue is the most preferred in such position, i.e. Proline. On the other side, the presence of positive charges in p6 and p7 reduces the interaction with DQ2, as observed by glutamine substitution with lysine side chains in alphaI (p6) and gamma I (p7). It is interesting to note that a positive charge in p5 seems to give a subtle improvement to the interaction of alpha II peptide with DQ2, as observed for leucine substitution with lysine. More detailed investigations give us the explanation at molecular levels of such effects of negative charges in anchor positions. In fact, p4 position is in proximity of the positive charge of lysine B69 of DQ2 molecule (see Figure 3). The distance between the glutamate in p4 position of a modified alpha II peptide and the positively charged residue of DQ2 is suitable for a favorable electrostatic interaction. Similarly, a repulsive interaction may occurs when the same position in the peptide is occupied by a residue with positive charge.

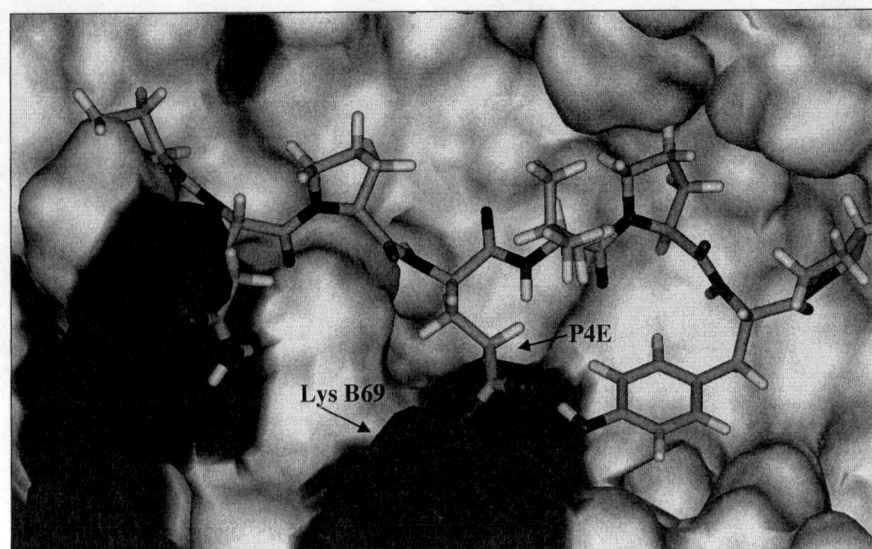

Figure 3. *Details of the interaction of alphaII peptide into DQ2 pocket. The glutamate side chain in position P4 of the peptide (P4E, stick representation) is very close to the positive charge of the LysB69 of DQ2-B chain (dark region on the DQ2 surface).*

4 CONCLUSIONS

Computational methods give results in good agreement with experimental date. Further investigations are ongoing in our group to evaluate the effects of other substitutions as well as with other gluten peptides with the aim to obtain more structural and functional information about the DQ2 – peptide recognition.

References

1. Sollid LM *Ann. Rev. Immunol.*, 2000, **18**, 53-81.
2. Sollid LM *Nat. Rev. Immunol.*, 2002, **2**, 647-655.
3. Quarsten H, Molberg O, Fugger L, McAdam SN, Sollid LM. *Eur. J. Immunol.*, 1999, **29**, 2506-2514.
4. Moustakas AK, van de Wal Y, Routsias J, Kooy YM, van Veelen P, Drijfhout JW, Koning F, Papadopoulos GK. *Int Immunol.*, 2000, **12**,1157-1166.
5. Arentz-Hansen H, Korner R, Molberg O, Quarsten H, Vader W, Kooy YM, Lundin KE, Koning F, Roepstorff P, Sollid LM, McAdam SN. *J. Exp. Med.*, 2000, **191**, 603-612.
6. Lee KH, Wucherpfennig KW, Wiley DC. *Nature Immunol.*, 2000, **2**, 501-507.
7. Facchiano AM, Stiuso P, Chiusano ML, Caraglia M, Giuberti G, Marra M, Abbruzzese A, Colonna G. *Protein Engineering*, 2001, **14**, 881-890.
8. Caporale C, Facchiano A, Bertini L, Leonardi L, Chilosi G, Buonocore V, Caruso C. *J Mol Model (Online)*, 2003, **9**, 9-15.
9. Altschul SF, Gish W, Miller W, Myers EW, Lipman DJ. *J. Mol. Biol.*, 1990, **215**, 403-410.
10. Sali A and Blundell TL. *J Mol Biol.* 1993 **234**, 779-815.
11. Laskowski RA, MacArthur MW, Moss DS and Thornton JM. *J. Appl. Cryst.* 1993, **26**, 283-291.

EVIDENCE OF *PANICUM MILIACEUM* AS A SAFE FOOD FOR COELIAC PATIENTS

F. Fusari[1], A. Petrini[1], M. De Vincenzi[2] and N. Pogna[3]

[1] CERMIS – Centro Ricerche e Sperimentazione per il Miglioramento Vegetale "N. Strampelli", Via Abbadia di Fiastra 3, Tolentino (MC), I-62029, E-mail: cermis@tin.it
[2] Laboratorio di Metabolismo e Biochimica Patologica, Istituto Superiore di Sanità, Viale Regina Elena 299, Rome, I-00161
[3] Istituto Sperimentale per la Cerealicoltura, Via Cassia 176, Rome, I-00191

1 ABSTRACT

In this work, toxicological and nutritional analyses provided evidence of the validity of *Panicum miliaceum*, a cereal species phylogenetically close to maize, as a safe food for people suffering from celiac disease. A collaboration between public and private institutions, from both Italy and other countries, has allowed to recover 23 genotypes of *Panicum miliaceum*, which have been grown in the field and studied for their main agronomic characteristics. Peptic-tryptic (PT) digests of ethanol soluble proteins from *Panicum miliaceum* have been submitted to agglutination tests on human myelogenous leukemia K562(S) cells, the agglutinating activity of the PT digests being strongly associated with toxicity for the coeliac intestine. The analysis revealed the absence of agglutination up to 2300 mg/l of PT digests as compared with 100% agglutination caused by 70 mg/l of PT digests from the positive check sample *Triticum aestivum* cultivar S. Pastore, suggesting the absence of toxicity of *Panicum miliaceum* for coeliac patients. Moreover, the nutritional analyses showed the absence of any inhibititory effect on the *in vitro* growth of CaCo-2 cells by the alcohol-soluble proteins of the *Panicum miliaceum* genotypes analysed here, even at protein concentrations as high as 10 g/l.

2 INTRODUCTION

Coeliac disease is a severe permanent alimentary pathology which causes a flattening of intestinal villi in genetically predisposed individuals (in Italy 1 : 100-180 inhabitants). The malabsorption, resulting from the damage to the small intestine epithelium in coeliac patients, causes a wide range of effects such as diarrhoea, weight loss, osteoporosis, anaemia, neuropathies and sterility.

At the present time, the prevention of pathological symptoms is possible only through the complete exclusion from the diet (for the rest of a patients' life) of food containing bread and durum wheat, spelt, Emmer, Einkorn, barley and rye, all of which contain toxic prolamins; maize and rice, on the contrary, are not toxic[1,2], and so it is probable that other cereal species phylogenetically close to maize, including proso millet (*Panicum miliaceum*), are not toxic for coeliac sufferers.

The aim of this study was to provide experimental evidence, through toxicological and nutritional analyses, on the validity of this crop as a safe food for coeliac patients.

3 MATERIALS AND METHODS

In order to increase the genetic variability of the subject matter, a collaboration with some public and private institutions, from both Italy and other countries, has been started and has enabled us to recover 23 genotypes of *Panicum miliaceum*.

All these varieties and populations have been grown in a field trial, without repetitions, using organic methods in 10 m² plots formed by 8 rows 8 metres long and an inter-row gap of 15 centimetres.

The following parameters were determined in the field: earing date, height from the ground to the last leaf, height from the ground to the base of cob, height of the plant (from the ground to the apex of cob), length of cob.

Further parameters were also measured on the grain harvest: unit production at 13% of humidity, hectolitric weight and 1000 kernel weight.

Table 1 *Characterization of 23 Panicum miliaceum genotypes*

Genotype code	Genotype	Earing date (days from July the 1st)	Height up to the last leaf (cm)	Height up to the base of cob (cm)	Plant height (cm)	Cob length (cm)	Production (t/ha at 13% of humidity)	1000 kernel weight (g)	Hectolitric weight (kg/hl)	% of agglutination (up to 2300 mg/l of PT digests)	Inhibition effect on CaCo-2 cells (up to 10 g/l)
1	BOSSI	10	89	93	119	26	3,96	0,59	77,8	0	absent
2	PELLICCIONI	10	87	93	115	22	3,71	0,58	79,6	0	absent
3	BIOLOGICO	10	97	106	132	27	1,64	0,62	69,8	0	absent
4	FLORAGRIA	22	98	105	129	24	4,75	0,58	77,5	0	absent
5	SALVIA	10	57	64	87	23	1,82	0,60	71,5	0	absent
6	MIGLIO BIANCO	3	31	37	51	14	0,30	0,65	73,1	0	absent
7	SUNRISE	3	40	45	62	17	0,80	0,64	69,0	0	absent
8	HUNTSMAN	10	41	48	65	18	1,37	0,62	64,5	0	absent
9	EARLYBIRD	3	40	48	63	14	1,33	0,66	73,6	0	absent
10	KINELSKOJE	3	63	63	79	16	0,47	0,79	76,0	0	absent
11	UNIKUM	3	55	63	90	28	1,13	0,60	77,4	0	absent
12	PAGLIERINO	3	64	69	87	18	1,61	0,61	73,5	0	absent
13	KORNBERGER M.	3	50	58	78	20	1,94	0,65	74,5	0	absent
17	VIR 160	3	25	38	51	13	0,31	0,63	-	0	absent
18	VIR 1991	22	74	80	104	24	2,14	0,54	73,5	0	absent
19	VIR 1992	10	88	96	123	27	2,81	0,55	78,6	0	absent
20	VIR 1993	3	65	72	95	23	2,76	0,51	73,6	0	absent
21	VIR 2033	3	67	73	96	23	2,55	0,55	74,5	0	absent
22	VIR 2198	31	85	88	118	30	1,64	0,62	64,4	0	absent
23	VIR 9181	22	90	99	121	22	3,69	0,57	80,3	0	absent
24	VIR 9774	3	49	51	73	22	2,53	0,58	77,9	0	absent
25	VIR 9783	3	51	57	76	20	1,65	0,61	67,2	0	absent
26	MIGLIO ROSSO	22	32	40	55	15	0,30	0,69	-	0	absent

The toxicological analysis on human myelogenous leukemia K562(S) cells was the agglutination test [3]; the agglutinating activity of prolamin peptides on K562(S) cells was found to be strongly associated with toxicity against the coeliac intestine[2]. Peptides were obtained through PT digestion of alcohol-soluble and acetic-acid-soluble proteins[4].

The nutritional characteristics were determined by a test on CaCo-2 cells[5] in order to verify the possible inhibitory effect of *in vitro* growth by the alcohol-soluble proteins of *Panicum miliaceum*.

A cytotoxic effect has been shown previously[5] whereby there was inhibition of the cell division by prolamin peptides on *in vitro* cultures of CaCo-2 cell lines, and the implication of these results for coeliac disease has been described.

4 RESULTS AND DISCUSSION

The agronomic and morphological characteristics of the 23 *Panicum miliaceum* genotypes are listed in Table 1.

The test on K562(S) cells indicated in all the samples the absence of agglutination of up to 2300 mg/l of PT digests (Table 1) as compared with 100% agglutination caused by 70 mg/l of PT digests from the positive check sample *Triticum aestivum* cultivar S. Pastore.

Alhough this test is indirect, it suggested a total absence of toxicity of *Panicum miliaceum* for coeliac patients.

The nutritional analyses showed the absence of any inhibitory effects on the *in vitro* growth of CaCo-2 cells even at protein concentration as high as 10 g/l (Table 1); so the alcohol-soluble proteins of the *Panicum miliaceum* genotypes analysed here do not produce anti-nutritional effects even at very high concentrations.

The positive results obtained through the agglutination test and the nutritional tests on CaCo-2 cells are encouraging for the continuation of studies on this interesting crop, which is little-known in developed countries.

References

1 B.S. Anand, J. Piris and S.C. Truelove, *Quaterly Journal of Medicine*, 1978, **185**, 101.
2 S. Auricchio, M. Cardelli, G. De Ritis, M. De Vincenzi, F. Latte and V. Silano, *Pediatric Research*, 1984, **18**, 1372.
3 M. De Vincenzi, M.R. Dessì, C. Giovannini, F. Maialetti and E. Mancini, *Toxicology*, 1995, **96**, 29.
4 M. De Vincenzi, M.R. Dessì, R. Luchetti, N. Pogna, R. Redaelli and G. Galterio, *Atla*, 1996, **24**, 39.
5 C. Giovannini, L. Maiuri and M. De Vincenzi, *Toxic. in vitro*, 1996.

EFFECTS OF RYE AND BARLEY IN COELIAC DISEASE

L. Sabbatella, S. Vetrano, M. Di Tola, C. Casale, M.C. Anania and A. Picarelli

Department of Clinical Sciences. University of Rome "La Sapienza", Viale del Policlinico 155 - 00161 Rome, Italy. (l.sabbatella@email.it)

1 INTRODUCTION

Coeliac disease (CD), defined as a permanent intolerance of the small bowel mucosa to the storage proteins of cereals, is one of the most common immunologically mediated gastrointestinal diseases in Europe[1] and the U.S.[2]. In this genetically related disease, an abnormal immune response to gliadin, so long considered the pivotal event in the pathogenesis of CD, initiates a cascade of, as yet, undefined events that leads to the typical tissue damage.

Coeliac disease is, in fact, usually diagnosed after histological analysis of biopsy samples of the intestinal mucosa showing a mucosal architecture characterised by villous flattening with crypt hyperplasia defined by villous height/crypt depth ratio below $3:1$[3].

The known deleterious constituents of cereals for CD patients are gliadins, secalins, ordeins and avenins, consequently the avoidance of cereals containing these proteins, has been advocated in the gluten-free diet (GFD)[4,5].

It is of interest that, although the culprit antigen of CD is gliadin, antiendomysial antibodies (EMA) and not antigliadin antibodies (AGA) have emerged as an outstanding tool in the screening and follow-up of CD patients because of their high values of sensitivity and specificity[6,7]. Moreover our recent investigations demonstrated that EMA are produced by the intestinal mucosa of CD patients. These studies showed that these antibodies, not present in the culture media of treated CD patients, are newly produced in these patients after in vitro challenge with gliadin or its peptides[8-10].

The organ culture method could be a useful tool to test the immunological effects of cereals on cultured intestinal mucosa from treated CD patients. In fact, we have previously used this system to show that avenin from oats lacks the capacity to induce immunological activation typical of CD[11]. This results have also been confirmed by several in vivo studies[12,13].

In order to better define the role of rye and barley in CD, we have tested whether these cereals are able to induce EMA production during in vitro culture of intestinal mucosa specimens from treated CD patients.

2 SUBJECTS AND METHODS

2.1 Patients

Twenty treated CD patients (8 males, 12 females, mean age 36.4 years, range 20-58 years) were enrolled in the study. Admission criteria included no histological signs of intestinal mucosa atrophy and the absence of EMA in culture supernatants after culture in medium alone. Only 11 out of the 20 patients enrolled met these criteria.

Fifteen patients (6 males, 9 females, mean age 34.5 years, range 21-60 years) affected by gastroenterological disease other than CD were also enrolled as disease control group.

Biopsy samples of duodenal mucosa were obtained from all these patients by means of oesophago-gastro-duodenoscopy for diagnostic purposes.

2.2 Ordein and secalin peptides

Gliadin peptide corresponding to 31-43 amino acid sequence of the α-gliadin (Leu-Gly-Gln-Gln-Gln-Pro-Phe-Pro-Pro-Gln-Gln-Pro-Tyr) was synthesised using the solid-phase method[14]. Peptide purity, higher than 99%, was verified by mass spectrometry (Maldi-MS analysis).

Peptic-tryptic digests (PT) of ordein and secalin were obtained by enzymatic sequential digestion of prolamine fractions extracted from pure varieties of barley (Arma-variety) and rye (variety 500 2G)[15, 16].

2.3 Biopsy culture

Duodenal mucosa biopsies from CD patients and disease controls were cultured for 72 hours at 37°C, one in the presence of 31-43 peptide of the α-gliadin (0.5 g/l), one in the presence of PT-ordein (2 g/l), one in the presence of PT-secalin (2 g/l) and one in medium alone. Culture supernatants were collected and stored at −70°C until used. Since different samples from the same patient were cultured with and without gliadin, ordein and secalin, each subject acted as their own internal control.

2.4 Antiendomysial antibodies detection

EMA were detected in undiluted culture supernatants on cryostat sections of monkey oesophagus by means of indirect immunofluorescence analysis using a commercially available kit (Eurospital, Trieste. Italia).

3 RESULTS

3.1 Detection of antiendomysial antibodies in culture supernatants

After in vitro gliadin challenge, all the 11 treated CD patients showed positive EMA results in culture supernatants of the mucosal biopsies. By contrast in the same patients, no EMA were detected in the supernatants of biopsy specimens cultured with medium alone

No EMA were detected in the culture media of biopsy samples, from the 11 treated CD patients, after in vitro challenge with PT-ordein and PT-secalin.

No EMA were detected in the culture media of the 15 control disease patients, irrespective of gliadin, ordein and secalin challenge (Table 1).

	Medium alone	31-43 (gliadin) (0.5 g/l)	PT-ordein (2 g/l)	PT-secalin (2 g/l)
Treated CD patients (n=11)	0/11	11/11	0/11	0/11
Disease controls (n=15)	0/15	0/15	0/15	0/15

Table 1. *EMA production in culture supernatants*

4 CONCLUSIONS

The injurious constituents for CD patients of wheat (gliadin), rye (secalin), barley (hordelin) and oats (avenin) are predominantly the alcohol-soluble protein fractions (prolamins). The amino acid sequences Pro-Ser-Gln-Gln and Gln-Gln-Gln-Pro are considered the toxic fractions of these prolamins. At least one of these sequences is present in the above-mentioned cereals[14].

Whether or not to include some cereals in the GFD of CD patients has been debated for many years. However, only in a few studies has the effect of these cereals on the small intestinal mucosa of CD patients been analysed; and almost all were performed to show beneficial effects of using oats in the GFD[12,13].

Previous studies demonstrated that EMA detection in culture supernatants of intestinal biopsies from treated CD patients, could be a suitable tool to test the immunological effect of gliadin[8-10]. We have already used this in vitro culture system to test the specific immunological effect, EMA production, of oats in CD[11].

In the present study we have demonstrated that EMA are detectable in culture supernatants only after culture with gliadin, whereas EMA were never detectable after culture in either medium alone or in medium with added PT-ordein or PT-secalin. Moreover, it may be argued that absence of EMA in those culture supernatants, despite the high concentration used (2 mg/ml) and the long culture time (72 hours) is proof of a non immunogenic effect of barley and rye in CD

In conclusion, our data suggest that in vitro challenge with PT-ordein or PT-secalin is unable to induce an immunological response typical of CD, evidenced by EMA production in treated CD patients. Therefore the results would suggest the safety of barley and rye consumption in the gluten-free diet of CD patients.

In addition, as we have previously shown for oats[11], this work demonstrates that the organ culture system can be used as a useful tool to test the immunologic effects of different cereals in coeliac disease patients.

References

1. C. Catassi, I.M. Ratsch, E. Fabiani, M. Rossini, F. Bordicchia, F. Candela, G.V. Coppa and P.L. Giorgi, *Lancet*, 1994, **343**, 200.
2. T. Not, K. Horvath, I.D. Hill, J. Partanen, A. Hammed, G. Magazzu and A. Fasano, *Scand J Gastroenterol*, 1998, **33**, 494.
3. M. Maki and P. Collin, *Lancet*, 1997, **349**, 1755.
4. P.G. Baker and A.E. Read, *Postgrad Med J*, 1976, **52**, 264.
5. A.S. Abdulkarim and J.A. Murray, *Aliment Pharmacol Ther*, 2003, **17**, 987.

6   M. Ferreira, S. Lloyd Davies, M. Butler, D. Scott, M. Clark and P. Kumar, *Gut*, 1992, **33**, 1633.
7   G. Corrao, G.R. Corazza, M.L. Andreani, P. Torchio, R.A. Valentini, G. Galatola, D. Quaglino, G. Gasbarrini and F. Di Orio, *Gut*, 1994, **35**, 771.
8   A. Picarelli, L. Maiuri, A. Frate, M. Greco, S. Auricchio and M. Londei, *Lancet*, 1996, **348**, 1065.
9   A. Picarelli, M. Di Tola, L. Sabbatella, M.C. Anania, T. Di Cello, R. Greco, M. Silano and M. De Vincenzi, *Scand J Gastroenterol*, 1999, **34**, 1099.
10 A. Picarelli, L. Sabbatella, M. Di Tola, S. Vetrano, C. Maffia, C. Picchi, A. Mastracchio, P. Paoluzi and M.C. Anania, *Clin Chem*, 2001, **47**, 1841.
11 A. Picarelli, M. Di Tola, L. Sabbatella, F. Gabrielli, T. Di Cello, M.C. Anania, A. Mastracchio, M. Silano and M. De Vincenzi, *Am J Clin Nutr*, 2001, **74**, 137.
12 E.K. Janatuinen, T.A. Kemppainen, P.H. Pikkarainen, K.H. Holm, V.M. Kosma, M.I. Uusitupa, M. Maki and R.J. Julkunen, *Gut*, 2000, **46**, 327.
13 E.K. Janatuinen, T.A. Kemppainen, R.J. Julkunen, V.M. Kosma, M. Maki, M. Heikkinen and M.I. Uusitupa, *Gut*, 2002, **50**, 332.
14 G. De Ritis, S. Auricchio, H.W. Jones, E.J.L. Lew, J.E. Bernardin and D.D. Kasarda, *Gastroenterology*, 1988, **94**, 41.
15 G. De Ritis, P. Occorsio, S. Auricchio, F. Gramenzi, G. Morisi and V. Silano, *Pediatr Res*, 1979, **13**, 1255.
16 S. Auricchio, G. De Ritis, M. De Vincenzi, P. Occorsio and V. Silano, *Pediatr Res*, 1982, **16**, 1004.

NUTRITIONAL COMPONENTS OF MILL STREAM FRACTIONS

L.D. Simmons[1], K.H. Sutton[1] and M.S. Noorman[2]

[1]New Zealand Institute for Crop and Food Research, Private Bag 4704, Christchurch, New Zealand
[2]Larensteinselaan 26a, P.O. Box 9001, 6880 GB Velp, The Netherlands

1 INTRODUCTION

The commercial flour milling of wheat grains results in the production of a considerable number of fractions or millstreams. These fractions are then blended to produce a commercial flour with specified characteristics. During the milling process there is a significant amount of segregation of the flour components at the molecular level.[1] This results in fractions containing varied concentrations of particular components, which may be utilised as a ready source of these materials if required. This work investigates the changes in distribution of nutritionally significant compounds, particularly antioxidants, that occur as a result of milling a number of disparate wheat varieties. Natural antioxidants in food are believed to be important agents in the reduction of diseases of ageing, if consumed in adequate amounts on a regular basis.[2-7] A knowledge of the distribution of these compounds within mill stream fractions could allow the production of material with enhanced nutritional performance or find uses for mill stream fractions that are currently under-utilised. The relationship between these compounds and the distribution of compounds such as proteins was also investigated, with a view to identifying alternative forms of component analysis to provide estimates of nutritional performance.

2 MATERIALS AND METHODS

Two wheat varieties from New Zealand, Sapphire and Monad, and two wheat varieties from Australia, Janz and Frame were used for the study. The wheats were milled as pure cultivars at the BRI pilot mill in Sydney, Australia.[1] Material from the each of the resulting mill streams was collected separately and a representative sample stored at −80°C until analysed. Each sample was measured for phytic acid level and total antioxidant activity. Phytic acid was measured using the method of Haug and Lantzsch,[8] and total antioxidant activity of each mill stream fraction was measured using the method of Miller and Rice-Evans.[9] Reagents and reference materials were sourced from Sigma.

3 RESULTS AND DISCUSSION

3.1 Phytic acid distribution

We found phytic acid was strongly segregated by the mill-streaming process in all four varieties, as shown in Fig. 1.

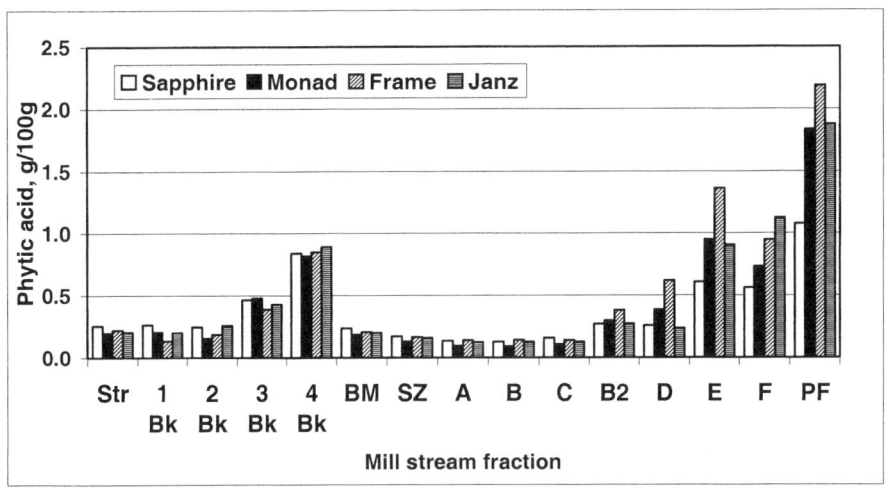

Figure 1 *Distribution of Phytic acid in mill-stream fractions*

Phytic acid (*myo*-inositol 1,2,3,5/4,6-hexakis[dihydrogenphosphate]) frequently occurs as phytin, a mineral storage material that is used to support seedling growth. Most of the phytate in wheat grains occurs in the aleurone layer.[10] During the milling process, most of these aleurone cells remain with particles of pericarp, so that phytate becomes concentrated in the bran fractions.[10] This effect can be clearly seen in Figure 1. Phytic acid has long been considered an anti-nutritional factor in cereals because of its negative effects on the bioavailability of iron, zinc and calcium. With a high density of negatively charged phosphate groups, phytic acid forms very stable complexes with mineral ions, rendering them unavailable for intestinal uptake.[11] However, studies have shown phytic acid to have anti-neoplastic properties in breast, colon, liver, leukaemia, prostate, sarcomas, and skin cancer.[12] It has been suggested that phytic acid acts as an antioxidant to inhibit the generation of reactive species from H_2O_2 by chelating minerals, resulting in the chemoprevention of cancer.[13] Phytic acid has been shown to contribute to the hypochloestrolemic properties of grain diets.[14]

3.2 Antioxidant activity

The distribution of total antioxidant activity, expressed as μmol Trolox equivalent antioxidant capacity per 100 grams, is shown in Figure 2. The data shows that antioxidant activity increases as colour darkens and bran content increases within each millstream fraction. Antioxidant activity is not uniform across wheat variety, cultivar Janz giving more activity than the other varieties, but the pattern of distribution of antioxidant activity is similar to that seen with the phytic acid; an increase across the four break streams, a drop with the initial reduction streams, and an increase with successive reduction streams culminating in the very high levels of the E, F, and polish-finisher (PF) streams. Comparing the antioxidant activity of each millstream fraction with that of the resulting straight-run flour (fraction Str in Figure 2) shows the segregating effect the milling process has on the distribution of these molecular level components. The levels of antioxidant activity in the lower reduction streams are similar to those seen in vegetables when assayed by the same system,[15] but are lower than those levels seen in berryfruit.[15]

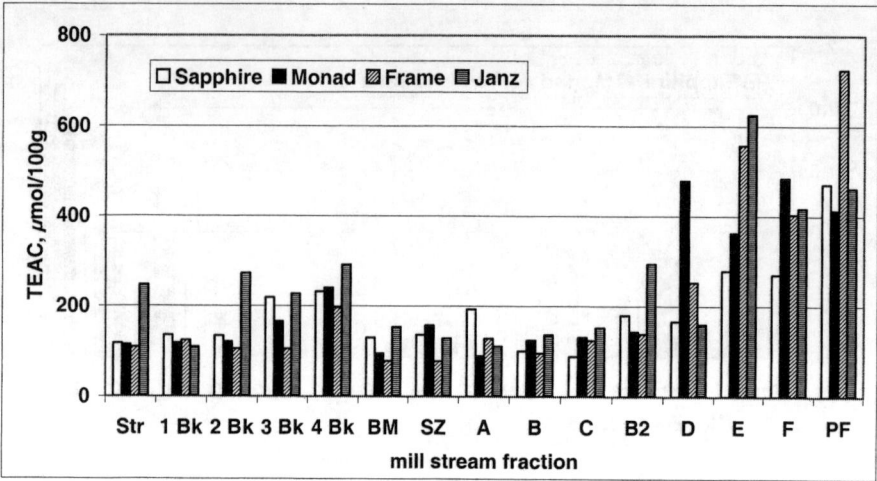

Figure 2. *Antioxidant activity of mill stream fractions, expressed as μmol/100g Trolox equivalent antioxidant capacity.*

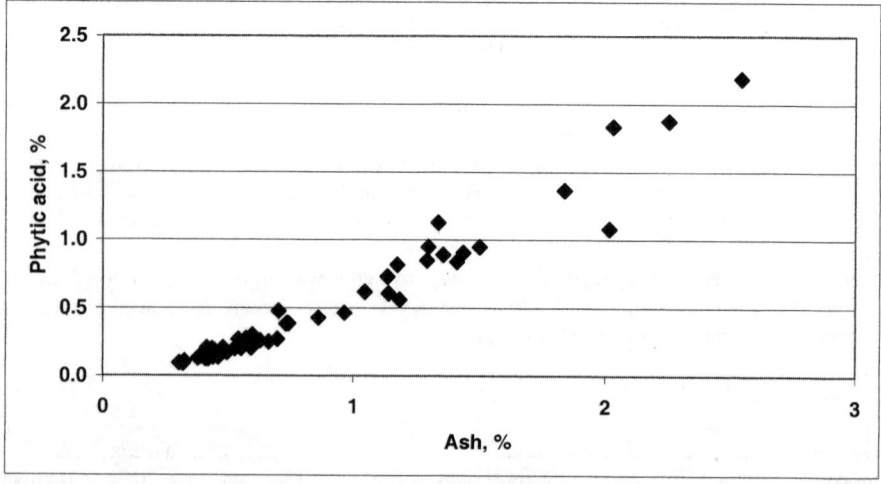

Figure 3. *Relationship of ash to phytic acid content in mill stream fractions.*

3.3 Relationship with other compounds.

The distribution of phytic acid within the millstream fractions showed a highly significant relationship with ash content (adjusted $r^2 = 0.95$, P<0.001)(Fig.3) and early eluting albumins and globulins (adjusted $r^2 = 0.92$, P<0.001), and a less strong relationship with Kent-Jones flour colour grade (adjusted $r^2 = 0.76$, P<0.001). The relationship with ash indicates the association of phytic acid with the aluerone layer in the wheat kernel,[10] and that this would be a good diagnostic feature to use when checking for phytic acid content. Phytic acid distribution showed little relationship to protein content, insoluble glutenin content, pentosan level or ß-glucan level (data not shown).

Total antioxidant activity showed a weaker relationship with ash content (adjusted $r^2 = 0.72$, P<0.001). This indicates that more than one compound can contribute to antioxidant activity, but that the source of at least some of these compounds lies in the aleurone and bran layers of the wheat kernel. The nature of these compounds is currently under study.

4 CONCLUSION

Wheat fractions were shown to have the equivalent antioxidant levels of vegetables. The milling process is an important way of concentrating these compounds, particularly in fractions that are of less economic importance in normal flour production. This may be a convenient way of providing a source of these compounds to augment other foods. Ash content was one simple way of estimating antioxidant activity in mill stream fractions.

References

1. L.D. Simmons and K.H. Sutton, *Proceedings of the 51st Australian Cereal Chemistry Conference*, 9-13 September 2001, ed. M.Wootton,75.
2. B.B. Maziya-Dixon, C.F. Klopfenstein and H.W. Leipold, *Cereal Chem.*, 1994, **71**, 359.
3. E.R. Greenberg and M.B. Sporn, *New Engl. J. Med.*, 1996, **334**, 1189.
4. E.B. Rimm, A. Ascherio, E. Giovannucci, D. Spiegelman, M.J. Stampfer and W.C. Willett, *JAMA*, 1996, **275**, 447.
5. L. Chatenoud, A. Tavani, C. Vecchia, D.R. Jacobs, E. Negri, F. Levi and S. Franceschi, *Int. J. Cancer*, 1998, **72**, 24.
6. A Wolk, J.E. Manson, M.J. Stampfer, G.A. Colditz, F.B. Hu, F.E. Speizer, C.H. Hennekens and W.C.Willett, *JAMA*, 1999, **281**, 1998.
7. D.R. Jacobs, L. Marquart, J. Slavin and L.H. Kushi, *Nutr. Cancer*, 1998, **30**, 85.
8. W. Haug and H-J. Lantzsch, *J. Sci.Food Agric.*, 1983, **34**, 1423.
9. N.J. Miller and C.A Rice-Evans, *Free Radical Research*, 1997, **26**, 195.
10. B. O'Dell, A.R. de Boland and S.R. Koirtyohann, *J. Agr. Food Chem.*, 1972, **20**, 718
11. H.W. Lopez, F. Leenhardt, C. Coudray and C. Remesy, *Int. J. Food Sci. Tech.*,2002, **37**, 727.
12. C.H.Fox and M. Eberl, *Complementary Therapies in Medicine*, 2002, **10**, 229
13. K. Midorikwa, M. Murata, S. Oikawa, Y. Hiraku and S. Kawanishi, *Biochem and Biophys. Research Comm.*, 2001, **288**, 552
14. B.B. Maziya-Dixon and C.F. Klopfenstein, *Cereal Chem.*, 1994, **71**, 539
15. C.E.Lister. Crop & Food Research Confidential Report 173, 2000,17.

FERMENTED WHEAT GERM EXTRACT IN THE TREATMENT OF COLORECTAL CANCER

R. Tömösközi-Farkas[1] and M. Hidvégi[2]

[1]Central Food Research Institute, Budapest, Hungary
[2]Jewish University, Budapest, Hungary

1 INTRODUCTION

The Hungarian Nobel-prized biochemist Albert Szent-Györgyi extensively studied various of the wheat plant for their anticarcinogenic effects and the biological activity of quinones which present in wheat germ. According to his theory, the two quinones, 2-methoxy-p-benzoquinone (2-MBQ) and 2,6-dimethoxy-p-benzoquinone (2,6-DMBQ) are likely to be responsible for the supposed biological properties of wheat germ, because similar quinones, together with ascorbic acid, are involved in a series of metabolic reactions of vital importance in which molecular oxygen is consumed. In wheat germ, 2-MBQ and 2,6-DMBQ appear in the form of glucoside. During fermentation of wheat germ with yeast, the quinones are released by the yeast glucosidase. The aim of our experiments was to examine the metastasis-inhibiting effects of this fermented wheat germ extract on different colorectal tumor lines[1,2].

2 METHODS AND RESULTS

2.1 Animals

In all experiments inbred lines of mice and rats from Semmelweis Medical University (Budapest) were used. They were maintained in plastic cages and were fed with rodent pellets and tap water *ad libitum*.

2.2 Tumor models

The following transplantable tumor lines, grown on mice or rats, were used in the experiments: C38 mouse colorectal tumor and HCR25 human colorectal cancer. C38 mouse colorectal carcinoma was mantained in C57B1/6 mice by subcutaneous transplantation. HCR25 human colon carcinoma xenograft was established from a moderately differentiated primary human colon carcinoma. The tumor was maintained in immunosuppressed CBA/CA mice. The technique of preparing single cell suspensions in these cases was similar. In our experiments the C38 and HCR25 tumor lines were used for establishing a spleen-liver metastasis model.

2.3 Treatment

MSC treatment was started 24 hours after tumor implantation. MSC was dissolved in water and administered by means of a gastric tube. The daily dose was 3 g/kg body weight per os administered in 0,1 ml of water. Control animals received tap water daily (0,1 ml), also via gastric tube.

These experiments were completed 20 (C38) and 51 (HCR-25) days after tumor inoculation, by means of exsanguination during anaesthesia.

In colon carcinogenesis experiments, carcinogenesis was induced by azoxy-methane (AOM). AOM was dissolved in physiologic saline and the animals were given three subcutaneous injections, (each 15 mg/kg body weight) at 1 week intervals.

2.4 Effect of MSC treatment alone

The results of the experiments in which only MSC treatment alone was applied are summarized in Figure 1. for the case of the HCR-25 human colon carcinoma. The MSC treatment decreased the amount of liver metastases. The number of metastases as compared to the control group of both the splenectomized MSC-treated animals was around 50% [1].

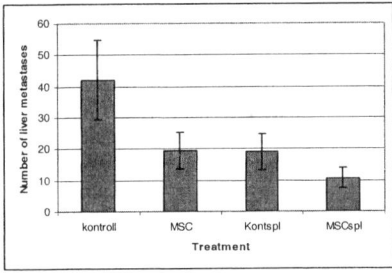

Figure 1 *Effect of MSC on liver metastases number in HCR25 human colorectal carcinoma*

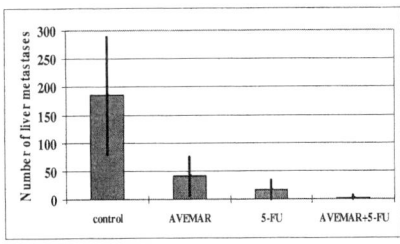

Figure 2 *Effect of therapeutic composition(MSC+5FU) on liver metastases of C38 colorectal carcinoma*

2.5 Combined used of MSC and cytostatics

The aim of this experiment was to find out how the daily treatment with MSC would influence the tumor growth and metastasis inhibiting effect of treatment with the well known antineoplastic agent (5-Fluoro-Uracil – 5FU) widely used in clinical oncology.

Figure 2. shows that a 20 days treatment of C38 colorectal carcinoma with the therapeutic composition of MSC and 5FU decreased the number of liver metastases synergically. The effect was significant[2,3].

2.6 Effect of MSC on carcinogenesis

This experiment demonstrated that the wheat germ extract prevents colonic cancer in laboratory animals. Ten rats served as untreated controls (group 1), whilst a second group was treated with AOM. In two additional groups MSC was administered as a tentative chemo-preventive agent. In group 3, animals started to receive MSC 2 weeks prior to the first administration of AOM and was given daily throughout the experimental period. In group 4 the basal diet and MSC were administered only.

The results of this experiment are presented in Table 1. No macroscopically lesions occurred in the intestinal tract of group 1 and 4. Out of 47 AOM-treated rats (group 2) 39 animals developed grossly identifiable polypoid tumors in the colon. In group 3, where MSC was administered before and following AOM doses the number of animals with neoplastic lesions in the colon significantly decreased. The average number of colon tumors per individual animal also significantly decreased upon administration of MSC. The chemopreventive effectiveness of the fermented wheat germ extract was estimated to be 70 per cent[4].

	No. of animals with colon tumors	Average number of colon tumors per animals	Average diameter of the colon tumors	Remark
Untreated controls ($n = 10$)	0/10	0/10	—	
Azoxymethane (AOM) ($n = 47$)	39/47 (83.0 %)	2.3 ± 0.21	2.35 ± 0.25	1 Wilms' tumor
MSC (AVEMAR) +AOM ($n = 29$)	13/29 (44.8 %)*	1.3 ± 0.17 **	2.21 ± 0.12	
MSC (AVEMAR) ($n = 9$)	0/9	0/9	—	

Table 1 *Chemopreventive effect of MSC (AVEMAR) = 100 - {100 x (0.448 x 1.3)/(0.83 x 2.3)} = 0.7 = 70 %, * P < 0.001, ** P < 0.004*

2.7 Clinical studies of MSC

This open-label cohort trial has compared anticancer treatments plus MSC (9g once daily) vs anticancer treatments alone in colorectal cancer patients, enrolled from three oncosurgical centres. Sixty-six colorectal cancer patients received MSC supplement for more than 6 months and 104 patients served as controls. No statistical differences were noted in the time from diagnosis to the last visit between the two groups. All patients were evaluated at base line, at the end of the first month and every 12 weeks thereafter. Evaluation included assessment of all measurable lesions by imaging techniques, laboratory tests, physical examination and data collection of treatment-related toxicities. End-point analysis revealed that progression-related events were significantly less frequent in the MSC group (Figure 3). Survival analysis showed significant improvements in the MSC group regarding progression-free and overall survivals probabilities (Figure 4)[5].

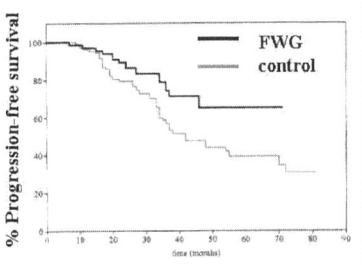

Figure 3. *Kaplan-Meier estimate of the cumulative probability of remaining free from disease progression in colorectal cancer patients*

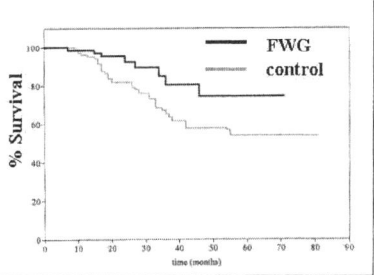

Figure 4. *Kaplan-Meier estimate of the cumulative probability of overall survival in colorectal cancer patients*

3 CONCLUSION

Both the animal model investigations and clinical studies proved that the wheat germ extract, in combination with surgery plus radio/chemotherapy, may inhibit overall tumour progression, including the formation of new metastases, and may prolong the survival of colorectal cancer patients.

References

1. M. Hidvégi, E. Rásó, R. Tömösközi-Farkas, S. Paku, K. Lapis, B. Szende, *Anticancer Res* 18: 2353-2358, 1998.
2. K. Lapis, B. Szende *Immunopharmacology* 41: 183-186, 1999.
3. M. Hidvégi, E. Rásó, R. Tömösközi-Farkas, B. Szende, S. Paku, L. Prónai, J. Bocsi, K. Lapis *Cancer Biother Radiopharm* 14: 277-289, 1999.
4. A. Zalatnai, K. Lapis, B. Szende, E. Rásó, A. Telekes, Á. Resetár, M. Hidvégi *Carcinogenesis* 22: 1649-1652, 2001.
5. F.Jakab, Y.Shoenfeld, Á. Balogh, M. Nichelatti, A. Hoffmann, Zs. Kahán, K.Lapis, Á.Máyer, P. Sápy, F.Szentpétery, A.Telekes, L. Thurzó, A.Vágvölgyi and M. Hidvégi *Brit.J. Cancer* 89, 465-469, 2003

Acknowledgment: The financial support of the National Scientific Research Fund (OTKA, F038335)and Ministry of Education (OM, BIO-00072/2000) is greatly appreciated.

Modifications Due to Parasite Attacks

SPECIFICITY OF ACTION OF THE WHEAT BUG (*NYSIUS HUTTONI*) PROTEINASE

D. Every[1], K. H. Sutton[1], P. R. Shewry[2], A.S. Tatham[2] and T. Coolbear[3]

1. NZ Institute for Crop & Food Research Limited, Private Bag 4704, Christchurch, New Zealand. 2. Rothamsted Research, Harpenden, Hertfordshire AL5 2JQ, UK. 3. Fonterra Research Centre, Private Bag 11029, Palmerston North, New Zealand.

1 INTRODUCTION

The New Zealand wheat bug (*Nysius huttoni*) injects a salivary proteinase into immature wheat kernels while feeding. This proteinase remains in flour made from bug-damaged wheat, and in dough digests gluten to produce slack, sticky dough and poor quality bread.[1] A micro-assay for *Nysius*-proteinase was developed using gluten as substrate[2], and the enzyme was purified and characterised as a serine proteinase[3]. Early attempts to develop an assay[2-4] showed that the enzyme did not act on several proteins and synthetic substrates commonly used for assay of proteases. The enzyme did, however, specifically hydrolyse the high molecular weight glutenin subunits (HMW-GS) of glutenin.[1,4] This paper describes enzyme specificity tests on a range of proteins and synthetic peptides, and reports the site of *Nysius*-proteinase action on HMW-GS 2, β-casein and κ-casein.

2 METHODS AND RESULTS

2.1 Methods

Nysius proteinase was extracted from wholemeal flour of NZ wheat that was seriously bug damaged by *Nysius huttoni*, and partially purified by ion exchange and hydrophobic interaction chromatography.[3] Papain and α-chymotrypsin (Sigma) were used as comparative controls in some assays. SDS-insoluble glutenin I, SDS-soluble glutenin II and gliadin were extracted[5] from vital wheat gluten (Bunge, Australia). HMW-GS 2 was isolated from flour by RP-HPLC.[6] Gelatine was from Davis Gelatine (NZ) Ltd. All other proteins were from Sigma. Protease action on protein substrates was measured by a radial diffusion in gel method[7], a SDS-protein gel assay[2], SDS-PAGE[1] and alkaline urea-PAGE[8] analysis of substrate and products liberated by proteolysis. Synthetic fluorogenic peptides were made by D. Harding (Massey University, NZ) and tested as protease substrates using a Shimadzu Fluorimeter. Peptides from proteolysis of purified HMW-GS 2, β-casein and κ-casein were isolated by RP-HPLC, sequenced by an automated Edman procedure and identified with peptide sequences in the original substrate proteins.[6,9]

Table 1. *Proteinase activity in the radial diffusion assay.*[a]

Substrate	*Nysius* proteinase	α-Chymotrypsin
Gelatine	-	+++
Casein	+-	++++
Azo-Casein	-	++++
Haemoglobin	-	++
Fibrin	-	+
Glutenin I	+-	+-
Glutenin II	+++	++
Gluten	+-	+

[a]Enzyme activity was a subjective estimate of the extent of radial hydrolysis of protein.
+- indicates weak activity that was difficult to assess.

Table 2. *Proteinase activity in the SDS-protein gel assay.*[a]

Substrate	*Nysius* proteinase	α-Chymotrypsin
Glutenin I	65000	480
Gluten	56000	590
Fibrin	0	1490
Hide powder	0	NT
Elastin	0	NT
Collagen	0	NT

[a]Enzyme activity was measured as Units/mg enzyme.[2] NT = Not tested.

2.2 Enzyme tests on protein and synthetic substrates

In the radial diffusion assay (Table1), glutenin II was the most reactive substrate for *Nysius* proteinase. Reactions of *Nysius* proteinase with gluten, glutenin II and casein were weak, probably because only a fraction of the constituent protein subunits were hydrolysed and the unhydrolysed protein subunits obscured visualisation of hydrolysed subunits. Other proteins did not react. α-Chymotrypsin hydrolysed all the proteins to variable extents.

The SDS-protein gel assay (Table 2), only used water insoluble proteins that could form gels in SDS-reagent to variable degrees. Hydrolysed protein became soluble in SDS-reagent and could not form a gel. In this assay, glutenin I and gluten were the only proteins hydrolysed by *Nysius* proteinase. Chymotrypsin hydrolysed all the proteins that were tested.

In the PAGE methods, the only proteins to be substantially hydrolysed by *Nysius* proteinase were the HMW-GS subunits of wheat, the M_r 75000 γ- and HMW secalins of rye, D hordein (the HMW subunit homologue) of barley, and the β- and κ-casein subunits of bovine milk.

The following synthetic fluorogenic substrates were not hydrolysed by *Nysius* proteinase: Pro-Gly-Gln-Aminoquinoline (AQ), Pro-Leu-Leu-AQ, Pro-Leu-Leu-Gln-AQ, Pro-Leu-Thr-Gln-AQ, Pro-Leu-Gly-Gln-AQ, Butyloxycarbonyl-Gln-AQ, Carbobenzoxy-Gln-AQ, Succinyl-Pro-Tyr-Gln-Gln-AQ, and Succinyl-Tyr-Tyr-Amino-4-methylcoumarin. As positive controls of the fluorometric assay, α-chymotrypsin hydrolysed

Table 3. *Proteinase activity determined by PAGE.* [a]

Substrate	*Nysius* proteinase	α-Chymotrypsin
Gluten	+ (HMW-GS bands only)	NT
Glutenin I	+ (HMW-GS bands only)	NT
Glutenin II	+ (HMW-GS bands only)	+ (all bands)
HMW-GS 2	+	NT
Gliadin	-	NT
HMW secalin of rye	+	NT
Rye proteins (total)	+ (HMW- & γ-seculins only)	NT
D hordein of barley	+	NT
Barley proteins (total)	+ (D-hordein only)	NT
Corn proteins (total)	-	NT
Haemoglobin	-	+
Bovine serum albumin	-	+
Papain	-	+
Cytochrome-C	-	+
Cytochrome-C oxidase	-	+
Casein (α- & β-Casein)	+ (β-Casein band only)	+ (all bands)
A-Casein	-	NT
B-Casein	+	NT
K-Casein	+-	NT
$α_{s1}$-Casein*	+-	NT
B-Casein*	+	NT
K-Casein*	+	NT

[a] Substrates marked * were analysed by Urea-PAGE, the rest were SDS-PAGE.
+ indicates disappearance or substantial reduction of stained bands.
+- indicates weak activity that was difficult to assess.

Table 4. *Cleavage sites in HMW-GS 2, β-casein and κ-casein cleaved by Nysius proteinase.*

Peptide	Likely cleavage site (↓) in sequence
Enzyme product of GS 2	↓GQP[a]-GYYP[b]TSLQQ-LGQGQQ-GYYPT-
Repeat block 1 of GS 2	-SGQ↓GQP -GYYP TSLQQ-LGQGQS-GYYPTSPQQ-
Repeat block 2 of GS 2	-PGQ↓GQP -GYYP TSPQQ-SGQGQP-GYYPTSSQQ-
Repeat block 3 of GS 2	-PGQ↓GQP -GYYP TSPLQ-SGQGQP-GYYLTSPQQ-
Repeat block 4 of GS 2	-PGQ↓GQP -GYYP TSPLQ-PGQGQP-GYDPTSPQQ-
Repeat block 5 of GS 2	-PGQ↓GQP -GYYL TSPLQ-LGQGQQ-GYYPTSLQQ-
Sequence 1 in β-casein	-PLLQ↓SYM-
Sequence 2 in β-casein	-PLTQ↓TPV-
Sequence 3 in β-casein	-PFAQ↓TQS-
Sequence 4 in β-casein	-PVPQ↓KAV-
Sequence 1 in κ-casein	-I PI Q↓YVL-
Sequence 2 in κ-casein	-SPAQ↓ILQ-

[a] P was mostly detected at this site, but Q was also detected.
[b] Both P and L were detected at this site.

Succinyl-Tyr-Tyr-Amino-4-methylcoumarin and papain hydrolysed Pro-Leu-Leu-AQ, but none of the other substrates.

2.3 *Nysius*-proteinase cleavage sites in proteins

Sequence analysis of the peptide products of *Nysius*-proteinase reaction on HMW-GS showed that the major peptide was from the repetitive domain, and corresponded to part of a hexapeptide followed by a nonapeptide, hexapeptide and nonapeptide (Table 4). There are, in fact, only five such blocks of repeats (6.9.6.9) in subunit 2, and in Table 4 they have been aligned with the sequence of the peptide product of the *Nysius*-proteinase reaction. None of the sequences correspond exactly to the peptide product. However, it is clear that the enzyme has cut within a hexapeptide, between glutamine (Q) and glycine (G). Sequence analysis of the peptide products of *Nysius*-proteinase reaction on β-casein and κ-casein also revealed that glutamine occupied the P1 position relative to the scissile bond at all cleavage sites (Table 4), but glycine did not occupy the P'1 position. For all three proteins, however, another preferred structural feature around the site of the scissile bond appears to be a proline (P) in the P3 or P4 position.

3 CONCLUSION

Nysius-proteinase is a very specific enzyme, which does not hydrolyse a range of proteins and synthetic substrates, some of which are commonly used in protease assays. These include low molecular weight glutenin subunits and gliadins of wheat, M_r 40000 γ- and ω-secalins of rye, barley proteins other than D hordein, corn proteins, gelatine, hide powder (keratin), fibrin, collagen, elastin, haemoglobin, bovine serum albumin, azo-casein, cytochrome-C, cytochrome-C oxidase and papain. The only proteins hydrolysed by *Nysius*-proteinase are the HMW-GS subunits of wheat, the M_r 75000 γ- and HMW secalins of rye, the D hordein of barley, and the β- and κ-casein subunits of bovine milk. Sequence analysis of the peptide products of *Nysius*-proteinase reaction for HMW-GS 2, β-casein and κ-casein revealed that glutamine occupied the P1 position relative to the scissile bond at all cleavage sites.

References

1 P.J. Cressey and C. L. McStay, *J. Agric. Food Chem.*, 1987, **38**, 357.
2 D. Every, *Anal. Biochem.*, 1991, **197**, 208.
3 D. Every, *J. Cereal Sci.*, 1993, **18**, 239.
4 W.H. Swallow and D. Every, *Cereal Foods World*, 1991, **36**, 505.
5 A. Graveland, P. Bosveld, W.J. Lichtendonk, H.H. Moonen and A. Scheepstra, *J. Sci. Food Agric.* 1982, **33**, 1117.
6 K.H. Sutton, *J. Cereal Sci.*, 1991, **14**, 25.
7 G.F.B. Schumacher and W.B. Schill, *Anal. Biochem.*, 1972, **48**, 9.
8 L.K. Creamer, *Bulletin Int. Dairy Fed.*, 1991, **261**, 14.
9 J.R. Reid, C.H. Moore, G.G. Midwinter and G.G. Pritchard, *Appl. Microbiol. Biotechnol.*, 1991, **35**, 222.

INHIBITION EFFECTS OF PLANT EXTRACTS ON THE PROTEASE ACTIVITY IN BUG (*Eurygaster* spp.) DAMAGED WHEATS

B. Olanca and D. Sivri

Hacettepe University, Faculty of Engineering, Food Engineering Department, Beytepe Campus, 06532, Ankara / TURKEY

1 INTRODUCTION

Some *Heteropterous* insects called "wheat bugs" (*Eurygaster* spp.) are the most important cereal pests in Turkey. Bug damage to wheat causes substantial losses to the yield and breadmaking quality of wheat.[1,2,3] Before harvest, the wheat bug injects saliva containing proteolytic enzymes into the grain. The proteolytic enzymes cause the breakdown of the gluten proteins during the breadmaking process. Dough prepared from bug-damaged wheat flour has an unusual consistency and produces loaves of low volume and unsatisfactory texture.[3,4,5,6] The rheological properties and breadmaking quality of the resulting flour are affected negatively when the wheat contains as little as 0.3-0.4% bug damaged kernels. At higher levels (>5%) it is not possible to use bug damaged wheat in the production of bread. In our preliminary study, natural inhibitors of wheat bug protease were investigated.[7] Food and feed legumes were preferred as the source of inhibitor due to high concentrations of serine protease inhibitors in their content. The inhibition effects of the crude plant extracts prepared from the legumes were tested by sodium dodecyl sulfate polyacrylamide gel electrophoresis (SDS-PAGE). The results showed that three plant extracts inhibited the bug protease activity. In the present study, the plant extract (PE) showing the highest inhibition of the bug protease activity was selected. Its inhibitory effect on the hydrolysis of glutenin proteins in four bug damaged wheat samples were investigated at different addition levels.

2 MATERIALS AND METHODS

2.1 Materials

Bug damaged wheat samples used in this study were grown at the experimental plots of Field Crops Improvement Center (Lodumlu, Ankara, Turkey). Hard red wheats (HRW1 and HRW2) represent strong gluten quality, and hard white wheats (HWW1 and HWW2) represent weak gluten quality. The levels of bug damage determined by a modified sedimentation test[8] were comparable among the cultivars. The protein contents (Nx5.7) for HRW1, HRW2, HWW1 and HWW2 determined by the Kjeldahl method were 12.2; 12.4, 12.4 and 11.3%, respectively.[9] Wheats were ground in a Buhler experimental mill into straight grade flour. The plant samples were prepared by a coffee grinder after air-drying. The flour of a hard red spring wheat (Katepwa) was used as a standard for electrophoresis.

2.2 Preparation of the Plant Extract

The finely ground plant (5.0 g) was extracted in 50 ml distilled water. The extraction was carried out for 3 hours at 4 °C on a magnetic stirrer. The suspension was centrifuged at 3600xg for 20 min and the supernatant was freeze-dried.

2.3 SDS-PAGE (Sodium Dodecyl Sulfate Polyacrylamide Gel Electrophoresis)

The bug damaged wheat samples (50 mg) were mixed with the plant extract (PE) at different ratios (10:1; 5:1 and 2:1, flour:PE). The samples were incubated with 200 µl distilled water for 2 hours at 37°C. After the incubation, the tubes were centrifuged at 2200xg for 3 min. Glutenin proteins for SDS-PAGE were prepared according to the 50% 1-propanol insoluble glutenin procedure of Fu and Sapirstein and used in SDS-PAGE[10].

2.4 Densitometry

Integrated optical densities of glutenin bands (IOD) were obtained by densitometric analysis of the electrophoregrams. Percentage decreases in IOD values of a glutenin bands were presented as relative bug protease activity (RPA%) and calculated with the following equation:

$$\text{RPA} (\%) = \frac{\text{IODc} - \text{IOD}}{\text{IODc}} \times 100 \quad (1)$$

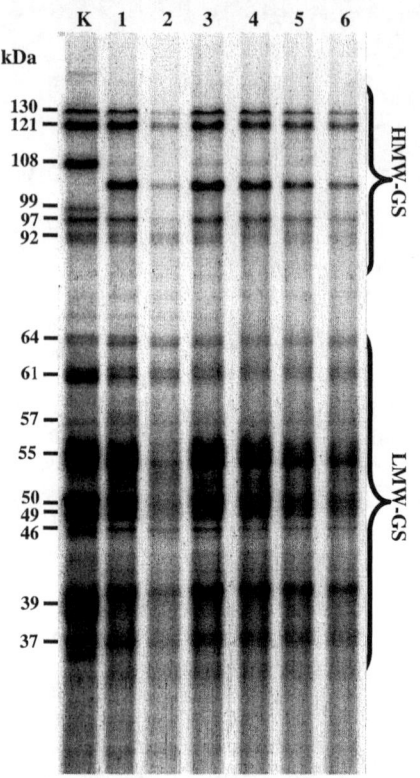

Figure 1 *The inhibition effects of the plant extract (PE) on the hydrolysis of glutenins in HRW1 cultivar*
(K) Katepwa,
(1) HRW1 (Control),
(2) HRW1*,
(3) HRW1 + PE (2:1),
(4) HRW1 + PE (2:1)*,
(5) HRW1 + PE (5:1)*,
(6) HRW1 + PE (10:1)*.
* incubation (37°C, 2 hr)

IOD_c: integrated optical density of glutenin bands (0hr)
IOD: integrated optical density of glutenin bands (2hr)

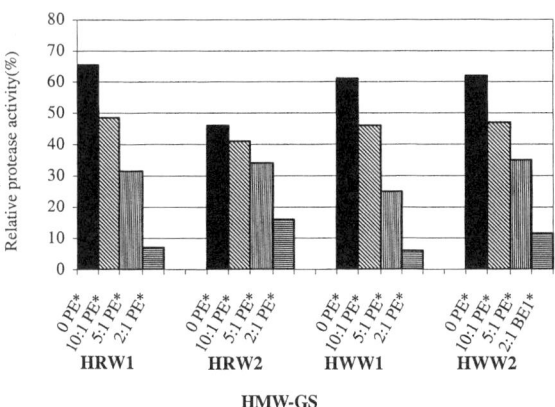
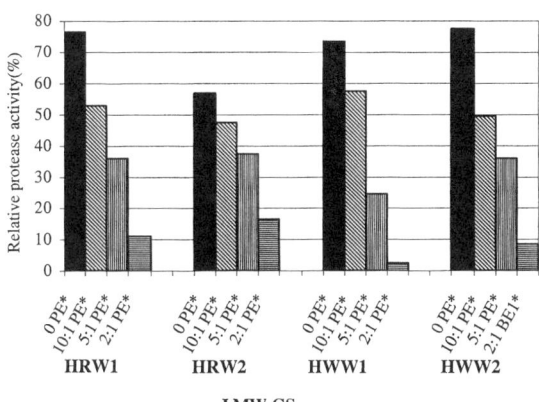
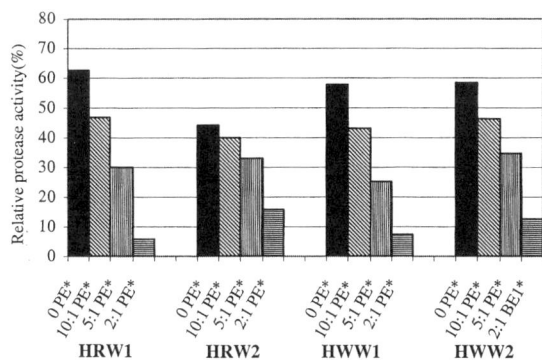

Figure 2 *Relative protease activity of wheat cultivars at various the plant extract (PE) addition levels*

3 RESULTS AND DISCUSSION

Glutenin proteins of the bug damaged wheat cultivars with added plant extract (PE) at different ratios (10:1; 5:1 and 2:1, flour:PE) were investigated by the SDS-PAGE method. A representative electrophoregram is given for HRW1 cultivar in Figure1.

SDS-PAGE results show that relative intensities of protein bands considerably decreased in the cultivars without PE (Fig.1, lanes 1, 2). This indicate that the insect proteases significantly affect the glutenins of bug damaged wheat samples with no inhibitors present. This result is agreement with our earlier studies[11].

However, after incubation (37°C, 2hr), intensities of the protein bands were greater in the samples with the inhibitor (PE) at all levels of addition (Fig.1, lanes 4, 5 and 6) than that of the samples with no inhibitor (Fig.1, lane 2). Therefore, it can be concluded that the PE decreased the proteolytic activity of the bug damaged wheats.

Relative protease activities on HMW, LMW and total glutenin subunits (GS) of the cultivars are shown in Fig. 2. RPA values of HRW1, HWW1 and HWW2 samples with no inhibitors were comparable for HMW-GS and LMW-GS. The lowest RPA was observed in HRW2 cultivar. HMW-GS were considerably affected by the bug protease for the all wheat cultivars. Higher specificity for the HMW-GS than for the LMW-GS confirms our earlier observations[12]. The inhibition effects of PE on the hydrolysis of total, HMW and LMW glutenins by the bug protease varied depending on the addition levels to the wheat samples. More than 50% decreases in the protease activities were observed for the total glutenin proteins in HRW1 and HWW1 due to the addition of the PE at a medium level (5:1). The addition of PE at the highest level (2:1) caused more than 80% inhibition of the bug protease activities in HRW1, HWW1 and HWW2 cultivars.

4 CONCLUSION

The results show that the hydrolysis of 50% 1-propanol-insoluble glutenin subunits due to bug proteases was significantly decreased by addition of the inhibitor. However, further work is required to determine the extent of this inhibitory effect on the rheological and baking properties of bug-damaged wheats.

References
1. F. Paulian and C. Popov, in *Wheat*, Ciba-Geigy, Basel, 1980, p. 69.
2. R.H. Miller and J.G. Morse, in *FAO Plant Production and Protection Paper*, No:138, Rome, 1996, p. 165.
3. B.R. Critchley, *Crop Protect.*, 1998, **17**, 271.
4. V.L. Kretovich, *Cereal Chem.*, 1944, **21**, 1.
5. K. Lorenz and P. Meredith, *Starch/Starke*, 1988, **40**, 136.
6. E. Karababa and A. Ozan, *N., J. Sci. Food Agric.*, 1998, **77**, 399.
7. B. Olanca, D. Sivri and H. Köksel, *ICC Conference 2002*, Budapest, Hungary, 2002, p. 160.
8. W.T. Greenaway, M.H. Neustadt and L. Zeleny, *Cereal Chem.*, 1965, **42**, 577.
9. Approved Methods of the American Association of Cereal Chemist, 8th ed., 1983, Method No: 46-12, The Association, St. Paul, MN, USA,.
10. B.X. Fu and H.D. Sapirstein, *Cereal Chem.*, 1996, **73**, 143.
11. D. Sivri, H. Köksel and W. Bushuk, N. Z. J. *Crop Hort. Sci*, 1998, **26**,11.
12. D. Sivri, H. Sapirstein, H. Köksel, and W. Bushuk, *Cereal Chem.*, 1999, **76**, 816.

Acknowledgments: This project was partially supported by the Hacettepe University Research Found under project number 01 01 602 004.

EFFECTS OF INTERCULTIVAR VARIATION ON THE GLUTEN PROTEINS AND RHEOLOGICAL PROPERTIES OF BUG (*Eurygaster* spp.) DAMAGED WHEATS

D. Sivri*, B. Olanca*, A. Atlı** and H. Köksel*

*Hacettepe University, Faculty of Engineering, Food Engineering Department
Beytepe Campus, 06532, Ankara / TURKEY
**Harran University, Faculty of Agriculture, Food Technology Department, 63040, Şanlıurfa / TURKEY

1 INTRODUCTION

In some wheat growing countries the functionality of gluten is adversely affected by proteases due to pre-harvest bug-damage to wheat. The damaged wheat is unsuitable for processing into bread because the dough produced from it becomes runny and sticky during kneading, fermentation and molding. Furthermore, the resulting bread has low volume and coarse texture.[1] It has been shown that strong bread wheat cultivars are less susceptible to adverse effects of bug protease in baking than the weak ones.[2] In this study, we extend earlier studies[2-3] to investigate the effects of bug protease on a range of bread wheat cultivars that represent different gluten qualities at various bug damage and protein content levels.

2 MATERIALS AND METHODS

Eight hard red winter (HRW) and eight hard white winter (HWW) wheat cultivars were selected to represent samples with a range of physical dough properties (Table 1). The difference between the SDS-sedimentation[4] (S) and modified sedimentation[5] (MS) test values was used to estimate bug-damage level and the samples were subdivided according to their damage levels (low, medium and high). Flour samples (40 mg) were incubated with distilled water (200 µl) for 2 hr at 35°C and 50% 1-propanol (v/v) insoluble glutenins were prepared for SDS-PAGE[6]. Unincubated samples were used as control. Gels were scanned by the Imager BL, (Biolab, Sweden) and quantified by GeneTools software (SynGene-Version 3.02). Relative proteolytic activity (RPA) was calculated using the following equation.

$$\text{Relative Protease Activity (\%)} = \frac{\text{IOD}_c - \text{IOD}}{\text{IOD}_c} \times 100 \qquad (1)$$

IOD_c: integrated optical density of protein bands (Unincubated)
IOD : integrated optical density of protein bands (Incubated for 2 hr)
Farinograph and alveograph test were performed according to the AACC methods[8].

Table 1: *Protein contents, sedimentation (S) and modified sedimentation (MS) values of the wheat cultivars*

Cultivars	Protein Contents*	S (ml)**	MS (ml)**	SV-MSV	Bug-Damage Level
HRW1	12.0	34	19	14	Medium
HRW2	12.0	37	23	15	Medium
HRW3	14.4	39	13	26	High
HRW4	14.9	32	7	25	High
HRW5	12.3	36	20	16	Medium
HRW6	12.3	42	20	22	High
HRW7	12.5	38	8	30	High
HRW8	15.1	34	5	29	High
HWW1	12.0	35	30	5	Low
HWW2	15.0	31	11	20	High
HWW3	14.1	39	10	29	High
HWW4	12.4	38	10	28	High
HWW5	11.4	31	20	11	Medium
HWW6	12.5	34	18	16	Medium
HWW7	14.5	33	5	28	High
HWW8	14.4	33	5	28	High

* (Nx5.7)
** (14% moisture basis)
HRW: Hard Red Winter Wheat
HWW: Hard White Winter Wheat

3 RESULTS AND DISCUSSION

3.1 SDS-PAGE Results

Quantification of SDS-PAGE analysis showed that the hydrolysis rates of glutenin proteins in HRW and HWW wheat samples were different (Figure 1). Although, HRW3, HWW3 and HWW8 samples had comparable protein contents and bug-damage levels, HRW3 showed the lowest relative proteolytic activity. High protein content HWW2 (15.1%) and low protein content HWW4 (12.4%) samples were comparable both in terms of damage level and RPA value for HMW-GS, indicating that effect of protein content on protein hydrolysis rate was small. HWW1 with the lowest bug damage level had higher RPA value than that of HRW3 with a high damage level suggesting that in some cases effects of bug damage level on the rate of hydrolysis were not significant in both HMW-GS and LMW-GS. In all cultivars RPA values of HMW-GS were higher than those of LMW-GS indicating that the bug protease had a high specificity for the HMW-GS. This result is in agreement with the earlier work.[3-4]

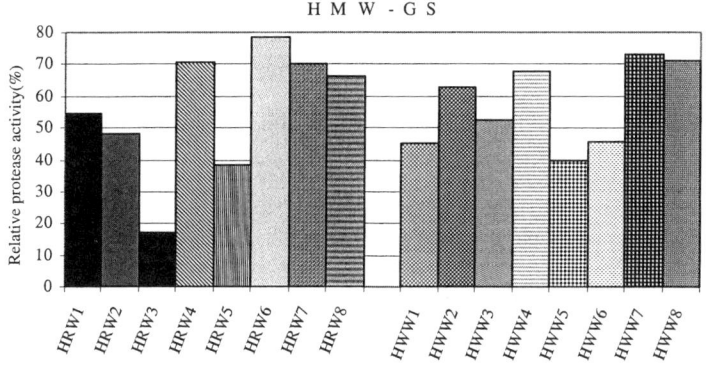
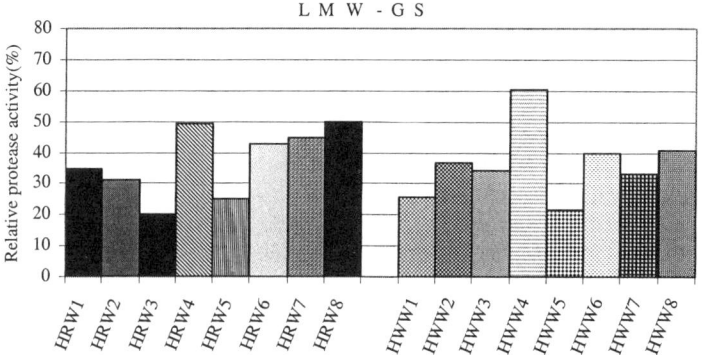
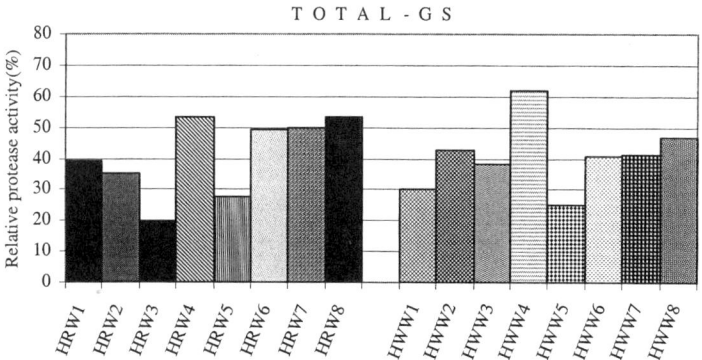

Figure 1 *Relative protease activities of bug damaged wheat cultivars on HMW-GS, LMW-GS and Total 50% 1-propanol insoluble glutenin*

HRW6 and HWW7 showed higher RPA values for HMW-GS than HRW7 and HWW8, respectively. However, there was a reverse trend in their RPA values on LMW-GS.

Therefore, it can be concluded that the rate of hydrolysis of HMW-GS and LMW-GS was affected from genotype.

Four cultivars (HRW1, HRW2, HRW5 and HWW6) were comparable in terms of protein content (12.0-12.5) and bug-damaged level (medium). However HRW5 had lower (<30%) and HWW6 had higher RPA values (~40%) for total glutenin proteins. Although in the present study HRW cultivars had generally higher HMW-GS contents than those of HWW cultivars, the results showed that HRW samples did not have a greater resistance to proteolysis by bug protease compared to HWW samples.

All of these results are consistent with our earlier results[3] indicating that there is a considerable level of intercultivar variation in the resistance of polymeric glutenin to the activity of bug protease. It seems that this intercultivar variation can not be explained by the differences in protein content, HMW-GS content and bug damage level. The nature of this different susceptibility of cultivars to bug protease might also be due to the differences in their HMW-GS subunit composition. Further investigations are required to explain the intercultivar variation in resistance of polymeric glutenin to the bug protease.

3. 2 Rheological Measurements

Although there were significant differences in all farinogram parameters of the samples, correlation coefficients between farinogram parameters and RPA values were low. At high bug damage levels, stability times decreased significantly. Mixing tolerance index and softening degree values increased from 100 to 400 Brabender Units indicating a highly significant weakening of gluten (results not presented). All alveograph parameters (W, P, G and L) were generally low for the all bug-damaged samples. The tenacity value (P) recognized as the indicator of dough resistance to deformation varied among the cultivars (results not presented). Significant correlations were found between deformation energy (W) and RPA values for HRW ($R^2= -0.62$) and HWW ($R^2= -0.77$) cultivars. Correlation coefficients calculated between extensibility (L) and RPA values of HRW and HWW cultivars were –0.70 and –0.54, respectively.

References

1   F. Paulian and C. Popov, in *Wheat*, Ciba-Geigy, Basel, 1980, p. 69.
2   D. Every, J. A. Farrell, M. W. Stufkens and A. R. Wallace, *J. Cereal Sci.*,1998, **27**, 37.
3   D. Sivri, H. Sapirstein, H. Köksel, and W. Bushuk, *Cereal Chem.*, 1999, **76**, 816.
4   Approved Methods of the American Association of Cereal Chemist, 8th ed., Method No: 46-12, No. 54-21, N0. 54-30, St. Paul, MN, USA, 1990.
5   W.T. Greenaway, M.H. Neustadt and L. Zeleny, *Cereal Chem.*, 1965, **42**, 577.
6   B.X. Fu and H.D. Sapirstein, *Cereal Chem.*, 1996, **73**, 143.

RELATIONSHIPS BETWEEN TIMING OF *Eurigaster maura* ATTACKS AND GLUTEN DEGRADATION IN TWO BREAD WHEAT CULTIVARS

P.Vaccino [1], M.Corbellini [1], A.Curioni [2], G.Zoccatelli [3], M.Migliardi [4] and L.Tavella [4]

[1] Istituto Sperimentale per la Cerealicoltura, via Mulino 3, 26866 S.Angelo Lodigiano (LO) Italy; vaccino@iscsal.it
[2] Dipartimento di Biotecnologie agrarie, Università di Padova (Italy)
[3] Dipartimento Scientifico e Tecnologico, Università di Verona (Italy)
[4] Di.Va.P.R.A. Entomologia e Zoologia applicate all'Ambiente, Università di Torino (Italy)

1 INTRODUCTION

Cereal bugs belonging to the genera *Aelia* (Pentatomidae) and *Eurygaster* (Scutelleridae) are among the main wheat pests in South-Eastern Europe, Western Asia and Northern Africa[1]. The species live on winter cereals and grasses, are univoltine and overwinter as adults. In Italy, infestations of *A. rostrata* have been reported in Southern regions[2] and in the Po Valley; recently, outbreaks of *E. maura* have been observed in Piedmont (North-Western Italy)[3,4], even if its population is usually well-controlled by egg parasitoids. Cereal bugs feed on ears, piercing the developing grains and injecting their saliva rich of proteolytic enzymes in the endosperm. Early attacks during the grain filling period can cause losses in kernel weight and germination ability, while later attacks, that may result associated to the formation of a discoloured area – the bug patch- on kernels around the point of stylet penetration (Figure 1), affect the baking quality.

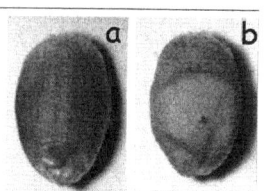

Figure 1 *Bread wheat kernels: sound (a), damaged (b).The bug patch surrounding the penetration point is evident.*

Flours derived from damaged kernels originate sticky doughs and poor bread due to the detrimental effect of proteases on gluten structure. Up to date, no sources of resistance are known, thus the only way of reducing the damage is the chemical bug control in the crop. To implement an efficient environment friendly control strategy there is a need to pinpoint the relationships between the timing of the bug attack and gluten degradation. The present work reports the results of a study carried out by caging plants of two bread wheat cultivars, characterized by different seed texture and bread-making quality, and introducing adults of *E. maura* in four periods corresponding to different grain filling stages: technological performance of flours from sound and affected kernels was analysed and biochemical investigations on gluten were performed by means of SDS gel electrophoresis and HPLC analyses of storage proteins.

2 MATERIALS AND METHODS

Two bread wheat cultivars, Centauro and Taylor, characterized by different seed texture and bread-making quality, were used: cv Centauro is an ordinary bread-making wheat with *soft* seed texture, cv Taylor is an improver wheat with *hard* texture.

The study was performed according to a completely randomized experimental design with five treatements and five replications. Plants of the two cultivars, grown in the same field, were caged and adults of *E.maura* were introduced and left for 10 days, at four stages of wheat development: heading (T1), early milk-ripe (T2), milk-ripe (T3) and late milk-ripe (T4). No insects were introduced in the T5 cages.

To assess the extent of bug damage, duplicate counts of 100 kernels for each sample were visually examined for the "bug patches". Protein content was determined by Near Infrared Reflectance according to AACC39-11. SDS sedimentation volume test was performed on wholemeal with the method of Every[5].

To analyse gluten protein degradation, micro-doughs were made by mixing 200 mg of flour with water at 30°C and incubating them for 30 min at 30°C. The glutenin components were extracted from the freeze-dried micro-dough (40 mg) and electrophoretically separated by SDS-PAGE (12.5%) according to Pogna et al [6]. For HPLC analyses, sample preparation was performed as described by Gupta et al.[7] Briefly, 20 mg of each sample were extracted for 30 min with 1 ml of 0.05 M phosphate buffer pH 6.9, containing 0.5% SDS and Complete protease inhibitor and centrifuged at 12,000g. Supernatants were collected (soluble fraction), while pellets were dissolved in the same buffer and sonicated for 30 seconds to solubilise the remaining proteins (insoluble fraction).

Chromatographic analyses were performed according to Batey et al.[8] by a Beckman Cromatograph (System Gold) using a Waters Protein-Pak 300 size exclusion column. Injection volume was 20 µl, running buffer was acetonitrile/water (1:1 v/v) containing 0.05 % TFA at 0.5 ml/min. Absorbance was measured at 214 nm.

Statistical analysis of data was performed by ANOVA (MstatC, 1991)[9].

3 RESULTS AND DISCUSSION

Significant differences among treatments were found for 1000 kernels weight, bug damage, protein content and SDS sedimentation volume (Table 1). The timing of the bug attack was clearly related to the amount of bug damage and to qualitative decline, shown by the decreasing of protein content and SDS sedimentation volume. Maximum technological damage occurred with insect attacks at the late milk-ripe stage (T4).

Previous results obtained by means of SDS gel electrophoresis of storage proteins showed that bug damage was associated with the degradation of some gliadin or glutenin components in damaged seeds. In some cases however, in spite of an evident technological damage no degradation of protein components was observed. This was explained by the latency of the proteases injected by the insects. The incubation of flour samples at 30°C for the activation of the proteases as suggested by Sivri et al.[10] was still not unequivocal, so a micro-method was set up by mixing at 30°C small quantities of flour with water.

Table 1 ANOVA and mean values of some qualitative parameters

	df	1000 kernel weight (g)	Bug damage (%)	Protein Content (% d.m.)	SDS Sed. Vol (mL)
ANOVA					
VARIETY (V)	1	ns	ns	**	**
TREATMENT (T)	4	**	**	**	**
VxT	4	**	ns	ns	ns
Mean					
Variety					
CENTAURO		38.8	2.5	12.1	55
TAYLOR		40.9	2.7	14.7	67
LSD (p<0.01)		ns	ns	0.9	3
Treatment					
T1		42.6	0.3	14.8	67
T2		38.5	0.6	14.2	65
T3		37.7	6.1	12.8	61
T4		35.7	5.6	11.9	49
T5		42.4	0.4	13.3	63
LSD (p<0.01)		1.9	2.2	1.4	6

The extraction of glutenin components by these micro-doughs, followed by one-dimensional SDS-PAGE, revealed progressive degradation (from T1 to T4) of two components of the high molecular weight glutenins in damaged samples (Figure 2, *), associated with the appearance of minor fragments with lower Mr (Figure 2, °), that should represent the product of the degradation.

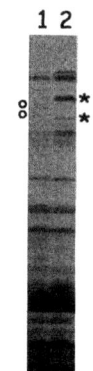

Figure 2 SDS-PAGE of glutenin components in the presence (lane 1) or absence (2) of proteolytic damage.

This result was confirmed by HPLC analyses. In Figure 3 the profiles obtained from a sound and a bug-damaged micro-dough sample of cv Taylor are reported. The soluble fraction profiles do not show any variation (compare –a- and –c- graphs); on the contrary a breakdown of the first peak of the insoluble fraction, representing the high Mr typical of the gluten network is evident in the damaged sample (compare–b- and–d- graphs).

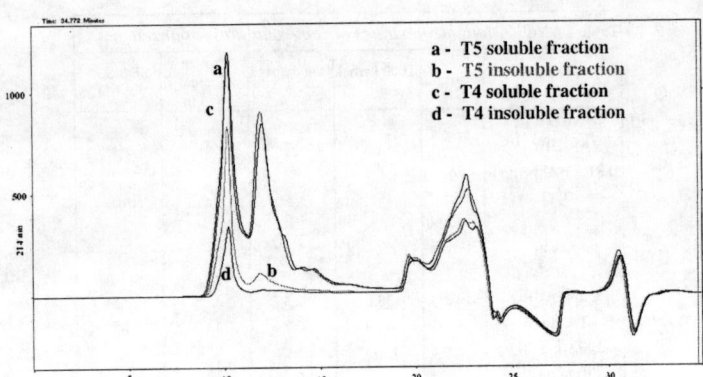

Figure 3 *HPLC profiles obtained from cv Taylor. T5, sound sample; T4, bug-damaged sample.*

A detailed description of HPLC analysis is reported in Zoccatelli *et al.* (present issue).

In the adopted experimental conditions maximum degradation of gluten aggregates was observed with bug attacks at the late milk-ripe stage of the grain filling period. This would suggest to focus chemical crop protection avoiding general treatments throughout the plant life-cycle.

References

1 F. Paulian and C. Popov, in: *Wheat,* ed. E. Hafliger, Ciba-Geigy, Basel, 1980, 69.
2 A. Spina, *L'Inf. Agr.*, 2000, **56**, 45.
3 M. Corbellini and P. Vaccino, *L'Inf. Agr.*, 2000, **56**, 39.
4 L. Tavella and M. Migliardi, *L'aratro*, 2000, **81**, 7.
5 D. Every, *New Zealand J. Crop and Hortic. Sci.*, 1992, **20**, 305.
6 N.E. Pogna, F. Mellini, A. Beretta and A. Dal Belin Peruffo, *J. Genet.& Breed.*, 1989, **43**, 17.
7 R. Gupta, K. Khan and F. MacRitchie, *J.Cer.Sci.*, 1993, **18**, 23.
8 I. L. Batey, R. B.Gupta. and F. MacRitchie, *Cereal Chem.*, 1991, **68**, 207.
9 MstatC, Michigan State University, USA, 1991.
10 D. Sivri, H. Koksel and W. Bushuk, *N. Z. J. Crop Hort. Sci.*, 1998, **26**, 117.

EFFECT OF *FUSARIUM* PROTEASES ON BREADMAKING PROPERTIES

D. Vázquez[1,2], S. Gonnet[1,3], M. Nin[1,4] and O. Bentancur[1,5]

[1] Mesa Nacional de Trigo, URUGUAY. [2] INIA La Estanzuela. CC 39173. Colonia. CP 70000. Uruguay. dvazquez@inia.org.uy [3] Laboratorio de Bioquímica. Facultad de Agronomía, Universidad de la República. Av.Garzón 780, Montevideo. CP 12900. Uruguay. [4] Departamento de Producción Vegetal. Facultad de Agronomía, Universidad de la República. Paysandú. CP 60000. Uruguay. [5] Departamento de Biometría, Estadística y Computación. Facultad de Agronomía, Universidad de la República. Paysandú. CP 60000. Uruguay.

1 INTRODUCTION

During the past few years, *Fusarium sp.* has caused important damage in the wheat producing area of Uruguay[1] and in many other countries[2]. The fungus causes loss in grain yield and is detrimental to industrial quality. The production of high levels of several mycotoxins by *Fusarium* makes the contaminated cereal unsuitable for either human consumption or animal feeding. At the same time, the fungus produces proteolytic enzymes, which may be active on gluten proteins. It was observed that some flours obtained from wheat damaged by *Fusarium*, even containing low levels of toxins, had poor breadmaking quality. The objective of this research was to evaluate the effect of *Fusarium* proteases on gluten proteins and its impact on breadmaking properties.

2 MATERIALS AND METHODS

2.1. Wheat samples

Samples of wheat grains were obtained from four different highly *Fusarium* affected farms, two from a good quality variety (INIA Boyero) and two from a regular quality variety (INIA Mirlo). They were cleaned by carefully handpicking affected grains. For each farm sample, "as is" grains were blended with clean grains in five different ratios (0:100, 25:75, 50:50, 75:25 and 100:0). All blends were done by duplicate, obtaining 40 samples (4 farms, 5 blends, 2 dups).

2.2. Methods

2.2.1. *Fusarium damage.* Percentage of *Fusarium* damaged kernels (FDK) was determined for each duplicate of each blend, reported as grams of affected kernels in 100g of sample.

2.2.2. *Flour quality.* Flours extracted from the 40 samples were analyzed for protein content (AACC method 46-11), wet and dry gluten (AACC method 38-12) and alveograms (AACC method 54-30)[3].

2.2.3. *Breadmaking*. French-type breads were baked following a typical Uruguayan procedure. Flour (100g) was mixed with yeast (2.2g), salt (2.5g) and water (adjusted amount). After 90min fermentation at 28°C, molding and proofing (90min), loaves were baked 20 min at 212°C. Volume, height and width of cooled bread were determined, and color was visually scored.

2.2.4. *Proteolytic activity*. Enzymes were extracted in pH 5.5 acetate buffer[4]. Several incubation pHs were tested, and pH 7.0 was chosen for being the optimum to measure the fungal activity (data not shown). Extract (0.5ml) was incubated with 0.4% azocasein (0.5 ml), in Tris-HCl (0.25M) buffer containing 10mM $CaCl_2$ (0.5ml) for 3 hours at 40°C. The reaction was stopped with 1M perchloric acid (0.5ml) and absorbance was measured at 337nm. Proteolytic activity was expressed as enzyme units (EU)/g of flour, where an EU is the amount of enzyme needed to produce an increase of one absorbance unit under the chosen experimental conditions.

2.2.5. *Statistical analysis*. Analysis of variance (ANOVA) was conducted using GLM procedure of SAS software (Version 8.0, 1999, SAS Institute, Cary, NC, USA). Structural equation models (path coefficient analysis) were conducted using CALIS procedure of the same software package.

3 RESULTS

FDK of "as is" samples ranged from 10.3% to 17.2%. Cleaned samples were as low as 0.7%, showing an efficient cleaning from highly damaged samples.

Table 1 shows ANOVA results for protein content, wet and dry gluten percentages and alveogram parameters. *Fusarium* damage did not influence significantly (P<0.05) either protein content or wet or dry gluten. The effect was significant on both alveogram W (deformation energy) and P (maximum resistance to extension), but not on L (extensibility). Figure 1 presents the relationship between W and FDK. W values of clean samples were lower than the characteristic values of the used varieties (data not shown), but W values of "as is" samples were up to 40% lower than those of clean samples.

Table 2 shows ANOVA results for breadmaking values. Breads were significantly (P<0.05) darker when the *Fusarium* damage increased. Breads shape was not significantly affected, but there was a detrimental trend. Crumb aspect was severely affected.

Although the proteolytic activity in cleaned samples was higher than in sound ones, it increased significantly (P<0.05) when the *Fusarium* damage increased. Structural equation

Table 1. *Effects of Fusarium damaged kernel on breadmaking properties*

Pr>F	Protein (%)	Wet gluten (%)	Dry gluten (%)	Alveo-gram W	Alveo-gram P	Alveo-gram L
INIA Mirlo, farm 1	0.037	0.450	0.532	0.024	0.024	0.515
INIA Mirlo, farm 2	0.510	0.992	0.990	0.017	0.055	0.485
INIA Boyero, farm 1	0.255	0.407	0.044	0.086	0.039	0.454
INIA Boyero, farm 2	0.394	0.002	0.050	0.008	0.036	0.227

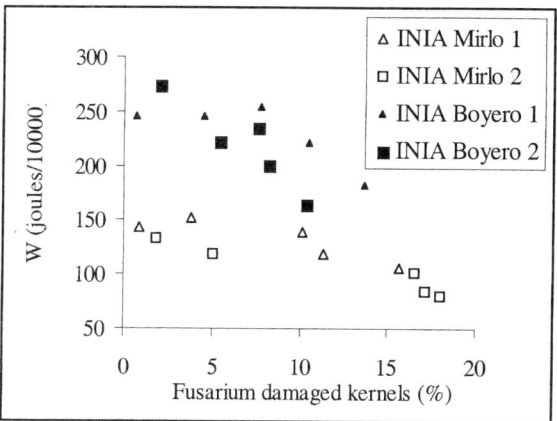

Figure 1. *Relationship between W and FDK*

Table 2. *Effects of Fusarium damaged kernel on breadmaking properties*

Pr>F	Bread volume (ml)	Crumb color (visual score)	Height/width ratio (dimensionless)
INIA Mirlo, farm 1	0.1240	0.0379	0.4652
INIA Mirlo, farm 2	0.3187	0.0411	0.2295
INIA Boyero, farm 1	0.4917	0.0199	0.2304
INIA Boyero, farm 2	0.3825	0.0108	0.1266

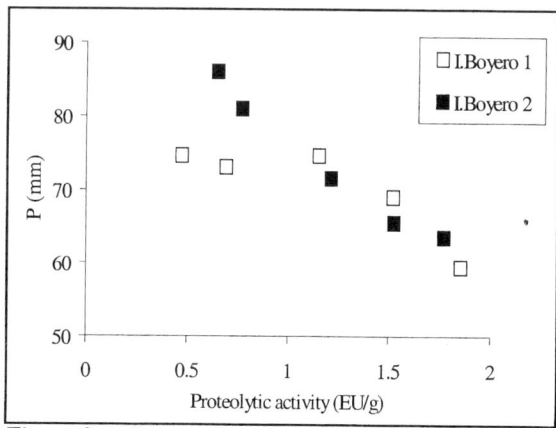

Figure 2. *Influence of proteolytic activity on alveogram P (resistance to extension).*

models revealed that increments in *Fusarium* damage have an indirect effect on W (r= -0.55) which was explained by the increase in proteolytic activity (r=0.93) that caused a decrease in P and W levels. This was explained by the direct effect of proteolytic activity onto P (r=-0.61, Figure 2). and the direct effect of P onto W (r=0.76).

4 CONCLUSIONS

Cleaning the original samples did not just decrease toxin levels, but also improved quality characteristics. However, cleaned grains are not as good as sound ones. The deterioration in dough rheological properties and breadmaking quality was associated with fungal proteolytic activity. It is proposed that the action of proteolytic enzymes on gluten deteriorates the structure, making the wheat unsuitable for breadmaking.

References

1. S. Pereyra, Serie Act. Dif. INIA, 2003, **312**,1
2. R.W.Stack, in *Fusarium Head Blight of Wheat and Barley*, ed. K.J.Leonard and W.R.Bushnell, The Phytopathological Society, St. Paul, MN, USA, 2003, ch.1, p.1.
3. AACC. Approved methods of the American Association of Cereal Chemists. Methods, The Association, St. Paul, MN, USA, 1993, methods 38-12, 46-11 and 54-30.
4. A. Pekkarinen, L. Mannonen, B.L. Jones and M.L. Niku-Paavola, J.Cereal Sci., 2000, **31**, 253.

Acknowledgments

Research supported by Facultad de Agronomía LIA 021, INIA-BID funds (Mesa Nacional de Trigo). Samples were provided by CALMER and CADYL. Special thanks to Patricia González, José Hernández, Daniela Ramallo, Claudia Fernández, María García, Lucía Usuca, Wilfredo Ibáñez, Silvia Pereyra and María José Bentancur for their technical support.

BREAKDOWN OF GLUTENIN POLYMERS DURING DOUGH MIXING BY *EURYGASTER MAURA* PROTEASE.

G. Zoccatelli[1], S. Vincenzi[2], M. Corbellini[3], P. Vaccino[3], L. Tavella[4], A. Curioni[2]

[1]Dipartimento Scientifico e Tecnologico, Università di Verona, Verona, Italy.
[2]Dipartimento di Biotecnologie agrarie, Università di Padova, via dell'Università, Legnaro (PD), Italy. E-mail: andrea.curioni@unipd.it
[3]Istituto Sperimentale per la Cerealicoltura, S. Angelo Lodigiano (LO), Italy.
[4]Di.Va.P.R.A. Entomologia e Zoologia applicate all'Ambiente, Università di Torino, Torino, Italy.

1 INTRODUCTION

Bug (a general term for insects belonging to the *genera Eurigaster.*, *Aelia* and *Nysius*) attacks during wheat kernel development impair the bread-making quality of the flour, giving sticky doughs and breads with a poor volume and texture.[1] The presence of very few bug-damaged kernels in a lot is sufficient to make the flour unsuitable for bread-making,[2] resulting in heavy economic losses. The damage is caused by the secretion injected by the insect into the kernel, which contains enzymes which hydrolyse the gluten proteins.[1] Here we study the molecular mechanisms involved in the detrimental effect on dough properties caused by the presence of few bug-damaged kernels.

2 MATERIALS AND METHODS

2.1 Wheat Samples

Wheat (var. Taylor) samples containing different percentages (0, 2.5, 5 and 11%) of bug (*Eurygaster maura*)-damaged kernels were used in this study. Flours and doughs taken at 5, 10 and 15 min of mixing in the farinograph were analysed. After sampling, doughs were freeze-dried and then reduced to a fine powder.

2.2 Rheological analyses

Farinographic and alveographic analyses were performed according the ICC methods 115-D-1972 and 121-1986, respectively.

2.3 Biochemical analyses

Protein extraction and Size-Exclusion (SE)-HPLC were done essentially as described by Gupta et al.[3] "Extractable" proteins were solubilised from flours and freeze-dried doughs (20 mg) with 1 ml of 50mM Na phosphate, pH 6.9, containing 0.5% SDS and a cocktail of protease inhibitors (Boehringer Mannheim, 1 tablet in 50 ml of buffer) (buffer A). "Un-

extractable" protein polymers were extracted from the residue by sonication for 30 sec in buffer A.
SDS-PAGE in reducing conditions and immunoblotting with HMW-GS- and prolamin-specific antibodies [4] were performed as previously described.[4] Gels were analysed by densitometry with the Multianalyst system (Bio-Rad)

3 RESULTS AND DISCUSSION

Flours deriving from wheat grains containing different percentages of bug (*Eurygaster maura*) damaged kernels were evaluated for technological performances by farinographic analysis. In comparison with an undamaged sample of the same variety, the flours deriving from bug-damaged kernels had poorer farinographic behaviours, showing, in particular, a lower mixing stability (Table 1). In addition, alveographic data indicated lower W values for the damaged samples (Table 1). These results confirm that bug attacks result in a worsening of the bread making quality.[2] However, the loss of rheological quality did not correlate with the percentage of damaged kernels found in the different samples.

Because the rheological behaviour of the wheat flour is largely affected by the molecular weight (MW) distribution of the glutenin polymers,[3] this parameter was studied by SE-HPLC for undamaged and damaged flours and dough samples taken after 5, 10 and 15 min of mixing in the farinograph. When an undamaged sample was analysed, a progressive decrease of the ratio between the SE-HPLC area of the un-extractable polymers (UEP), corresponding to the glutenin polymers with a "large" MW and that of the extractable polymers (EP), which corresponds to the glutenin polymers with a "low" MW[3] was noted during mixing (Figure 1). This shift of the MW distribution towards lower values corresponds to the well-known de-polymerisation of the glutenin polymers occurring a as consequence of mixing.[5, 6] The same analyses performed with bug-damaged flour samples showed that the MW distribution of the glutenin polymers was similar to that observed for the undamaged one (Figure 1), indicating that the bug attack did not modify the gluten proteins in the flour. In contrast, compared to the undamaged samples, a dramatic drop of the UEP/EP ratio occurred in the doughs derived from bug-damaged kernels after the first minutes of mixing (Figure 1), indicating a much faster glutenin de-polymerisation.

Table 1. *Characteristics of bug-damaged and undamaged wheat samples*

	Undamaged	Damaged		
Damaged kernels (%)	0	5	11	6
Protein (%)	15.9	16.7	16.4	17.2
Peak Time (min)	14.7	7.0	5.9	4.9
Stability (min)	17.8	14.1	8.4	5.6
Softening (FU)	0	38	107	154
P (mm)	108	84	54	41
L (mm)	103	126	112	85
P/L	1.05	0.66	0.49	0.48
W (J 10^{-4})	440	326	147	73
	A		B	

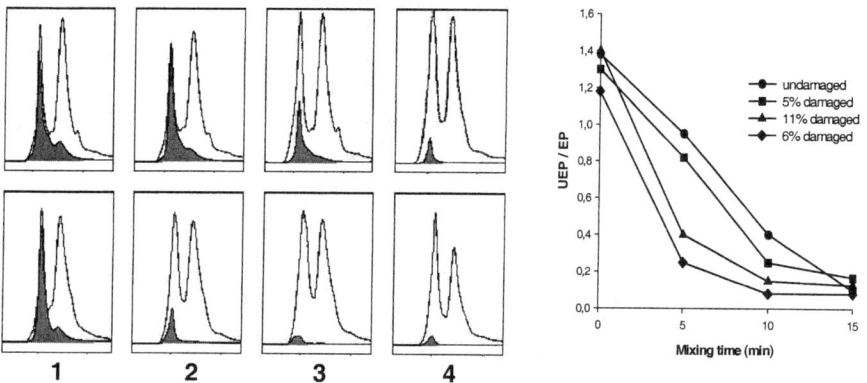

Figure 1 *(A) SE-HPLC profiles of extractable (white) and un-extractable (gray) proteins from the undamaged (top) and one of the damaged (bottom) samples. 1: flours; 2, 3, and 4: doughs at 5, 10 and 15 minutes of mixing, respectively. (B) Variation of the ratios between the areas of the chromatographic peaks corresponding to un-extractable (UEP) and extractable (EP) protein polymers extracted from undamaged and damaged flours and doughs at 5, 10 and 15 minutes of mixing*

These results indicate that the bug attack brings about its effects only after processing the wheat flour in a dough by reducing the degree of polymerisation of the gluten proteins. This can explain the low mixing stabilities and W values (table 1) of damaged wheat samples.

In order to determine the reason for the observed de-polymerisation, the proteins of the different samples were comparatively analysed by SDS-PAGE and densitometric scanning of the gels (not shown). The patterns obtained for the proteins of the undamaged and damaged flours were identical (Figure 2A, lanes 1 and 3), confirming the results obtained by SE-HPLC. In contrast, the analysis of the proteins extracted from the doughs taken at increasing times of mixing in the farinograph showed a certain progressive degradation of the HMW-GS in the damaged samples (Figure 2A, lanes 3-6), whereas the undamaged ones did not show any difference compared to the flours (Figure 2A, lanes 1 and 2). These results were confirmed and strengthened by immunoblotting experiments using antibodies specific for the HMW-GS and for the prolamin group (HMW-GS, LMW-GS and gliadins).[4] In fact, among the gluten proteins, the HMW-GS were preferentially hydrolysed during mixing, giving rise to immuno-reactive fragments of higher electrophoretic mobility (Figure 2B), whereas both the LMW-GS and gliadins seemed to be more resistant to the degradation (not shown). These results confirm previous findings.[7,8]

In order to distinguish the effect of mixing from that of hydration, the flours were incubated in an excess of tap water without mixing and the insoluble proteins, recovered by centrifugation, were analysed by both SDS-PAGE and immunoblotting. The observed breakdown of the HMW-GS in the bug-damaged flours, but not in the undamaged ones, indicated that it is the presence of water that induces HMW-GS degradation (not shown). Moreover, the water extract of a bug-damaged flour was able to cause the degradation of the HMW-GS of a sound wheat sample (var. Gladio) (not shown). All these degradation phenomena were totally abolished in the presence of a cocktail of protease inhibitors.

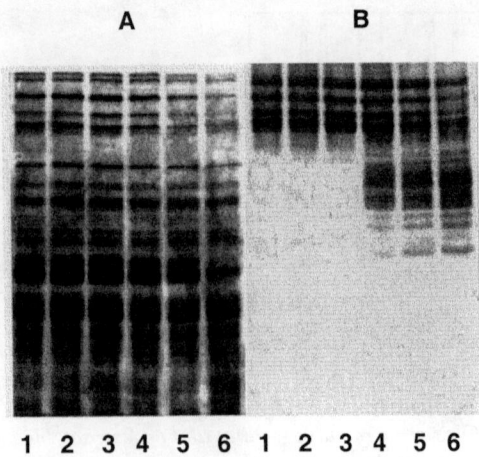

Figure 2 *SDS-PAGE of un-damaged flour (lanes 1) and dough after 15 min of mixing (lanes 2), damaged flour (lanes 3) and dough samples, taken after 5 (lanes 4), 10 (lanes 5) and 15 (lanes 6) min of mixing. Part A: Coomassie staining . Part B: immunoblotting with an antibody specific for HMW-GS*

Finally, the HMW-GS degradation increased when the pH of the incubation mixture was raised from 5 to 9 (not shown).

A joint consideration of all these results leads to the conclusion that the bug injects in the wheat kernel a water-soluble basic proteinase which reduces the molecular size of the glutenin polymers only after hydration of the flour by acting preferentially on the HMW-GS. Since these subunits are likely to constitute the "linear" portions of the glutenin complex, it is clear that only few proteolytic cleavages (corresponding to the low enzyme activity deriving from the presence of few bug-damaged kernels) on the HMW-GS are sufficient to have a dramatic effect on the polymer size and, as a consequence, on the rheological behaviour of the dough.

References

1 D. Every, *J. Cereal Sci*, 1992, **16**, 183.
2 E. Karababa, A.E. Ozan, *J. Sci. Food Agric.*, 1998, **77**, 399.
3 R.B. Gupta, K.W. Shepherd and F. MacRitchie, *J. Cereal Sci.*, 1993, **18**, 23.
4 A. Curioni, A. D.B. Peruffo, G. Pressi and N. E. Pogna, *Cereal Chem.*, 1991, **68**, 200.
5 K. Tanaka and W. Bushuk, *Cereal Chem.*, 1973, **50**, 590.
6 P.L. Weegels, R.J. Hamer, J.D. Schofield, *J. Cereal Sci.*, 1997, **25**, 155.
7 P.J. Cressey, C.L. McStay, *J. Sci. Food Agric.,* 1987, **38**, 357.
8 D. Sivri, H. Köksel, *New Zeal. J. Crop Hort.*, 1998, **26**, 117.

Non-Food Uses

A WAY TO IMPROVE THE WATER RESISTANCE OF GLUTEN-BASED BIOMATERIALS: PLASTICIZATION WITH FATTY ACIDS

M. Pommet[1,2*], A. Redl[1], M.H. Morel[2] and S. Guilbert[2]

[1] Amylum Europe N.V., Burchtstraat 10, 9300 Aalst, Belgium
[2] Unité de Technologie des Céréales et Agropolymères, ENSA.M/INRA, 2 Place Viala, 34060 Montpellier Cedex 1, France
* present address

1 INTRODUCTION

Biodegradable material showing interesting physical properties can be obtained by mixing wheat gluten with plasticizer, thanks to the thermoplastic behavior of gluten[1,2]. Hydrophilic compounds, such as water or glycerol, are usually used as plasticizer. Nevertheless, the use of such compounds impairs the water resistance of the material[3,4]. As wheat gluten carried both polar and non-polar amino acids, we decided to investigate its plasticization with amphiphilic compounds. Saturated fatty acids (FAs) were chosen. A common hydrophilic head (carboxylic group) defines the series, while the hydrophobic zone (carbon chain) increases, thus resulting in a variation of the hydrophobicity degree.

2 METHODS AND RESULTS

2.1 Mixing process

Materials were obtained by mixing vital wheat gluten with a saturated FA (with an even number of carbon from 6 to 18) in a proportion of 65/35. Mixing was performed in a two blade counter-rotating batch mixer turning at 100 rpm and recording torque and product temperature. The mixing chamber was regulated at different temperatures from 40 to 120°C.

The analysis of torque curves recorded during mixing will provide some knowledge concerning the plasticization mechanisms of wheat gluten by FAs. Briefly, a lag phase precedes a quick torque increase to a maximum. This torque development during mixing is associated with a change in consistency from a powder/liquid dispersion to a cohesive material with viscoelastic properties. The lag phase duration then depends on the ability of the constituents to interact together in given mixing conditions. Torque evolution of gluten/FA blends within mixing time was found to depend on the FA and on the regulation temperature as shown in Figure 1.

Firstly, the lag phase increases all the more as the carbon chain of the FA is long or as the regulation temperature is low. On the one hand, hydrophobicity degree and steric hindrance increase with FA carbon chain length. On the other hand, molecular mobility increases with temperature. In addition, higher temperatures result in an unfold of gluten proteins, thus exposing hydrophobic sites. Consequently, this lag phase can be related to wettability and diffusion characteristics as well as interaction possibilities. In addition, the time needed to reach the maximum torque can be described with an exponential relationship as a function of the carbon number of the FAs. Other exponential relationships were found for the other tested regulation temperatures, meaning that the FAs display a series behavior[5].

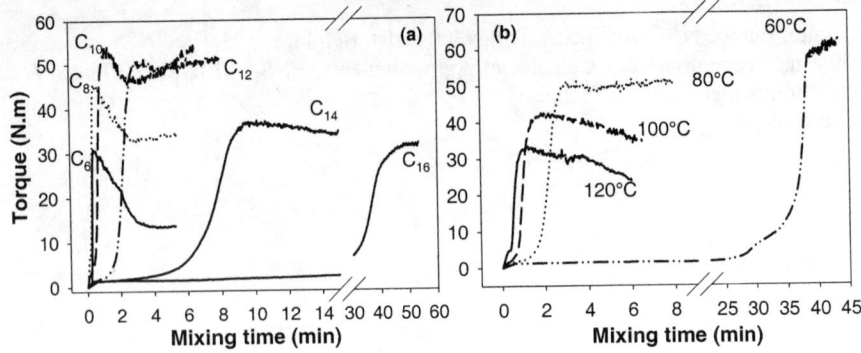

Figure 1 *Torque evolution during the mixing process (100rpm); (a): mixing of wheat gluten with different FAs (35% w/tw, identified by their number of carbons) at the regulation temperature of 80°C; (b): mixing of wheat gluten with lauric acid (C_{12}, 35% w/tw) at different regulation temperatures*

Secondly, the maximum torque values, reflecting the materials viscosity, increase all the more as the temperature is low which is a well-known behavior. However, at a given regulation temperature, it increases with FA carbon number until 10, and then decreases. As the plasticizer content was constant in weight, its mole content decreases with the increasing molecular mass of the FAs, resulting in more viscous materials. The low maximum torque values observed for the blends containing FA with more than 10 carbons may then be due to a lubricant effect, suggesting that those materials were not homogeneous, displaying a free part of FA.

2.2 Compatibility between Wheat Gluten and FAs

While wheat gluten and FA are compatible, they are in interaction; the FA is consequently unfreezable. But above the compatibility limit, the FA in excess is free and its melting point can then be detected. To quantify the free part of the FA, differential scanning calorimetry (DSC) measurements were investigated on gluten/FA blends and on FAs alone. The results, as well as scanning electron microscopy analyses, confirmed the presence of FA exudation for the gluten/FA blends with FA of more than 10 carbons as shown in Figure 2.

The compatibility limit values, calculated from the melting enthalpies, could also be predicted from the number of carbon of the FA, which is another proof of the series behavior of the FAs as gluten plasticizer[5].

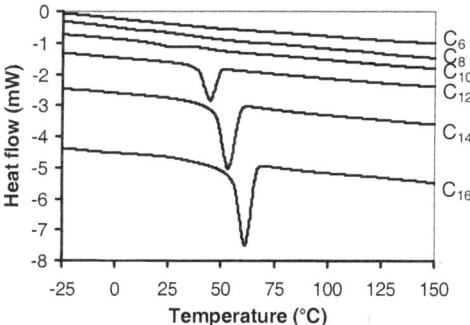

Figure 2 *DSC thermograms of wheat gluten-based materials plasticized with different FAs identified by their number of carbons (adapted from ref. 5)*

2.3 Study of the Plasticizing Effect of FAs

Dynamical mechanical thermal analyses were performed on gluten/FA blends mixed at the regulation temperature of 80°C in order to have a better knowledge of their plasticizing effect. According to the observed reduction of wheat gluten Tg, the plasticizing effect of FAs was evidenced[5]. Moreover, taking into account the effective FA moles involved in plasticizing interactions, the Tg results, obtained for gluten blends with different FAs, could be modeled by a unique Couchman-Karasz equation, which confirmed again the FA series behavior[5,6]. As shown on Figure 3, the plasticizing effect of FAs on wheat gluten was found to be in-between those of water and glycerol in the molar range tested.

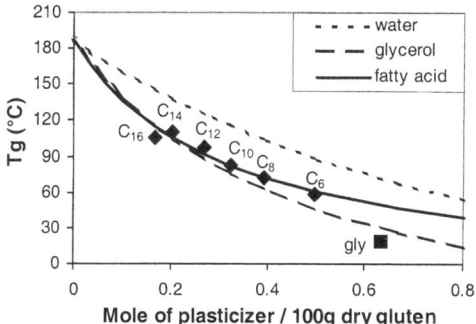

Figure 3 *Glass transition temperature of gluten-based materials plasticized with different FAs (identified by their number of carbons) or glycerol (gly) as a function of the plasticizer mole content. Lines are Tg prediction from Couchman-Karasz equation (from ref. 5)*

2.4 Water Vapor Permeability

Water vapor permeability measurements were performed on films obtained by pressure molding of the materials plasticized with the different FAs. The results were compared to a gluten-based film plasticized with glycerol and presented in Figure 4. Although the gluten networks were less reticulated in gluten films plasticized with FAs compared to those plasticized with glycerol, the water vapor permeability values were more than three times lower[5].

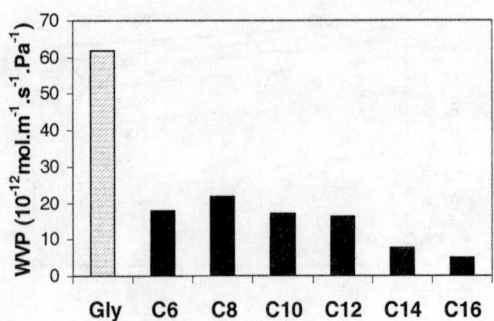

Figure 4 *Water vapor permeability (WVP) values of wheat gluten-based films plasticized with glycerol or different FAs identified by their number of carbons*

3 CONCLUSION

Saturated FAs were found to be effective gluten plasticizers and to behave as a homogeneous series. Consequently, the time needed to obtain a cohesive material by mixing, the compatibility limits between gluten and FAs, as well as the intensity of gluten Tg depress due to plasticization could be predicted by different relationships. This may be related with the amphiphilic nature of FAs: the carboxylic group would be the plasticizing function, while the hydrocarbon chain would act as an antiplasticizer. As the carbon chain gets longer, the compatibility between gluten and the FAs decreases, limiting the plasticization effect. Even small carbon chain length FAs, when used as gluten plasticizer, greatly improve the barrier properties of resulting materials. This amphiphilic plasticization opens new ways for the improvement of gluten-based biodegradable materials properties.

References

1 B. Cuq, N. Gontard and S. Guilbert, *Cereal Chem.*, 1998, **75(1)**, 1.
2 A. Redl, M.H. Morel, J. Bonicel, S. Guilbert and B. Vergnes, *Rheol. Acta*, 1999, **38(4)**, 311.
3 A. Gennadios, C.L. Weller and R.F. Testin, *Cereal Chem.*, 1993, **70(4)**, 426.
4 N. Gontard, S. Guilbert and J.L. Cuq, *J. Food Sci.*, 1993, **58(1)**, 206.
5 M. Pommet, A. Redl, M.H. Morel and S. Guilbert, *Polymer*, 2003, **44(1)**, 115.
6 P.R. Couchman and F.E. Karasz, *Macromolecules*, 1978, **11**, 117.

WHEAT GLUTEN BASED BIOMATERIALS: ENVIRONMENTAL PERFORMANCE, DEGRADABILITY AND PHYSICAL MODIFICATIONS

S. Domenek, M.-H. Morel and S. Guilbert

Cereal Technology and Agropolymers Laboratory, ENSA – INRA Montpellier, 2 place Viala, 34060 Montpellier, France

1 INTRODUCTION

Environmental concerns have driven research to develop new biodegradable polymers based on renewable resources. Wheat gluten is a promising raw material, because it unites technological advantages such as viscoelasticity and availability in large quantities at competitive prices.

The most important requirements for boimaterials are overall environmental benefit, biodegradability, non-toxicity, and functionality for the desired application.

In this paper, we investigated these requirements by conducting a preliminary life cycle analysis (LCA) of wheat gluten, testing biodegradability of several gluten materials and testing their functionality with respect to mass transport.

2 MATERIALS AND METHODS

The preliminary life cycle analysis (LCA) of gluten (from cradle to gate) was conducted following the methods of the ISO 14040 family. The standards divide the analysis into four steps, which are (i) goal and scope definition, (ii) life cycle inventory, (iii) life cycle assessment, and (iv) interpretation. The scope of wheat gluten was limited to from wheat farming (cradle) to commercial available gluten powder (gate) due to lack of a commercialized material application.

The preparation of cast wheat gluten films was previously described.[1] It consisted of chemical reducing disulfide bonds and spreading the film forming solution onto a Plexiglass surface. For the preparation of hot-moulded wheat gluten films. gluten-glycerol (35 w/w %) blend or wet gluten powder (19.40 % H_2O) was pressed in a heating press for various times at different temperatures and 150 bar.[2,3] Mixed gluten blends were prepared with the help of a two-blade counter rotating measuring mixer at 100 rpm according to Redl et al. (2003).[4] The fabrication conditions are given in Table 1.

The central control parameter of gluten materials is the percentage of aggregated, SDS-insoluble protein (F_i), which was measured by size-exclusion chromatography according to Domenek et al. (2002).[5] The F_i-values of the different materials are given in Table 1.

The international standard ISO 14852 method was chosen to evaluate biodegradability of wheat gluten materials in liquid medium. . Biological mineralization of the ground material was assessed by the quantity of CO_2 produced during biodegradation. Microbial inhibition was measured with the same test set-up by investigation of retardation of degradation of cellulose blended with ground gluten materials. Biodegradability in farmland soil was probed in a laboratory scale test that consisted of digging gluten samples into soil at constant water content (75 % of the soil's maximal water retention capacity) and temperature (20 °C). Degradation was assessed as sample weight loss with time.[3]

The enzymatic digestion of gluten materials with pepsin was measured with the ninhydrine method according to Prochazkova et al. (1999).[6]

The transport measurement of bovine serum albumin (BSA) through wheat gluten membranes was carried out with a diffusion cell test set-up, applying a concentration gradient of 100 μmol/L BSA over the swollen gluten membrane. BSA occurrence in the receiver compartment was measured by reverse phase HPLC.[2]

Table 1 *Preparation data of wheat gluten materials and corresponding degree of protein aggregation (F_i)*

Mat.	Preparation	F_i [%]	Mat.	Preparation	F_i [%]
1	casting	0	7	film glycerol 150 °C 20'	61
2	mix.[a] 30' 107°C 100 min^{-1}	64	8	film H$_2$O[c] 70 °C 10'	28
3	film glycerol[b] 100 °C 2'	28	9	film H$_2$O 80 °C 10'	35
4	film glycerol 100 °C 60'	49	10	film H$_2$O 90 °C 10'	52
5	film glycerol 120 °C 35'	84	11	film H$_2$O 100 °C 10'	72
6	mix. 86' 37 °C 10 min^{-1}	4.1	12	film H$_2$O 120 °C 10'	87

[a]*mixing with plasticizer glycerol;* [b]*pressed films plasticizer glycerol;* [c]*pressed films plasticizer water*

3 RESULTS AND DISCUSSION

3.1 Environmental Performance and Degradation

Preliminary investigation of environmental performance of wheat gluten in the impact categories, energy use and green house gas emissions, revealed an advantage of gluten compared to synthetic materials (Table 2). The environmental benefit of gluten is likely to continue during the secondary transformations necessary to manufacture a commercial product.

Table 2 *Environmental performance of different biopolymers in two impact categories*

Polymer	Energy use[a] [10^3 kJ/kg]	GHG emissions[b] [kg CO$_2$-eq]	Analysis scope
Low density polyethylene	80.6[c]	5.04[c]	Incineration
Starch	18.5[c]	1.08[c]	Cradle to gate
Thermoplastic starch	18.9[c]	1.10[c]	Cradle to gate
Mater-Bi foam grade	32.4[c]	0.89[c]	Composting
Polylactic acid	57[c]	3.84[c]	Incineration
Polyhydroxyalcanoates, various processes	66-573[c]	not analyzed	not analyzed
Wheat gluten, prel. data	10 (-17)	0.72	Cradle to gate

[a]*Use of energy accumulated over all actions relating to the process within the analysis scope;* [b]*Cummulation of green house gas (GHG) emissions;* [c]*Data from Patel, [2002]*[7]

Non-Food Uses

The gluten network structure can be modified by physical treatments causing crosslinking of the protein network without chemical additives. The crosslinking reaction can be monitored with the help of solubility studies in SDS. A variation of the percentage of SDS-insoluble protein (F_i) between 0 and 90% can be achieved with the help of different technological treatments (Table 1).

The effect of such an important change in the protein network on the material's biodegradability was investigated in liquid and in solid culture. All gluten materials were fully degradable within 36 days in liquid culture and no significant variation between the degradation curves was found (Figure 1a). Gluten material degradation in farmland soil confirms the result. After 50 days, all materials had completely vanished.

The microbial inhibition test carried out on strongly crosslinked gluten materials showed that during the mineralization process no formation of toxic metabolites, which slow the cellulose degradation, was observed (Figure 1b).

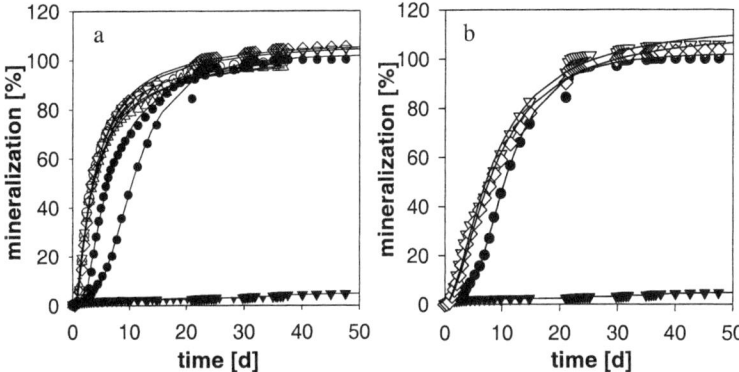

Figure 1 (a) *Degradation of wheat gluten materials in liquid culture in comparison to the degradation behaviour of cellulose: material 1 (+), material 2 (O), material 3 (△), material 4 (∇), material 5 (◇), cellulose (positive reference, ●), and negative reference (▼);* (b) *Microbial inhibition test on cellulose (●) blended with material 4 (∇) or material 5 (◇)*

3.2 Effect of physical Modification on Mass Transport Properties

The material's enzymatic digestibility is closely related to microbial degradation and important for mass transport through gluten films. Enzymatic digestion tests *in vitro* confirmed that the decrease of the protein network hydrolysis rate became significant only at very high crosslinking degrees (> 72 % F_i, cf. Table 3).

Furthermore, gluten membranes constitute an effective barrier against the physical transport of large molecules, such as BSA the size of which is comparable to digestive enzymes. The transport curves (Figure 3) show that it is not of diffusive type. BSA passes the membrane through membrane damages originating during the experiment. This stochastic process is mirrored in irreproducible lag-times of the transport (Figure 3, repeat experiments of material 9). The mass transfer rates of BSA, however, are reproducible and drop with increasing F_i (Figure 3), which indicates a damage size dependent on the crosslinking degree of the matrix.

Table 3 *Percentage of hydrolysis of the gluten film samples heat-treated at different temperatures. Multiple range test for the hydrolysis in function of the treatment temperature*

Mat.	Sample	hydrolysis[%]	σ [%][c]
gluten	vital gluten	7.03[a]	0.23
8	film H$_2$O 70 °C 10'	7.11[a]	0.24
9	film H$_2$O 80 °C 10'	6.69[a]	0.24
10	film H$_2$O 90 °C 10'	6.43[a,b]	0.24
11	film H$_2$O 100 °C 10'	6.53[a]	0.27
12	film H$_2$O 120 °C 10'	5.77[b]	0.23

[a,b] *groups a and b are statistically different at the 95 % confidence level;* [c] *σ standard deviation (10 repeats)*

4 CONCLUSION

The structure of wheat gluten materials can be modified with physical treatments resulting in a large range of attainable degrees of protein aggregation. High F_i have a positive effect on the gluten membrane barrier properties towards large molecules. The structural changes do not alter the biodegradability and do not cause toxic effects during material degradation, which makes gluten a promising candidate for fabrication of biomaterials.

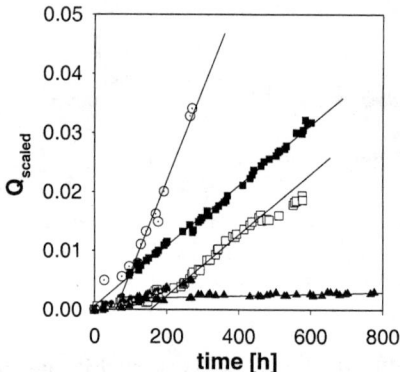

Figure 3 *BSA transport through wheat gluten films subjected to different heat-treatments: material 8 (○), material 9a (□), repeat experiment 9b (■), and material 12 (▲)*

References

1 N. Gontard, S. Guilbert, and J. Cuq, 1993, *J. Food Sci.*, **58**, 206.
2 S. Domenek, L. Brendel, M.-H. Morel and S. Guilbert, *Cereal Chem.*, 2003a, in print.
3 S. Domenek, P. Feuilloley, J. Gartraud, M.-H. Morel and S. Guilbert, *Chemosphere*, 2003b, in print.
4 S. Domenek, M.-H. Morel, J. Bonicel and S. Guilbert, *J. Food Chem. Agric.*, 2002, **50**, 5947.
5 S. Prochazkova, K. Varum and K. Ostgaard, *Carbohydr. Res.*, 1999, **320**, 82.
6 A. Redl, M.-H. Morel and S. Guilbert, 2003, *J Cereal Sci.*, **38**, 105.
7 M. Patel, C. Bastioli, L. Marini and E. Würdinger, in *Biopolymers*, Wiley-VCH, 2003, in print.

GLUTEN FILMS: EFFECT OF COMPOSITION AND PROCESSING ON PROPERTIES AND STRUCTURE

Y. Popineau, C. Mangavel and J. Guéguen

Institut National de la Recherche Agronomique, Unité de Recherche sur les Protéines Végétales et leurs Interactions, BP 71627, 44316 Nantes cedex 3, France

1 INTRODUCTION

Sustainable development will create a strong demand for materials issued from renewable resources which could overcome the applications of synthetic polymers. Among other agricultural products, wheat storage proteins are a possible basis for such material. They are currently extracted at the industrial scale and their natural properties make them candidates to prepare plastic films. For this purpose, they must be processed in the presence of plastifying molecules by a wet process (casting) or a dry process (thermo-moulding). Various factors are susceptible to modify the structure of the proteins and to alter the properties of the films. As a part of a collaborative European research programme (Gluten Biopolymer, FAIR CT 96 1979) the influence on the film properties of the following parameters was studied: the prolamin composition and aggregation, the type and concentration of plasticizing molecules, the thermal conditions of the process. The structure of gluten proteins and the mechanism of film formation were investigated as well.

2 MATERIAL AND METHODS

Industrial gluten was provided by Amylum (Aalst, Belgium). Flours of four wheat varieties (NSA2, ULI03, Gazul and Farak) were provided by ULICE (Riom, France). Gluten was extracted at the pilot scale from these flours by washing a dough and freeze-drying. Gliadin and glutenin-rich fractions were extracted from the industrial gluten in the presence of diluted acetic acid and freeze dried [1]. Protein composition of the glutens and fractions were determined by sequential extraction in SDS-containing buffers and SE-HPLC [2]. Thin film (80 μm) were prepared by casting as described [2,3]. Shortly, gluten or fraction were dispersed at pH 11 in sodium hydroxide and mixed with the required amount of plasticizer. The mixture was spread on a plate and dried under controlled conditions. Thermo-moulding was performed on blends of dry gluten and plasticizer at 120°C and 200 bars for 15 min in a heated press. The films were equilibrated at 60% of relative humidity before analysis. The changes of protein structure induced by processing were followed by determining protein extractability and by Fourier-transform infra-red spectroscopy [4]. Extension properties were analysed at 20°C and 60% RH on 5A-type specimens according to the ISO-527-2 standard as reported previously [3].

3 RESULTS AND DISCUSSION

3.1 Effect of protein composition

Films were prepared in the same conditions of protein and glycerol contents by casting with glutens extracted from four varieties. The glutens differed largely by their HMW-GS

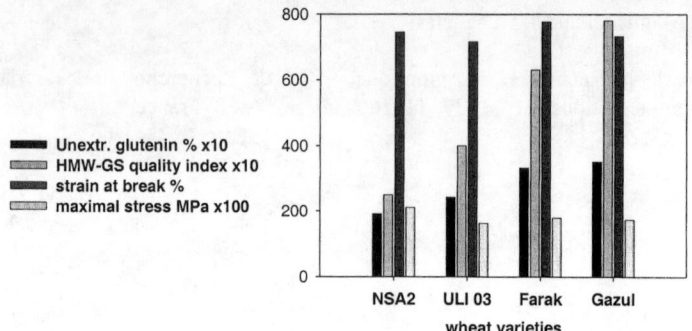

Figure 1 *Effect of gluten composition on the mechanical properties of the films*

quality index as well as by their content of unextractable glutenin (Figure 1) and by their rheological properties. However, the corresponding films showed no significant differences in their strain at break and maximal stress. Experiments performed on gliadin-rich and glutenin-rich fractions confirmed that the initial aggregation of prolamin had only a slight influence on the film properties. The maximal stress of the glutenin-rich film was only doubled when compared to that of the gliadin-rich film (2.7 vs 1.1 MPa), whereas the strain at break was not altered. Modifications of the plasticizer/protein ratio (0.2-0.6 w/w) resulted in far wider variations of the properties of the films. It was thus possible to vary strain at break between 0 and 900% and the maximal stress between 1 and 13 MPa.

3.2 Effect of plasticizer type and content

The effect of four hydrophilic plasticizers of the ethyleneglycol series differing by their number of carbons (2-8) was analysed. With all plasticizer, the plasticizing effect (*i.e.*

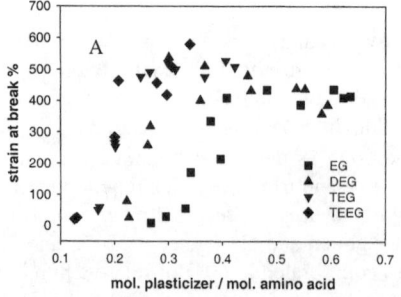
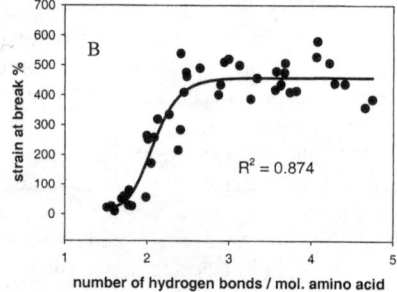

Figure 2 *Effect of various concentrations of ethyleneglycol (EG), di- (DE), tri (TEG), and tetraethyleneglycol (TEEG) on the strain at break of gluten films*

increase of strain at break, decrease of maximal stress) was positively linked to the plasticizer concentration. When the extension properties were expressed in function of the plasticizer/amino-acid molar ratio (Figure 2a), the longer the carbon chain the higher the plasticizing effect. However, all the data were fitted on the same curve when the number of H bonds potentially shared with the macromolecules was taken into account (Figure 2b). This showed that the plasticization of the film was due to the decrease of prolamin/prolamin interactions, because interchain prolamin H bonds were replaced by plasticizer/prolamin H bonds. This increased the mobility of the polypeptide chains. The results showed also that above 3 mol of H bond/mol of amino acid, additional plasticizer has no effects on the film mechanical properties, suggesting that above this ratio, the protein nerwork is «saturated» with plasticizer molecules. Beyond 4-5 mol of H bond/mol of amino acid, the film lost its coherence and could no longer be handled.

3.3 Effect of the processing conditions

Films were prepared by casting (gluten dispersed in NaOH solution) with a drying at 70°C for 1 hour, and by thermo-moulding (low hydrated plasticizer/gluten blend) with heating at 120°C under a pressure of 200 bars. The two processes resulted in films differing in their extension properties (Figure 3). The films obtained with the wet process showed a wide

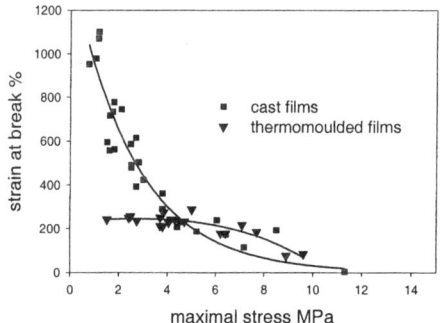

Figure 3 *Extension properties of gluten protein films prepared by the casting or the thermo-moulding processes with various concentrations of plasticizers*

range of variation of the strain at break - maximal stress value pairs in function of the plasticizer content, the highest elongation reaching about 1100%. On the other hand, with thermo-moulding (dry state, high temperature and pressure) the same range of maximal stress was obtained, but the respective variations of strain and stress did not follow the same law and the strain at break did not exceed 250%. The higher temperature applied during thermo-moulding could explain a part of the decreased elongation, by enhancing the covalent crosslinking of the polypeptide chains. Also the initial unfolding of the gluten proteins in the alkaline suspension (casting) could help the inclusion of plasticizer molecules into the network.

3.4 Changes of protein structure

The changes of protein structure were followed during the drying of an alkaline gliadin suspension in the presence of plasticizer (casting process). According to the drying temperatures (in the range 25-90°C) the extractability and secondary structure of the proteins of the dry film were modified (Figure 4). When the film was dried at 25°C neither

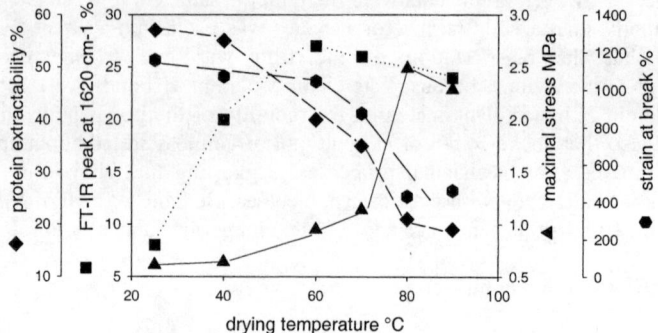

Figure 4 *Effect of the drying temperature of the casting process on the structure of the protein in the dry film*

extractability nor secondary structure were altered. At 40°C only a large increase of the β-structure percentage was observed. At 60°C the proportion of β-structures reached its maximum and extractability decreased. This corresponded to a significant increase of the maximal stress. Above 60°C the extractability decreased drastically as did the strain at break, whereas the maximal stress increased abruptly. This indicated that below 60°C the formation of the film is dominated by non-covalent aggregation of the proteins involving the appearance of a higher proportion of intermolecular H–bonded β-sheets (FT-IR peak at 1620 cm^{-1}). Above 60°C covalent crosslinking takes place through SH/SS exchanges and new SS bonds. This alters largely the extension properties of the film.

4 CONCLUSION

Gluten is a convenient biopolymer to develop protein-based plastic materials. The mechanical properties of thin films were only marginally depending on variations of the gluten protein composition. The formation and the extension properties of the protein network in the film are based on intermolecular H–bonded β-sheets and intermolecular SS bonds. The processing conditions are the prominent factors influencing the film properties. The extension properties are modulated by the ability of hydrophilic plasticizers to share H bonds with the protein network. So, film sensitivity to RH is a limit for some applications.

References

1 S. Bérot, S. Gautier, M. Nicolas, B. Godon and Y. Popineau, *Int. J. Food Sci and Technol.*, 1994, **29**, 489.
2 C. Mangavel, J. Barbot, E. Bervas, L. Linossier, M. Feys, J. Guéguen and Y. Popineau, *J. Cereal Sci.*, 2002, **36**, 157.
3 C. Mangavel, J. Barbot, J. Guéguen and Y. Popineau, *J. Agric. Food Chem.*, 2003, **51**, 1447.
4 C. Mangavel, J. Barbot, Y. Popineau and J. Guéguen, *J. Agric. Food Chem*, 2001, **49**, 867.

Non-Gluten Components

A LIPID TRANSFER PROTEIN FROM FARRO (*TRITICUM DICOCCON* SCHRANK) AND COMMON WHEAT (*TRITICUM AESTIVUM* L. CV. CENTAURO)

A. Capocchi[1], D. Fontanini[1], L. Galleschi[1], L. Lombardi[1], R. Lorenzi[1], F. Saviozzi[1] and M. Zandomeneghi[2]

[1]Dipartimento di Scienze Botaniche, Università di Pisa, via L. Ghini 5, I-56126-Pisa, Italy (lgalles@dsb.unipi.it)
[2]Dipartimento di Chimica e Chimica Industriale, Università di Pisa, via Risorgimento 35, I-56126, Italy

1 INTRODUCTION

Hulled wheats were among the first cultivated cereals. After a long period of decline in their agricultural use, social, cultural and economical reasons have made them popular once again. Among these reasons are the - frequently unsubstantiated - claims about their beneficial properties for human health, especially as dietary substitutes for the common wheat products. In fact, it has being suggested that these grains might have a lower allergenic potential than common cultivated wheats. Among potentially allergenic factors in wheat products are the non-specific lipid transfer proteins (nsLTPs), a protein class that has been recognized as a pan-allergen (1). Their allergenicity has been linked to sensitization and elicitation of IgE-mediated allergy via the gastrointestinal tract (GI) (2). We have purified nsLTPs from the soft wheat *Triticum aestivum* L. cv. Centauro and from the hulled wheat *Triticum dicoccon* Schrank (farro) bran. We also have investigated the resistance of purified nsLTP1 to the GI acidic proteolytic environment by pepsin digestion in simulated gastric fluid, and the ability of possible digestion products to bind antibodies against barley nsLTP1.

2 METHODS AND RESULTS

2.1 Purification of *T. aestivum* and *T. dicoccon* nsLTPs

Bran nsLTPs extraction and purification was based on our modification of a method by Charvolin et al. (3). Purification was achieved by chromatography (cation exchange, gel filtration, semipreparative and analytical RP-HPLC). At the end of the purification three peaks eluting at 25.5%, 27% and 30% acetonitrile, respectively, were obtained from the *T. aestivum* sample (a, b and c on Fig. 1A); two peaks eluting at 26.5% and 32.1% acetonitrile, respectively, were obtained from the *T. dicoccon* sample (d and e on Fig. 1B). The RP-HPLC fractions were analyzed for purity on 15% SDS-PAGE according to Laemmli (4)(Fig. 2A) and the nsLTP1 protein bands were identified by immunoblotting using antibodies against barley (*Hordeum vulgare* L. cv. Morex) nsLTP1 (Fig. 2B). The soft wheat peak eluting at 25.5% acetonitrile and the farro peak eluting at 26.5% acetonitrile, showed an apparent molecular weight (Mr) of 7.2 kDa on SDS-PAGE (Fig.

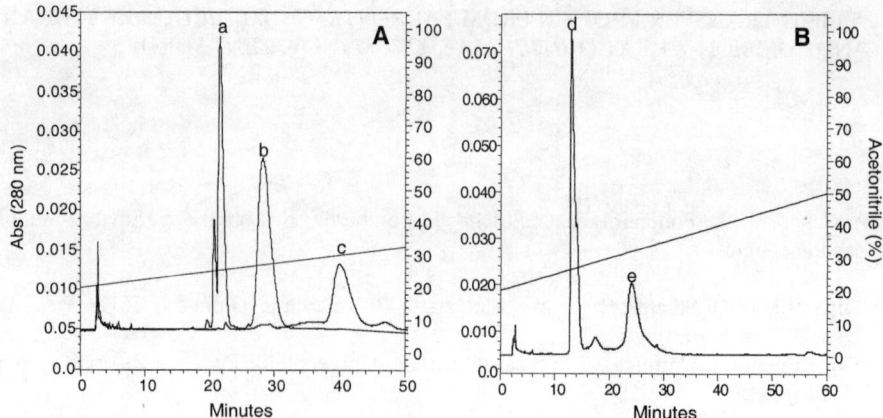

Figure 1 *Final purification of soft wheat and farro nsLTPs on analytical C-18 RP-HPLC. A, Soft wheat sample separation on water:acetonitrile:0.05% TFA gradient (0.25% acetonitrile per minute). B, Farro sample separation on a water:acetonitrile:0.05% TFA gradient (0.5% acetonitrile per minute).*

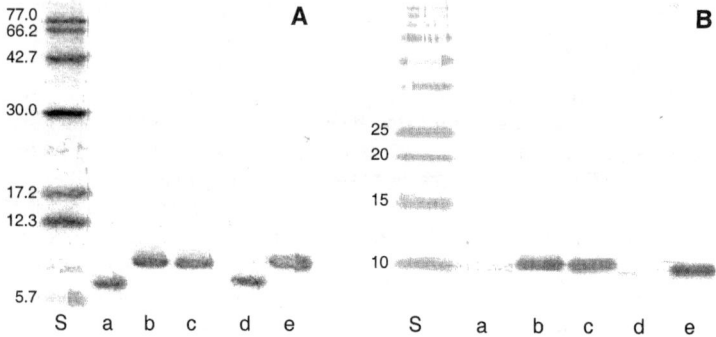

Figure 2 *SDS-PAGE and western blotting of the nsLTP samples purified from soft wheat and farro. A, 15% SDS-PAGE of the soft wheat (lanes a, b and c) and farro (lanes d and e) fractions obtained by RP-HPLC. B, Immunoblots of the fractions a-e from figure 2A. Lane S, Molecular weight standards.*

2A, a and d) and gave no cross-reaction with nsLTP1 antibodies (Fig. 2B, a and d). The 27% and 30% acetonitrile-eluting peaks from soft wheat, and the 32.1% acetonitrile-eluting peak from farro had a Mr of 9.2 kDa on SDS-PAGE (Fig. 2A, b, c and e) and reacted with nsLTP1 antibodies (Fig. 2B, b, c and e).

Based on HPLC retention times, Mr data and the lack of reactivity with nsLTP1 antibodies, the 7.2 kDa proteins could be identified as putative nsLTP2 and the 9.2 kDa proteins were assigned to the nsLTP1 class.

2.2 Stability of Soft Wheat and Farro nsLTP1 to Pepsin Digestion

nsLTPs from several plant sources have been recognized as food allergens that trigger IgE-mediated immune reactions. Experimental data indicate that these allergens are resistant to the acidic and proteolytic environment of the stomach and might reach, in an allergenic form, the gut mucosa where they come into contact with the immune system (1). Hence, to investigate nsLTPs potential for eliciting an IgE-mediated allergic reaction, it is conceivable that its ability to resist pepsin digestion be tested. In order to assess the chance that hydrolysis products retain epitopes for IgE binding, the peptides obtained from the pepsin hydrolysis should maintain the ability to bind anti-nsLTP antibodies. In our study, the purified nsLTP1 from soft wheat and farro, were tested in hydrolysis assays that used different pepsin concentrations. The hydrolysis products were analyzed by electrophoresis and western blots that used antibodies against barley nsLTP1 to verify the presence of nsLTP1 epitopes on possible digestion products.

2.2.1. nsLTP1 Digestion Reaction. The stability of purified soft wheat and farro nsLTP1 to pepsin digestion was assessed using the simulated gastric fluid (SGF) described by the US Pharmacopeia (4) as enzyme source. The SGF contained 3.25 mg/mL of pepsin dissolved in 30 mM NaCl, pH 1.2. Purified nsLTPs were dissolved in water to give a final concentration of 10 mg/mL.

The digestion reactions were carried out at 37°C in mixtures that contained a nsLTP1:pepsin ratio of 1:1, 1:6, 1:13 and 3:1. At times: 0, 15, 60 and 120 min, 41 µL-aliquots were taken from the reaction mixtures and added with 5X Laemmli-SDS sample buffer that contained 24% of 1.5 M Tris-HCl, pH 8.8. The aliquots were immediately frozen in dry ice, and heated at 100°C for 3 minutes before being separated by electrophoresis.

2.2.2. Analysis of the Reaction Products. The hydrolysates from the digestion reactions were analyzed on 15% SDS-PAGE gels (Fig. 3). The hydrolysates were analyzed compared to controls in which nsLTP1 or pepsin were incubated in 30 mM NaCl, pH 1.2, for 120 min under the reaction conditions. Duplicates of the SDS-PAGE gels were also analyzed by western blotting using polyclonal antibodies against barley nsLTP1 (Fig. 3). The results showed very little SGF hydrolysis of soft wheat nsLTP1 at higher nsLTP1:pepsin ratios (1:1, 3:1); this hydrolysis became apparent only when the ratio nsLTP1:pepsin was lowered to 1:6 and 1:13. As indicated by the western blots of the 1:13 reaction products, nsLTP1 retained its antigenicity even after 120 min of incubation in SGF. On the other hand, farro nsLTP1 was far more sensitive to SGF, as shown by the SDS-PAGE gels and immunoblots of Figure 3. Farro nsLTP1 hydrolysis was already apparent in the 1:1 reaction and its rate increased progressively as the pepsin concentration was increased. The protein was completely digested after 120 min in the 1:13 reaction and consequently lost its antigenicity as indicated by the lack of reaction with barley nsLTP1 antibodies. Lindorff-Larsen and Winther (6) have stated the high stability of barley nsLTP1 subjected to pepsin hydrolysis in mixtures with nsLTP1:pepsin ratio of 3:1 for 60 min. Soft wheat and farro nsLTP1 also showed a significant resistance to pepsin hydrolysis in these conditions. However, it is noteworthy that in our experiments soft wheat and farro nsLTP1 resistance to pepsin digestion was found to be dependant by the relative amount of enzyme; when the pepsin concentration in the reaction mixture was raised, the amount of hydrolysis was also increased. This was particularly evident when comparing the hydrolysis susceptibilities of the nsLTP1 from the two species.

3 CONCLUSION

Although the allergenicity of some cereal LTPs is still under debate, it has been suggested that their resistance to proteolysis could be an indication of their involvement in food allergies (2) as a mean to escape the gastric environment in an antigenically active form.

Due to its LTP1 low resistance to SGF digestion compared to that of soft wheat, farro could be an interesting food alternative to common wheats.

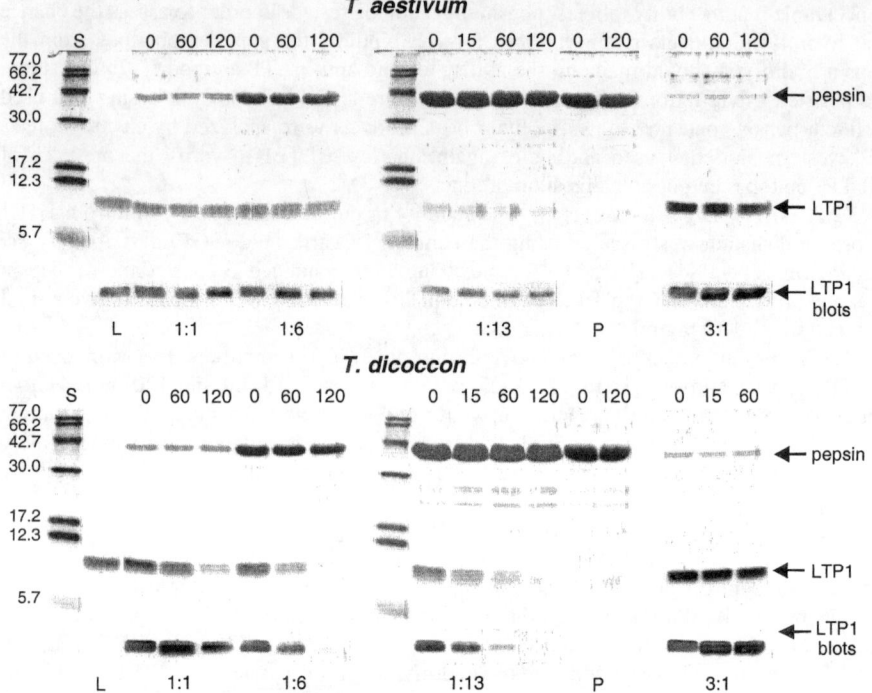

Figure 3 *15% SDS-PAGE and immunoblots analyses of soft wheat and farro LTP1 digests in SGF. Analyzed samples were taken at 0, (15), 60 and 120 min from reaction mixtures containing LTP1:pepsin ratios of 1:1, 1:6, 1:13 and 3:1. Lanes L, LTP1 incubated for 120 min in 30 mM NaCl, pH 1.2, under the reaction conditions. Lanes P, pepsin incubated for 0 and 120 min in 30 mM NaCl, pH 1.2, under the reaction conditions. Lanes S, Molecular weight markers. The numbers at the top of the gels indicate LTP1 digestion reaction times. The numbers at the bottom of the gels indicate the LTP1:pepsin ratio.*

References

1 R. Asero et al., *Int. Arch. Allergy Immunol.*, 2000, 122, 20.
2 R. Van Ree, *Biochem. Soc. Trans.*, 2000, 30, 910.
3 D. Charvolin et al., *Eur. J. Biochem.*, 1999, 264, 562.
4 U.K. Laemmli, *Nature*, 1970, 227, 680.
5 Board of Trustees. Simulated Gastric Fluid. In *The United States Pharmacopeia* 23, The National Formulary, 18, The United States Pharmacopeial Convention, Inc., Rockville, MD, 1995, p 2053.
6 K. Lindorff-Larsen and J.R. Winther, *FEBS Letters*, 2001, 488, 145.

EVIDENCE OF A GENE CODING FOR GRAIN SOFTNESS PROTEIN (*GSP-1a*) ON THE LONG ARM OF CHROMOSOME 5D IN *Triticum aestivum*

L.Gazza, A.Niglio, F.Nocente and N.E. Pogna

Department of Applied Genetics, Experimental Institute for Cereal Science, Via Cassia,176 - 00191 Rome, Italy

1 INTRODUCTION

The 2S protein family of wheat endosperm includes three main components, *i.e.* puroindoline a (pin a), puroindoline b (pin b) and grain softness proteins (GSPs)[1]. Minor variations in cDNA sequences have defined GSP-1a, -b and -c proteins, which show about 40% amino acid sequence homology with puroindolines[2]. Pin a and pin b are known to bind polar lipids, in particular those of starch granule membrane and because of amino acid homology between GSP-1 and puroindolines, it is probable that GSP-1 has the same behaviour. Each of the short arm of group 5 chromosomes of bread wheat (*T. aestivum* L.) was found to contain a GSP-1 encoding gene. On chromosome 5DS this gene is closely linked at the *Ha* (= Hardness) locus and to the *Pina-D1* and *Pinb-D1* loci coding for pin a and pin b, respectively. GSP-1 genes have never been detected in tetraploid durum wheat (*T. turgidum* spp. *durum*). In this work, durum wheat genotypes, in which *Pina-D1* and *Pinb-D1* loci had been introduced by allosyndetic recombination, were crossed with the commercial durum wheat cv. Colosseo and the progeny analysed for GSP-1 composition.

2 METHOD AND RESULTS

2.1 Development of durum wheat cultivars with soft kernel texture

The substitution line Langdon 5D(5B) has been crossed with the mutant line of durum wheat cv. Cappelli lacking the *Ph1* locus (allele *ph1c*)[3] . The absence of the *Ph1* locus in the progeny of this cross facilitates pairing and allosyndetic recombination between chromosomes 5D and 5B. The tetraploid inbred line RIL1 with soft texture was selected as described before[4] (Figure 1). This line was crossed as the male parent with the durum wheat cv. Colosseo and the resulting F_2 progeny were analysed for the presence of *Pina-D1* and *Pinb-D1* by PCR amplification with puroindoline specific primers, to confirm the presence of the recombinant 5B chromosome[4].

2.2 Assessment of the presence of *GSP-1a* gene on the long arm of chromosome 5D

F$_2$ plants from the cross between durum wheat cv. Colosseo and the tetraploid inbred line RIL 1 were analyzed by PCR using *GSP1-a* specific primers which anneal to a region of GSP-1a molecule with no homology with puroindolines, the sense-strand primer being : 5'-CGGATGGTTTTGGGGAAT-3' and the anti-sense strand primer being : 5'-GTCTGCACCGTTCTGGCTTTA-3'. Reactions were performed in 50μl containing 300ng of genomic DNA, isolated from leaves as described by Della Porta et al.[5], 10 pmol of each primer, 100μM of each dNTP, 1X Taq DNA polymerase buffer (Amersham), 2.5 U of Taq DNA polymerase (Amersham). The samples, denatured at 94°C for 3 min before adding Taq, were submitted to 35 cycles of 1 min denaturation at 94°C, 1'30" annealing at 53°C, and 1'30" elongation at 72°C. A final cycle with an extension of 5 min at 72°C completed the reactions. As expected, progeny possessing *Pina-D1* and *Pinb-D1* all gave an amplification product of 316bp in size (Figure 2) but, surprisingly, progeny lacking puroindoline-encoding loci turned out to be positive for *GSP-1a* (lanes without stars in Figure 2).

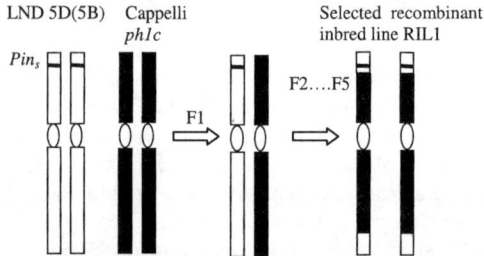

Figure 1 *Development of tetraploid inbred lines with soft texture*

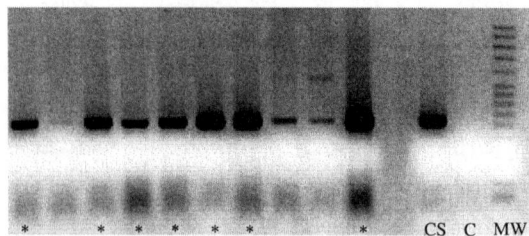

Figure 2 *PCR amplification with GSP-1a specific primers on F$_2$ progeny resulting from cv. Colosseo x RIL 1. Stars indicate lines with Pina-D1 and Pinb-D1 genes. CS: Chinese Spring; C: Colosseo; MW: Molecular Weight marker (50bp ladder).*

Since recombination events between *GSP-1a* and *Pina-D1*/*Pinb-D1* on the short arm of chromosome 5D are expected to be very rare, we supposed the presence of an additional *GSP-1* gene on chromosome 5D. This gene, provisionally called GSP-1a^L (L=long arm), likely occurs on the long arm of recombinant 5B chromosome shown in Figure 1. The amplification product of 316bp obtained by PCR on CS-DT5DL, a Chinese

Spring ditelosomic line lacking the short arm of chromosome 5D, using primers specific for *GSP-1a* (mentioned above), proved to be identical to *GSP-1a* in its nucleotide sequence (Figure 3); the absence of any amplification product by PCR on CS-N5DT5B (Figure 3), a nullisomic line lacking chromosome 5D, using the same primers, confirmed our hypothesis.

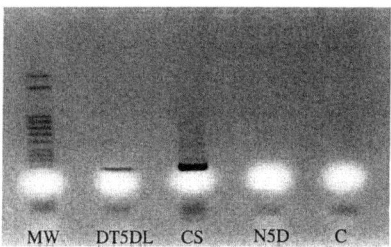

Figure 3 *PCR amplification with primers for GSP-1a on genomic DNA from Chinese Spring (CS), CS-DT5DL, CS-N5DT5B (N5D) and Colosseo (C).*

Furthermore, using primers which were able to amplify in low stringency conditions (T.a.=50°C) both *Pin a* and *GSP-1a* genes (Figure 4), it was possible to isolate tetraploid progeny which possess only the GSP-1a^L gene (lanes without stars in Figure 4). The primers used, which anneal in a region of strong homology between *Pin a* and *GSP-1a*, were: sense-srand primer: 5'-GCACCAAAACACACTGACAACA-3' and anti-sense strand primer: 5'-TCACCAGTAATAGCCAATAGTG-3'; the reactions were performed as described above at the T.a. mentioned. The sequencing of PCR fragments eluted from agarose gel with NucleoSpin Extract (Machenery-Nagel) and processed in duplicate on a Perkin Elmer ABI Prism 377 DNA sequencer, using both PCR primers, confirmed the amplification of both genes.

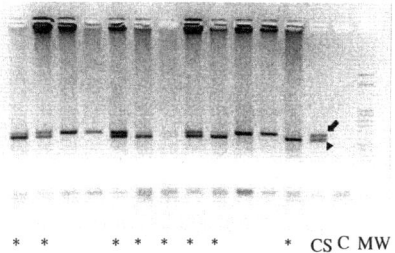

Figure 4 *PCR amplification with primers annealing in the region of homology between the GSP-1a (arrow) and Pin a (arrowhead) in F2 progeny from the cross between cv. Colosseo x RIL 1. Stars indicate plants with Pina-D1 and Pinb-D1. CS: Chinese Spring; C: Colosseo; MW: Molecular Weight marker (50bp ladder).*

3 CONCLUSION

Results indicate that among the F_2 progeny from the cross between durum wheat cv.Colosseo and tetraploid inbred line RIL 1, there are three different genotypes with regard to recombinant 5B chromosome (Figure 5):

i) *Pina-D1+Pinb-D1+GSP-1a^S*(S=Short)*+GSP-1a^L* (L=Long) (the different intensity of PCR response in lanes with stars in Figure 2, likely reflects the presence of two copies of gene *GSP-1a*).

ii) *Pina-D1+Pinb-D1+GSP-1a^S*

iii) *GSP-1a^L* (the minor intensity of PCR amplification product in lanes without stars in Figure 2 is likely due to the presence of a single copy of the *GSP-1a* gene).

Experiments are in progress to analyse recombinant 5B chromosomes in the F_3 progeny. In particular, genotype (iii), because of the absence of *Pin$_s$*-encoding loci, should be investigated to better understand the localization and the function of GSP-1a protein, likely involved in kernel hardness.

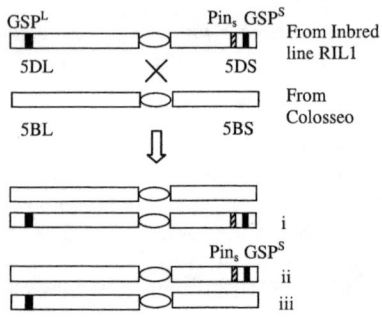

Figure 5 *Parental (i) and recombinant (ii, iii) 5B chromosomes produced by the F_1 progeny of the cross between durum wheat cv.Colosseo and tetraploid inbred line RIL 1. Pin$_s$ (hatched): puroindolines encoding loci; GSP$_s$(black): GSP-1a on the short(^{S}) and on the long (^{L}) arm.*

References

1 M. Turner, Y. Mukai, P. Leroy, B. Charef, R. Appels and S. Rahman, *Genome,*1999, **42**, 1242.
2 C. Jolly, G. Glenn and S. Rahman, *PNAS*, 1996, **93**, 2408.
3 B. Giorgi, *Mut. Breed. News.*, 1978, **11**, 4.
4 L.Gazza, A. Niglio, E. Mei, E. De Stefanis, D. Sgrulletta and N. Pogna, in *Proceedings* of 2nd International Workshop *"Durum Wheat* and *Pasta Quality"*, Rome 19-20 Nov. 2002, p. 285.
5 S. Dellaporta, J. Wood and J.Hicks, *Plant Mol.Biol. Rep.*,1983, **1**, 19.

RELATIONSHIP BETWEEN SEQUENCE POLYMORPHISM OF *GSP-1* AND PUROINDOLINES IN *TRITICUM AESTIVUM* AND *AEGILOPS TAUSCHII*

A.N. Massa[1] and C.F. Morris[2]

[1] USDA ARS Western Wheat Quality Laboratory & Washington State University, Pullman, WA, USA
[2] USDA ARS Western Wheat Quality Laboratory, E-202 Food Science & Human Nutrition Facility East, P.O. Box 99164-6394, Washington State University, Pullman, WA, 99164-6394 USA

1 INTRODUCTION

Grain Softness Protein-1 (Gsp-1) was coined by Rahman and colleagues[1] and has evolved to define a set of homoeologous genes which are closely related to the indolines–most notably, puroindoline a and b; and closely co-located in chromosome 5DS at the *Hardness* locus. Indolines appear in most taxa of the Triticeae and Aveneae tribes. They are members of the same cysteine-rich protein family as the CM proteins and non-specific Lipid Transfer Proteins. Our primary interest in Gsp-1 stems from the unique properties of this family of proteins, and their utility in phylogenetic studies.

Hexaploid wheat (*Triticum aestivum* L.) was formed some 7,000-10,000 years ago through one or more rare hybridization events between a wild AABB tetraploid species and a DD genome-bearing *Aegilops tauschii* (Coss.). Consequently, genetic diversity of modern wheat is limited by this "catastrophic" bottleneck of evolution, and the specific genetic composition of those few individual *Ae. tauschii* plants of long ago. Surveys of hundreds of hexaploid wheat varieties have indicated limited diversity in the puroindoline genes. Of the seven variants identified to date, 6 are single-nucleotide polymorphisms (SNPs) and one is a null form which produces no mRNA or protein. All seven variants result in hard-texture kernel phenotype. Surveys of *Ae. tauschii* accessions have indicated a much greater degree of nucleic acid and protein sequence polymorphism for the two puroindoline genes compared to that found in wheat. Surveys of *Gsp-1* gene sequences have not been conducted in *Ae. tauschii*. Consequently, we undertook the present study, which involved 50 accessions of *Ae. tauschii* drawn from the collection of the Wheat Genetics Resource Center, Kansas State University, and provided by Prof. Bikram Gill. The results will contribute to the understanding of what present-day *Ae. tauschii* accession is most closely related to the donor of the D-genome of modern wheat, and provide greater insight into the molecular evolution of the *Hardness* locus in wheat.

2 MATERIALS AND METHODS

Seeds of 50 accessions of *Ae. tauschii* were obtained from the Wheat Genetics Resource

Center, Kansas State University, Manhattan, KS, USA from Prof. Bikram Gill. Accessions were those originally collected by Kihara et al.[2] in the center of diversity and included the range of morphological variation of the species. Two seeds per accession were planted; seedlings were vernalized at 4°C for five weeks. Plants were grown in a glasshouse to maturity using routing cultural practices.

Genomic DNA was isolated from young leaves (Ae. tauschii) or single seeds (cvs. Chinese Spring and Yecora Rojo) according to the methods of Murray and Thompson[3] and Morris and Massa,[4] respectively. Full-length (495 bp) Gsp-D1 gene sequences were produced via PCR and the following primers which were based on Gsp-D1 sequence.[5]

Forward: 5'-TGGCCTCATCTCATCTTTCA-3'
Reverse: 5'-GCTCACCAATGGAAGCTACA-3'

```
Gsp-D1a  MKTFFLLAFL  ALVVSTAIAQ  YAEVPSPAAQ  APTADGFGEW  VAIAPSASGS  ENCEEEQPKV
Gsp-D1b  ..........  ..........  ..........  ..........  ..........  ..........
Gsp-D1c  ..........  ..........  .A........  ..........  ..........  ..........
Gsp-D1d  ..........  ..L.......  .A........  ..........  ..........  ..........
Gsp-D1e  ..........  ..........  .A........  ..........  ..........  ..........
Gsp-D1f  ..........  ..........  .A........  ..........  ..........  ..........
Gsp-D1g  ..........  ..........  .A........  ..........  ...V......  ..........
Gsp-D1h  ..........  ..........  .A........  ..........  ...V......  ..........

Gsp-D1a  DSCSDYVMDR  CVMKDMPLSW  FFPRTWGKRS  CEEVRNQCCK  QLRQTTSRCR  CKAIWTSIQG
Gsp-D1b  ..........  ..........  ..........  ..........  ......P...  ..........
Gsp-D1c  ..........  ..........  ..........  ..........  ......P...  ..........
Gsp-D1d  ..........  ..........  ..........  ..........  ......P...  ..........
Gsp-D1e  ..........  ..........  ..........  ..........  ......P...  ..........
Gsp-D1f  ..........  ..........  ..G.......  ..........  ......P...  ..........
Gsp-D1g  ..........  ..........  ..G.......  ..........  ......P...  ..........
Gsp-D1h  ..........  ..........  ..G.......  ..........  ....T.P...  ..........

Gsp-D1a  DLSGFKGLQQ  GLKARTVQTA  KSLPTQCNID  PKFCNIPITS  GYYL
Gsp-D1b  ..........  ..........  ..........  ..........  ....
Gsp-D1c  ..........  ..........  ..........  ..F.......  ....
Gsp-D1d  ..........  ..........  ..........  ..........  ....
Gsp-D1e  ..........  ..........  ..........  ..........  ....
Gsp-D1f  ..........  ..........  ........L.  ..........  ....
Gsp-D1g  ..........  ..........  ..........  ..........  ....
Gsp-D1h  ..........  ..........  ..........  ..........  ....
```

Figure 1 *Translated amino acid sequence diversity of* Gsp-D1 *genes from* Aegilops tauschii *and* Triticum aestivum *cv. Chinese Spring. Periods ('.') represent the same codon and amino acid present in* Gsp-D1a; *otherwise the letter indicates the non-conservative or conservative change at the nucleotide level.*

Reactions were performed in 25 μL containing 100 ng of genomic DNA, 10 pmol of each primer, 250 μM of each dNTP, 1X Taq DNA polymerase reaction buffer, 0.5 unit of Taq DNA polymerase (Promega, Madison, WI) and 1.5 mM of $MgCl_2$. Annealing temperature for both forward and reverse primers was maintained at 58°C. PCR products were analyzed on 1.5% (w/v) agarose gels, stained with ethidium bromide, and visualized using UV light. PCR products were purified from dNTPs and oligonucleotide primers using Exonuclease I and Shrimp Alkaline Phosphatase (ExoSAP-IT, UBS, Cleveland, OH), and sequenced directly with amplification 5' primers. Singleton variants were confirmed by using a second plant of the accession and sequencing the genomic DNA PCR products from both plants in both directions. All sequences were aligned using ClustalW algorithm.[6]

Table 1 Distribution of Grain Softness Protein-1 (*Gsp-D1*) gene alleles among 50 accessions of *Aegilops tauschii*.

Gsp-D1 allele	Taxonomic group[a]	Number of accessions	Per cent of taxa	Accession number[b]
Gsp-D1b	*tauschii*	9	60	TA1583, TA1675, TA10092, TA10094, TA10095, TA10101, TA10130, TA10131, TA10132
	typica	4	29	TA2460, TA2475, TA2483, TA2485
	meyeri	1	14	TA2529
Gsp-D1c	*strangulata*	9	82	TA2377, TA2450, TA2452, TA2454, TA2464, TA2467, TA2468, TA2470, TA2472
	tauschii	2	13	TA10085, TA10086
	typica	1	7	TA2369
Gsp-D1d	*typica*	5	36	TA2536, TA2394, TA2419, TA2487, TA2458
	tauschii	1	7	TA10110
	anathera	3	100	TA2374, TA2419, TA2436
	strangulata	1	9	TA2462
Gsp-D1e	*typica*	4	29	TA2512, TA2572, TA2574, TA2495
	tauschii	3	20	TA10090, TA10098, TA10105
	meyeri	1	14	TA2527
Gsp-D1f	*meyeri*	1	14	TA1649
	strangulata	1	9	TA2455
Gsp-D1g	*meyeri*	2	29	TA1599, TA2378
Gsp-D1h	*meyeri*	2	29	TA1691, TA2530

[a] *typica*, *anathera* and *meyeri* refer to *Ae. tauschii* subsp. *tauschii* var. *typica*, var. *anathera*, and var. *meyeri*; *strangulata* refers to *Ae. tauschii* subsp. *strangulata*.
[b] Wheat Genetics Resource Center accession identifier.

3 RESULTS AND DISCUSSION

Full-length *Gsp-D1* gene sequence was obtained for cv. Chinese Spring, which was assigned the '*a*' "wild-type" allele designation, following convention[7] (Figure 1). Yecora Rojo also had this exact sequence. Future studies will determine the extent of sequence diversity within *T. aestivum* for *Gsp-D1*, as well as *Gsp-A1* and *Gsp-B1*. A low level of polymorphism within *T. aestivum* may be expected at the D-locus.[1]

Among the 50 *Ae. tauschii* accessions, a total of 7 polymorphic nucleotide positions throughout the *Gsp-D1* gene were found, resulting in a total of 4 amino acid differences

(Figure 1). Each unique gene sequence was assigned an allele (*i.e.* '*b*' through '*h*'), and was recorded in the 2003 annual supplement of the *Catalogue of Gene Symbols in Wheat*.[8] None of these *Gsp-D1* gene sequences matched exactly that found in Chinese Spring. Consequently, one interpretation is that none of these *Ae. tauschii* accessions embody the specific ancestral D-genome donor of wheat. The *Gsp-D1b* allele was most similar to Chinese Spring, differing by only 1 nucleotide and 1 amino acid.

The distribution of the various alleles among the subspecies and botanical varieties of *Ae. tauschii* are provided in Table 1. Alleles *b, c, d* and *e* were relatively common with 8 to 14 accessions each, whereas alleles *f, g* and *h* had 2 accessions each. The allelic variation within subsp. *tauschii* was greater than the observed in subsp. *strangulata*, which is also in agreement with the morphological variation observed for subsp. *tauschii*.

Overall, these results on the sequence diversity of *Grain Softness Protein-1* in *Ae. tauschii* open up new perspectives for studying the phylogenetic relationships among members of this taxon, modern wheat and other members of the *Triticeae* and *Aveneae* tribes. The sequence similarity and the close co-localization of *Gsp-D1* with *Puroindoline a-D1* and *Puroindoline b-D1* provide further intrigue.

REFERENCES

1 C.F. Morris, *Plant Molec. Biol.*, 2002, **48**, 633.
2 H. Kihara, K. Yamashita and M. Tanaka, in *Cultivated Plants and their Relatives. Results of the Kyoto University Scientific Expedition to Korakoram and Hindukush*, ed. K. Yamashita, Kyoto University, 1955, **1**, 41.
3 M.G. Murray and W.F. Thompson, *Nucl. Acids Res.*, 1980, **8**, 4321.
4 C.F. Morris and A.N. Massa, *Cereal Chem.*, 2003, **80**, in press.
5 M. Turner, Y. Mukai, P. Leroy, B. Charef, R. Appels and S. Rahman, *Genome*, 1999, **42**, 1242.
6 J.D. Thompson, T.J. Gibson, F. Plewnicak, F. Jeanmougin, and D.G. Higgins, *Nucl. Acid Res.*, 1997, **24**, 4876.
7 R. McIntosh, G.E. Hart, K.M. Devos, M.D. Gale and W.J. Rogers, *Proceedings of the 9th International Wheat Genetics Symposium, vol. 5, Catalogue of Gene Symbols for Wheat*, 1998, August 2-7, 1998, Saskatoon, SK, Canada, Univ. Extension Press, Univ. Saskatchewan, Canada, 235 pp.
8 R. McIntosh, K.M. Devos, J. Dubcovsky, C.F. Morris and W.J. Rogers, *The 2003 Supplement of the Wheat Gene Catalogue*, 2003, published on-line at: http://wheat.pw.usda.gov/ggpages/wgc/2003upd.html.

Author Index

Abonyi T., 323
Addeo F., 375
Alberghina G., 304
Albertini M., 231
Allen H., 89
Alvarez J.B., 117
Amato M.E., 304, 308
Ames N.A., 148, 192
Ames N.P., 132
Anania M.C., 398
Anderson O.D., 6, 10, 18
Andersson A., 180
Aramini M., 152
Arendt E.K., 379
Atl A., 421
Augustin M., 337

Bagulho A.S., 113
Balmer Y., 169
Bancel E., 30, 38, 173
Banerjee R., 162
Basman A., 345
Bason M.L., 219
Batey I.L., 337
Baticz O., 267, 323
Battais F., 383
Bedő Z., 66, 140, 267, 323
Békés F., 22, 85, 109, 140, 215, 219, 267, 285, 323
Bellavita I., 70
Belton P.S., 54, 203
Ben Hamida J., 173
Benedettelli S., 158
Bentancur O., 429
Bhandari D.G., 312
Bharadwaj A., 162
Birzele B., 327
Blakeney J.L., 219
Blechl A.E., 6, 10
Bonomi F., 387
Branlard G., 30, 38, 173
Bregitzer P., 6
Brismar K., 288
Brites C., 113

Buchanan B.B, 169
Butow B.J., 85

Caballero L., 117
Caboni M.F., 259
Cai N., 169
Cannell M., 14
Capocchi A., 453
Carcea M., 121
Carollo V., 18
Carrillo J.M., 113, 184
Casale C., 398
Çelik S., 74
Ceriotti A., 26
Cerletti P., 349
Cervantes F., 156
Chambon C., 38
Chao S., 18
Chiaraluce R., 316
Ciclitira P.J., 371
Clarke F.R., 148, 192
Clarke J.M., 132, 148, 192
Colaprico G., 152
Colonna G., 391
Consalvi V., 316
Coolbear T., 413
Corbellini M., 425, 433
Costantini S., 391
Cubadda R., 259
Cunsolo V., 34
Curioni A., 425, 433

D'Egidio M.G., 387
D'Ovidio R., 10, 22, 136, 316
Dang J.M.C., 219
Dardevet M., 30
Day L., 337
De Ambrogio E., 81
De Vincenzi M., 395
Dekova T., 125, 129
Delcour J.A., 292
Della Cristina P.A., 26
Denery-Papini S., 383
Dessalegn T., 105
Dexter J.E., 132

Di Gennaro S., 70
Di Tola M., 398
Dobraszczyk B.J., 231, 255, 300
Doherty A., 14
Domenek S., 243, 443
Don C., 177, 223, 285
Dornez E., 292
Douliez J.P., 383
Dumur J., 30, 38
Dupont F.M., 42

Edwards K.J., 3
Edwards N.M., 132
Ellis H.J., 371
Engel W., 371
Every D., 341, 413

Facchiano A.M., 391
Farisei F., 70
Ferrante P., 136
Ferranti P., 375
Ficca A.G., 70
Fisichella S., 304, 308
Fontanini D., 453
Foti S., 34
Franchi E., 121
Fraser J., 371
Friedt W., 144
Fu B.X., 239
Funatsuki H., 46
Funatsuki W.M., 46
Fusari F., 395

Gale K.R., 85
Gallagher E., 379
Galleschi L., 453
Gazza L., 457
Gecheff K., 125
Georgiev S., 125, 129
Gianibelli M.C., 58, 62, 81, 89, 109, 132, 140, 275
Gonnet S., 429

Gonzalez-Santoyo H., 156
Goodwin J., 14
Gormley T.R., 379
Griffin W.B., 341
Griffiths-Jones S., 367
Gruber H., 50
Guéguen J., 447
Guerrieri N., 121, 320, 349
Guilbert S., 243, 439, 443
Guóth A., 323
Gutser R., 188

Halford N.G., 54
Hamer R.J., 177, 223, 285, 361
Haraszi R., 140, 219, 267
Hassani M.E., 58
He Z.H., 97
Hedlund U., 247
Helmerich G., 353
Henton S.M., 196
Hidvégi M., 406
Holdworth M.J., 3
Hormes J., 327
Houshmand S., 148
Hurkman W.J., 169

Iafelice G., 259
Iametti S., 387
Ikeda T.M., 101
Izydorczyk M.S., 296

Jamieson P.D., 196
Jenkins J.A., 367
Johansson E., 180, 288
Jones H.D., 3, 14
Juhász A., 62, 66, 89, 140, 323

Kanny G., 383
Kasarda D.D., 42
Kato A., 46
Khan K., 239
Kieffer R., 235, 251
Killermann B., 50, 144
Kindler A., 267

Király I., 323
Knox R.E., 148, 192
Koehler P., 188, 327, 353, 357
Koen E., 105
Köksel H., 74, 345, 421
Kolster P., 211
Kovacs M.I.P., 239
Kuktaite R., 180, 288

Labuschagne M.T., 105
Lafiandra D., 10, 34, 81, 109, 136, 152, 158, 275, 304, 308, 316
Láng L., 267
Larroque O.R., 81, 85, 89, 109, 132, 140, 323
Larsson H., 180, 247, 288
Laudencia-Chinguanco D., 18
Lazo G.R., 18
Lefebvre J., 207
Lenton J.R., 3
Li W., 255, 300
Lichtendonk W.J., 223
Lin J., 6
Lindhauer M.G., 271
Liu L., 97
Lombardi L., 453
Lookhart G., 177
Lorenzi R., 453

Ma W., 85
MacRitchie F., 177, 227
Maesmans G., 231
Majoul T., 173
Mamone G., 375
Mangavel C., 447
Mann G., 215, 285
Mantarro D., 304, 308
Marconi E., 121, 259
Margiotta B., 109, 152
Marion D., 383
Martín L.M., 117
Martínez M.C., 184
Martre P., 196
Marttila S., 288
Masci S., 10, 34, 136, 152, 316
Massa A., 461

McCaig T.N., 132
McCarthy D.F., 379
Melck D., 375
Merlino M., 30
Messia M.C., 121, 259
Meuser F., 271
Migliardi M., 425
Millar S.J., 312
Mills E.N.C., 54, 367
Modrow H., 327
Moneret-Vautrin D.A., 383
Morel M.-H., 243, 439, 443
Morell M.K., 85, 89, 21
Morgenstern M.P., 263
Morris C.F., 461
Muacho M.C., 113
Muccilli V., 34

Naeem H., 177
Nagamine T., 101
Nagy I.J., 66
Nakamura H., 93
Napier J.A., 22
Newberry M.P., 263
Ng P.K.W., 345
Nguyen S.B., 6
Niglio A., 457
Nin M., 429
Nishio Z., 46
Nocente F., 457
Noorman M.S., 402

O'Brien K., 6
Olanca B., 417, 421
Oliver J., 89
Orsi A., 26

Pagani M.A., 387
Palermo A., 304, 308
Panichi D., 70
Pastori G., 14
Patacchini C., 10, 136, 152, 316
Peña R.J., 97, 156
Petrini A., 395
Pflüger L.A., 158
Picarelli A., 398

Plijter J.J., 223
Poerio E., 70
Pogna N.E., 395, 457
Pollock E., 371
Pommet M., 243, 439
Popineau Y., 207, 383, 447
Porter J.R., 196
Prange A., 327
Preston K.R., 296
Prieto-Linde M.L., 180, 288
Pyle D.L., 331

Rakszegi M., 267
Redl A., 243, 439
Richmond J.C., 312
Rodríguez-Quijano M., 184
Rossi M., 391
Rousseau C., 207
Ruiz M., 184

Sabbatella L., 398
Saito K., 46
Saletti R., 34
Salvatorelli S., 121
Saruyama H., 46
Savage A., 54
Savarino A., 304, 308
Saviozzi F., 453
Scarlata G., 304, 308
Schober T.J., 379
Schofield J.D., 231, 331
Schurgers B., 292
Scossa F., 10

Secundo F., 349
Sedlmeier M., 50
Shariflou M.R., 58
Sharp P.J., 58, 109
Shewry P.R., 3, 14, 22, 54, 367, 413
Simmons L.D., 263, 341, 402
Singh H., 227
Sivri D., 417, 421
Sliwinski E.L., 211
Smewing J., 231
Solomon R.G., 58
Sparks C., 14
Sparvoli F., 26
Sreeramulu G., 41
Stevenson S.G., 296
Sutton K.H., 341, 402. 413

Tabiki T., 46
Tafuro F., 375
Takács I., 66
Takata K., 46
Tamás L., 58, 66
Tanaka C.K., 169
Tatham A.S., 413
Tavella L., 425, 433
Tömösközi S., 267, 323
Tömösközi-Farkas R., 406
Tosi P., 22
Triboï E., 173, 196
Trivisonno M.C., 259
Tsiami A.A., 331

Unbehend L., 271

Uthayakumaran S., 275

Vaccino P., 425, 433
Vaishnav P.P., 162
van Vliet T., 211, 361
Varga J., 267
Vázquez D., 279, 429
Vázquez J.F., 184
Vensel W.H., 42, 169
Veraverbeke W.S., 292
Vetrano S., 398
Vincenzi S., 433
von Tucher S., 188

Wang M., 361
Watts B., 279
Wellner N., 54
Wieser H., 188, 235, 371
Wilde P.J., 300
Wong J.H., 169
Woods S.M., 239
Wrigley C.W., 337
Wu H., 14

Yahata E., 46
Yamauchi H., 46
Yano H., 101
Yaşar F., 74
Yordanov Y., 129
You S., 296

Zandomeneghi M., 453
Zardi M., 387
Zimmermann G., 144
Zoccatelli G., 425, 433

Subject Index

Antibodies
  HMW-GS specific, 50

Biomaterials based on wheat gluten
  characterisation of, 443, 447
  plasticization with fatty acids, 439

Distribution area of bread wheat, 93
Doubled haploids
  use for analysis of dough properties, 85, 89, 144, 192

Genetic variation
  of bread wheat, 125, 129
  of spelt, 117
Genomics
  use of ESTs to analyse wheat seed proteins, 18, 62
Gliadins
  ω-type: characterisation and evidence for post-translational cleavage, 42
  effect of high-pressure and temperature on their extractability, 235
  interaction with polysaccharides, 349
  molecular modeling, 74
  naturally present chimeric γ-type genes, 66
  peptides involved in celiac disease, 371, 375
  relation to quality, 140, 144
  surface, rheological properties and localisation of purified gliadins in the dough, 300
Gluten allergies and intolerances
  α-gliadins peptides, 371
  allergy to LTP proteins, 383
  complex between gluten peptides and HLA-DQ2 molecule, 391
  effect of rye and barley, 398
  gluten-free bread, 379
  proteins involved in type I allergy, 367
  use of immature grains, 387
  use of *Panicum miliaceum* as a safe food, 395

Glutenins *see also* HMW-GS *and* LMW-GS
  effect of high-pressure and temperature on their extractability, 235
  particle size, 177, 285
  surface, rheological properties and localisation of purified glutenins in the dough, 300
Glutenin polymers
  breakdown due to bug protease, 433
  effect of ascorbic acid, 341
  effect of dough mixing, 223
  effect of HMW-GS, 285
  FFF fractionation and MALLS for the determination of polymer size, 296
  influence of cultivar, environment and dough treatment, 180
  influence of growing conditions, 177
  in transgenic wheats, 10, 22
  in durum wheat, 81, 132
  in relation to quality, 105, 109, 288, 323
  of wheat in comparison to barley proteins, 331
  structural studies of, 308

Heterologous expression
  of LMW-GS, 136, 316
  of y-type HMW-GS of *T. tauschii*, 58
HMW-GS
  correlation with quality parameters, 81, 85, 89, 97, 109, 144, 156, 158, 162
  development of NILs, 152
  effect on dough properties, 207
  effect on glutenin particle size, 285
  influence of sulfur fertilisation, 188
  novel alleles, 81
  of *T. tauschii*, 58
  purification of, 304
  structural studies of, 308
  transformation with, 6

LMW-GS
  aggregation of in vitro synthesised, 26

characterisation of genes and
  polypeptides, 62, 66, 101, 136
chimeric genes, 66
correlation with quality parameters, 46,
  144, 156, 158
influence of dough properties, 89, 97
influence of sulfur fertilisation, 188
mass spectrometry of, 34
molecular modeling, 74
structural analyses of, 316
transformation with, 10, 22

Mass spectrometry
  as a tool in the study of celiac disease,
    375
  of ω-gliadins, 42
  of LMW-GS, 34, 38

Non-gluten components
  association with quality, 337
  confocal scanning laser microscopy of
    lipids in dough, 300
  grain softness protein (GSP-1), 457
  influence of puroindolines, 113
  lipid transfer poteins (LPT), 383, 453
  polysaccharide-gliadin interaction, 349
  puroindolines, 461
Nutritional aspects
  components of mill stream fraction,
    402
  fermented wheat germ extract, 406

Proteins
  content prediction by portable FT-NIR,
    312
  protease inhibitors, 70
  sulphur content, 327
  surface hydrophobicity, 320
  wheat bugs proteases, 413, 417, 421
Proteomics
  as a tool to study genetic and molecular
    aspects of gluten, 30, 34, 38
  response to development and high
    temperature, 169, 173

Quality *see also* Rheology
  assessment and improving by
    biotechnological approaches, 3,
    6, 10, 14, 22, 54, 58

association with non-protein
  components, 337
association with QTL, 148
correlation of HMW-GS with, 58, 81,
  323
correlation of LMW-GS with, 10, 22,
  46
effect of wheat bugs proteases, 421,
  425, 429, 433
influence of environmental factors,
  184, 188, 192, 196
influence of glutenins and
  puroindolines, 113
of durum wheat, 121, 275
of wheats containing HMW-GS
  transgenes, 6
of wheat containing LMW-GS
  transgenes, 10, 22
use of SE-HPLC as a prediction test,
  105, 109, 132

Rheology *see also* Quality
  application of polymer concepts, 227
  changes of gluten organisation during
    dough formation, 251
  development and distribution of gas
    cells, 255
  dough forming properties in different
    conditions, 271
  effect of ascorbic acid, 341, 357
  effect of carbon atom and thermal
    stability of proteins, 239
  effect of dough mixing, 223
  effect of HMW-GS, 207
  effect of high-pressure and
    temperature, 235, 243
  effect of large deformations, 211
  effect of pentosan, 361
  effect of phospholipids, 353
  effect of salt and pH, 247
  effect of transglutaminase, 345
  extensional measurements, 215
  factors influencing extensibility, 279
  influence of micro-scale gluten-starch
    separation on protein
    agglomeration, 292
  mechanisms of stability of bubble
    expansion, 231

relationship between dough
    components and dough
    development during mixing, 263
theory of viscoelasticity, 203
use of the different mixers, 219, 267
vital wheat gluten, 259

Transgenic wheat
    transformation with HMW-GS genes,
        6, 207
    transformation with LMW-GS genes,
        10, 22

DATE DUE

DUE DATE SUBJECT TO CHANGE
IF A RECALL IS REQUESTED

SEP 2 8 2005 Rtnd SS	AUG 5 - 2005